Ultrasonic Guided Waves

Ultrasonic Guided Waves

Special Issue Editor

Cliff Lissenden

MDPI • Basel • Beijing • Wuhan • Barcelona • Belgrade • Manchester • Tokyo • Cluj • Tianjin

Special Issue Editor
Cliff Lissenden
University Park
USA

Editorial Office
MDPI
St. Alban-Anlage 66
4052 Basel, Switzerland

This is a reprint of articles from the Special Issue published online in the open access journal *Applied Sciences* (ISSN 2076-3417) (available at: https://www.mdpi.com/journal/applsci/special_issues/UGW).

For citation purposes, cite each article independently as indicated on the article page online and as indicated below:

LastName, A.A.; LastName, B.B.; LastName, C.C. Article Title. *Journal Name* **Year**, *Article Number*, Page Range.

ISBN 978-3-03928-298-2 (Pbk)
ISBN 978-3-03928-299-9 (PDF)

Contents

About the Special Issue Editor

Cliff Lissenden is a professor of engineering science and mechanics at Penn State. He joined the Department of Engineering Science and Mechanics in 1995 and gained a joint appointment in Acoustics in 2011. He is an ASME Fellow and was the founder and director of the Ben Franklin Center of Excellence in Structural Health Monitoring. His current research is mainly focused on the nondestructive characterization of materials using ultrasonic guided waves and is broadly applicable to metals, composites, concrete, rock, and bone. Characterization systems use piezoelectric, magnetostrictive, electromagnetic, and laser-based transduction of ultrasonic waves for nondestructive testing, inspection, and monitoring, often in harsh environments. The basic physics employed are that the wave speeds depend on the elastic properties, that acoustic impedance mismatches cause wave scattering, and that material nonlinearity distorts the waveform. Current applications include robotic nondestructive inspection of stress corrosion cracking, structural integrity analysis of bonded joints in composites, detection of incipient damage with nonlinear ultrasonics, process monitoring of additive manufacturing with laser ultrasonics, cloaking critical infrastructure from earthquakes, and characterization of bone healing.

Editorial

Applied Sciences Special Issue: Ultrasonic Guided Waves

Clifford J. Lissenden

Department of Engineering Science and Mechanics, Pennsylvania State University, University Park, PA 16802, USA; lissenden@psu.edu

Received: 9 September 2019; Accepted: 10 September 2019; Published: 15 September 2019

The propagation of ultrasonic guided waves in solids is an important area of scientific inquiry due primarily to their practical applications for the nondestructive characterization of materials, such as nondestructive inspection, quality assurance testing, structural health monitoring, and for achieving material state awareness. This Special Issue of the journal covers all aspects of ultrasonic guided waves (e.g., phased array transducers, meta-materials to control wave propagation characteristics, scattering, attenuation, and signal processing techniques) from the perspective of modeling, simulation, laboratory experiments, or field testing. In order to fully utilize ultrasonic guided waves for these applications, it is necessary to have a firm grasp of their requisite characteristics, which include being multimodal, dispersive, and comprised of unique displacement profiles through the thickness of the waveguide.

The majority of the manuscripts in this Special Issue report on waveguides that are pipes and plates, but rods, rail, multi-strand wire, human bone, and half-spaces are also investigated. Hakoda et al. [1] provide experimental results on an aluminum plate, demonstrating how wave propagation can be manipulated by changing the boundary conditions. Specifically, surface-mounted resonators block the propagation of the S0 Lamb mode at low frequencies. Forbidding the propagation of the S0 Lamb mode turns out to be very different from forbidding the propagation of the A0 Lamb mode but can be explained in terms of the boundary conditions, which provide an alternative approach to meta-materials. The remaining 18 manuscripts pertain directly to the detection of defects within the waveguide in order to enable the assessment of the structural integrity.

While the purpose of pipes is to transport or circulate fluids, their hollow cylindrical geometry forms an efficient waveguide for longitudinal, torsional, and flexural ultrasonic waves. Torsional waves are often preferred for long-range defect detection because the wave energy does not leak into the fluids inside the pipe. Six manuscripts in the Special Issue investigate ultrasonic guided waves in pipes for the purpose of defect detection. Rose et al. [2] show that an array of fiber Bragg grating (FBG) sensors multiplexed along an optical fiber can receive torsional guided waves in a pipe, thus improving the range of piezoelectric and magnetostrictive actuators because reflections do not have to propagate back to the actuator. Pedram et al. [3] demonstrate the use of split-spectrum processing to improve the signal-to-noise ratio (SNR) of axisymmetric torsional waves in pipes and investigate its limitations. They show a 30 dB increase in SNR. Mahal et al. [4] first propose a method to detect defects in pipes from axisymmetric torsional waves having a low signal-to-noise ratio by leveraging the signals received from individual elements in the ring array transducer. Then, they [5] propose an adaptive leaky normalized least-mean-square filter to reduce the effect of unwanted flexural waves on the detection of axisymmetric waves. In a third paper [6], the same authors describe the exploitation of the power spectrum to distinguish torsional waves from flexural waves in pipes for more reliable defect detection. Taking a different approach, Wu and Lee [7] propose applying a multilayer perceptron neural network to detect leaks in pressurized natural gas pipelines, based upon acoustic emissions from the leaks. Accurate localization depends on the group velocity of the flexural waves generated during the acoustic emission events.

The Lamb waves and shear-horizontal (SH) waves that propagate in traction-free homogeneous isotropic plates are often used for defect detection when access is limited or to cover a large area. Hakoda and Lissenden [8] demonstrate how the analysis of partial waves provides new perspectives on the propagation characteristics of elastodynamic guided waves in plates as well as other types of waveguides. Fan et al. [9] employ wavenumber filtering to remove Lamb modes other than the A0 mode to detect flat bottom holes in a plate. Wu et al. [10] propose a methodology for nondestructive inspection of cylindrical storage tanks' base plates whereby a transducer is scanned around the perimeter of the base plate. The generated fundamental SH waves propagate in the radial direction of the base plate. Defects scatter the waves, and reflections are received by the transducer. A synthetic aperture focusing technique based on the so-called exploding reflector model is applied to image the tank bottom. Wu et al. [11] then demonstrate a 12 dB improvement in the amplitude of SH waves generated by a magnetostrictive transducer when a soft magnetic patch is applied, which reduces the resistance of the magnetic circuit. This optimization of the magnetic circuit is also applicable to transducers that send and receive torsional waves in pipes. Crespo et al. [12] propose an experimental method to construct the Lamb wave and SH wave dispersion curves for a plate using a transmitter and two receivers, which requires no knowledge of the material properties. The paper by Rucka et al. [13] completes the applications to plates by investigating the integrity of the adhesive in single-lap joints between aluminum plates by images processed from the wavefield acquired by scanning laser Doppler vibrometer and the weighted root mean square.

Axial transmission of ultrasound in human cortical bone is under development as a biomarker for early detection of osteoporosis and fracture risk. While the geometry of human bones, such as the tibia, is far from the geometry of plates, it is common to use Lamb waves as surrogates for waves in cortical bones. Okumura et al. [14] developed a rapid high-resolution technique to determine the wavenumbers of propagating waves in surrogates for human bone.

One manuscript addresses Rayleigh waves propagating along the surface of a half-space. Wang et al. [15] provide forward and inverse analyses of Rayleigh wave scattering from surface breaking defects based on reflection coefficients in order to predict the shape of realistic surface-breaking flaws.

The final four manuscripts investigate defect localization for long-range inspection of different types of one-dimensional waveguides. Zhang et al. [16] considered long-range ultrasonic guided waves in a steel square bar and analyzed the non-detection zone for flaws located near the back wall. Xing et al. [17] propose a single mode extraction algorithm to precisely localize defects in rail. Zhang et al. [18] analyze guided waves, including second harmonic generation associated with the contact acoustic nonlinearity, in multi-strand cable where transducers are coupled to only two strands in the cable. Finally, Chang and Moon [19] investigate accurate group velocity of electromagnetic waves in dispersive waveguides, such as insulated cables, in which defect localization is performed using reflectometry.

Funding: This research received no external funding.

Acknowledgments: I would like to thank all of the authors who submitted manuscripts to this special issue, as well as the reviewers, who helped to improve the quality of the manuscripts.

Conflicts of Interest: The author declares no conflicts of interest.

References

1. Hakoda, C.; Lissenden, C.; Shokouhi, P. Clamping resonators for low-frequency S0 lamb wave reflection. *Appl. Sci.* **2019**, *9*, 257. [CrossRef]
2. Rose, J.; Philtron, J.; Liu, G.; Zhu, Y.; Han, M. A hybrid ultrasonic guided wave-fiber optic system for flaw detection in pipe. *Appl. Sci.* **2018**, *8*, 727. [CrossRef]
3. Pedram, S.; Mudge, P.; Gan, T. Enhancement of ultrasonic guided wave signals using a split-spectrum processing method. *Appl. Sci.* **2018**, *8*, 1815. [CrossRef]

4. Mahal, H.; Yang, K.; Nandi, A. Detection of defects using spatial variances of guided-wave modes in testing of pipes. *Appl. Sci.* **2018**, *8*, 2378. [CrossRef]
5. Nakhli Mahal, H.; Yang, K.; Nandi, A. Improved defect detection using adaptive leaky NLMS filter in guided-wave testing of pipelines. *Appl. Sci.* **2019**, *9*, 294. [CrossRef]
6. Nakhli Mahal, H.; Yang, K.; Nandi, A. Defect detection using power spectrum of torsional waves in guided-wave inspection of pipelines. *Appl. Sci.* **2019**, *9*, 1449. [CrossRef]
7. Wu, Q.; Lee, C. A modified leakage localization method using multilayer perceptron neural networks in a pressurized gas pipe. *Appl. Sci.* **2019**, *9*, 1954. [CrossRef]
8. Hakoda, C.; Lissenden, C. Using the partial wave method for wave structure calculation and the conceptual interpretation of elastodynamic guided waves. *Appl. Sci.* **2018**, *8*, 966. [CrossRef]
9. Fan, G.; Zhang, H.; Zhang, H.; Zhu, W.; Chai, X. Lamb wave local wavenumber approach for characterizing flat bottom defects in an isotropic thin plate. *Appl. Sci.* **2018**, *8*, 1600. [CrossRef]
10. Wu, J.; Tang, Z.; Yang, K.; Wu, S.; Lv, F. Ultrasonic guided wave-based circumferential scanning of plates using a synthetic aperture focusing technique. *Appl. Sci.* **2018**, *8*, 1315. [CrossRef]
11. Wu, J.; Tang, Z.; Yang, K.; Lv, F. Signal strength enhancement of magnetostrictive patch transducers for guided wave inspection by magnetic circuit optimization. *Appl. Sci.* **2019**, *9*, 1477. [CrossRef]
12. Hernandez Crespo, B.; Courtney, C.; Engineer, B. Calculation of guided wave dispersion characteristics using a three-transducer measurement system. *Appl. Sci.* **2018**, *8*, 1253. [CrossRef]
13. Rucka, M.; Wojtczak, E.; Lachowicz, J. Damage imaging in lamb wave-based inspection of adhesive joints. *Appl. Sci.* **2018**, *8*, 522. [CrossRef]
14. Okumura, S.; Nguyen, V.; Taki, H.; Haïat, G.; Naili, S.; Sato, T. Rapid high-resolution wavenumber extraction from ultrasonic guided waves using adaptive array signal processing. *Appl. Sci.* **2018**, *8*, 652. [CrossRef]
15. Wang, B.; Da, Y.; Qian, Z. Forward and inverse studies on scattering of rayleigh wave at surface flaws. *Appl. Sci.* **2018**, *8*, 427. [CrossRef]
16. Zhang, L.; Yang, Y.; Wei, X.; Yao, W. The study of non-detection zones in conventional long-distance ultrasonic guided wave inspection on square steel bars. *Appl. Sci.* **2018**, *8*, 129. [CrossRef]
17. Xing, B.; Yu, Z.; Xu, X.; Zhu, L.; Shi, H. Research on a rail defect location method based on a single mode extraction algorithm. *Appl. Sci.* **2019**, *9*, 1107. [CrossRef]
18. Zhang, P.; Tang, Z.; Lv, F.; Yang, K. Numerical and experimental investigation of guided wave propagation in a multi-wire cable. *Appl. Sci.* **2019**, *9*, 1028. [CrossRef]
19. Chang, S.; Moon, S. Compensation for group velocity of polychromatic wave measurement in dispersive medium. *Appl. Sci.* **2017**, *7*, 1306. [CrossRef]

Article

Clamping Resonators for Low-Frequency S0 Lamb Wave Reflection

Christopher Hakoda, Cliff J. Lissenden * and Parisa Shokouhi

Department of Engineering Science and Mechanics, Pennsylvania State University, University Park, Pennsylvania 16802, PA, USA; cnh137@psu.edu (C.H.); parisa@engr.psu.edu (P.S.)
* Correspondence: lissenden@psu.edu

Received: 30 November 2018; Accepted: 9 January 2019; Published: 12 January 2019

Abstract: A recent elastic metamaterial study found that resonators that "clamp" a plate waveguide can be used to create a frequency stop-band gap. The result was that the resonator array can prohibit the propagation of an A0 Lamb wave mode. This study investigates whether the concept can be extended to S0 Lamb wave modes by designing resonators that can prohibit the propagation of S0 Lamb wave modes in a 1-mm aluminum plate waveguide at 50 kHz. The frequency-matched resonators did not reduce the transmitted signal, leading to the conclusion that the design concept of frequency-matched resonators is not always effective. On the other hand, the resonators designed to clamp the upper surface of the plate were very effective and reduced the transmitted signal by approximately 75%.

Keywords: metamaterial; resonator; low-frequency; lamb wave

1. Introduction

The field of metamaterials has emerged over the past 20 years and inspired numerous innovations in optics, electromagnetics e.g., [1–3] and more recently in acoustics e.g., [4–6]. Phononic crystals and acoustic metamaterials are composites engineered to mold elastic wave dispersion through scattering or local resonances to achieve desired spectral and phase properties such as band gaps around specific frequencies. Phononic crystals are composed of a periodic array of scatterers. The mechanism for the formation of band gaps in phononic crystals is a Bragg-like scattering of waves with wavelength comparable to the dimension of the period of the crystal or the crystal lattice constant [7]. This characteristic limits the use of phononic crystals as low-frequency filters. However, low-frequency filters can be achieved by using a different class of acoustic metamaterials that comprise a series of locally resonating units which are not necessarily periodic. Liu et al. [8] demonstrated that such assemblies are capable of muting waves of wavelengths that are orders of magnitude longer than the lattice constant. By tuning the properties of the individual units and their arrangement, the metamaterial can act as a low-frequency band-gap filter (blocking the incoming elastic energy within certain frequency bands) or wedge (converting the mode and diverting the incident wave). The studies of resonators that followed have been mostly parametric in nature and focused on designing elements with strong frequency-coupled characteristics that will yield band gaps about a given center frequency [9–11].

As noted above, the field of elastodynamic metamaterials is relatively new, but nonetheless presents a wide variety of resonator design methodologies [9–13]. An experiment conducted by Rupin et al. [14] found that cylindrical rods adhered to a plate waveguide could be used to prohibit the propagation of the A0 Lamb wave mode at low frequencies (<11 kHz). Soon thereafter, a paper by Williams et al. [15] sought to determine what characteristic of the cylindrical rods contributed to the A0 Lamb wave mode band gap. They found that the rods vibrated in an axial mode which applied a "clamping"-like effect to the waveguide. That is, the rods would vibrate in such a way that the out-of-plane displacement at their bases was minimized. They also compared the frequencies at which the minimal displacements occurred with the band gaps reported in Rupin et al.'s paper and found a good agreement between the two spectra.

The spectrum detailed in Williams et al.'s paper was calculated using a vibration model subjected to a harmonic forcing function. Upon further investigation, we found that the frequencies at which minimum displacement occurred were equal to the resonant frequencies of an axially vibrating, fixed-free beam. Since these resonant frequencies have well-known solutions, this finding presented itself as a very straightforward method for designing sub-wavelength resonators for the A0 Lamb wave mode to create band gaps around specific frequencies. However, this approach was only shown to be effective for low-frequency A0 Lamb wave modes which have primarily out-of-plane displacement. The purpose of this article is to extend this concept to design resonators that prohibit the propagation of low-frequency S0 Lamb wave modes and to determine its applicability at higher frequencies. To accomplish this, two resonator designs, one designed based on a frequency matching approach and the other based on the sole concept of clamping, were tested to see how well they can impede the propagation of a 50 kHz S0 Lamb wave. These resonators are hereafter referred to as the frequency-matched resonators and the clamping resonators, respectively. While the design of the frequency-matched resonators is straightforward (based on resonance), the theoretical basis for the design of the clamping resonators is more involved and requires the full attention of a separate publication. We believe that a rational design protocol is possible and will use the experimental results provided herein to demonstrate it in a subsequent publication. Thus, in this article we simply provide finite-element simulations to qualitatively confirm the experimental findings.

2. Experimental Setup

For the experiment, the consistency of various experimental parameters was prioritized to obtain a comparison between the resonator types. That is, various experimental parameters are the same regardless of the resonators being tested. This includes the transducer positions, waveguide dimensions, resonator spacing/positioning, the area of contact between the resonator and plate, resonator material properties, and overall resonator mass. The only difference between tests is the shape of the resonators.

The guided waves that will be used for the test are the S0 Lamb wave mode with a 50 kHz 5-cycle pulse, which corresponds with a wavenumber of 58.1 rad/m. At these low frequencies, this mode corresponds with a wave-structure (i.e., displacement profile) that is effectively planar and consists of in-plane displacement through the thickness of the plate waveguide. The S0 Lamb wave mode was excited and received using shear transducers with the polarization aligned with the wave-propagation direction. A schematic of the setup, including transducer placement and resonator positions, is shown in Figure 1. Three rows of ten resonators were used for the resonator array and adhered to the surface of the waveguide using cyanoacrylate adhesive. The spacing of the resonator array shown in Figure 1b is based on how close together the clamping resonators could be positioned.

The main method of assessing the resonators will be to compare two-dimensional fast Fourier transforms (2DFFTs) of the signal transmitted past the resonator array [16]. The receiving transducer was moved away from the resonator array in 2 cm increments. This was done 20 times to measure an A-scan at 20 positions along the expected wave-propagation direction. For the 2DFFT, zero-padding was applied to increase the number of bins in the wavenumber spectrum, but no zero-padding was applied for the frequency spectrum.

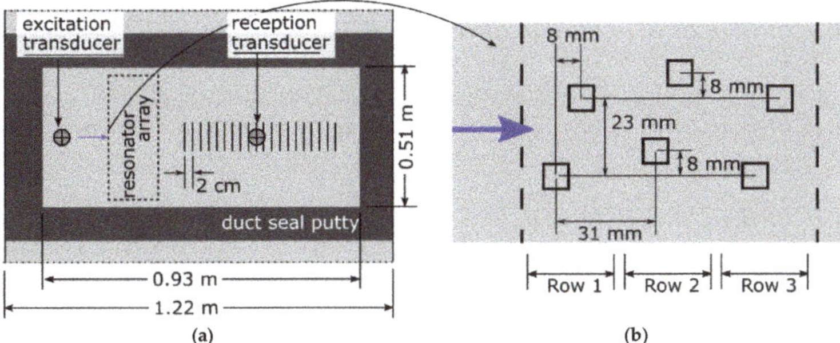

Figure 1. Schematic (not to scale) of experimental setup: (**a**) top-view of the transducer positions with respect to the resonator array, and (**b**) a zoomed in view of the resonator array that shows how the resonators are positioned with respect the wave-propagation direction. The squares are the contact areas between the resonators and the plate waveguide.

3. Frequency-matched Resonators

Analogous to Rupin et al.'s [14] axial rod design, the frequency-matched resonator was designed to vibrate at 50 kHz in a flexural mode under fixed-free boundary conditions. The fixed boundary condition was expected to apply the desired clamping effect described in Williams et al. [15]. The refined design of these frequency-matched resonators consists of a 7.9 × 7.9 mm square rod that is 23.4 mm long. Figure 2a shows the 2DFFT of a baseline measurement which was taken prior to bonding the resonators to the plate waveguide. The peak at a wavenumber of −58.1 rad/m and a frequency of 53.9 kHz shown in Figure 2a indicates that the S0 mode is being excited.

The Frequency-matched Resonators were positioned on the waveguide as shown in Figure 1 and adhered to the plate waveguide. The A-scans described in section 2 were measured and used to calculate the 2DFFT shown in Figure 2b.

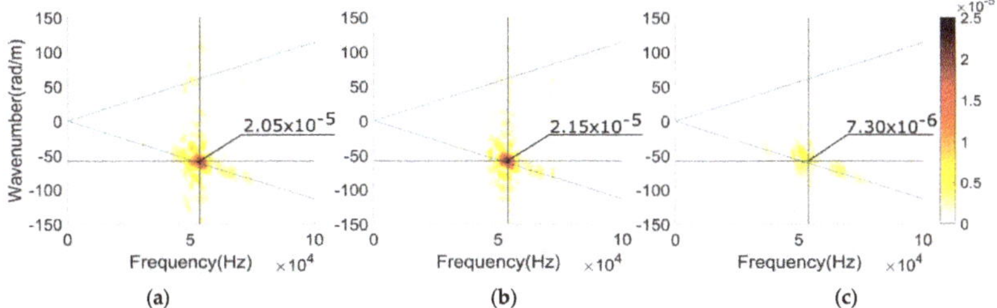

Figure 2. The 2DFFT of the transmission measurements for (**a**) the Baseline, (**b**) the Frequency-matched Resonator and (**c**) the Clamping Resonator. The diagonal lines are the dispersion curves for the forward and backward propagating S0 Lamb wave modes. The vertical and horizontal lines show where the frequency- and wavenumber-spectrum slices shown in Figure 6 are obtained. The peak values are noted in each figure.

When comparing Figure 2a to Figure 2b, the frequency-matched resonators do not reduce the transmitted S0 signal. This is believed to be due to the resonators vibrating differently than expected. That is, unlike the axial vibration described in Williams et al., the flexural vibration of the frequency-matched resonator is likely to cause unwanted out-of-plane displacement.

The frequency-matched resonators are an attempt at designing resonators based on resonant frequency, but as shown, it was not an effective approach for this application. In the next section, we propose a clamping resonator that was designed to effectively clamp the S0 wave's in-plane motion without introducing unwanted out-of-plane displacements. That is, we focused on how the resonator clamps the surface of the waveguide.

4. Clamping Resonators

The clamping resonators (shown in Figure 3) were designed using symmetry as a means to reduce the unwanted displacement components. The resonators are composed of a center cube with a side length of 7.9 mm, and four cylindrical "arms" that are 4.8 mm in diameter and 15 mm long. These components are joined using cyanoacrylate adhesive so that the resonator shown in Figure 3 is built. The arms perpendicular to the wave-propagation direction are meant to clamp the in-plane displacement, while the arms parallel to the wave-propagation direction are meant to reduce the unwanted out-of-plane displacement. All these arms undergo flexural vibration, presumably at the same frequency as the propagating wave, to achieve this effect. The guiding principles of the conceptual design were achieved through a combination of theoretical modeling and finite-element analysis, which will be described in more detail in a subsequent publication.

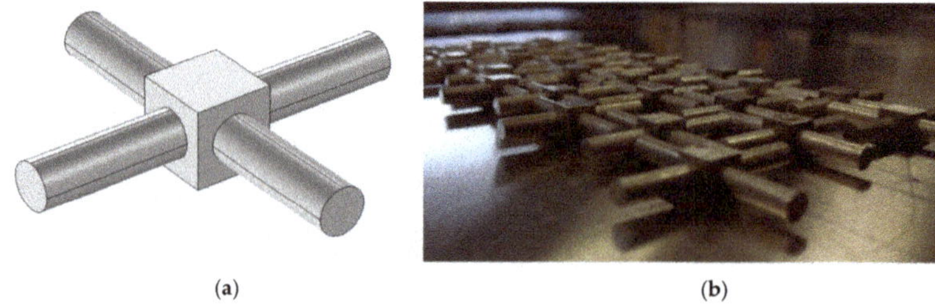

(a)   (b)

Figure 3. (a) Isometric view of a clamping resonator, and (b) the 30 clamping resonators in an array according to the spacing shown in Figure 1b.

4.1. Transmission beyond the Clamping Resonator Array

A cursory look at Figure 2c reveals that compared to the baseline results shown in Figure 2a, the clamping resonators noticeably reduce the amplitude of the transmitted S0 Lamb wave mode. By extracting slices of data from Figure 2 that are parallel to the frequency or wavenumber axis we can plot the frequency spectrum, or the wavenumber spectrum as shown in Figure 4a and b, respectively. These spectra reveal that the clamping resonator reduces the S0 Lamb wave mode by about 75% of the baseline measurement. On the other hand, the frequency-matched resonator unexpectedly causes a very slight increase in the transmitted S0 mode.

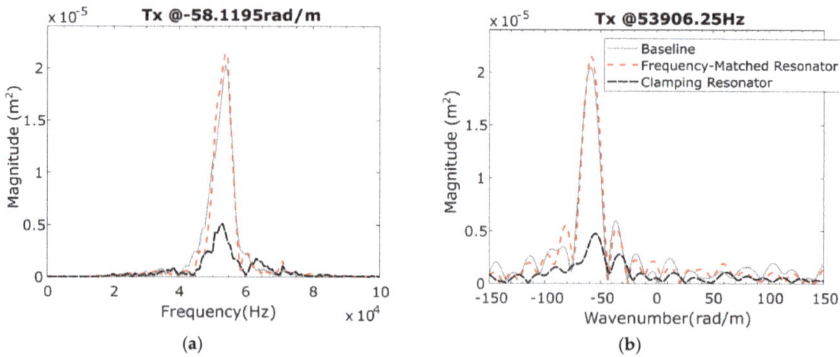

Figure 4. (a) The frequency spectra of the transmitted signal, which were extracted from the 2DFFTs shown in Figure 2 at a wavenumber of −58.1 rad/m; (b) The wavenumber spectra of the transmitted signal, which were extracted from the 2DFFTs shown Figure 2 at a frequency of 53.9 kHz.

Examples of the transmitted waveforms for the baseline and clamping resonators are shown in Figure 5. The comparison shows that the wave packet received beyond the clamping resonators has a significantly reduced amplitude at its center (around 200 µs). The inclusion of the clamping resonators caused the measured signals to vary slightly more than the signals from the baseline and frequency-matched resonator measurements; presumably due to scattering from the resonators. We note that putty was used to reduce edge reflections, but it did not completely attenuate the wayward waves, which could explain why some noise remains in the A-scans shown in Figure 5. The 2DFFT was used to help reduce unwanted signals reflecting off the edges of the plate by organizing the components of the A-scan measurements by wavenumber and frequency.

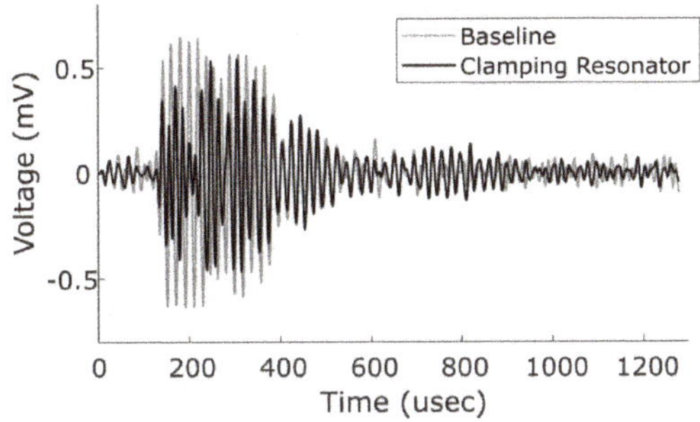

Figure 5. A-scans measured 69 cm from the transmitting transducer. The A-scan (black) from the plate with clamping resonators has a reduced amplitude around 200 µs when compared to the baseline (grey). These A-scan examples were some of the noisier signals that were measured.

4.2. Reflection from the Clamping Resonator Array

To determine whether the clamping resonators reflect or attenuate the incident S0 Lamb wave, the transducer setup was changed to measure and calculate the 2DFFT of possible reflections as shown in Figure 6a. The incident/reflection A-scans were used to calculate the 2DFFT shown in Figure 6b. Extracting a slice of the data parallel to the wavenumber axis at a frequency of 53.9 kHz yields the wavenumber spectrum shown in Figure 6c. The wavenumber spectrum indicates that about 62% of the incident wave was reflected from the resonator array. It should be noted that the plot shown in Figure 6c is not directly comparable to those in Figure 4b because different transducer arrangements were leveraged.

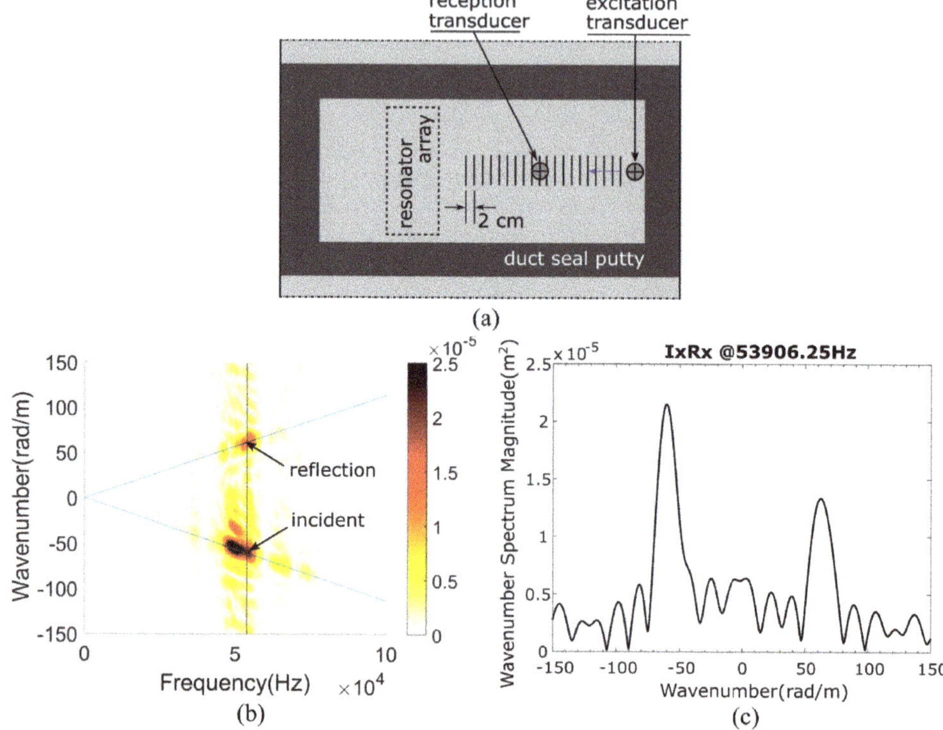

Figure 6. (**a**) Schematic of transducer positioning with respect to the resonator array for measuring reflected waves; (**b**) The 2DFFT of the incident/reflection measurements. The diagonal blue lines are the dispersion curves for the forward and backward propagating S0 Lamb wave. The vertical black line is where the wavenumber-spectrum slice data at a frequency of 53.9 kHz are obtained for (**c**) the wavenumber spectrum.

4.3. Finite-Element Simulations

Frequency domain finite-element analysis [17] was performed in order to provide confidence in the experimental results presented in Figures 2 and 4. This simulation assumes a plane wave solution and uses the frequency domain by assuming $e^{i\omega t}$ dependence, hence for a forward traveling wave the wavenumber k_x is negative-valued. Perfect matching layers are used on the ends of the model to prevent "end-wall" reflections. The remaining parts of the domain are divided into buffer, excitation, incident/transmit, and "affected" regions. The purpose of this model is to evaluate the transmission and reflection qualities of reflectors defined in the "affected" region. A similar model was used in Hsu [18], but we apply a body-load excitation instead of line excitation to excite a single mode. The displacement component u_x is plotted along the x-axis in Figure 7. The y-direction width of the model was chosen to correspond with the spacing of the resonators, which equates to a periodically repeating array of resonators in the $\pm y$-directions since periodic boundary conditions are applied to these edges. The clamping resonators are effective at prohibiting the propagation of much of the S0 Lamb wave, while the frequency-matched resonators are completely ineffective. The wavenumber spectrum was calculated using the complex-displacements measured from the transmitted wavefield and is shown in Figure 8. At the wavenumber of the S0 mode we see that the frequency-matched resonators only reduce the spectrum by approximately 7%, while the clamping resonators reduce it by 90%.

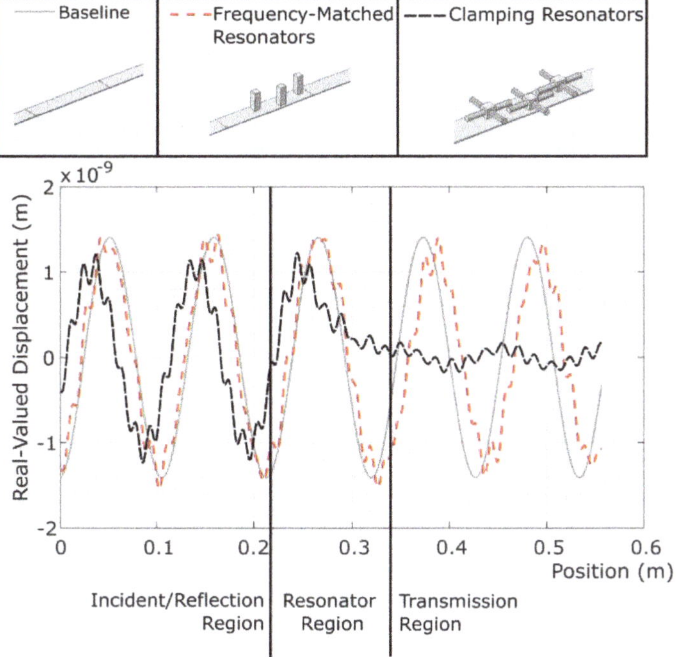

Figure 7. Finite-element predictions of the *x*-component displacement of an S0 Lamb wave at 50 kHz plotted along the *x*-axis showing that the clamping resonators block wave propagation, while the frequency-matched resonators do not.

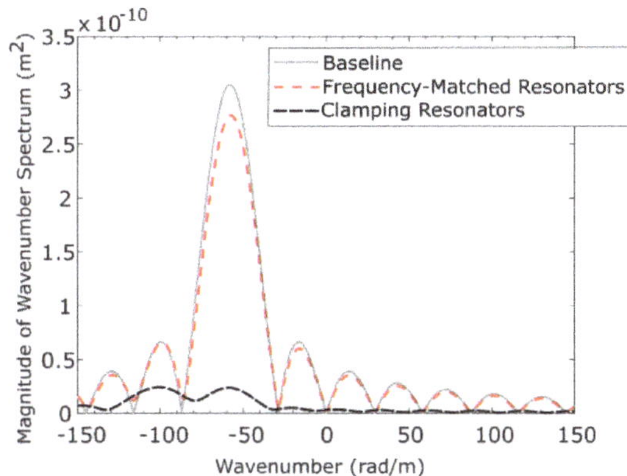

Figure 8. Finite-element-predicted wavenumber spectrum of the guided waves that were allowed to transmit past the resonator array. This spectrum is expected to only contain forward (+*x*) traveling waves.

5. Conclusions

Based on Williams et al.'s [7] identification of the clamping phenomenon, we sought to investigate whether the concept's applicability could be extended to S0 Lamb wave modes. Two approaches for designing a resonator were used based on the concepts discussed by Williams et al. These approaches resulted in the frequency-matched resonators and the clamping resonators (based on the clamping concept). The clamping resonators performed well by decreasing the amplitude of the transmitted wave by about 75%. When measuring the reflection from the clamping resonators, 62% of the incident wave was found to be reflected. The reason these two percentages do not equal each other is provisionally attributed to scattering from the resonator array and beam spread. In comparison, the frequency-matched resonators did not reduce the S0 Lamb wave amplitude. While the results for the clamping resonators used in these experiments are limited to the S0 Lamb mode at 50 kHz, these results and those of Williams et al. suggest that by designing a resonator based on the concept of clamping the waveguide, a metamaterial capable of prohibiting the propagation of a low-frequency A0 or S0 Lamb wave is possible. This also suggests that these resonators could be mode-specific by using different resonator designs for each mode. Moreover, a more detailed description of the basis for the concepts used to design the clamping resonators will be provided along with the associated theoretical modeling effort, which is still ongoing. Future work will focus on the frequency stop-bandwidth.

Author Contributions: Conceptualization, P.S., C.L. and C.H.; investigation, C.H.; writing—original draft preparation, C.H.; writing—review and editing, C.H., C.L. and P.S.; supervision, C.L.; project administration, P.S.; funding acquisition, P.S.

Funding: This research was funded by a seed grant from the College of Engineering at The Pennsylvania State University, "Engineering a giant metamaterial: a band-stop seismic/blast filter to shield critical civil infrastructure".

Conflicts of Interest: The authors declare no conflict of interest.

References

1. Pendry, J.B. Negative refraction makes a perfect lens. *Phys. Rev.* **2000**, *85*, 3966–3969. [CrossRef] [PubMed]
2. Cubukcu, E.; Aydin, K.; Ozbay, E.; Foteinopoulou, S.; Soukoulis, C.M. Negative refraction by photonic crystals. *Nature* **2003**, *423*, 604–606. [CrossRef] [PubMed]
3. Schurig, D.; Mock, J.J.; Justice, B.J.; Cummer, S.A.; Pendry, J.B.; Starr, A.F.; Smith, D.R. Metamaterial electromagnetic cloak at microwave frequencies. *Science* **2006**, *314*, 977–980. [CrossRef] [PubMed]
4. Norris, A.N.; Shuvalov, A.L. Elastic cloaking theory. *Wave Motion* **2011**, *48*, 525–538. [CrossRef]
5. Farhat, M.; Guenneau, S.; Enoch, S. Ultrabroadband elastic cloaking in thin plates. *Phys. Rev. Lett.* **2009**, *103*, 1–4. [CrossRef] [PubMed]
6. Stenger, N.; Wilhelm, M.; Wegener, M. Experiments on elastic cloaking in thin plates. *Phys. Rev. Lett.* **2012**, *108*, 1–5. [CrossRef] [PubMed]
7. Deymier, P.A. (Ed.) *Acoustic Metamaterials and Phononic Crystals*; Springer: Berlin, Germany, 2009.
8. Liu, Z.; Zhang, X.; Mao, Y.; Zhu, Y.Y.; Yang, Z.; Chan, C.T.; Sheng, P. Locally resonant sonic materials. *Science* **2000**, *289*, 1734–1736. [CrossRef] [PubMed]
9. Bilal, O.R.; Hussein, M.I. Trampoline metamaterial: Local resonance enhancement by springboards. *Appl. Phys. Lett.* **2013**, *103*, 111901. [CrossRef]
10. Wu, T.C.; Wu, T.T.; Hsu, J.C. Waveguiding and frequency selection of lamb waves in a plate with a periodic stubbed surface. *Phys. Rev. B* **2009**, *79*. [CrossRef]
11. Rupin, M.; Roux, P. A multi-wave elastic metamaterial based on degenerate local resonances. *J. Acoust. Soc. Am. Express Lett.* **2017**, *142*, EL75–EL81. [CrossRef] [PubMed]
12. Li, Y.; Zhu, L.; Chen, T. Plate-type elastic metamaterials for low-frequency broadband elastic wave attenuation. *Ultrasonics* **2017**, *73*, 34–42. [CrossRef] [PubMed]
13. Zhao, J.; Bonello, B.; Boyko, O. Focusing of the lowest-order antisymmetric lamb mode behind a gradient-index acoustic metalens with local resonators. *Phys. Rev. B* **2016**, *93*, 174306. [CrossRef]
14. Rupin, M.; Lemoult, F.; Lerosey, G.; Roux, P. Experimental demonstration of ordered and disordered multiresonant metamaterials for lamb waves. *Phys. Rev. Lett.* **2014**, *112*, 234301. [CrossRef] [PubMed]
15. Williams, E.G.; Roux, P.; Rupin, M.; Kuperman, W. Theory of multiresonant metamaterials for A0 lamb waves. *Phys. Rev. B* **2015**, *91*, 104307. [CrossRef]
16. Alleyne, D.N.; Cawley, P. The interaction of lamb waves with defects. *IEEE Trans. Ultrason. Ferroelectr. Freq. Control* **1992**, *39*, 381–397. [CrossRef] [PubMed]

Appl. Sci. **2019**, 9, 257

17. Chillara, V.K.; Ren, B.; Lissenden, C.J. Guided wave mode selection for inhomogeneous elastic waveguides using frequency domain finite element approach. *Ultrasonics* **2016**, *67*, 199–211. [CrossRef] [PubMed]
18. Hsu, J. Low-frequency forbidden band in phononic crystal plates with Helmholtz resonators. *Jpn. J. Appl. Phys.* **2011**, *50*. [CrossRef]

 *applied sciences*

Article

A Hybrid Ultrasonic Guided Wave-Fiber Optic System for Flaw Detection in Pipe

Joseph L. Rose [1,2]**, Jason Philtron** [1,*]**, Guigen Liu** [3]**, Yupeng Zhu** [3] **and Ming Han** [3]

1   FBS, Inc. dba Guidedwave, 450 Rolling Ridge Drive, Bellefonte, PA 16823, USA;
    jlresm@engr.psu.edu or jrose@gwultrasonics.com
2   Department of Earth and Engineering Sciences, Pennsylvania State University, University Park,
    PA 16802, USA
3   Department of Electrical and Computer Engineering, Michigan State University, East Lansing,
    MI 48824-1226, USA; guigenliu@hotmail.com (G.L.); yup.zhu@gmail.com (Y.Z.); mhan@egr.msu.edu (M.H.)
*   Correspondence: jphiltron@gwultrasonics.com

Received: 11 April 2018; Accepted: 2 May 2018; Published: 5 May 2018

Abstract: The work presented in this paper shows that Fiber Bragg Grating (FBG) optical fiber sensors can potentially be used as receivers in a long-range guided wave torsional-mode pipe inspection system. Benefits over the conventional pulse-echo method arise due to reduced total travel distance of the ultrasonic guided wave reflections, since reflections from defects and structural features do not need to propagate a full round trip back to the transmitting collar. This is especially important in pipe configurations with high attenuation, such as coated and buried pipelines. The use of FBGs as receivers instead of conventional piezoelectric or magnetostrictive elements also significantly reduces cabling, since multiple FBG receivers can be placed along a single optical fiber which has a diameter on the order of only around 100 μm. The basic approach and sample results are presented in the paper. Additionally, a brief overview of some topics in ultrasonic guided waves is presented as a background to understand the inspection problem presented here.

Keywords: ultrasonic guided waves; fiber optics; fiber Bragg grating; nondestructive testing; structural health monitoring; pipe inspection

---

1. Introduction

Guided wave ultrasound is increasingly being used to solve inspection problems in a wide variety of applications. This is because guided waves can be used to inspect over long distances from a single probe position, offer complete volumetric coverage, and can inspect hidden/inaccessible structures, such as those under water, coatings, insulation, and soil. Before discussing the merger of ultrasonic guided waves with fiber optic technology, the major topic in this paper, a brief review of some of our first accomplishments on various topics in ultrasonic guided waves is presented. This illustrates the background leading to the topic of this paper.

Two textbooks on the subject of wave mechanics fundamentals with emphasis on guided waves for nondestructive testing (NDT) and structural health monitoring (SHM) are presented by Rose, [1] 1999, [2] 2014. Some selected developments over the last few decades include the following. The value of computing and selecting specific wave structures from the phase velocity dispersion curves was initially pointed out by Ditri, Rose, and Chen [3] 1991, where defects close to a surface were found with wave structure having lots of energy near the surface. Since then paying attention to wave structure has led to solutions for many important problems in NDT and SHM. Pilarski, Ditri, and Rose in [4] 1993, demonstrated the ability to locate wave structure profiles with in-plane displacement on the surface of a waveguide, which has played a significant role in ice detection and the impact of water loaded surfaces on many NDT and SHM problems. A variety of problems in the aircraft industry are covered by Rose and Soley [5] 2000. Such topics as lap joint, tear strap and honeycomb inspection are discussed. A method of focusing ultrasonic guided waves in pipe was introduced by Rose, Sun, Mudge, and Avioli [6] 2003. The phased array focusing approach improves detection sensitivity, axial and lateral resolution and even penetration power. The ability to generate dispersion curves for structures having an arbitrary cross section was introduced by Hayashi, Song, and Rose [7] 2003. Many new applications with guided waves became possible and in particular a breakthrough in treating guided wave propagation in rails. Guided wave tomography concepts and several sample problems were coved by Gao, Shi, and Rose [8] 2005. The critical problem of rail inspection under shelling for transverse crack detection in the rail head along with possibilities of web and base defect detection was presented by Lee, Rose, and Cho [9] 2008. Adhesive bondline inspection breakthroughs was reported for certain substrate systems by Puthillath and Rose [10] 2010. Utilization of magnetostrictive technology advancements for the generation and utilization of guided waves was reported by Van Velsor, Royer, Owens, and Rose [11] 2013. Guided wave mode and frequency tuning and optimization is discussed in Philtron and Rose [12] 2014, making it possible to find defects in a variety of different structures more completely. A phased array system for plate inspection was developed and reported by Rose, Borigo, Owens, and Reese [13] 2017. This technique is responsible for amazingly fast inspection of corrosion type defects in large plate-like structures employing radar type circumferential or sector scans. Other NDT applications for guided waves, for example, include oil-tank weld detection using EMAT-generated shear-horizontal guided waves [14] 2017, fundamental torsional mode detection of angled cut flaws [15] 2016, and liquid level displacement measurements using magnetostrictive sensors [16] 2017.

Our background in ultrasonic guided waves thus prepared us for a merger with concepts presented by Perez, Cui, and Udd [17] 2001. Perez et al. [17] showed the possibility of ultrasonic wave interaction with fiber Bragg gratings (FBG) in an optical fiber. To date, FBG ultrasonic sensors have been extensively investigated [18] 2008. The use of the fundamental shear horizontal mode in plates and torsional mode in pipes is of interest for ultrasonic inspection due to the nondispersive nature and insensitivity to water loading that these modes offer. However, the majority of FBG ultrasonic sensors have been used to detect Lamb-type guided waves which are quite different from the shear horizontal (SH) waves in plates and torsional waves in pipes, there are only a few papers experimentally describing the usage of FBG sensors for SH wave detection [19–21] but with inclusive results. The mechanism governing the interaction of an FBG sensor with SH waves in plates has not been thoroughly reported until our work was recently published by Liu, Philtron, Zhu, Rose, and Han [22] 2018. Our results in [22] confirm both theoretically and experimentally the feasibility of using FBG sensors for SH wave detection with a maximum reception efficiency around $45°$ incidence, which shows a completely different angular dependence in comparison to Lamb-type wave detection.

In this current paper, taking advantage of the remote sensing (attenuation of the optical signal in an optical fiber is ultralow) and multiplexing capability of FBG sensors, the work is expanded upon our previous work in [22] to demonstrate how the technique could be used to extend the range in medium and long-range pipe inspection applications using torsional guided waves. The use of the fundamental shear horizontal mode in plates and torsional mode in pipes is of interest for ultrasonic inspection due to the nondispersive nature and insensitivity to water loading that these modes offer. Since long-range guided wave pipe inspection is critical and taking place all over the world on thousands of miles of pipe, the work reported in this paper was intended to investigate if improvements in inspection reliability and cost reduction would be possible with a hybrid ultrasonic guided wave-FBG approach. Some earlier work on this subject was reported by Hu, Zhao, and Li [23] 2012 and Ray, Srinivasan, Balasubramaniam, and Rajagopal [24] 2017. Hu et al. [23] 2012 performed a modeling study suggesting feasibility to integrate optical fiber sensors into pipeline inspection using the L(0,2) mode, but an experiment was not reported. Ray et al. [24] 2017 describe the use of fiber optic sensors to detect changes in the guided wave features of the L(0,2) mode to assess cross sectional irregularities such as pipe eccentricity. Obviously, their work does not address torsional wave detection. The work reported in this paper will illustrate how FBG receivers could be implemented as part of a hybrid ultrasonic pipe inspection technology using torsional guided waves.

2. Hybrid Ultrasonic—Fiber Optic System Concept

Commercially available systems for ultrasonic guided wave pipe inspection can quickly and reliably scan large sections of pipeline for flaws. However, there are certain situations where increased range of an inspection system is desired. This may occur in such situations as for highly attenuative pipe, such as buried, coated, or generally corroded lines, or in situations where access is limited. In these situations, use of an alternate method, such as fiber optic sensors, in conjunction with conventional ultrasonic guided wave inspection can be beneficial to increase inspection range into the non-inspected region. Fiber optic sensors may be attached to a pipeline during installation and then left in place for use during periodic inspections. The fiber optic sensors could be placed near the end of the range of a conventional pulse-echo inspection system, allowing for increased range by as much as 50%, if the optical system has a similar signal-to-noise ratio. This can be achieved because the reflected wave energy from a flaw need not travel the full return distance to the actuator, but only back to the remote fiber optic sensor to be detected.

This concept is illustrated in Figure 1. The conventional ultrasonic collar generates guided waves in the pipe at location 0. Remote fiber optic sensor(s) are placed at a distance X from the collar, which should be near, but not beyond, the effective inspection range of the conventional system. The inspection range of the hybrid system is then extended such that a flaw at a distance Y from the collar is now detectable from the original collar location through use of the fiber optic sensor. Note that the optical system could be collocated with the ultrasonic equipment, or not. One key difference between optical fiber and conventional metal cable carrying ultrasonic signals is that the loss of the optical signal is very low in comparison.

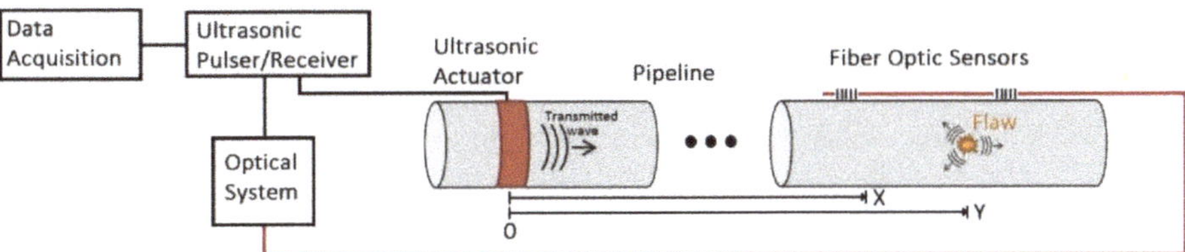

Figure 1. Global hybrid ultrasonic guided wave and fiber optic inspection system concept. The range of a conventional collar at position 0 can be increased beyond position X using a fiber optic sensor attached near position X. This allows the hybrid system to detect a flaw at a further distance, position Y.

Consider an example coated pipeline with high attenuation. It is possible that a conventional inspection distance is only 50 feet. Flaws beyond 50 feet from the collar are not detected because potential flaw reflections are hidden within the noise. However, if a fiber optic sensor was installed near the 50-foot location, the inspection range could be increased by 25 feet, to a total of 75 feet. Assuming constant attenuation along the pipe, the ultrasonic wave energy would decrease the same amount traveling from 50 feet back to the collar, or from 50 to 75 feet and back to the fiber optic sensor. If the fiber optic sensor has a similar signal-to-noise ratio as the ultrasonic collar, the effective inspection range can be increased by as much as 50%. The fiber optic sensors could be attached and left in place (under coatings) during pipeline installation, saving significant time and cost from inspection and preparation costs to expose the pipe for collar application, particularly for the case where access must be dug in the middle of a buried section of pipeline. Note that the use of a hybrid system could be costly due to the need for extra components to perform optical sensing such as the laser, controller, photodetector, etc. However, in certain cases where it can achieve improved performance, such as in cases where increased inspection range provides inspection of previously un-inspectable or expensive or hard-to-inspect areas, use of a hybrid system will be worthwhile. Future developments will also lead to reduced cost of optical components. Generally, it is expected that the fiber sensors will be used in a similar method as for conventional piezoelectric and magnetostrictive sensors, i.e., to detect reflections from flaws. Although it may be possible to detect flaws between a sender and receiver, flaw detection is much easier and more straight forward when analyzing a reflected wave signal from a flaw.

Note also that additional SMS collar(s) could be installed at location X, etc., to extend the inspection range, but these collars are generally more expensive than optical fiber, would require many, long cables (typically 16, one per channel) with a remotely used data acquisition system, and there may not physically be access space to install these collars. The use of fiber optic sensors also allows for many sensors along a single optical cable.

3. Materials and Methods

An experiment was conducted on a 21-foot section of 6-inch NPS schedule 40 carbon steel pipe. A segmented magnetostrictive (SMS) collar was placed 3 feet from the pipe end and used to generate torsional axisymmetric ultrasonic guided waves in the pipe. Generation of ultrasound using the magnetostrictive effect is described in Ref. [25] 1993. The SMS collar was constructed of copper meandering coils printed on flexible circuit boards with a design frequency of 64 kHz, which were placed over a 2-inch-wide strip of iron cobalt (FeCo) that was bonded to the circumference of the pipe. Olympus UltraWave LRT hardware (OSSA, Waltham, MA, USA) and custom software were used to generate torsional mode ultrasonic guided waves and collect data. This sending setup, and the SMS coil design, is identical to the one used and described in Ref. [11]. Optical FBG sensors were initially collocated with the collar, on the top and bottom of the pipe, and used to receive ultrasonic guided wave energy. The optical fiber sensors were oriented at 45° relative to the pipe axis and directly attached to the pipe surface using clear Scotch® tape. Tape was used because a temporary method for sensor bonding was desired so that the fiber optic sensor could be easily and quickly moved to various locations during experimental testing. Epoxy or superglue (cyanoacrylate) would be a more appropriate long-term choice for permanent sensor installation. The 45° orientation was previously shown to result in maximum reception of shear horizontal-type guided wave energy in plates [22] 2018. Ref. [22] also gives a theoretical justification for why this angle achieves the highest amplitude response to shear-type guided waves. Two flaws were introduced in the pipe at 13 feet and 10 feet from the pipe end. A schematic of the pipe is illustrated in Figure 2.

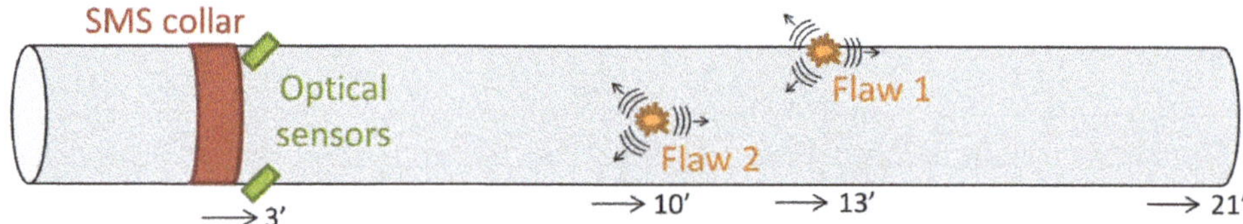

Figure 2. Schematic of the 21-foot pipe specimen. The SMS collar and the initial FBG sensor locations were 3 feet from the pipe end. Flaws were introduced at 13 feet and 10 feet from the pipe end. The side with flaw 2 will be referred to as the right side. Number and locations of the optical fiber sensors vary in the different experiments.

A schematic of the test equipment is shown in Figure 3. A laptop was used for data acquisition and custom software was used to control the UltraWave pulser/receiver hardware. The pulse was sent to the collar to generate ultrasonic guided waves in the pipe. One or more fiber optic sensors were coupled to the pipe and a distributed feedback (DFB) laser was used with a locking system to keep the laser wavelength locked onto the spectral notch of the optical fiber sensor, and a photodetector was used to convert the light signal into a voltage signal for recording. Different from the benchtop narrow linewidth laser used in our previous work [22], the much more portable and low-cost DFB lasers (A1905LMI, Avanex, New York, NY, USA) were used in this paper with moderate compromise in signal-to-noise ratio (due to the larger noise of a DFB laser). Furthermore, a two-channel system was used in this work to demonstrate the multiplexing capability of the FBG sensing system (see Figure 3), i.e., one DFB laser and FBG pair works at 1550 nm wavelength while the other pair works at 1545 nm. A dense wavelength division multiplexer (DWDM) was used to optically separate the two channels and the ultrasonic signals received by the two FBG sensors are acquired simultaneously. Additional details describing each single-channel setup are found in Ref. [22] 2018. Although the two-channel system is enough for the demonstration of multiplexing capability, when using this type of optical setup additional lasers could be added for additional simultaneous reception by fiber sensors connected to other channels of the DWDM but it is not the focus of this paper. Note that the type of optical fiber sensor used in this study is a Fabry-Perot interferometer formed by a pair of tandem chirped fiber Bragg gratings, each with a 5-mm grating length, abbreviated as CFBG-FPI and referred to in this work simply as an optical fiber sensor. Any reasonable optical fiber Bragg grating type sensor may be used to detect torsional guided waves; this sensor type was selected due to its relatively high signal-to-noise ratio, although it has other properties such as the ability to withstand relatively large static strains in the structure while still using its narrow spectral notch pattern [26] 2017.

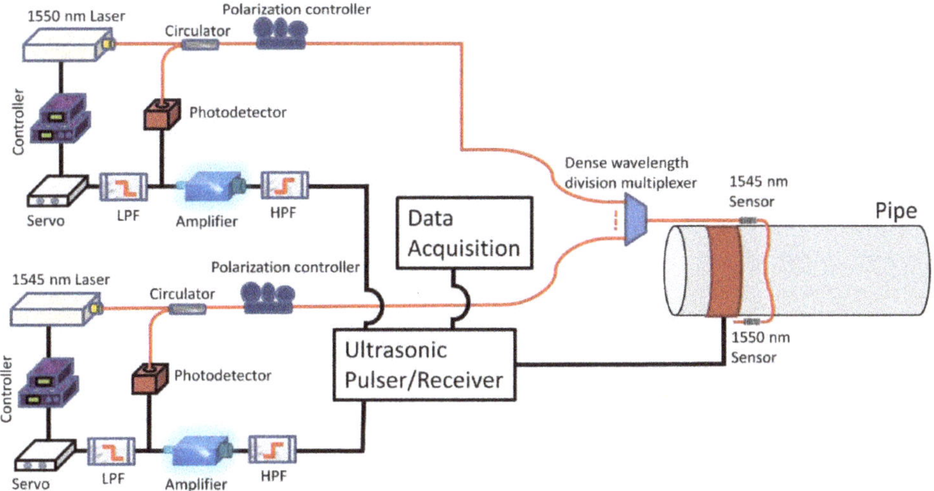

Figure 3. Schematic of the integrated ultrasonic-optical system setup. A servo controller was used to keep the laser wavelength locked to the slope of the FBG spectrum for each channel. The optical signal was converted to a voltage signal and recorded using conventional ultrasonic hardware.

4. Results

4.1. Confirmation of Torsional Mode and Flaw Detection

Figure 4 shows a frequency scan (F-Scan) plot of data collected using ultrasound generated by the SMS collar and received by the optical fiber sensor attached on the bottom of the pipe at 3 feet from the pipe end. The F-Scan plot shows data from a series of A-scans (amplitude scans) with different pulse center frequencies vs. distance, with amplitude shown in color. The dead zone, pipe end reflections, and flaw indications are indicated. These two flaws are located at 7 feet and 10 feet from the sensor and on the top and right side of the pipe, respectively. The flaws are 0.5-inch diameter and 0.25-inch deep drilled holes, which represent a simulated 2% CSA (pipe cross-sectional area) corrosion flaw.

Reflections from a reverse wave are also shown in Figure 4. Although the SMS collar was activated with directional control, in which two axially-separated coils are used with a time delay to cancel the backward-travelling wave, the cancellation is not perfect, and therefore extra reflections are shown at 3 feet and 13 feet, and to a lesser extent near 10.5 feet, from the collar location. These signals would not be as prevalent in a longer length of pipeline, since a 100% reflector (cut pipe end) would not be present. Also, these reverse wave reflections could be suppressed if two axially-separated optical fiber sensors are used and a directional control routine is applied on the received signals.

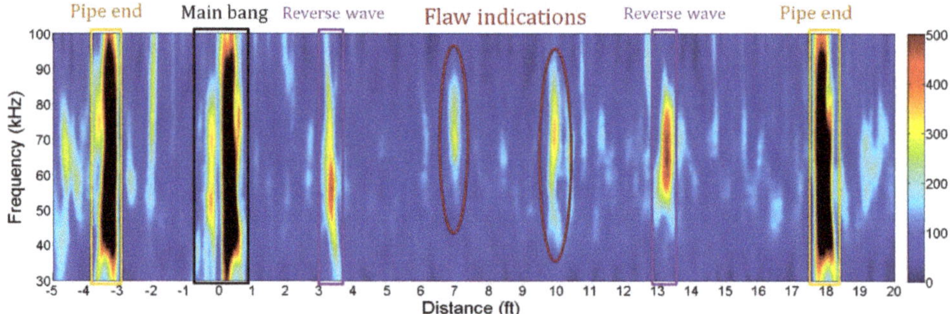

Figure 4. F-Scan generated using an FBG sensor directly bonded to the bottom side of the pipe at 3 feet from the pipe end, collocated with SMS sending collar. Flaw indications are shown at 7 feet and 10 feet from the sensor. Pipe end reflections are shown at −3 feet and 18 feet. Indications at 3 feet and 13 feet are due to waves traveling from the SMS collar in the reverse direction and bouncing off the near end of the pipe at −3 feet, a result of imperfect directional control. The optical fiber sensor can clearly be used for flaw detection.

Even though the FBG sensor is effectively sensing at only a small portion (approximately a "point") along the circumference, as opposed to sensing around nearly the full circumference as is achieved with a conventional long-range ultrasonic collar, it is clear from Figure 4 that positive flaw detection occurs over a range of frequencies. The ability of the "point" optical fiber sensor to detect a flaw reflection is dependent on the relative locations of the flaw and sensor, as will be discussed further below.

For comparison, F-Scan data using solely the SMS collar in pulse-echo mode is displayed in Figure 5. The two 2% CSA flaws are clearly detectable at 7 and 10 feet. Also, there is a higher signal-to-noise ratio for this setup, which is typically seen for magnetostrictive sensors. Use of different optical components, such as a laser with lower noise, could achieve a similar signal-to-noise ratio. Note also that the dual axially-separated coils allow for directional control and also a suppression algorithm on the received wave, suppressing the reverse waves shown in Figure 4, which is not possible for a single fiber optic sensor but could be achieved with multiple fiber optic sensors, as noted above. The experimental pipe tested here does not highlight the advantages of the hybrid sensing method, since this application of a short pipe of low attenuation would not benefit from range extension. The overall goal of this work is to prove that detection with a hybrid system is possible and describe future applications where it could prove beneficial.

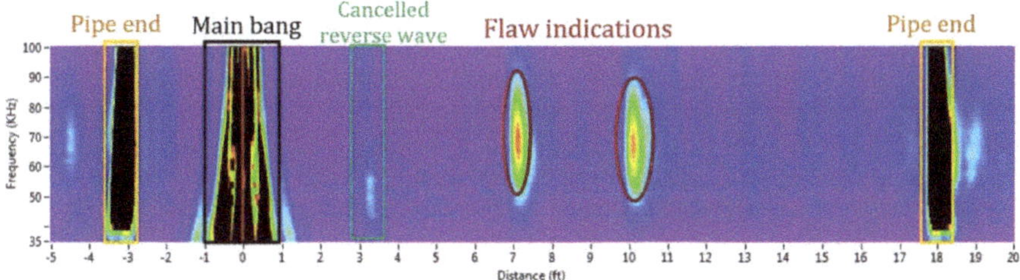

Figure 5. F-Scan generated using the SMS collar in pulse-echo mode at 3 feet from the pipe end. Flaw indications are shown at 7 feet and 10 feet from the sensor. Pipe end reflections are shown at −3 feet and 18 feet. There is a higher signal-to-noise ratio for this setup. Note also that the dual coils allow for directional control of the received wave, suppressing the reverse waves shown in Figure 4.

4.2. Progressive Damage Test in SHM Mode

Hybrid ultrasonic-optical data was also collected in a structural health monitoring (SHM) mode. In SHM mode, data was collected on the undamaged pipe, and then after each successive growth in flaw size. Figure 6 shows enveloped waveforms from several SHM states when the second flaw, located 7 feet from the sensor, was at 0%, 0.5%, 1% and 2% CSA. As shown, the reflection amplitude increases as the flaw size increases. Additionally, the reflection amplitude from the flaw at 10 feet from the sensor remains relatively constant, because the size of this flaw remains the same. Note that the 0.5% and 1% CSA flaws represent a 0.25-inch diameter hole with 0.125-inch depth and a 0.25-inch diameter hole with 0.236-inch depth, respectively.

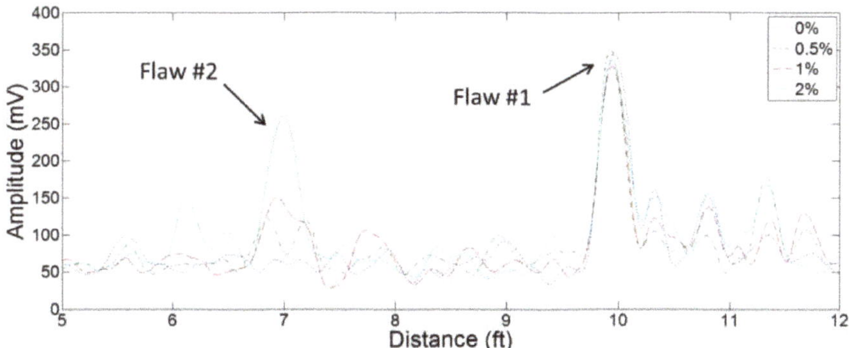

Figure 6. Waveform data (amplitude envelope) for SHM states with flaw growth of 0 to 2% CSA for the flaw located 7 feet from the optical fiber sensor. Waveforms are shown for an excitation frequency of 76 kHz using the SMS collar and a single direct-bonded FBG optical fiber sensor used as a receiver. As the flaw grows, the reflected ultrasonic wave energy from the flaw increases. The optical fiber sensor can clearly be used for flaw sizing.

4.3. Individual FBG Sensor NDT

It is clear from Figures 4 and 6 that the ultrasonic-optical approach can be effective for flaw detection. This section assesses the ability of a point-like receiver to reliably detect reflections from a particular flaw location. What is found is that if the flaw is relatively far from the receiving sensor, then detection is possible at all circumferential locations (see Figure 7). However, if the flaw is relatively close to the receiving sensor, then detection is less reliable as it is more dependent on the circumferential location of the FBG relative to the flaw (see Figure 8). This result will be dependent on the scattering field from the flaw at small sensor-flaw distances, which will vary according to flaw geometry and ultrasonic wavelength. However, at further distances, for circular and many corrosion-like flaws, the reflected wave will be detectable at nearly any circumferential location due to the wrapping of the reflected wave path circumferentially around the pipe. Additionally, sweeping inspection frequency to generate the F-scan further improves the likelihood of detection.

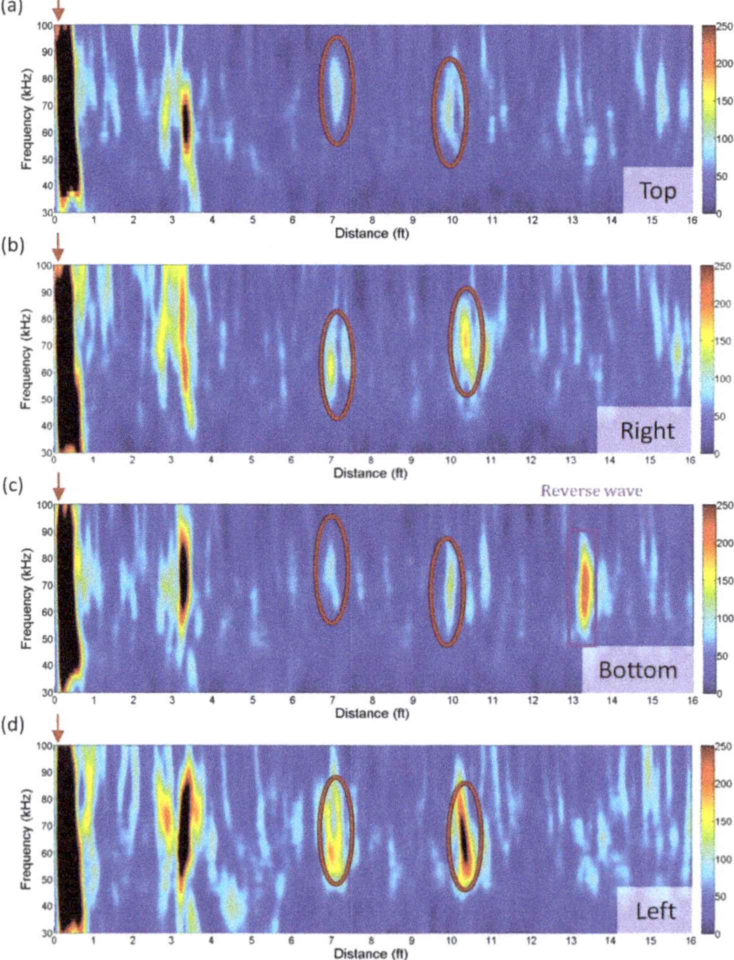

Figure 7. F-Scans collected using a direct-bonded FBG receiver collocated with the sending collar at several circumferential locations: (**a**) top; (**b**) right side; (**c**) bottom; and (**d**) left side of the pipe. Since the receiver is relatively far from the flaws, both flaws (red ovals) are clearly detectable at all circumferential locations. Flaw indications are marked by the red ovals. Arrows indicate the axial location of the fiber optic receiver. Distance is indicated from source/receiver position. When the receiver is relatively far from the flaw, i.e., multiple pipe circumferences, flaw detection is reliable at any circumferential location. Note that the indication in (**c**) at 13 feet (rectangle) is due to a reverse wave reflection.

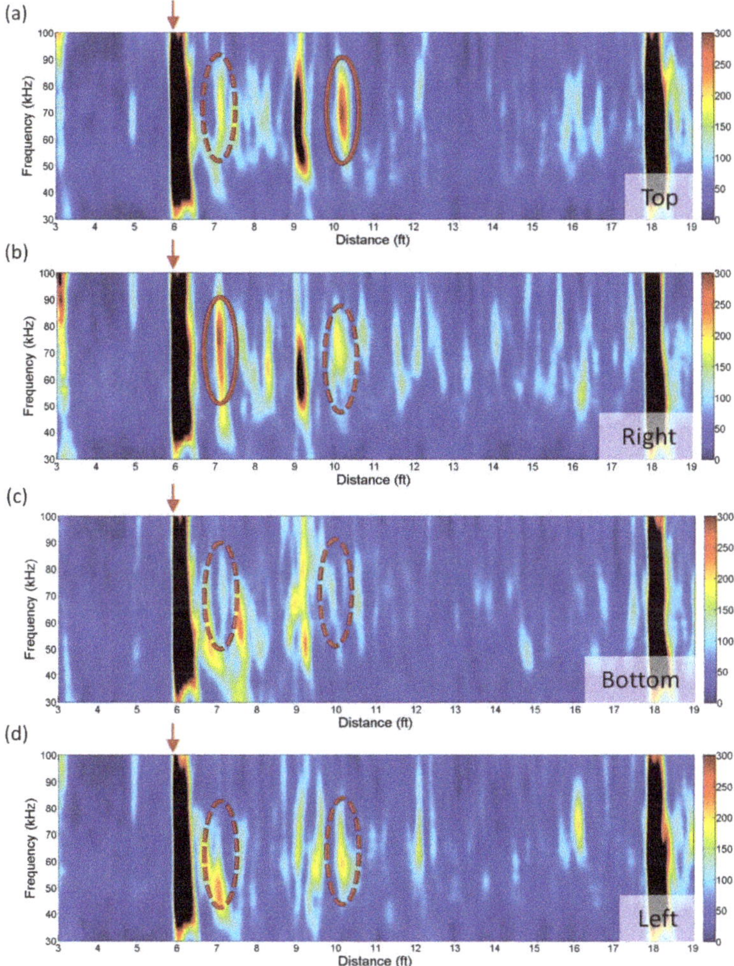

Figure 8. F-Scans collected using a direct-bonded FBG receiver 6 feet in front of the sending collar at several circumferential locations: (**a**) top; (**b**) right side; (**c**) bottom; and (**d**) left side of the pipe. Since the receiver is relatively close to the flaws, the best flaw detection occurs directly in line with the flaw location, while the detection at other circumferential locations is unreliable. Clear flaw indications are marked by the red ovals and less clear or undetectable flaw indications are marked by dashed red ovals. Arrows indicate the axial location of the fiber optic receiver. Distance is indicated from the sending collar. When the receiver is near the flaw, but not at the same circumferential location, flaw detection is not reliable, due to the use of a point-like sensor.

Figure 7 shows F-Scans collected with optical fiber sensors collocated with the collar at four distinct circumferential positions (top, right, bottom, and left sides). The axial distance to the two flaws is 7 feet and 10 feet, respectively. Compared to the pipe circumference, this axial distance is relatively large compared to the pipe circumference, at 4 and 5.8 times the circumference, respectively. At this axial distance, sensors at all four circumferential positions detect both flaw reflections. It is believed that detection is possible at all circumferential locations because the scattered wave generated by the interaction of the initial axisymmetric wave and the flaw has had sufficient time to spiral around the pipe by the time the waves return to the receiving sensor locations.

Figure 8 shows F-Scans collected with optical fiber sensors located 6 feet in front of the collar at four separate circumferential positions. In this setup, the axial distance to the two flaws is 1 foot and 4 feet, respectively. This is a relatively small axial distance between the sensors and flaws at 0.6 and 2.3 times the circumference, respectively. In this case, the sensor located on the opposite side of the pipe from the flaw has the lowest reflected wave amplitude from that position. In both cases, the sensor on the opposite side does not effectively detect the flaw, and the reliability of the other sensor locations for flaw detection is intermittent. This may be due to the smaller amount of energy in the scattered wave field travelling at high angles relative to the axis of the pipe, and also the natural angular reception efficiency of the optical sensor which decreases as incident angle increases relative to the pipe axis, with a maximum efficiency at 45°. The exception is for the sensor at the same circumferential location (directly in line) with the flaw. The sensor on the top of the pipe clearly detects the flaw at 10 feet, and the sensor on the right of the pipe clearly detects the flaw at 7 feet. These results suggest that to have 100% detection coverage of the pipe using point-like sensors, including at axial distances close to the sensors, an array of optical fiber sensors is preferred—either circumferentially around the pipe, axially around the pipe, or a combination of the two. An array of sensors can detect flaws regardless of the closeness of the optical sensors to the flaws.

4.4. Sparse Array NDT

As mentioned at the end of the previous section, to increase flaw detection reliability, a sparse array of optical fiber sensors could be situated at different circumferential and/or axial locations along the pipe. This would allow detection of a flaw at any location along a pipe, because if it is close to one sensor it will also be sufficiently far from other sensors to allow for reliable detection. Several sparse arrays were simulated by moving the optical fiber sensors to different locations on the test pipe. An optical fiber sensor was located at 0 feet, 3 feet, 6 feet and 9 feet from the sending collar on the top, left, bottom, and right sides at each of those axial locations. Data was collected at two locations at a time until data from all 16 locations was collected. Data from multiple positions is then combined to allow for an assessment of a variety of sensor array geometries.

In the present work, two optical fiber sensor arrays are considered. Both spiral down the pipe so that there is one sensor located on the top, right, bottom, and left sides of the pipe, each at a different axial location. Group 1 contained sensors on the left, bottom, right, and top, and group 2 contained sensors on the right, top, left, and bottom of the pipe at axial locations of 0 feet, 3 feet, 6 feet and 9 feet, respectively. This is shown schematically in Figure 9. These groups were selected such that group 1 had sensors directly in front of the two flaws, while the sensors closest to the flaws in group 2 are on the opposite side of the pipe (and therefore those two sensors are not expected to easily detect the flaws). Note that the flaws are located on the top and right sides of the pipe at 10 feet and 7 feet from the sending collar, respectively.

To combine the four waveforms collected via the four FBGs in each array into a single "group waveform", the individual waveforms are processed by suppressing known position-dependent signals including the dead zone, the direct wave arrival, and the reverse pipe end reflection. When these events are present in the signals, the amplitude is reduced to a value of 20 mV, chosen because it is close to the noise level. The amplitude envelope of each of the four waveforms is also shifted in time to correspond to their respective positions along the pipe axis. After time-shifting, the waveforms are added, and the average is calculated. Flaws are more easily detectable in the group waveform than in any individual waveform from a single FBG. Additionally, the combined waveform is more robust than data from a single location since we are unlikely to miss a flaw due to a particular flaw position relative to a single sensor. Note that in general, the arrival times of these suppressed features would be known *a priori* since the sensor locations are known. This suppression algorithm also adds a jaggedness to the waveforms, as the suppression is applied as a step function. For the data shown in

Figure 10, this jaggedness occurs for the approximate length of the initial wave pulse (1 foot) as the wave passes each sensor location: 1–2 ft, 3–4 ft, 6–7 ft and 9–10 ft.

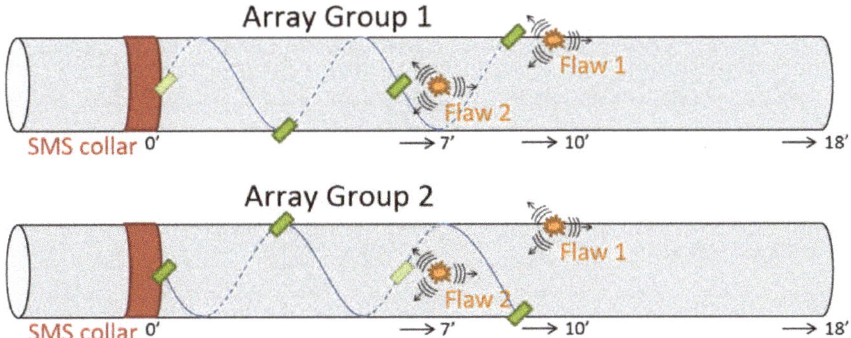

Figure 9. Schematic showing sensor locations for array groups 1 (**above**) and 2 (**below**). Fiber optic sensors are indicated by green rectangles and the optical fiber is indicated by the blue line. Dashed lines indicate the optical fiber and sensor on the far side of the pipe. The pipe is being viewed from the right side (flaw 2 side).

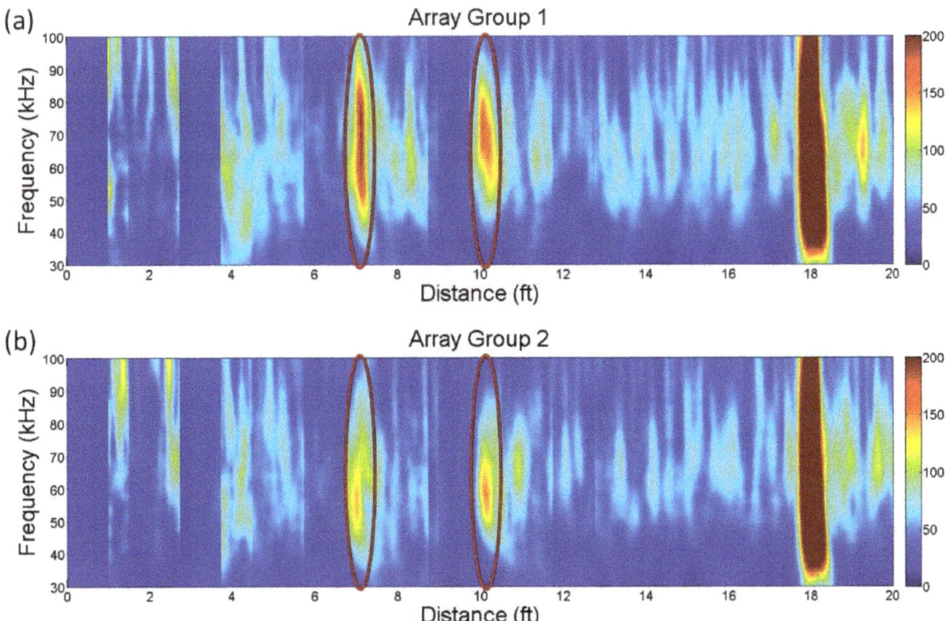

Figure 10. F-Scans of the processed group waveforms for both (**a**) group 1 and (**b**) group 2 fiber optic sensor arrays. Flaws (red ovals) are clearly detected at 7 feet and 10 feet from the sending collar across a wide range of frequencies. The far pipe end reflection is detected at 18 feet. Note that groups 1 and 2 both detect the flaws even though group 2 has multiple sensors near flaws in non-ideal locations for efficient flaw detection.

Figure 10 shows F-Scans of the processed group waveforms for both (a) group 1 and (b) group 2 fiber optic sensor arrays over 30 to 100 kHz frequencies generated with the SMS collar. The combined group 1 and group 2 signals both show good flaw detection capabilities. Both FBG sensor groups detect the flaws at 7 feet and 10 feet from the collar location, even though multiple sensors in group 2 are on the far side of the pipe from the flaw to which they are nearest. This confirms that the potential to miss a flaw based on its relative location to a single sensor is greatly diminished using this approach versus placing a single fiber sensor at a single axial location. The array approach can increase detection confidence. Creating a group signal from sensors at 4 circumferential locations at a single axial location can also give a similar result. Depending on inspection requirements and access limitations, different implementations of a sensor array may be desired. Additionally, the array group signals appear to be more stable over a wider frequency range, which makes for more reliable interpretation by the inspector. Due to these characteristics, the F-Scans in Figure 10 are easier to interpret and more reliable for flaw detection than the F-Scans shown in Figures 7 and 8, for a single fiber sensor at a single axial and circumferential location.

5. Discussion

It is believed that the work presented here represents the first study of flaw detection in pipes using an ultrasonic-optical system and torsional ultrasonic guided waves. Torsional wave reception using FBG sensors was successful, and damage detection, monitoring, and localization were demonstrated. Additionally, one method for increased confidence for flaw detection was demonstrated, in which signals from a sparse array of optical fiber sensors were combined.

A hybrid ultrasonic-optical inspection system has a wide variety of potential applications. One application is with permanently-mounted long-range ultrasonic guided wave pipe inspection collars. The optical fiber could be far from the excitation area and can increase the effective inspection range from a single sending collar position. Optical fiber sensors may also be used in places with limited access for traditional collars. The optical cable is very thin (as little as 80 μm diameter for bare fiber), so it can easily be mounted under coatings. Additionally, the optical fiber sensors can withstand harsh environments such as high temperature, radiation, and high levels of electromagnetic interference (EMI). Finally, the optical fiber sensors have a much wider bandwidth than traditional piezoelectric or magnetostrictive sensors.

Future work should focus on the development of ruggedized and miniaturized optical components and the demonstration of system effectiveness in field tests. Tests demonstrating an increase in inspection range on buried pipe could open avenues for additional applications. Optical fiber sensors provide unique advantages over traditional sensor types which will encourage their use in challenging inspection problems in the future.

Author Contributions: Joseph L. Rose and Jason Philtron contributed the ultrasonic guided wave and hybrid aspect of the research work including theory, experiment, and the write up. Guigen Liu, Yupeng Zhu, and Ming Han contributed fiber optic Bragg grating experimental portion of the work and assisted in manuscript revision.

Funding: The work carried out by J. L. Rose and J. Philtron was funded by Naval Research Laboratory grant number N0001417RX00182. The work carried out by G. Liu, Y. Zhu, and M. Han was carried out primarily at University of Nebraska-Lincoln and funded by the Office of Naval Research under Grant N000141712819 and Grant N000141410456.

Acknowledgments: The authors would like to thank Bob Brown and James Tagert at the Naval Research Laboratory for making this work possible. The authors would also like to acknowledge Russell Love and Cody Borigo for assistance with software and the experimental setup.

Conflicts of Interest: The authors declare no conflict of interest.

References

1. Rose, J.L. *Ultrasonic Waves in Solid Media*; Cambridge University Press: New York, NY, USA, 1999.
2. Rose, J.L. *Ultrasonic Guided Waves in Solid Media*; Cambridge University Press: New York, NY, USA, 2014.

3. Ditri, J.J.; Rose, J.L.; Chen, G. Mode Selection Guidelines for Defect Detection Optimization Using Lamb Waves. In Proceedings of the 18th Annual Review of Progress in Quantitative NDE, Brunswick, ME, USA, 28 July–2 August 1991; Plenum: New York, NY, USA, 1991; Volume 11, pp. 2109–2115.

4. Pilarski, A.; Ditri, J.J.; Rose, J.L. Remarks on Symmetric Lamb Waves with Dominant Longitudinal Displacements. *J. Acoust. Soc. Am.* **1993**, *93*, 2228–2230. [CrossRef]

5. Rose, J.L.; Soley, L.E. Ultrasonic Guided Waves for Anomaly Detection in Aircraft Components. *Mater. Eval.* **2000**, *50*, 1080–1086.

6. Rose, J.L.; Sun, Z.; Mudge, P.J.; Avioli, M.J. Guided Wave Flexural Mode Tuning and Focusing for Pipe Inspection. *Mater. Eval.* **2003**, *61*, 162–167.

7. Hayashi, T.; Song, W.J.; Rose, J.L. Guided Wave Dispersion Curves for a Bar with an Arbitrary Cross-section, a Rod and Rail Example. *Ultrasonics* **2003**, *41*, 175–183. [CrossRef]

8. Gao, H.; Shi, Y.; Rose, J.L. Guided Wave Tomography on an Aircraft Wing with Leave in Place Sensors. In *Review of Quantitative Nondestructive Evaluation*; AIP: New York, NY, USA, 2005; Volume 24, pp. 1788–1794.

9. Lee, C.M.; Rose, J.L.; Cho, Y. A Guided Wave Approach to Defect Detection under Shelling in Rail. *NDT E Int.* **2008**, *42*, 174–180. [CrossRef]

10. Puthillath, P.; Rose, J.L. Ultrasonic Guided Wave Inspection of a Titanium Repair Patch Bonded to an Aluminum Aircraft Skin. *Int. J. Adhes. Adhes.* **2010**, *30*, 566–573. [CrossRef]

11. Van Velsor, J.; Royer, R.; Owens, S.; Rose, J.L. A Magnetostrictive Phased Array System for Guided Wave Testing and Structural Health Monitoring of Pipe. *Mater. Eval.* **2013**, *71*, 1296–1301.

12. Philtron, J.H.; Rose, J.L. Mode Perturbation Method for Optimal Guided Wave Mode and Frequency Selection. *Ultrasonics* **2014**, *54*, 1817–1824. [CrossRef] [PubMed]

13. Rose, J.L.; Borigo, C.; Owens, S.; Reese, A. Rapid Large Area Inspection from a Single Sensor Position: A Guided Wave Phased Array Scan. *Mater. Eval.* **2017**, *75*, 671–678.

14. Chen, L.; Choi, W.R.; Lee, J.G.; Kim, Y.G.; Moon, H.S.; Bae, Y.C. Oil-Tank Weld Detection Using EMAT. *Int. J. Hum. Robot.* **2017**, *14*, 1750008. [CrossRef]

15. Kim, Y.G.; Chen, L.; Moon, H.S. Numerical Simulation and Experimental Investigation of Propagation of Guided Waves on Pipe with Discontinuities in Different Axial Angles. *Mater. Eval.* **2016**, *74*, 1168–1175.

16. Chen, L.; Kim, Y.G.; Bae, Y.C. Long Range Displacement Measurements Systems Using Guided Wave. *Int. J. Fuzzy Log. Intell. Syst.* **2017**, *17*, 154–161. [CrossRef]

17. Perez, I.; Cui, H.-L.; Udd, E. Acoustic emission detection using fiber Bragg gratings. In Proceedings of the SPIE's 8th Annual International Symposium on Smart Structures and Materials, Newport Beach, CA, USA, 4–8 March 2001; International Society for Optics and Photonics: Bellingham, WA, USA, 2001.

18. Wild, G.; Hinckley, S. Acousto-ultrasonic optical fiber sensors: Overview and state-of-the-art. *IEEE Sens. J.* **2008**, *8*, 1184–1193. [CrossRef]

19. Li, F.; Murayama, H.; Kageyama, K.; Shirai, T. Guided wave and damage detection in composite laminates using different fiber optic sensors. *Sensors* **2009**, *9*, 4005–4021. [CrossRef] [PubMed]

20. Harish, A.V.; Ray, P.; Rajagopal, P.; Balasubramaniam, K.; Srinivasan, B. Detection of fundamental shear horizontal mode in plates using fibre Bragg gratings. *J. Intel. Mater. Syst. Struct.* **2016**, *27*, 2229–2236. [CrossRef]

21. Ray, P.; Rajagopal, P.; Srinivasan, B.; Balasubramaniam, K. Feature-guided wave-based health monitoring of bent plates using fiber Bragg gratings. *J. Intel. Mater. Syst. Struct.* **2017**, *28*, 1211–1220. [CrossRef]

22. Liu, G.; Philtron, J.H.; Zhu, Y.; Rose, J.L.; Han, M. Detection of Fundamental Shear Horizontal Guided Waves Using a Surface-bonded Chirped-fiber-Bragg-grating Fabry-Perot Interferometer. *IEEE J. Lightwave Technol.* **2018**, *36*, 2286–2294. [CrossRef]

23. Hu, Y.; Zhao, M.; Li, S. Pipeline Defect Detection Based on Ultrasonic Guided Wave Technique Using Fiber Bragg Gratings. In Proceedings of the International Conference on Pipelines and Trenchless Technology (ICPTT), Wuhan, China, 19–22 October 2012; pp. 996–1010.

24. Ray, P.; Srinivasan, K.; Rajagopal, P. Fiber Bragg Grating-based Detection of Cross Sectional Irregularities in Metallic Pipes. In Proceedings of the SPIE, 25th International Conference on Optical Fiber Sensors, Jeju, South Korea, 24–28 April 2017; pp. 1–4.

25. Trémolet de Lacheisserie, E. *Magnetostriction: Theory and Applications of Magnetoelasticity*; CRC Press: Boca Raton, FL, USA, 1993; Volumes 339–352, pp. 359–361.
26. Zhang, Q.; Zhu, Y.; Luo, X.; Liu, G.; Han, M. Acoustic Emission Sensor System Using a Chirped Fiber-Bragg-grating Fabry-Perot Interferometer and Smart Feedback Control. *Opt. Lett.* **2017**, *42*, 631–634. [CrossRef] [PubMed]

 *applied sciences*

Article

Enhancement of Ultrasonic Guided Wave Signals Using a Split-Spectrum Processing Method

Seyed Kamran Pedram [1,*], **Peter Mudge** [1] **and Tat-Hean Gan** [1,2]

[1] TWI Ltd., Granta Park, Great Abington, Cambridge CB21 6AL, UK; peter.mudge@twi.co.uk (P.M.); Tat-Hean.Gan@brunel.ac.uk (T.-H.G.)
[2] Brunel University London, Institute of Materials and Manufacturing, Kingston Lane, Uxbridge UB8 3PH, UK
* Correspondence: kamran.pedram@twi.co.uk

Received: 10 August 2018; Accepted: 29 September 2018; Published: 3 October 2018

Abstract: Ultrasonic guided wave (UGW) systems are broadly utilised in several industry sectors where the structural integrity is of concern, in particular, for pipeline inspection. In most cases, the received signal is very noisy due to the presence of unwanted wave modes, which are mainly dispersive. Hence, signal interpretation in this environment is often a challenging task, as it degrades the spatial resolution and gives a poor signal-to-noise ratio (SNR). The multi-modal and dispersive nature of such signals hampers the ability to detect defects in a given structure. Therefore, identifying a small defect within the noise level is a challenging task. In this work, an advanced signal processing technique called split-spectrum processing (SSP) is used firstly to address this issue by reducing/removing the effect of dispersive wave modes, and secondly to find the limitation of this technique. The method compared analytically and experimentally with the conventional approaches, and showed that the proposed method substantially improves SNR by an average of 30 dB. The limitations of SSP in terms of sensitivity to small defects and distances are also investigated, and a threshold has been defined which was comparable for both synthesised and experimental data. The conclusions reached in this work paves the way to enhance the reliability of UGW inspection.

Keywords: signal processing; SNR; split-spectrum processing; ultrasonic guided waves

---

1. Introduction

Long-range ultrasonic testing (LRUT), also known as guided wave testing (GWT) is an advanced non-destructive testing (NDT) method that utilises ultrasonic guided wave (UGW) signals for the inspection. This inspection could be applied to any large complex structures such as pipes, rails, cables, etc. for defect detection. This method is widely utilised for the inspection of pipelines that mainly contain oil and gas products, and it has the ability to screen long distances (up to 50 m in each direction) from a single location to identify defects in the structure (e.g., corrosion, erosion) [1–3]. GWT often operates at a low-frequency range (20–100 kHz) (compared to conventional ultrasonic testing (UT), which operates at MHz range) to transmit the waves using one or more rings of dry-coupled transducers around the circumference of the pipe, which are pneumatically forced against the surface. These waves propagate within the pipe wall along the pipe's main axis, and scattering occurs when the waves encounter discontinuities in wall thickness. The transducers are used to record these changes to obtain information about the presence and characteristics of the features within a pipe [4].

In order to reduce the effect of dispersion and to achieve a good resolution between features, a tone burst signal is employed for the transmission of the signal, as shown in Figure 1a.

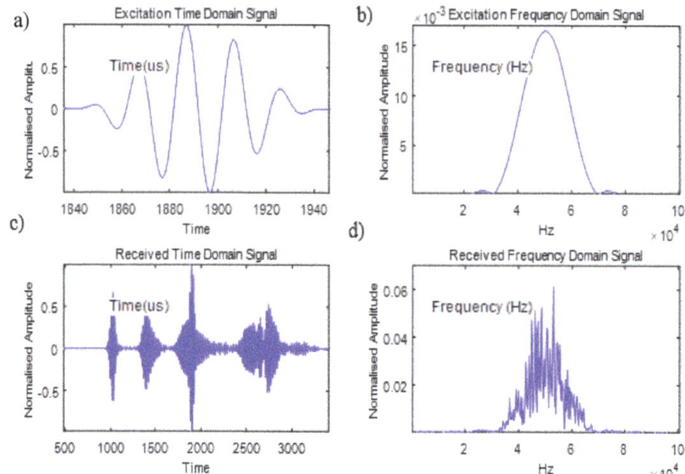

Figure 1. guided wave testing (GWT) signals: an excitation (**a**) time domain and (**b**) frequency domain signal, received (**c**) time domain and (**d**) frequency domain signal.

This is a 50 kHz 5-cycle Hann windowed excitation signal, with a frequency response as shown in Figure 1b, which are generated in this work using MatLab software. A typical response is illustrated in Figure 1c, which consists of a number of peaks that correspond to reflections from structural features under investigation (e.g., defects and welds). In addition, its frequency response is displayed in Figure 1d, exhibiting the same frequency bandwidth as the input signal [5].

It is ideal to generate an axisymmetric wave mode to promote non-dispersive propagation; however, the interaction of the guided wave signal with non-axisymmetric features within the pipeline can cause mode conversion. This results in the generation of dispersive wave modes (DWM) that travel with different velocities according to the different frequency components in the signal [6]. Hence, the energy spreads over space during propagation, and compromises the ability to distinguish echoes from closely spaced reflectors. This lack of spatial resolution leads to coherent noise and reduces the sensitivity of the inspection. In order to enhance the sensitivity and the SNR of such signals, it is vital to minimise the presence of coherent noise. Dispersion is one of the main sources of coherent noise; hence, the aim of this work is to reduce the effect of dispersive wave modes.

There are many researchers who study the effect of dispersion in GWT [7–14]. Wilcox [7] developed a method for reversing the effect of dispersion by using the knowledge of wave mode characteristics to map signals from the time to the distance domain, and then reversed the effect of DWM and restored them to undispersed pulses. Zeng and Lin [8] investigated the dispersion pre-compensation method using chirp-based narrowband excitation signals to compress the time duration of received wave packets during the extracting process. They employed the benefits of chirp excitation by utilising previous knowledge of the dispersion curve and the propagation distance. Other researchers such as Xu et al. [9–11] employed dispersion compensation (DC) to analyse the propagation behaviours of the signals. However, most of them required the knowledge of the propagation distance in advance. Toiyama and Hayashi [12] combined the DC method with a pulse compression (PuC) algorithm by employing a chirp signal. They considered a scenario of a single wave mode without introducing the quantitative SNR enhancement. The combination of DC with PuC is utilised by Yucel et al. [13,14] to enhance the SNR of UGW response employing a broadband maximal length sequence excitation signal. The result showed that the technique was successful for highly dispersive flexural wave modes, but it was not that effective for longitudinal wave modes that are not dispersive. Mallet et al. [5] considered cross-correlation and wavelet de-noising algorithms for reduction of the effect of dispersion in UGW. He claimed that neither of these methods was suitable for the reduction of coherent noise, as both methods removed the smaller amplitudes regardless of whether or not they were signal or noise.

Newhouse et al. [15] considered split-spectrum processing (SSP) in the field of NDT to enhance the SNR by splitting the signal's response into a set of sub-band signals. A theoretical basis for the selection of filter bank parameters was investigated by Karpur et al. [16]. They proposed an equation to predict SNR enhancement by compounding a number of frequency diverse signals. The result showed that some parameters achieved a larger value than expected, which could be the result of using the Gaussian function for filtering (because of its simplicity) while the calculation was based on the Sinc function. Shankar et al. [17] employed a polarity thresholding (PT) algorithm for the detection of a single target. They showed how sensitive the SNR enhancement was to the selection of filter bank parameters. Saniie et al. [18] investigated the performance of order statistic (OS) filters in conjunction with SSP in the context of ultrasonic flaw detection, to improve the flaw-to-clutter ratio of backscattered signals. It has been claimed that the OS filter performs well where the flaw and clutter echoes have good statistical separation in a given quartile. However, its performance deteriorated with the contamination from unwanted statistical information.

Gustafsson and Stepinski [19] adapted an SSP method using an artificial neural network (ANN) to implement PT for UT signals. In order to allow the relative importance of the different sub-bands to be taken into account, weighting factors were added to the input signal. The results showed better performance than, PT but only for one particular sample. Gustafsson [20] then extended the method by employing both the filter bank and the non-linear processing as an ANN for SSP. This method was time-consuming, although the results indicated that the ANN could "eliminate most of the noise". Rubbers and Pritchard [21] developed a complex-plane SSP (CSSP) method, which was a modified version of SSP for the ultrasonic inspection of castings, and improved the SNR for a number of conventional UT techniques. It utilised an additional mathematical dimension to improve the result, while maintaining linearity in both the amplitude and the energy content of a defect signal. Moreover, they gave an overview of SSP methods [22] with a variety of SSP reconstruction algorithms and parameters. However, they claimed that as the amplitude of the processed signal is non-linear, it does not allow for the sizing of flaws, hence the use of this method is limited.

Rodriguez et al. [23,24] proposed a new filter bank design for SSP, based on the use of variable bandwidth filters, where filters were equally spaced in frequency and their energy gain equalised. They utilised stationary models for the grain noise in the presence of a single defect. They claimed that a frequency multiplication (FM) algorithm gave the greatest SNR enhancement. They stated that the number of filter bands compared to other algorithms is reduced; hence, it reduced the system complexity. However, this technique was not evaluated for non-stationary models, highly dispersive material, or a model with multiple defects. Syam and Sadanandan [25] employed a combination of SSP and order statistic filters to reduce the effect of reverberation for flaw detection in conventional UT using a wideband signal. He stated that by processing the multiple echoes corresponding to a set of transmitted signals, the effect of microstructure reflections could be suppressed with respect to the flaw echo. However, this method tested for a simulated signal only, without revealing the value of the SSP parameters.

Overall, it is clearly shown by some of the cited literature [7–14] that their techniques required prior knowledge of the dispersion curve and/or propagation distance in order to perform well. Other cited literature [15–25] indicated that the successful implementation of the SSP technique was highly sensitive to the selection of filter bank parameters. Although most of these papers claimed that they enhanced the SNR of the signal's response, the enhancement was mainly achieved for conventional UT. However, those parametric values are not suitable for use in GWT that contain a combination of axisymmetric and non-axisymmetric wave modes with different phase velocities. Hence, a full study is required to find the optimum filter bank parameters for SSP in terms of its capacity to provide such improvements in GWT. To the best of the authors' knowledge, prior to this study, apart from Mallet et al. [5], no one else has investigated the use of SSP in guided wave testing.

An analysis of SSP with application to GWT was conducted in our previous paper [26] for reducing the effects of DWMs in the signal response, and the optimum parameters have been proposed to enhance the SNR and spatial resolution of such signals. However, the limitations of SSP for use in GWT are still unclear, and they need further investigation. Hence, in this paper, the optimum parameters that were identified in the previous work are utilised for deeper investigation to improve the SNR, and to enhance the spatial resolution further by investigating the limitations of SSP mainly in two areas: (i) in terms of defect sensitivity and (ii) in terms of minimum distance between two features (e.g., weld and defect). Thus, the core concept explored in this work is to address these issues. Such a parametric study has not been undertaken in the field of GWT prior to this work.

In order to do this, a synthesised signal has been created to identify the limitations of SSP. The limitations have been tested, evaluated, and identified synthetically in a controlled environment. Then, in order to validate the effectiveness of the technique, laboratory experiments have been carried out on an eight-inch pipe utilising the Teletest-guided wave system [27]. It is shown that the proposed method reduces the presence of coherent noise and improves the SNR by up to 30 dB. In addition, the limitations of SSP have been identified and a threshold has been defined, below which the temporal resolution will be reduced.

The paper is organised as follows: Section 2 describes the theory and concept of SSP, including the implementation and selection of filter bank parameters for guided wave testing. Sections 3 and 4 provide details and discussion of SSP for synthesised and experimental testing, and finally Section 5 concludes the paper.

2. Split-Spectrum Processing (SSP)

2.1. Theory of SSP

SSP is an advanced signal processing technique that was initially developed from the frequency agility techniques used in radar [28]. This method was then considered for SNR enhancement in NDT applications such as conventional UT, to reduce grain scatter in the received signal. A significant amount of research has been undertaken over the last few decades in this area with respect to the reduction of non-random noise (coherent noise) in NDT applications, due to ultrasonic scattering. The application of SSP in GWT is relatively new and, to the best of the authors' knowledge, this technique and its limitations have not been previously investigated in this field.

SSP splits the spectrum of a received signal in the frequency domain using a bank of bandpass filters to generate a set of sub-band signals at incremental centre frequencies. These sub-band signals are normalised and then subjected to a number of possible non-linear processing algorithms to generate an output signal. Figure 2 illustrates a block diagram for SSP with its step-by-step implementation. It shows that the input time domain signal, $x(t)$, is transformed into the frequency domain, $X(f)$, using a fast Fourier Transform (FFT), and is then filtered by a bank of band-pass filters. Subsequently, the outputs from the filter banks, $X_k(f)$ ($k = 1, 2, \ldots, n$), are converted back into the time domain using the inverse FFT (IFFT) and normalised by a weighting factor, w_k. Then one of the recombination algorithms will be employed to combine these non-linear signals to produce the output signal $y(t)$. In general, SSP application shows a great potential to reduce those signal components that vary across a frequency range, in particular dispersive wave modes, and to suppress the regions of the signal of interest that are constant in that frequency range [29].

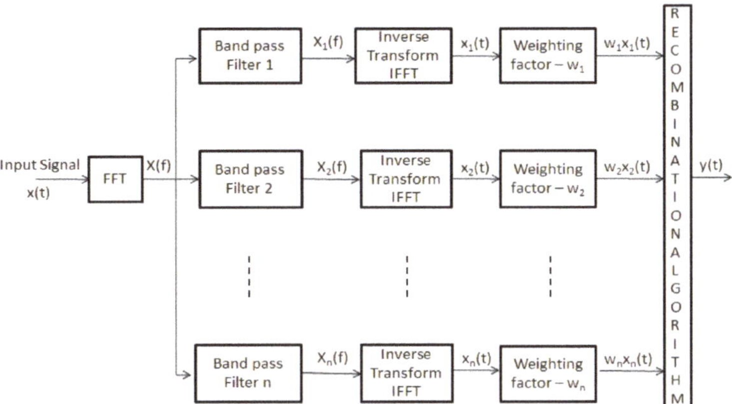

Figure 2. Split-spectrum processing (SSP) block diagram.

2.2. SSP Filter Bank Parameters

The SSP filter-bank parameters were first investigated by trial-and-error for NDT applications as they were processed. However, this was not very practical for field inspection as there are typically large amounts of data to analyse; therefore, researchers sought to find the optimum values for NDT methods. Hence, the optimum values have been proposed, developed and examined for UT techniques. However, these values are not suitable for GWT due to the long duration and narrow bandwidth of the signal that operates in the kHz range, whereas traditional UT operates in the MHz range. Therefore, further investigation was required to find an optimum value suitable for GWT. As a result, the rules and key factors of parameter selection for conventional UT have been reviewed and then modified for use in GWT. Figure 3 illustrates the scheme of SSP filtering. The parameters that need to be quantified are listed below with a brief description:

1. The total bandwidth for processing (B); this needs to be large enough such that the reflections of the signal from features in the specimen are constant across this range, and the reflections from coherent noise vary. If the bandwidth is too large, then it may cause the features to be lost, as at least one of the filter outputs will not contain the feature signal. Hence it reduces the spatial resolution in the processing. In general, narrowband waveforms were used as the excitation signals to reduce the effect of unwanted wave modes, and to suppress the dispersion effect in the GWT response. Hence, the bandwidth of the transmitted signal could be employed as the total bandwidth of processing.

2. The filter separation (F); is the distance between the sub-band filters. Karpur et al. [16] claimed that the optimum spectral splitting could be attained by using the frequency-sampling theorem, whereby the spectrum of a time-limited signal can be reconstructed from sample points in the frequency domain separated by 1/T Hz, where T is the total duration of the signal. Note that the Gaussian filter is employed for calculating the filter bank in practice, due to its simplicity, whereas the Sinc function was utilised for actual calculation. Thus, the filter separation could be calculated as F = 1/T.

3. The sub-band filter bandwidth (B_{filt}) is the width to be used for each filter in the filter bank. It was recommended [15–17] that the value of the sub-band filter bandwidth needs to be set at three to four times the filter separation. It should be noted that a bandpass filter could reduce the temporal resolution of the signal. This is because reducing the bandwidth of a time-limited signal will increase its duration. This means that the SSP filter bank needs to be selected precisely, otherwise it could lead to a reduction in temporal resolution, as the pulses that correspond to reflections from features spread out in time and mask one another. Moreover, the correlation between

adjacent sub-bands could be affected by the overlap of the filters. This means the correlation increases with an increase in overlaps. On the other hand, little or no overlap could lead to loss of information. It is notable that the noise in adjacent filters needs to be uncorrelated and the features should be correlated. Hence, the overlap needs to be selected somehow to minimise the correlation between coherent noise regions in adjacent sub-bands without losing information.

4. The filter crossover point (δ); or a cutoff frequency at the edges, is a boundary in the frequency response at which the energy flowing through the structure starts to decrease rather than passing through.

5. The number of filters (N); is the number of sub-band filters that is required to be selected in order to enhance the SNR and spatial resolution by correlating the signal of interest and minimising the correlation between the coherent noise region in the adjacent sub-bands signal. Overall, these parameters are dependent on each other, which means that their values have a direct effect on other parameter values. Therefore, it is necessary to search for the optimum parameters and to select them appropriately. As an example, increasing the number of filters (N) would be required to increase the total bandwidth (B), or to reduce the filter separation (F), or a combination of both. Thus, as is shown in Figure 3 the number of filters (N) could be calculated as below:

$$N = \frac{B}{F} + 1 \tag{1}$$

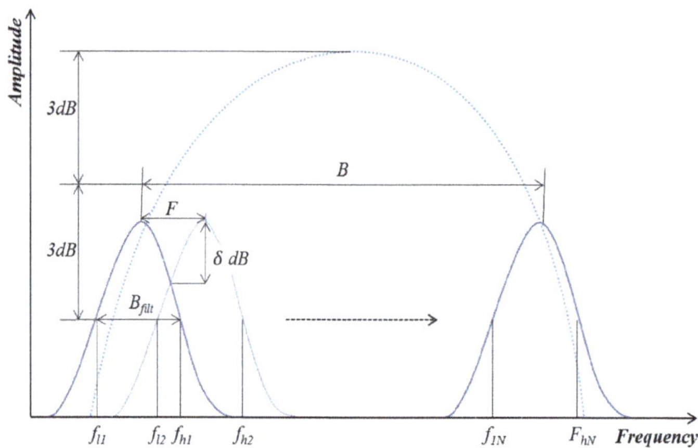

Figure 3. Filter bank parameters of SSP.

2.3. Recombination Algorithms of SSP

There are numerous SSP recombination algorithms that could be employed for use in GWT to reduce the effect of coherent noise which have been described in the literature [16,22]. In this paper, the two most common ones that obtained the highest SNR and spatial resolution in the previous work are explained in more detail:

The polarity thresholding (PT) that is expressed as:

$$y_{PT}[m] = \begin{cases} x[m] & \text{if all } x_i[m] > 0, \quad i = 1,\ldots, n \\ x[m] & \text{if all } x_i[m] < 0, \quad i = 1,\ldots, n \\ 0 & \text{Othrewise} \end{cases} \tag{2}$$

where y_{PT} is the output result of the signal that is obtained after processing at m, n is the number of filter band signals, x[m] is the unprocessed signal, and $x_i[m]$ care the sub-band signals. This method looks at the signal sub-bands at each sample time, and if the samples are all negative or all positive, then the output is the unchanged input signal. Otherwise, the output is zero. This has the effect of only passing the time samples where the polarity is not affected by the frequency. Thus, those sections of the signal that are highly frequency-dependent must be removed. However, the amplitude of the signal of interest needs to be larger than the amplitude of the coherent noise response, whereas if the noise signal has greater amplitude, then it will change the signal's sign.

Polarity thresholding with minimisation (PTM), which is defined as:

$$y_{PTM}[m] = \begin{cases} \min(x[m]) & \text{if all } x_i[m] > 0, \quad i = 1, \ldots, n \\ \min(x[m]) & \text{if all } x_i[m] < 0, \quad i = 1, \ldots, n \\ 0 & \text{Othrewise} \end{cases} \quad (3)$$

PTM is the combination of the minimisation and the PT algorithm where the output is the minimum amplitude of PT when there is no change in polarity. This method takes the minimum points that the PT algorithm has passed in order to suppress the noise even further than is obtained by the PT algorithm. Since the variance of the points containing noise is generally larger than those that are containing the signal of interest, the use of the minimisation algorithm reduces those that are the result of noise. However, this method is less effective when the noise level is larger than the actual signal's amplitude; thus, reducing the noise level will significantly reduce the signal amplitude in certain sub-bands, and gives minimum values for the PTM's output in that region.

2.4. Implementation of the Filter Bank

A MatLab program has been written that takes an unprocessed signal in the time domain and converts it to the frequency domain. It then filters the signal using a Gaussian bandpass filter to generate a set of sub-band signals, and applies the recombination algorithms into these sub-bands. The input is the signal to be filtered, with the upper and lower 3 dB cut-off frequencies. Therefore, the lower cut-off frequency f_l, and the higher cut-off frequency f_h, for each sub-band filter, are calculated as:

$$f_{l_n} = \begin{cases} f_{min} - \frac{B_{filt}}{4} & n = 1 \\ f_{l_{n-1}} + F & n = 2, 3, \ldots N \end{cases} \quad (4)$$

$$f_{h_n} = f_{l_n} + B_{filt} \quad n = 1, 2, \ldots N \quad (5)$$

where F is the filter separation, N is the number of filters, f_{min} is the lower cut-off frequency of B, and B_{filt} are the sub-band filters. The lower cut-off frequency for the first sub-band f_{l_1}, needs to cover the start point of the signal. The selections of these values are inspired by the values that have been employed in UT, and then adjusted for the use in GWT using the brute force search algorithm. Table 1 shows the optimum values of SSP that were proposed in the previous paper [26]. These values are employed in this paper to find the limitations of SSP. The performance of the proposed technique is quantified by measuring the SNR, the spatial resolution, and the defect sensitivity of the output signal.

Table 1. Recommended values for SSP parameters.

SSP Parameters	Symbols	Recommended Values
Total bandwidth	B	99% of total energy
Sub-band filter bandwidth	B_{filt}	B/11
Filter crossover point	δ	$B_{filt}/3.5$
Filter separation	F	1 dB
Number of filters	N	$B/F + 1$

3. Signal Analysis

This section shows how this methodology has been evaluated. Initially, signals have been simulated using the technique that was presented by Wilcox [7] to represent the guided wave reflections in a pipe with known dispersion characteristics. This allows the SSP limitation to be found under controlled conditions. Secondly, experimental data was collected from an eight inch pipe that was equivalent to the simulated pipe. This gives two benefits: (i) it allows the approach to be tested under more realistic conditions; and (ii) the reusable SSP parameters across equivalent pipes could be evaluated.

Signal Synthesis

The signal synthesis was utilised to generate the propagation of the dispersive wave modes in time/space, based on applying a frequency-dependent phase shift to the wave packet of interest via DISPERSE software (developed by Imperial College, London, UK) [30]. The core concept of this section was to identify the limitation of SSP in terms of finding the smallest defect size that could be detected, and to find the distance limitation when the location of the defect is close to a dominant feature with a high amplitude, such as a weld. In order to achieve that, the axisymmetric T(0,1) wave mode is excited with a 10-cycle pulse and a centre frequency of 44 kHz, using the synthesised model. It is assumed that there are only two reflection echoes (i) from the defect, and (ii) from the pipe's end. Since the proposed method has already proved its capability to remove the DWM [26], here, it is assumed that only the T(0,1) wave mode is reflected from the features on the pipe. According to Böttger et al. [31], there is a linear relationship between the amplitude of the reflected signal of T(0,1) and the cross-section area (CSA) of its defect. Hence, the attenuation of T(0,1) is linear, which means that if 10% of the excited signal reflects from the defect, then the rest of the energy (90%) will reflect from the pipe end.

The set-up of this experiment is shown in Figure 4. The distances of the defect and pipe end are X = 3 m and X = 4.5 m from the excitation point, respectively. It is assumed that the defect reflects 6% of the total energy and rest of the energy (94%) is reflected by the pipe end, as displayed in Figure 5a. The reflections from the defect were reduced gradually in order to find the smallest defect size that can be recognised by the proposed method. These are illustrated in Figure 5, where the defect sizes are gradually reduced from 6% CSA to 1% CSA. The results demonstrate that the proposed method has the potential to detect defects down to 1% CSA when the distance between two features is 1.5 m. Then, in order to find the distance limitation, it is assumed that a 10% CSA defect is 1 m from the pipe end and it is moved towards the pipe end in steps of 0.1 m as illustrated in Figure 4. Results in Figure 6a–d clearly demonstrate that the defect was recognisable until the distance from the pipe end was around 0.7 m. After that, as shown in Figure 6f, the resolution was gradually reduced until it reaches 0.5 m. Afterwards, the defect reflection started to superimpose on the pipe end reflection, and it travelled below the limit of the resolution. Therefore, according to this result, SSP with the proposed parameters can detect small amplitudes that are close to the dominant amplitude only when the distance between them is greater than 0.5 m. Note that this is the result for 10 cycles of excitation signal with a defect size of 10% CSA and a centre frequency of 50 kHz.

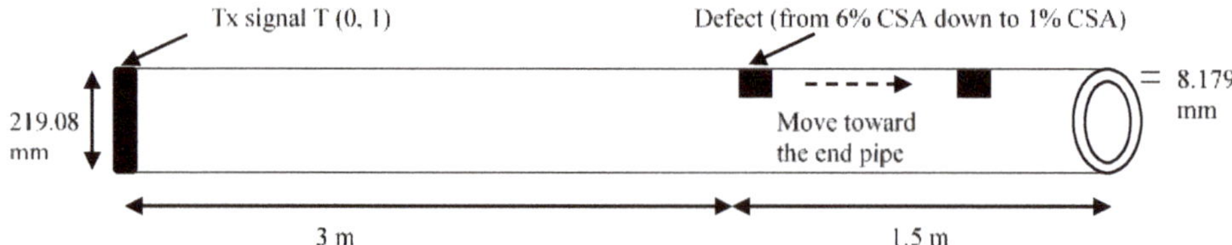

Figure 4. Synthesised setup for an eight inch pipe with a wall thickness of 8.179 mm and an outside diameter of 219.08 mm.

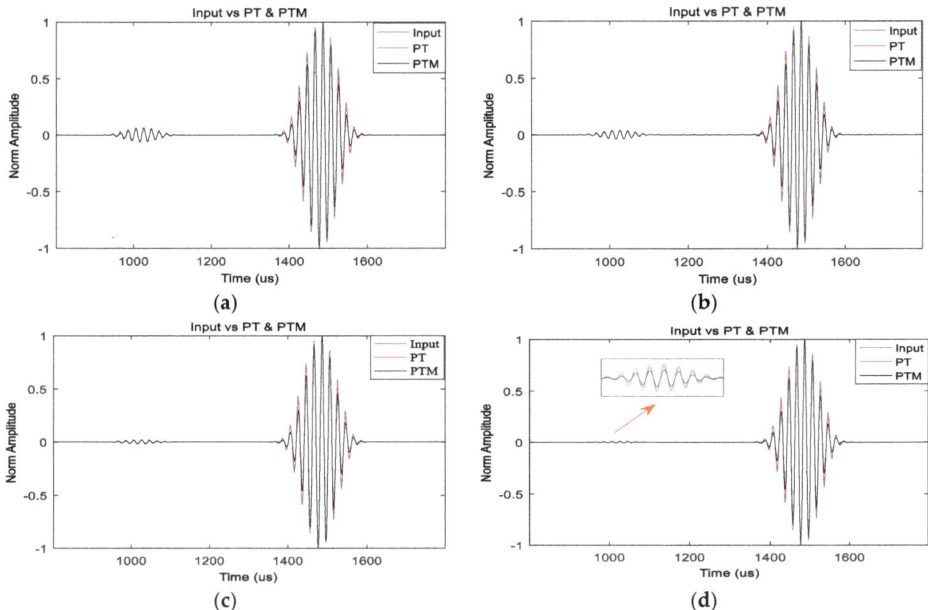

Figure 5. Results for the synthesised signal before and after applying SSP ((polarity thresholding & Polarity thresholding with minimisation)PT & PTM). The defect and the pipe end are located at X = 3 m and X = 4.5 m from the excitation signal. The defect sizes are (**a**) 6% cross-section area (CSA); (**b**) 4% CSA; (**c**) 2% CSA; and (**d**) 1% CSA.

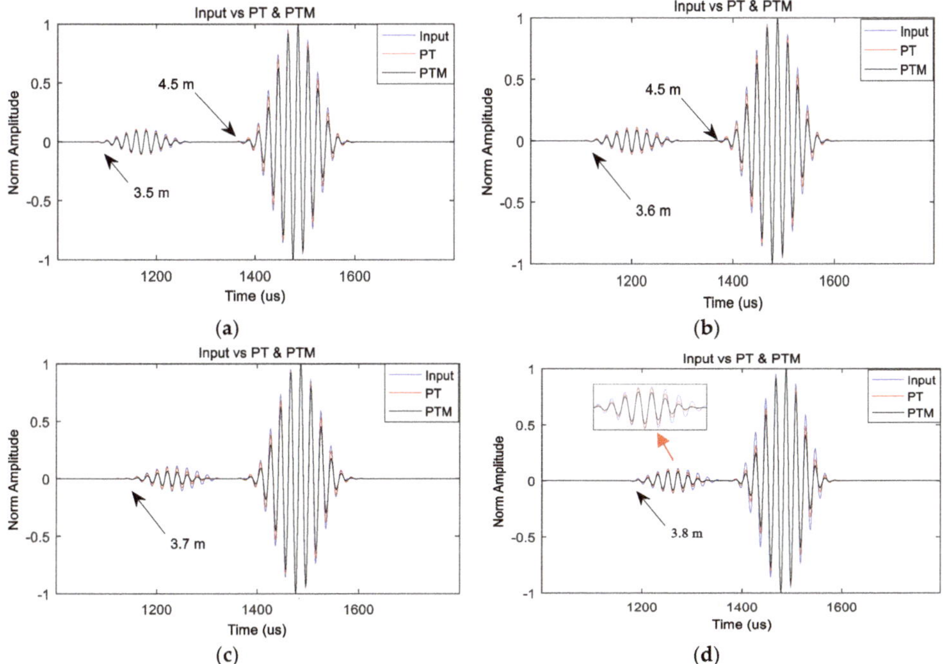

Figure 6. *Cont.*

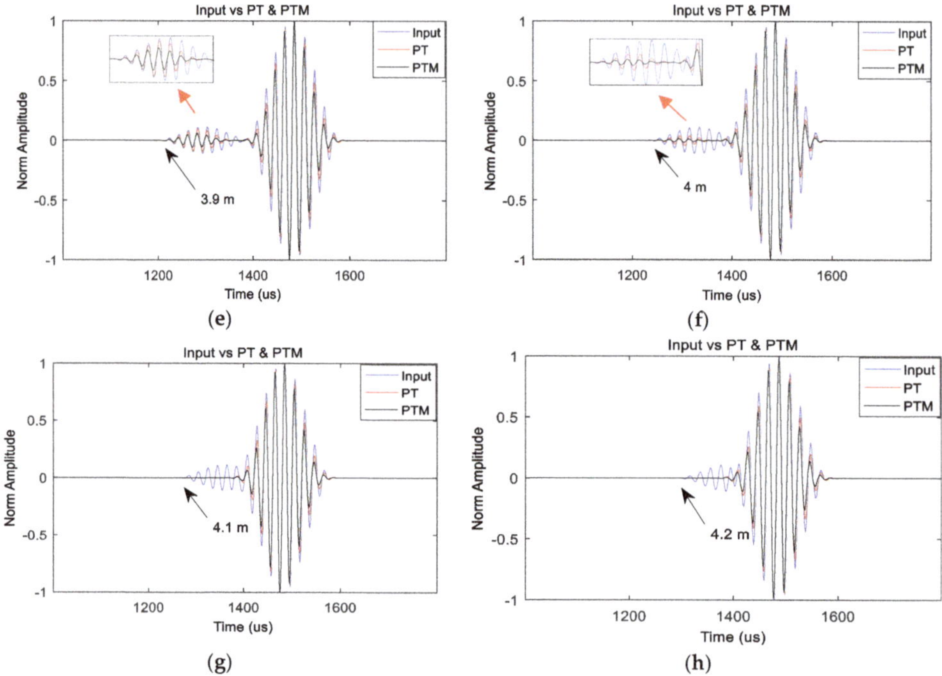

Figure 6. Results for the synthesised signal before and after applying SSP (PT & PTM). The defect (X = 3.5 m) is moved towards the pipe end (X = 4.5 m) in steps of 0.1 m. The defect distances are X= (**a**) 3.5 m; (**b**) 3.6 m; (**c**) 3.7 m; (**d**) 3.8 m; (**e**) 3.9 m; (**f**) 4 m; (**g**) 4.1 m; (**h**) 4.2 m.

Furthermore, as shown in Figure 6, the results of PT and PTM algorithms are almost identical while the distance between the two peaks is greater than 0.8 m. However, the PT algorithm gives a better temporal resolution when the distance is around 0.5 m (Figure 5f). In these regions, the PT identify defects partially, whereas the PTM loses the information completely, and finally, when the distance is less than 0.4 m, both algorithms miss the defect. The result of the synthesised signal showed that SSP application has the potential to reduce the level of coherent noise significantly, due to the presence of DWM, hence enhancing the SNR and the spatial resolution of signals.

In addition, a threshold (0.4 m distance between two features) was defined as a distance limitation, below which threshold the temporal resolution will be reduced. However, further work is required for the different scenarios. Furthermore, in order to validate the outcome of these synthesised results, SSP was applied to the experimental data that was gathered in the lab. These validations are described next.

4. Experimental Validation

In this section, two experiments were carried out in the laboratory to validate the proposed method for the reduction of DWM, and to enhance the spatial resolution in GWT.

4.1. Experiment #1: One Saw Cut Defect

The first experiments were conducted in the lab using a nominal eight-inch steel pipe, 6 m long, with a wall thickness of 8.28 mm and an outer diameter of 219.08 mm. The setup for the experiment is illustrated in Figure 7. The signal was excited/received (Tx/Rx) using a Teletest system to transmit a 10-cycle Hann window modulated tone burst of T(0,1) wave mode. The ring spacing between the transducers was 30 mm.

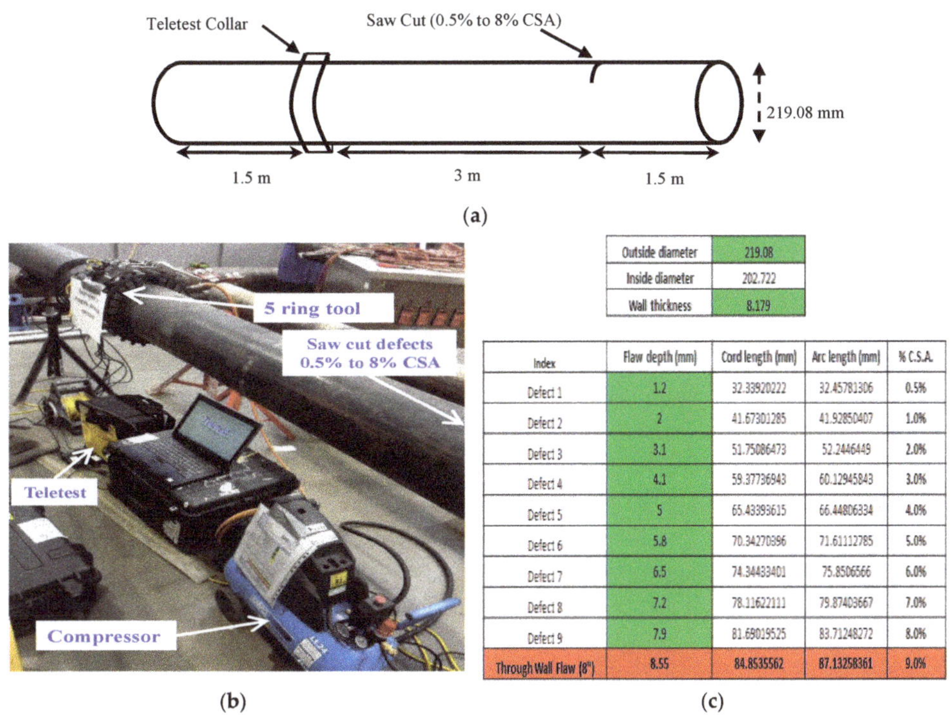

Outside diameter	219.08
Inside diameter	202.722
Wall thickness	8.179

Index	Flaw depth (mm)	Cord length (mm)	Arc length (mm)	% C.S.A.
Defect 1	1.2	32.33920222	32.45781306	0.5%
Defect 2	2	41.67301285	41.92850407	1.0%
Defect 3	3.1	51.75086473	52.2446449	2.0%
Defect 4	4.1	59.37736943	60.12945843	3.0%
Defect 5	5	65.43393615	66.44806334	4.0%
Defect 6	5.8	70.34270396	71.61112785	5.0%
Defect 7	6.5	74.34433401	75.8506566	6.0%
Defect 8	7.2	78.11622111	79.87403667	7.0%
Defect 9	7.9	81.69019525	83.71248272	8.0%
Through Wall Flaw (8″)	8.55	84.8535562	87.13258361	9.0%

Figure 7. Experimental setup up for an eight inch steel pipe with a wall thickness of 8.179 mm and an OD of 219.08 mm (**a**,**b**), and (**c**) its flaw size plan.

The frequencies that give the best results for this particular pipe size according to the dispersion curve are 27 kHz, 36 kHz, 44 kHz, 64 kHz, and 72 kHz. Therefore, the data was collected at these frequencies for analysis. The Teletest Collar was placed at 1.5 m away from the near pipe-end, and a saw cut defect was introduced 1.5 m away from the far pipe-end. The size of the defect was incrementally increased from 0.5% CSA to 8% CSA in nine steps. The flaw size plan is displayed in Figure 7c. In order to reduce incoherent noise, the collection was repeated 512 times, and the received signals were averaged. The sampling frequency of the received signal was set to 1MHz. Since PT algorithm gave the best result for synthesised signal, it was used in this experiment with the same filter bank parameters that were employed for the synthesised signal as illustrated in Table 1.

As mentioned earlier, the experiment was run for different frequencies, and the result of each frequency was compared with the conventional model that is currently employed in the Teletest system. However, the comparison for 44 kHz is presented in greater detail in this paper. The results indicated that defects that are smaller than 2% CSA are almost impossible to identify before and/or after applying the proposed technique. Therefore, the investigation and comparison were confined to when the defect size was greater than 2% CSA. It is notable that the current sensitivity for reliable detection of the Teletest system was 9% CSA, which is equivalent to 5% amplitude reflection.

Figure 8 shows the zoom-in plot around the defects area from 0.5% CSA to 8% CSA using MatLab software for both the unprocessed (blue trace) signal and the SSP (red trace) signal. The figure confirms that the SSP technique using optimum parameters enhances the defect sensitivity down to 2% CSA, which was hidden below the noise level. As shown in Figure 8, it was still difficult to identify a 3% CSA defect with the conventional techniques, whereas the proposed method removed all of the surrounding noise, and only the defect's reflection remained. Hence, the defect was easily noticeable and it could be identified with confidence. This size flaw was typical of that which can be challenging to reliably detect with GWT systems. Therefore, as a result of this experiment, SSP demonstrates great potential to enhance the SNR, and to increase the sensitivity and spatial resolution of signal response, and it is able to identify defects down to 2% CSA.

SNR Calculation

In order to quantify the improvements shown by the proposed technique, the SNR enhancement was calculated as:

$$\text{SNR} = 20 \times \log_{10}\left(\frac{S}{N}\right) \tag{6}$$

where S is the maximum amplitude of the defect's reflection, and N is the root mean square (RMS) value of the noise region around the defect. The SNR of the unprocessed signal was 7.8 dB when the defect size was 2% CSA, and 13.25 dB when the defect size was 3% CSA. The results presented in Table 2 show that the SSP algorithms enhanced the SNR by 34.9 dB and 42.9 dB for these cases respectively. Furthermore, a comparison of the amplitude of the pipe-end reflection and the defect reflection was undertaken in order to evaluate how well the SSP maintains the amplitude of the signal of interest. The results indicate that when the size of the defect was 3% CSA or greater, there were no significant amplitude changes that occurred for the proposed method.

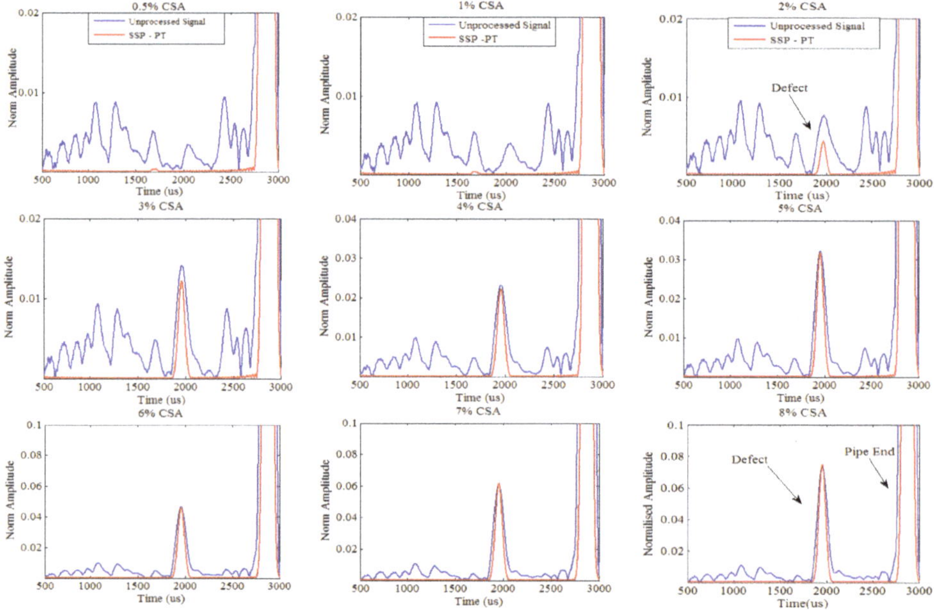

Figure 8. Zoom in around the defect area from 0.5% CSA up to 8% CSA.

Table 2. Signal-to-noise ratio (SNR) enhancement of experimental signal.

Signal	2% Defect	3% Defect
SSP	34.9 dB	42.9 dB
Unprocessed	7.8 dB	13.3 dB

Figure 9 illustrates the result of the unprocessed signal (blue trace) and the SSP algorithm (red trace) for the above experiment when there was no defect (baseline), up to when the defect size was 8% CSA. The results clearly illustrated that the performance of the proposed technique massively improved the SNR of the GWT response compared to the unprocessed data, achieving around 30 dB improvement for 44 kHz. However, as mentioned earlier, the comparison commenced when the defect size was at least 2% CSA. Hence, in order to clarify it, a dotted line was added at 2% CSA.

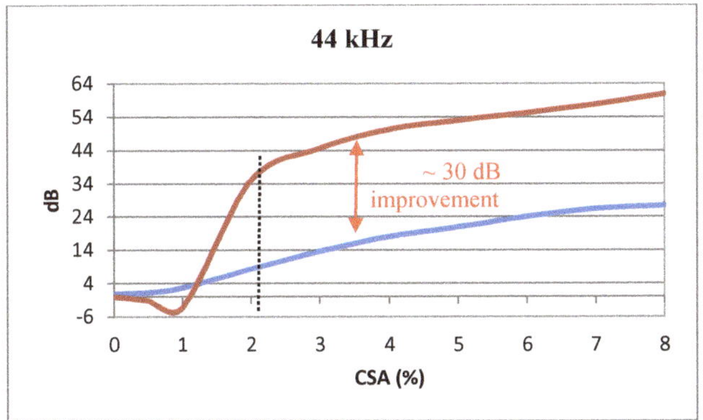

Figure 9. SNR calculation–peak amplitude of the defect (S) to the root mean square (RMS) value of the noise region (N) for 44 kHz.

4.2. Experiment #2: Two Saw Cut Defects

The aim of this test was to investigate the distance limitation of SSP where a small feature (i.e., defect) is close to a dominant feature (e.g., weld). The initial investigation started with the optimum parameters utilised for a single saw cut defect to evaluate the outcome of SSP distance limitation.

An experiment was carried out on the same eight-inch pipe that was utilised for the previous experiment. In this experiment, another saw cut defect was added to the pipe at a location that was 48 cm from the far pipe end, as shown in Figure 10. The size of the defect was incrementally increased from 1% CSA to 8% CSA. Note that another 8% CSA saw cut defect already existed 1.5 m from the far pipe end. All of the setups, the tool location, and the frequency test region were exactly the same as the previous exercise. However, only the results relating to 44 kHz were presented here because, according to the previous experiment, this frequency gave the best SNR enhancement.

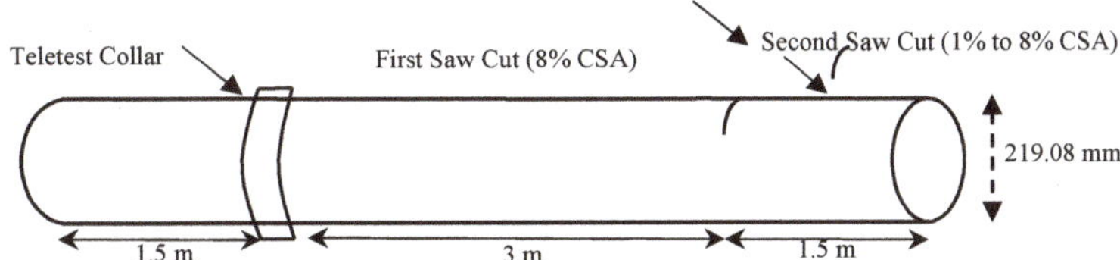

Figure 10. Experimental setup for the same eight inch pipe (Figure 7) with two saw cut defects.

The proposed method was applied to the collected data gathered from this experiment. It was observed that the initial result obtained with these parameters was not as successful, and it was only able to identify defects down to 4% CSA. Hence, the brute force search algorithm was applied to find the optimum parameter values for this scenario, again in order to improve the performance of the defects, and the capability to find smaller defects. As a result, optimum values were discovered that gave the chance to detect defects down to 2% CSA for a second defect. The unprocessed signal, and the signal after applying SSP with the new optimum parameters are presented in Figure 11. This figure clearly demonstrates that the data after applying SSP are tidier in general, and in terms of defect detection, down to 2% CSA is noticeable. However, it can be seen that the coherent noise is hardly reduced by the new parameters, and SNR is slightly improved. Therefore, it is confirmed that there is a trade-off between detecting small features next to a dominant signal and improving the SNR. In addition, this result confirms the result of the synthesised analysis where it was stated that the distance limitation of SSP to identify adjacent features is around 50 cm when using a 10-cycle Hann windowed at 44 kHz.

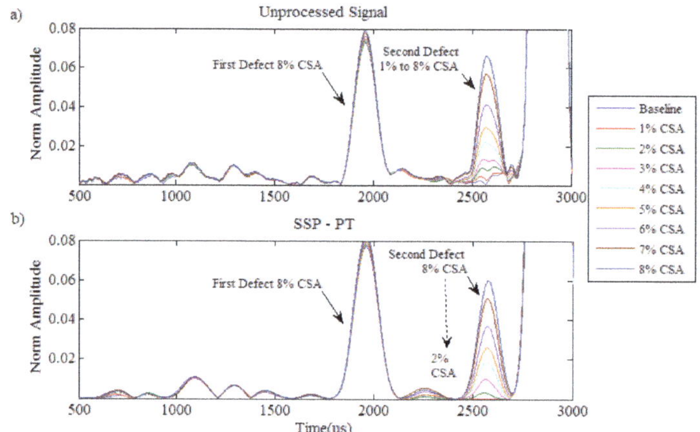

Figure 11. Zoom in result with two defects. (**a**) Unprocessed signal; (**b**) SSP signal.

4.3. Discussion

The limitations of SSP were first investigated for synthesised signals using the two most common SSP recombination algorithms, polarity thresholding (PT) and PT with minimisation (PTM). The synthesised signals were utilised to find the limitations of SSP by measuring the SNR and spatial resolution of the received signal. Results showed that the SSP with PT algorithm was able to identify defects down to 1% CSA, and stated that 50 cm is the minimum distance for the SSP to identify a defect next to the pipe end. Then, in order to validate the synthesised results, two experimental tests have been carried out in the lab on the same sized pipe.

Two experiments were carried out on an eight inch pipe, 6 m long, with a wall thickness of 8.28 mm and an outer diameter of 219.08 mm, using a pulse-echo technique. The Teletest system was utilised to transmit a 10-cycle Hann window-modulated tone burst of T(0,1) with a centre frequency of 44 kHz. In the first experiment, a saw cut defect was created 1.5 m from the far pipe-end. Nine saw cut defects were created, the sizes of which gradually increased from 0.5% CSA to 8% CSA. The results demonstrate that the SNR was improved by approximately 30 dB compared to the unprocessed signal. The results indicated that defects smaller than 2% CSA cannot be identified both before and after SSP. This is due to the sensitivity of the system. Therefore, the investigation and comparison were conducted only when the defect size was greater than 2% CSA. It was shown that the 2% CSA defect's reflection was almost masked by the coherent noise level, and that the identification of responses using conventional signal interpretations was not feasible. However, SSP removed all the surrounding coherent noise but the defect's reflection. Therefore, this provided a good evidence that SSP has the potential to identify defects down to 2% CSA, and to enhance the spatial resolution.

In the second experiment, a new saw cut defect was created, which varied in size from 1% CSA to 8% CSA in each test, in addition to the already existing 8% CSA defect. The aim was to validate the distance limitation that was observed for the synthesised exercise. Therefore, the defects were created 48 cm from the far pipe-end. The results illustrated that defects up to 4% CSA were detectable with the same filter bank parameters as those used for the previous scenarios. However, in terms of identifying smaller defects, the parameters needed to be modified, and this was undertaken using a brute force search algorithm. As a result, new filter bank parameters were introduced to identify defects down to 2% CSA. However, this was achieved at the cost of reducing the SNR enhancement.

The above algorithm and methodology explain a means of identifying potentially reusable parameters of SPP for the application of suppressing dispersive wave modes in GWT. It should be noted that pipes of the similar geometry and material share the same guided wave characteristics. The relative rates of dispersion between the desired/undesired wave modes will be the same for equivalent pipes. Regulation in the oil and gas sectors means that pipelines are manufactured from a limited set of standard geometries and materials. Hence, equivalent pipes are commonplace. Therefore, it is anticipated that once the set of parameters have been identified for a specific pipe, they can be reused effectively for similar or slightly different pipes. This would mean it would not be required to run the brute force search algorithm for every inspection. However, the proposed technique was tested in a limited trial, and in order to build a signal processing toolbox (automated), more field data analysis is required to be investigated.

5. Conclusions

A novel solution based on signal processing is proposed in this work to address the problem of coherent noise in guided wave testing using the SSP technique. The main concern was to identify the limitations of SSP in terms of sensitivity and resolution when two features are close to each other.

Therefore, a synthesised signal has been created to identify the limitations of SSP in terms of establishing the smallest defect that can be detected, and to establish the resolution limitation when the location of a defect is close to a dominant feature. It was demonstrated that the SSP technique with optimum parameters successfully identified defects down to 2% CSA, and enhanced the SNR of the received signal. In addition, a threshold of 50 cm has been defined, below which the temporal resolution will be significantly reduced. The outcome was then experimentally validated for an eight-inch pipe containing two saw cut defects using the Teletest system, where it achieved a comparable result. To sum up, it is demonstrated throughout this work that the proposed method using a polarity thresholding algorithm has the potential to improve the sensitivity and spatial resolution of GWT in terms of SNR (by up to 30 dB), by detecting smaller defects (down to 2% CSA) with a resolution threshold of 50 cm between the two features. Thus, this work shows that SSP, as implemented here, could be applied for pipeline inspection using the GWT technique.

Further work on this subject should focus on validation of this technique with field data, which are usually more complex and contain different types of defects. Furthermore, the sensitivity of this algorithm could be investigated for coated and buried pipelines, where the attenuation rates are sufficiently high that they cause a major reduction in guided wave test capability.

Author Contributions: S.K.P. conducted the synthesised modelling, analysis, programming, experiments, and drafted the manuscript; P.M. helped edit the manuscript; T.-H.G. contributed to reviewing the manuscript.

Funding: This research was funded by the National Structural Integrity Research Centre (NSIRC), managed by TWI with the partnership of Brunel University.

Acknowledgments: The authors gratefully acknowledge TWI and PI Ltd. for granting access to the Teletest equipment.

Conflicts of Interest: The authors declare no conflict of interest.

References

1. Rose, J.L. An introduction to ultrasonic guided waves. In Proceedings of the 4th Middle East NDT Conference and Exhibition, Manama, Bahrain, 2–5 December 2007.
2. Rose, J.L. Ultrasonic guided waves in structural health monitoring. *Key Eng. Mater.* **2004**, *270–273*, 14–21. [CrossRef]
3. Catton, P. Long Range Ultrasonic Guided Waves for the Quantitative Inspection of Pipelines. Ph.D. Thesis, Brunel University, Uxbridge, UK, 2009.
4. Fateri, S.; Lowe, P.S.; Engineer, B.; Boulgouris, N.V. Investigation of ultrasonic guided waves interacting with piezoelectric transducers. *IEEE Sens. J.* **2015**, *15*, 4319–4328. [CrossRef]

5. Mallett, R.; Mudge, P.; Gan, T.; Balachandra, W. Analysis of cross-correlation and wavelet de-noising for the reduction of the effects of dispersion in long-range ultrasonic testing. *Insight-Non-Destr. Test. Cond. Monit.* **2007**, *49*, 350–355. [CrossRef]

6. Wilcox, P.; Lowe, M.; Cawley, P. The effect of dispersion on long-range inspection using ultrasonic guided waves. *NDT E Int.* **2001**, *34*, 1–9. [CrossRef]

7. Wilcox, P.D. A rapid signal processing technique to remove the effect of dispersion from guided wave signals. *IEEE Trans. Ultrason. Ferroelectr. Freq. Control* **2003**, *50*, 419–427. [CrossRef] [PubMed]

8. Zeng, L.; Lin, J. Chirp-based dispersion pre-compensation for high resolution Lamb wave inspection. *NDT E Int.* **2014**, *61*, 35–44. [CrossRef]

9. Xu, K.; Ta, D.; Moilanen, P.; Wang, W. Mode separation of Lamb waves based on dispersion compensation method. *J. Acoust. Soc. Am.* **2012**, *131*, 714–722. [CrossRef] [PubMed]

10. Xu, K.; Ta, D.; Hu, B.; Laugier, P.; Wang, W. Wideband dispersion reversal of lamb waves. *IEEE Trans. Ultrason. Ferroelectr. Freq. Control* **2014**, *61*, 997–1005. [CrossRef] [PubMed]

11. Xu, K.; Liu, C.; Ta, D. Ultrasonic guided waves dispersion reversal for long bone thickness evaluation: A simulation study. In Proceedings of the 2013 35th Annual International Conference of the IEEE Engineering in Medicine and Biology Society (EMBC), Osaka, Japan, 3–7 July 2013; pp. 1930–1933.

12. Toiyama, K.; Hayashi, T. Pulse compression technique considering velocity dispersion of guided wave. *Rev. Prog. Quan. Nondestr. Eval.* **2008**, *975*, 587–593.

13. Yucel, M.; Fateri, S.; Legg, M.; Wilkinson, A.; Kappatos, V.; Selcuk, C.; Gan, T.-H. Coded Waveform Excitation for High Resolution Ultrasonic Guided Wave Response. *IEEE Trans. Ind. Inform.* **2016**, *12*, 257–266. [CrossRef]

14. Yucel, M.; Fateri, S.; Legg, M.; Wilkinson, A.; Kappatos, V.; Selcuk, C.; Gan, T.-H. Pulse-compression based iterative time-of-flight extraction of dispersed ultrasonic guided waves. In Proceedings of the 2015 IEEE 13th International Conference on Industrial Informatics (INDIN), Cambridge, UK, 22–24 July 2015; pp. 809–815.

15. Newhouse, V.; Bilgutay, N.; Saniie, J.; Furgason, E. Flaw-to-grain echo enhancement by split-spectrum processing. *Ultrasonics* **1982**, *20*, 59–68. [CrossRef]

16. Karpur, P.; Shankar, P.; Rose, J.; Newhouse, V. Split spectrum processing: Optimizing the processing parameters using minimization. *Ultrasonics* **1987**, *25*, 204–208. [CrossRef]

17. Shankar, P.M.; Karpur, P.; Newhouse, V.L.; Rose, J.L. Split-spectrum processing: Analysis of polarity threshold algorithm for improvement of signal-to-noise ratio and detectability in ultrasonic signals. *Trans. Ultrason. Ferroelectr. Freq. Control* **1989**, *36*, 101–108. [CrossRef] [PubMed]

18. Saniie, J.; Oruklu, E.; Yoon, S. System-on-chip design for ultrasonic target detection using split-spectrum processing and neural networks. *Trans. Ultrason. Ferroelectr. Freq. Control* **2012**, *59*, 1354–1368. [CrossRef] [PubMed]

19. Gustafsson, M.G.; Stepinski, T. Adaptive split spectrum processing using a neural network. *Res. Nondestruct. Eval.* **1993**, *5*, 51–70. [CrossRef]

20. Gustafsson, M.G. Towards adaptive split spectrum processing. *IEEE Ultrason. Symp.* **1995**, *1*, 729–732.

21. Rubbers, P.; Pritchard, C.J. Complex plane Split Spectrum Processing: An introduction. *NDT Net* **2003**, *8*, 1–8.

22. Rubbers, P.; Pritchard, C.J. An overview of Split Spectrum Processing. *J. Nondestruct. Test.* **2003**, *8*, 1–11.

23. Rodríguez, A. A new filter bank design for split-spectrum algorithm. *NDT* **2009**, *1*, 1–7.

24. Rodriguez, A.; Miralles, R.; Bosch, I.; Vergara, L. New analysis and extensions of split-spectrum processing algorithms. *NDT E Int.* **2012**, *45*, 141–147. [CrossRef]

25. Syam, G.; Sadanandan, G.K. Flaw Detection using Split Spectrum Technique. *Int. J. Adv. Res. Electr. Electron. Instrum. Eng.* **2014**, *3*, 8118–8127.

26. Pedram, S.K.; Fateri, S.; Gan, L.; Haig, A.; Thornicroft, K. Split-Spectrum Processing Technique for SNR Enhancement of Ultrasonic Guided Wave. *Ultrasonics* **2018**, *83*, 48–59. [CrossRef] [PubMed]

27. Eddyfi Ltd. 2018. Available online: http://www.teletestndt.com/ (accessed on 10 July 2018).

28. Beasley, H.R.; Eric, W. A Quantitative Analysis of Sea Clutter Decorrelation with Frequency Agility. *IEEE Trans. Aerosp. Electron. Syst.* **1968**, *4*, 468–473. [CrossRef]

29. Pedram, S.; Haig, A.; Lowe, P.; Thornicroft, K.; Gan, L.; Mudge, P. Split-Spectrum Signal Processing for Reduction of the Effect of Dispersive Wave Modes in Long-range Ultrasonic Testing. *Phys. Procedia* **2015**, *70*, 388–392. [CrossRef]

30. Pavlakovic, B.; Lowe, M.; Alleyne, D.; Cawley, P. Disperse: A general purpose program for creating dispersion curves. In *Review of Progress in Quantitative NDE*; Springer: New York, NY, USA, 1997; pp. 185–192.

31. Bottger, W.; Schneider, H.; Weingarten, W. Prototype EMAT system for tube inspection with guided ultrasonic waves. *Nucl. Eng. Des.* **1987**, *102*, 369–376. [CrossRef]

 *applied sciences*

Article

Detection of Defects Using Spatial Variances of Guided-Wave Modes in Testing of Pipes

Houman Nakhli Mahal [1,2,*], **Kai Yang** [3] **and Asoke K. Nandi** [1]

[1] Department of Electronic and Computer Engineering, Brunel University London, Uxbridge UB8 3PH, UK; asoke.nandi@brunel.ac.uk
[2] NSIRC, Granta Park, Cambridge CB21 6AL, UK
[3] TWI, Granta Park, Cambridge CB21 6AL, UK; kai.yang@twi.co.uk
* Correspondence: houman.nakhlimahal@brunel.ac.uk

Received: 19 October 2018; Accepted: 19 November 2018; Published: 24 November 2018

Abstract: In the past decade, guided-wave testing has attracted the attention of the non-destructive testing industry for pipeline inspections. This technology enables the long-range assessment of pipelines' integrity, which significantly reduces the expenditure of testing in terms of cost and time. Guided-wave testing collars consist of several linearly placed arrays of transducers around the circumference of the pipe, which are called rings, and can generate unidirectional axisymmetric elastic waves. The current propagation routine of the device generates a single time-domain signal by doing a phase-delayed summation of each array element. The segments where the energy of the signal is above the local noise region are reported as anomalies by the inspectors. Nonetheless, the main goal of guided-wave inspection is the detection of axisymmetric waves generated by the features within the pipes. In this paper, instead of processing a single signal obtained from the general propagation routine, we propose to process signals that are directly obtained from all of the array elements. We designed an axisymmetric wave detection algorithm, which is validated by laboratory trials on real-pipe data with two defects on different locations with varying cross-sectional area (CSA) sizes of 2% and 3% for the first defect, and 4% and 5% for the second defect. The results enabled the detection of defects with low signal-to-noise ratios (SNR), which were almost buried in the noise level. These results are reported with regard to the three different developed methods with varying excitation frequencies of 30 kHz, 34 kHz, and 37 kHz. The tests demonstrated the advantage of using the information received from all of the elements rather than a single signal.

Keywords: signal processing; defect detection; spatial domain; array analysis; pipeline inspection; ultrasonic guided waves (UGWs)

---

1. Introduction

Pipelines are the main means of transferring oil and gas. They are usually installed in a hostile environment and suffer from defects such as corrosion or mechanical damage. To prevent the failure of pipelines, there is a need to inspect them in service in order to ensure their structural integrity. One of the many methods for the non-destructive testing of pipes is ultrasonic guided waves (UGW). This technology uses sets of transducer arrays to generate a low-frequency ultrasound signal. This ensures that elastic waves are guided through the boundary of the pipe; hence, the name guided waves [1]. Therefore, from one test point, the long-range inspection of pipelines is possible. However, the interoperation of UGW signals tends to be difficult due to the existence of the coherent noise in the tests.

Despite using arrays of transducers, UGW testing systems use a general routine process that outputs a single monodirectional signal in a time domain [2–5]. The generated signal is then used by inspectors to report any anomalies above the regional signal noise level [6]. To allow an ease of inspection, many types of research have been defined to use the characteristics of guided waves to increase the signal-to-noise ratio (SNR), increasing the defect detection accuracy.

One of the widely used techniques in the signal processing of guided waves is the work of Wilcox et al., which uses previously calculated dispersion curves in order to compensate for the effect of dispersion for the wave mode of interest in a certain propagation distance [7]. In 2013, Zeng et al. [8] used the basis of the dispersion compensation technique and introduced a novel method to design waveforms in order to pre-compensate for a certain propagation distance. Various researchers investigated the pulse compression approach, which consisted of exciting a known coded sequence and match-filtering the results in order to detect the known sequence [9–12]. Mehmet et al. [9,10] introduced an iterative dispersion compensation for removing the dispersion effect of guided-wave propagation. In 2016, Lin et al. [11] reported on the performance of different excitation sequences in terms of their ability to achieve δ-like correlation on Lamb waves. Most recently, Malo et al. [12] developed a two-dimensional compressed pulse analysis in order to enhance the achieved SNR of each wave mode. Coded waveform approaches have also been implemented in the enhancement of SNR in the inspection of Long Bones using guided waves [13]. In the pulse compression approaches, the results are always a trade-off between the spatial resolution and SNR; for having a high-energy, δ-like correlation result, the wavelength must be increased. Furthermore, the transfer function of the excitation system can affect the accuracy of the coded waveforms. These two factors mean that the initial waveform design and the accuracy of inspection using this method depend on the testing conditions. Furthermore, if non-dispersive regions of wave modes are used, there will be no need to use dispersion compensation algorithms.

Another method for filtering the guided waves signal is the work of Pedram et al. in Split-Spectrum Processing (SSP) [14,15], whereby through selecting the optimum filter parameters, the coherent noise of the test cases can be removed. However, automatic selection of the optimum filtering parameter is not possible at present. These parameters can be varying with regard to central excitation frequencies and the pipe geometries. On a similar note, frequency sweep examination, which was developed by Fateri et al. [16] in 2014, uses a similar concept in a way that defect features should exist in all of the frequency bands. Based on this fact, a novel method was developed that processes signals with different excitation frequencies, and is able to distinguish between the dispersive and non-dispersive wave modes that exist in the test. This method has never been tried in the pipes before, and requires many different excitations to be calculated on the same test subject. As it is currently known, due to backward cancellation algorithms and an overall system transfer function [4], it is not perfect, and not all of the frequencies can be used for optimum excitation (it can be varying depending on the pipe and tool set used). This will affect the accuracy of the algorithm if it is converted for pipeline inspection.

In 2015, Ching-Tin Ng [17] investigated signal processing approaches for the enhancement of results produced by guided-waves signals, in which he reported that Hilbert transform, Gabor wavelet transform, and discrete wavelet decomposition are able to improve the detectability of the results. UGWs have also been used for monitoring purposes. In 2015, Dobson et al. implemented independent component analysis (ICA) combined with previously developed temperature compensation methods [18,19] to enable the successful guided-wave monitoring of pipelines. In 2017, Chang et al. [20] introduced the idea of an adaptive sparse deconvolution method, with the aim of distinguishing the overlapping echoes of an ultrasonic guided-waves signal.

All of the algorithms have been developed with one purpose in mind: the improvement of the SNR to increase the possibility of the detection of a defect signal. However, the full potential of the system has not yet been established. The signal information in these systems are received from three rings of (at least) 24 transducers, but in the mentioned work, there is only one signal, which is achieved after the propagation routine of the device by summing the phase-delayed version of all of the transducers together. This routine was created to reduce the received flexural signals and have a unidirectional inspection [4,5]. Nonetheless, the received informatics can be processed in a better method, i.e., to use all of the available sensors for detecting the defect location rather than changing the information into a single time-domain signal.

On the other hand, array processing techniques that use individual signals are generally used for imaging and focusing applications [21–27]. However, these algorithms are computationally expensive, and might require the individual excitation of transducers [26], or a phased-array focusing of the transducers [23], rather than synthetic focusing, where all of the transducers in the array are excited at the same time [21]. This adds to the complexity of the interpreting the results. Furthermore, the goal of these algorithms is imaging a segment of the pipe rather than the detection of the defect location.

In this paper, we propose a new algorithm to solve the problem of detections, which can possibly be combined with all of the above-mentioned methods for SNR enhancement. The aim of this algorithm is to illustrate the possibility of using all of the individual transducers for defect detection instead of using a single signal achieved after the general routine. This way, both the spatial and temporal characteristics of the signal can be examined in more depth, which can lead to more accurate results. Furthermore, unlike imaging techniques, it uses normal excitation sequences, and thus can be applied to all of the data sequences, even if they are achieved in the past.

Section 2 explains the nature of noise on the guided-waves signal and the wave formation with regards to unique features. In Section 3, a novel method is proposed that uses the characteristic of axisymmetric waves to allow the detection of the defect signal. Validation was done by laboratory trials using two separate defect locations. The first defect is in a region where less coherent noise exists in the baseline with a varying size of 2% and 3% of the pipe's cross-sectional area (CSA), and the second defect is located in higher spatial noise with a varying size of 4% and 5% CSA. The setup of these tests and the results are reported in sections 4 and 5, respectively. In the last section, a summary of current achievements as well as the further works is reported.

2. Background Theory

Guided waves are dispersive and multimodal [2]. Dispersion causes the energy of a signal to spread out in space and time as it propagates [7]. Guided wave modes are categorized based on their displacement patterns (mode shapes) within the structure. Three main families of waves exist in pipes, which are longitudinal, torsional, and flexural. Longitudinal and torsional waves are axisymmetric waves, while flexural waves are non-axisymmetric. The popular nomenclature used for them are in the format of $X(n,m)$, where X can be replaced by the letters L for longitudinal, T for torsional, and F for flexural waves; n shows the harmonic variations of displacement and stress around the circumference, and m represents the order of existence of the wave mode [28].

Due to the sinusoidal and spatial variations of flexural waves, these coherent noises in the inspection can be reduced by adding the signals received from linearly spaced transducers around the circumference of the pipe. The resultant effect is the cancellation of non-axially symmetric flexural waves and the enhancement of axially symmetric torsional or longitudinal waves. However, cancellation is not ideal because of the imperfections of the real-world devices. These factors include the non-linear spacing of the transducer's placement, the variability of the transducer's frequency response, the number of transducers used, or the tolerances of the electronic test device. Furthermore, a scattering of an ideal axisymmetric wave with a non-axisymmetric feature would also result in the generation of other wave modes, which is commonly known as a wave mode conversion phenomenon [29–32]. Thus, inspecting guided waves without the interference of these coherent noises is near impossible.

For decades, the inspectors have been trained to interpret these resultant time-domain signals to extract the defect location. Nevertheless, the problem of detection can be defined in another way, i.e., the linearity and symmetry between the phases of the received waveforms. For better understating of this matter, in the following section, different conditions of waves arrived at spatially are demonstrated using a synthesized model.

2.1. Wave Formation

The scattered wave would consist of the excited axisymmetric and its associated flexural waves. Depending on the feature shapes, the energy distribution of each of the created wave modes would be different. As an example, a perfectly axisymmetric feature in the ideal scenario would result in the generation of a perfect axisymmetric response with no other flexural waves. In these scenarios, if the axisymmetric wave is stronger than the coherent regional noises in the area, an anomaly is reported by the inspectors. However, since it is a visual interpretation, in some scenarios, the resultant mixture of flexurals can easily lead to false alarms due to their coherence to the expected received signal. Furthermore, the axisymmetric wave can be smaller than the local noise level, and therefore would not be detected. These are some of the main challenges faced when inspecting the temporal domain.

Figure 1 shows the spatial arrival of the signals received from two rings of 32 source points. These signals are generated by the finite element model (FEM) [30]. In this setup, ideal source points are assumed, which leads to the generation of a perfect $T(0,1)$ wave mode in all of the figures except (b) and (i), where the frequency response of each transducer is variable.

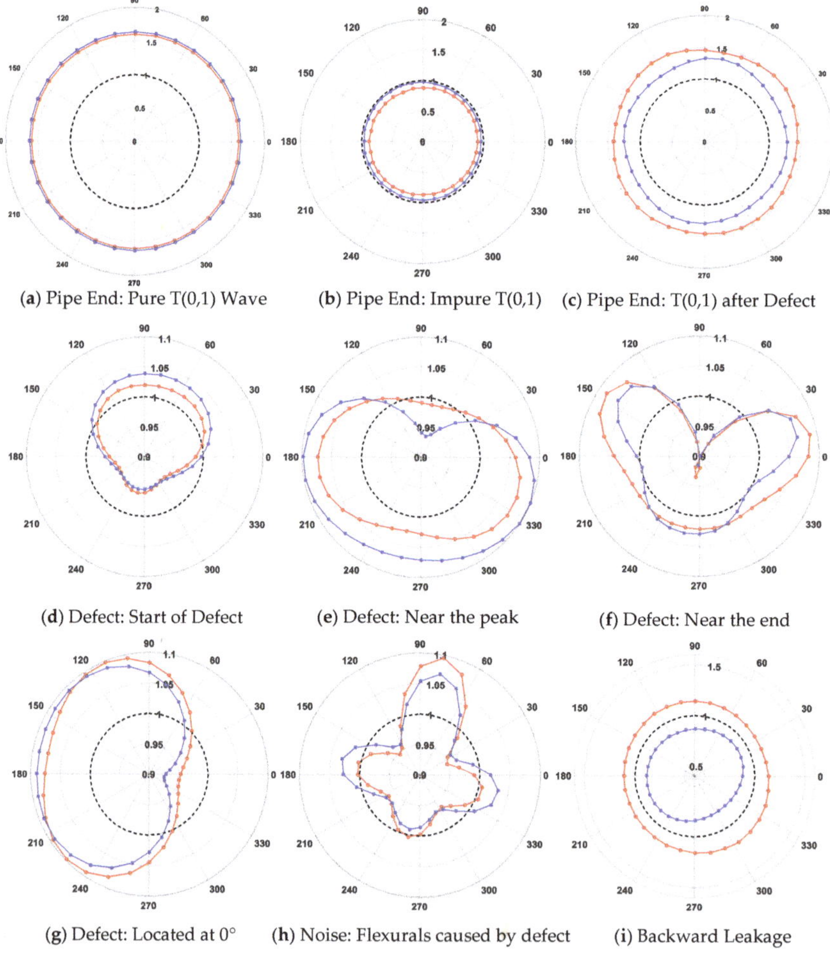

(**a**) Pipe End: Pure T(0,1) Wave   (**b**) Pipe End: Impure T(0,1)   (**c**) Pipe End: T(0,1) after Defect

(**d**) Defect: Start of Defect   (**e**) Defect: Near the peak   (**f**) Defect: Near the end

(**g**) Defect: Located at 0°   (**h**) Noise: Flexurals caused by defect   (**i**) Backward Leakage

Figure 1. Spatial signal reception from two rings of 32 source points from various features where the red, blue, and black dotted lines are showing the first ring, second ring, and the reference offset.

The red and blue signals are showing the first and second ring, respectively, and the reference is illustrated by the black dotted line. Thus, the values inside the reference would mean a negative phase, and values higher than the circle would mean a positive amplitude of the individual transducer.

2.1.1. Pipe End

The received signals from the pipe end are shown in Figure 1a–c. As can be seen in (a), a pure $T(0,1)$ wave is generated, since both the excitation sequence and the feature are perfectly axisymmetric. However, in the cases of (b) and (c), where the flexurals are created by imperfect excitation and from wave mode conversion, respectively, the received response is the superposition of $T(0,1)$ and flexural waves. The scattering from the pipe end would result in all of the energy of the wave to be reflected. Thus, a strong proportion of the received signal energy corresponds to the $T(0,1)$ wave, and small variations are caused by flexural waves.

2.1.2. Defect

In cases (d)–(g) of Figure 1, the received waves is generated from a non-axisymmetric defect with a cross-sectional area size of 3%. The shown signals are different sequences from various times with respect to the temporal domain signal. The capture times are shown in Figure 2b where the red line is representing case (d), the blue line is representing case (e), and the green line is representing case (f). Case (g) is captured at the same time as case (e), but the generated wave is from a defect located at $0°$ (the right side of the pipe), while the other cases were from a defect angled at $90°$ (top side). Four important points are illustrated from the figures:

1. Unlike the pipe end, defects that are non-axisymmetric features will inherently generate flexural waves. Nonetheless, the strongest energy will be associated with the $T(0,1)$ wave.
2. The end of the signal envelope will contain more flexural waves compared to the start of the envelope. At the start of the envelope, at most one cyclic variation will exist in the spatial domain, which would represent a low-order flexural wave mode. Nonetheless, near the end of the envelope, higher order variations (three or more) are starting to increase. This is because after the center of the envelope, the power of the $T(0,1)$ wave is starting to decrease, and the higher order flexural with a lower travel speed is beginning to arrive.
3. The strongest energy of the $T(0,1)$ wave is associated with the peak of each cycle in the temporal domain, where most of the transducers will have the same phase.
4. When the angle of the defect is changed, the direction of the arrival changes, but the phases and the overall formation of the wave remain the same.

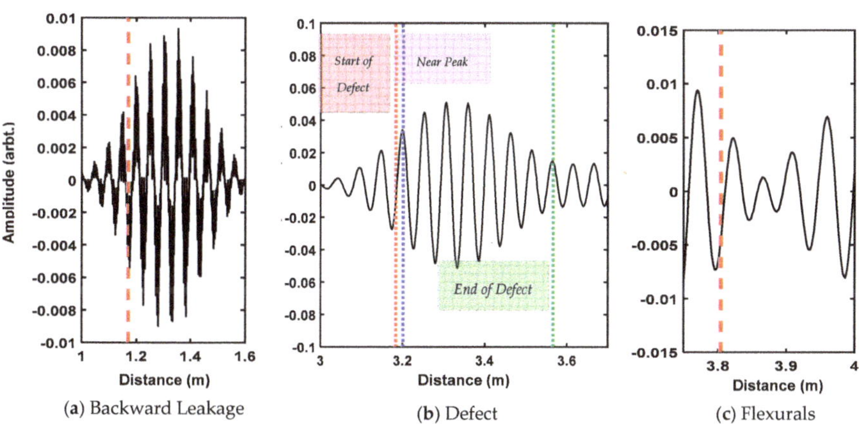

(a) Backward Leakage   (b) Defect   (c) Flexurals

Figure 2. Temporal domain of cases (d)–(i) from Figure 1

2.1.3. Coherent Noise

The spatial arrival of the flexural noises created by the wave mode conversion and the backward leakage due to the imperfect excitation is shown in Figure 1h,i, respectively. Their corresponding time domain signals are shown in Figure 2b,c. As can be seen, for the flexural signal, the reception point phases are highly non-linear, which is completely different from the case of the defect signal.

Backward cancellation is not always perfect, due to the fixed ring spacing and the variance along the received waves from each transducer. In the temporal domain, these waves might still be detected, but when one looks at the phases between the two rings, it is observable that the two rings would have the exact opposite phases. Thus, it would be easier to detect signals received from the opposite direction in the spatial domain.

3. Methodology

In the previous section, it was demonstrated that in the case of flexural waves, high-order sinusoidal variation along the pipe's circumference will be created. In ideal scenarios, where a perfect axisymmetric wave hits a perfect axisymmetric feature, the received amplitude in each transducer must be the same. However, due to the imperfections of the transducers and the testing conditions, there will be variance along each of the received amplitudes from the transducers; nonetheless, they all will have the same phase. Similar to the pipe ends, defects would generate a significant amount of torsional mode. However, they also do create flexurals with respect to the defect size and location. For signals coming from the backward direction, it is observed that the phases would practically be the inverse of each other for each transducer.

Based on these observations, by detecting the linearity between the phases of the sensors, the axisymmetric waves can be detected. Since not all of the phases would become the same in the case of defect signals, a measure of similarity between the phases must be defined, which can then be the threshold for detection.

3.1. General Routine

The first step in the general propagation routine of guided wave devices is to cancel the signal received from the backward direction. For doing so, the signals received from Ring 2 are initially subtracted from Ring 1 and Ring 3, and then the results are phase-delayed with respect to the tool set's ring distancing and the excitation wave mode dispersion curve ($T(0, 1)$ in this case). The results are stored in the Signals matrix:

$$\text{Signals}[i,\ n] = \begin{cases} \text{PhaseDelayed}(\text{Ring1} - \text{Ring2}), \text{ for } i \leq NS/2 \\ \text{PhaseDelayed}(\text{Ring3} - \text{Ring2}), \text{ for } i > NS/2 \end{cases} \tag{1}$$

where n represents the sample number, i is the sensor number, and NS represents the total number of achieved signals, which is twice the number of transducers in each ring. This method will result in $T(0, 1)$ components to be overlapped. For the signals received from the forward test direction, this overlap will be positive, and for the backward direction, the amplitudes are inverted, which allows the unidirectional inspection. The temporal domain signal, which is generally used for inspection, is achieved by taking the mean of the Signals:

$$\text{temporal}[n] = \frac{1}{NS} \sum_{i=1}^{NS} \text{Signals}[i,\ n] \tag{2}$$

3.2. Initialization

All of the transducers have the same phase when the sum of all of the signals from the same-phase transducers is equal to the sum of all of the signals. Therefore, the measure of similarity of the phases is defined by taking the difference between the sum of all of the signals and the sum of in-phase signals received from each transducer. The sign of the temporal domain demonstrates which sign is the greater power of iteration. Based on this sign, a mask is created for each transducer as follows:

$$\text{mask}[i, n] = \begin{cases} 1, & \text{sign}(\text{Signals}[i, n]) = \text{sign}(\text{temporal}[n]) \\ 0, & \text{else} \end{cases}. \tag{3}$$

Using this mask, in-phase signals with the temporal domain of each sample can be identified, and their sum can be calculated as:

$$\text{inPhase}[n] = \left| \sum_{i=1}^{NS} \text{mask}[i, n] * \text{Signals}[i, n] \right| \tag{4}$$

Afterward, the similarity can be defined using:

$$\text{comparisonValue}[n] = \text{inPhase}[n] - \left| \sum_{i=1}^{NS} \text{Signals}[i, n] \right| \tag{5}$$

3.3. Threshold Selection

In cases where a significant proportion of iterations energy is from the $T(0, 1)$ wavemode, the comparisonValue will be zero. However, as shown previously, defect signals will contain a mixture of flexural and torsional wave modes, leading this value to be non-zero. Hence, there is a need for defining a threshold for detecting the defects. In this section, three methods are presented defining this threshold.

3.3.1. Method One

In this method, a static constant is defined by the inspector. If the comparisonValue is less than this defined threshold, the sample would be marked as one, which means that a feature is detected; otherwise, it would be marked as zero:

$$\text{Result}[n] = \begin{cases} 1, & \text{comparisonValue}[n] \leq \text{Threshold}_{\text{One}} \\ 0, & \text{else} \end{cases} \tag{6}$$

For small signals, this formula would not be accurate, since inherently, their sum would be bigger than the threshold; thus, a compensation factor, 0.001, is added to prevent outliers from being detected in the noise region. This is an acceptable level, since the defect signals smaller than 0.001 would not be detected by the analog to digital converters of the device due to their limited discretization range. If the amplitude of the signal in the temporal domain is greater than this compensation factor, the algorithm would work as normal, else it would be zero:

$$\text{Result}'[n] = \begin{cases} \text{Result}[n], & |\text{temporal}[n]| > \text{compensation} \\ 0, & \text{else} \end{cases} \tag{7}$$

The disadvantages of this method are that firstly, a user input is required, and secondly, the value is fixed for all of the features, disregarding the existing characteristics in each sample.

3.3.2. Method Two

To use signal characteristics more effectively, the amplitude of the inPhase signals can be used to define the threshold, since a greater inPhase signal can be the result of a greater linearity of the phases. By taking a percentage of the inPhase signals, a higher threshold can be assigned to such cases. In this method, a fixed percentage value needs to be set by the inspector, and the threshold for each iteration would be adaptively changed:

$$\text{Threshold}_{\text{Two,V1}}[n] = \text{inPhase}[n] * \text{percentage} \tag{8}$$

Another improvement to this algorithm would be to consider the effect of comparisonValue as well. A higher comparisonValue means that there is a higher chance that the signal is not linearly received by all of the transducers. This effect is significant in the case of signals received from the backward direction. In each iteration, the comparisonValue can be subtracted from the inPhase, which would decrease the resultant threshold value for backward propagating or non-axisymmetric waves:

$$\text{Threshold}_{\text{Two,V2}}[n] = (\text{inPhase}[n] - \text{comparisonValue}[n]) * \text{percentage} \tag{9}$$

3.3.3. Method Three

To eliminate the human error, the requested input from the user must be eliminated. The phases are the main factor in detecting the results; the percentage can be defined based on the ratio of the in-phase sensors over the total number of sensors:

$$\text{percentage}[n] = \frac{1}{\text{NS}} \sum_{i=1}^{\text{NS}} \text{mask}[i, n] \tag{10}$$

In no iteration is the value of mask zero; it always have a positive offset, which means that the percentagecannot be replaced directly into Equation (8). Nonetheless, this offset can be removed by subtracting the mean of the calculated percentage for all the samples from each individual sample:

$$\text{Average} = \frac{1}{\text{N}} \sum_{n=1}^{\text{N}} \text{percentage}[n] \tag{11}$$

where N is the total number of samples in the test. The final considered percentage for each sample would be:

$$\text{percentage}_{\beta}[n] = \text{percentage}[n] - \text{Average} \tag{12}$$

Doing so removes the need for user input, and provides a more accurate thresholding with regard to the phases of the signals in each sample; the higher number of the same phase signals means a higher percentage value for the sample, which leads to a higher chance of detection. The formula for the first version (V1) and second version (V2) of this method can be achieved by substituting the new percentage achieved from Equation (12) in Equation (8) or Equation (9):

$$\text{Threshold}_{\text{Three,V1}}[n] = \text{inPhase}[n] * (\text{percentage}[n] - \text{Average}) \tag{13}$$

$$\text{Threshold}_{\text{Three,V2}}[n] = (\text{inPhase}[n] - \text{comparisonValue}[n]) * (\text{percentage}[n] - \text{Average}) \tag{14}$$

3.4. Limits Definitions

Three main detection limits exist in these algorithms. (1) The first if the minimum, which is the minimum value required for each method to detect at least one cycle of the defect signal. (2) Second is the maximum, which is the maximum allowable value for the methods to detect the defect without any outliers. By moving toward the maximum value, the detection length increases; however, the accuracy reduces as the chance of detecting outliers increases. Thus, for comparing the methods together, an average value is defined using the minimum and maximum. This value is the best possible compromise between the achieved defect length and the possibility of detecting false alarms. For method three, the average value is calculated using Equation (11), and minimum and maximum values are achieved by manually replacing the average in Equation (14). For both of the other methods, the average value is calculated manually from the achieved minimum and maximum values in the tests. Also, the *Safe zone* margins are defined as the difference between the minimum and maximum, which illustrate the values where manual inputs will produce a result without any false alarms.

3.5. Program Flowchart

The program flowchart is shown in Figure 3. As can be seen, the three stages of general routine, initialization, and result generation are the same in all of the methods. However, the threshold value that is used in the result stage is selected based on methods one, two, or three, which are explained in the thresholding section.

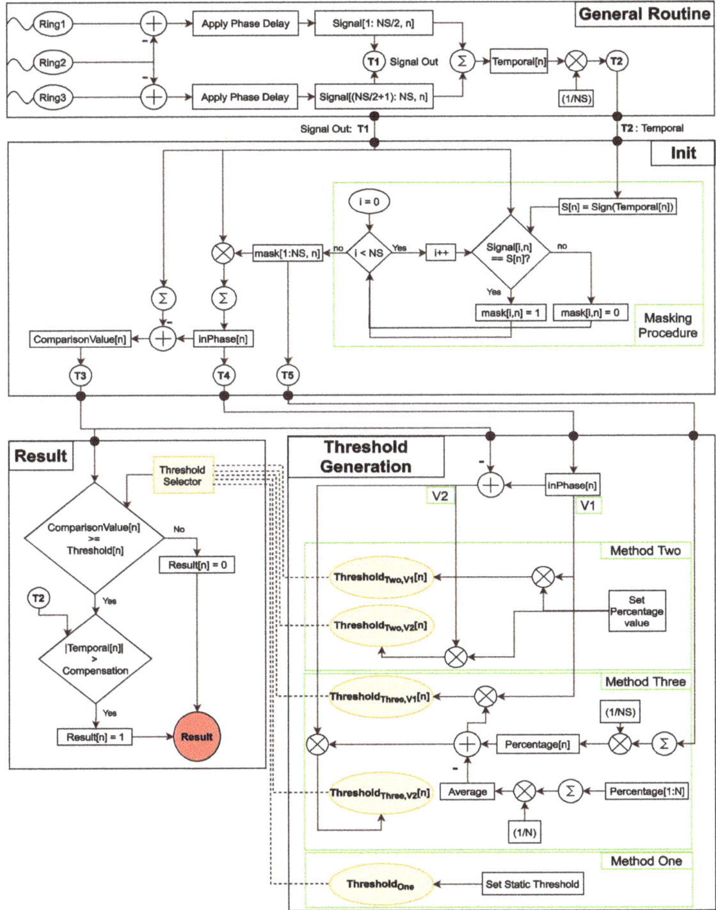

Figure 3. System flowchart.

4. Experimental Setup

The Teletest device [33], manufactured by Teletest branch of Eddify technology family in Cambridge (UK), used in order to inspect an eight-inch schedule 40 steel pipe with a length of six meters. The test tool included three rings of 24 linearly spaced transducers with 30-mm spacing, which are located 1.5 m away from the back end of the pipe. The main aim of these trials is to assess the ability of the algorithm in highly noisy scenarios and investigate the dependency of each method to the excitation frequency.

In the first set of tests, two defects of 3% and 2% CSA have been created; these are located 4.5 m from the back end of the pipe, which generates signals with less/almost the same energy as the noise regions. In order to confirm the ability of this thresholding method in detecting another defect in the same pipe, defects with CSA sizes of 4% and 5% have been created after the initial trials, which are located 5.5 m away from the back end of the pipe. Figure 4 shows the signal reception routes of this setup.

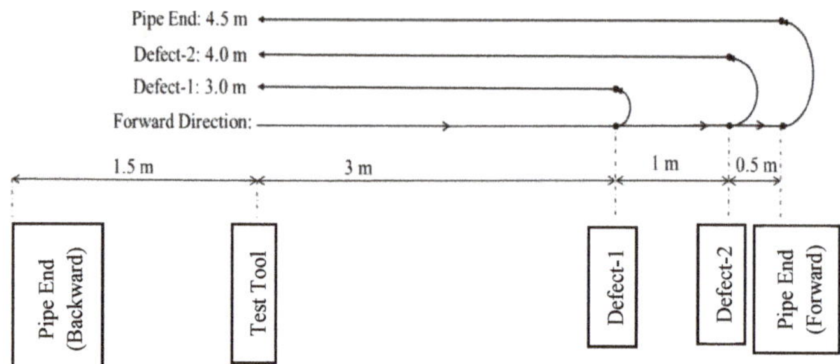

Figure 4. Signal reception routes.

The excitation sequence is a 10 cycle Hann-windowed sine waves with frequencies of 30 kHz, 34 kHz, and 37 kHz. The selected frequencies are considered the optimum excitation sequences with regard to their resultant forward propagating energy achieved using the backward cancellation algorithm [30]. A summary of the settings used for the first set of tests (first defect), which are numbered from one to six, are shown in Table 1; for the second set of tests, they are numbered from seven to 12 (second defect), and are shown in Table 2.

Table 1. Summary of the first set of tests conditions (first defect). CSA: cross-sectional area.

Test (#)	Defect CSA (%)	Frequency (kHz)
1		30
2	3	34
3		37
4		30
5	2	34
6		37

Table 2. Summary of the second set of tests conditions (second defect).

Test (#)	Defect CSA (%)	Frequency (kHz)
7		30
8	5	34
9		37
10		30
11	4	34
12		37

Figure 5 shows the baseline signals achieved from the same pipe with no defects, with excitation frequencies of (a) 30 kHz; (b) 34 kHz; and (c) 37 kHz. As can be seen, even though there are no defects in the pipe, the baseline signal is highly polluted by the coherent noise. The first defect signal is expected to be received approximately three meters away from the tool test, between the first and second red dotted lines. The second defect is expected to be located approximately four meters away, between the third and fourth red dotted lines. Three major conclusions can be drawn with regard to the baselines:

1. Defect 1: The signal with lowest amount of existing coherent noise in the baseline are 34 kHz, 30 kHz, and 37 kHz, respectively.
2. Defect 2: The lowest amount of existing coherent noise in the baseline belongs to 37-kHz, 34-kHz, and 30-kHz signals, respectively

3. All of the respective frequencies have a higher amount of noise in the Defect-2 region.

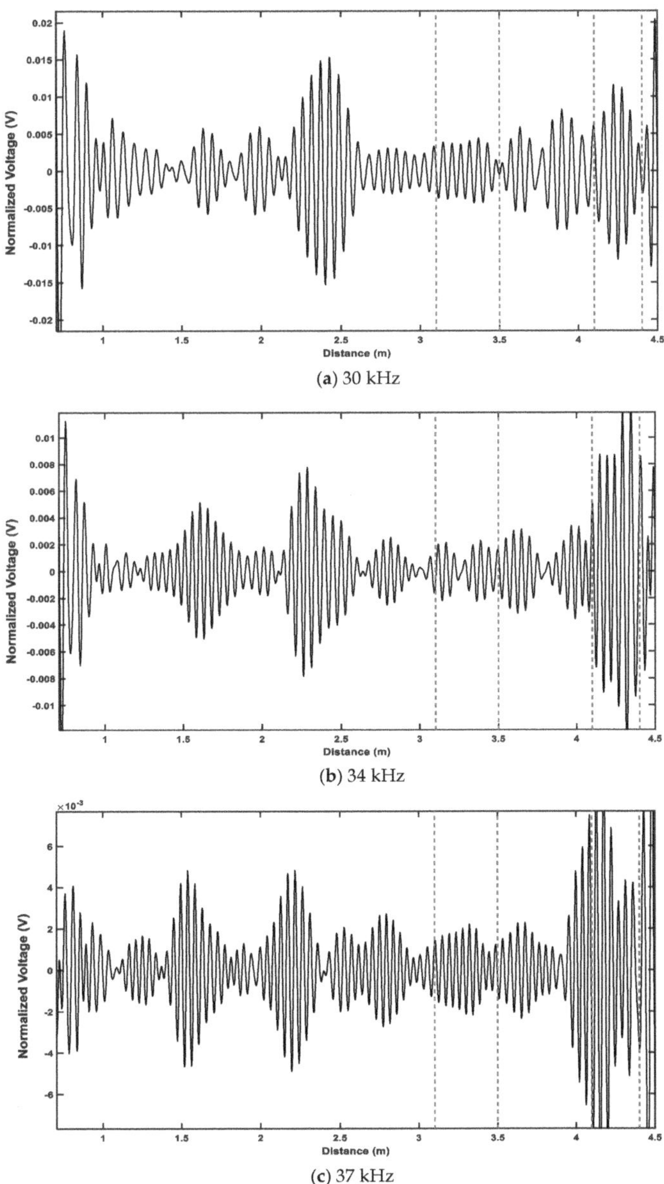

(**a**) 30 kHz

(**b**) 34 kHz

(**c**) 37 kHz

Figure 5. Baseline signals of pipe with no defect with frequencies of (**a**) 30 kHz; (**b**) 34 kHz; and (**c**) 37 kHz. The first marked region (3–3.5 m) is the location of the first defect, and the second marked region (4.1–4.5) is the location of the second defect.

It is expected that the thresholds are required for detecting regions with lower existing coherent noises, which means that overall, Defect 2 will be more difficult to detect.

5. Results

In this section, the limits achieved for detecting the defect signal using each method is reported. The SNR of the defects in each test is demonstrated in Table 1. These values are generated using the following equation:

$$\text{SNR} = 20\log_{10}\left(\frac{\text{energy(Signal)}}{\text{energy(Noise)}}\right) \tag{15}$$

The SNRs are calculated with respect to the overall noises in the region. In the case of test numbers 1–6, only the energy of Defect 1 and the pipe end signal is removed for calculating the energy of the noise; however, for test numbers 7–12, as well as the Defect-1 and pipe end signals, the Defect-2 signal is also removed for calculating the noise. As demonstrated in Table 3, the defect signals are within the noise level, and identification of them using the inspection routine would not be accurate, especially for a 2% CSA defect. Nonetheless, the overall SNR of Defect 2 is much higher due to both the increment of CSA size, and the existing coherent noises of the region.

Table 3. Calculated signal-to-noise ratio (SNR) with respect to the coherent noises in the inspection (in dB).

Test #	1	2	3	4	5	6	7	8	9	10	11	12
SNR	2.97	8.93	10.36	−2.65	−0.98	−3.43	18.57	27.63	28.05	14.73	24.00	23.10

5.1. Example Test Case

Figure 6 shows the Teletest signal of Test 5 before processing, where the defect's signal region is in between the two dotted red lines. The SNR of a defect in this test is −0.98 dB, which means that it is completely buried in the noise. The noise signal is calculated by removing the torsional waves (pipe end and defect signal) from the shown signal in Figure 6. However, since Defect 2 is not present in this test, the location in between 4–4.5 m is also considered as noise.

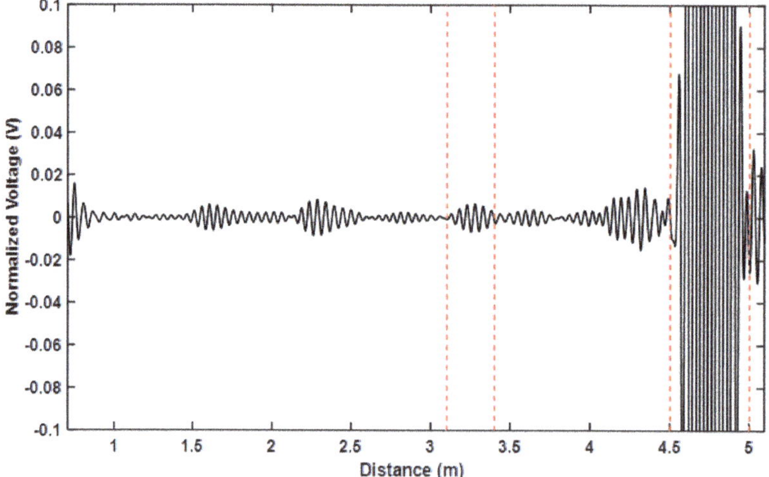

Figure 6. Example signal achieved using general routine, tested by 34 kHz where the first two red dotted lines shows the defect region, and the third and fourth ones show the pipe end.

Figure 7 shows the assigned values to the first (black line) and the second version (red line) of thresholding, which is the inPhase and subtraction of the comparisonValue from the inPhase, respectively. As can be seen both in the backward leakage signal and the noise region, the decrease in the assigned threshold is more than the defect and pipe end signals. In method two, a static percentage of either of the mentioned values is considered the threshold for each iteration; however, in method three, this percentage is assigned using the ratios of inPhase signals subtracted from its mean, as shown in Figure 8. The calculated threshold for each iteration using the average value of method three is shown in Figure 9. In the same figure, if method one was used, the threshold value would be a constant line for all of the iterations. The processed signal calculated from method three is illustrated in Figure 10, where it shows the locations at which the comparisonValue is higher than the threshold history. As expected, the pipe end is completely detected, since it is a complete axisymmetric feature, but only the center part of the defect could be detected.

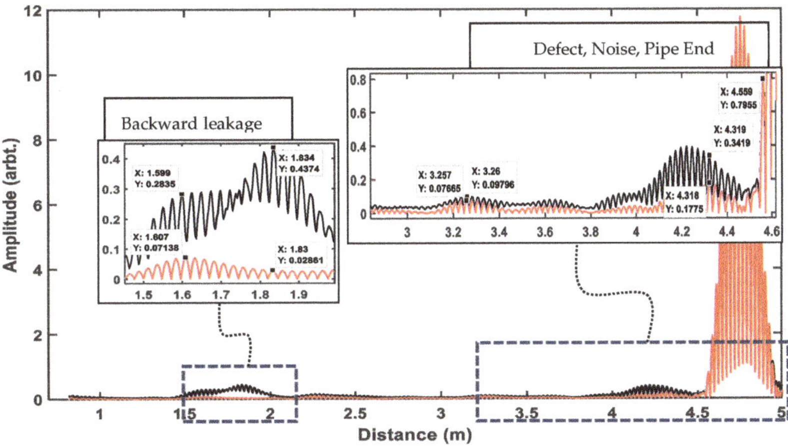

Figure 7. Comparison of the first and second thresholding versions on the Test 5 signal where the two zoomed regions are showing the backward leakage, defect, noise, and pipe end, respectively. The black and red signals are showing the first and second versions, respectively.

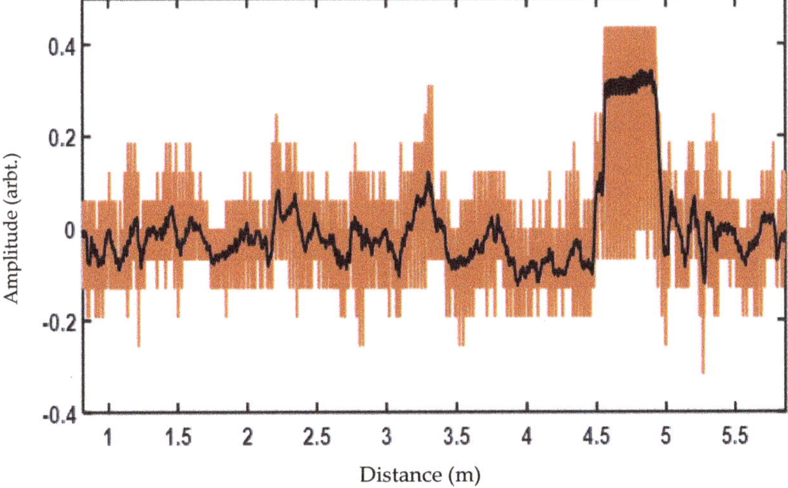

Figure 8. Calculated percentage of sensors with same phase, where the offset (mean values) are removed (orange); the black signal is showing the filtered version for better illustration.

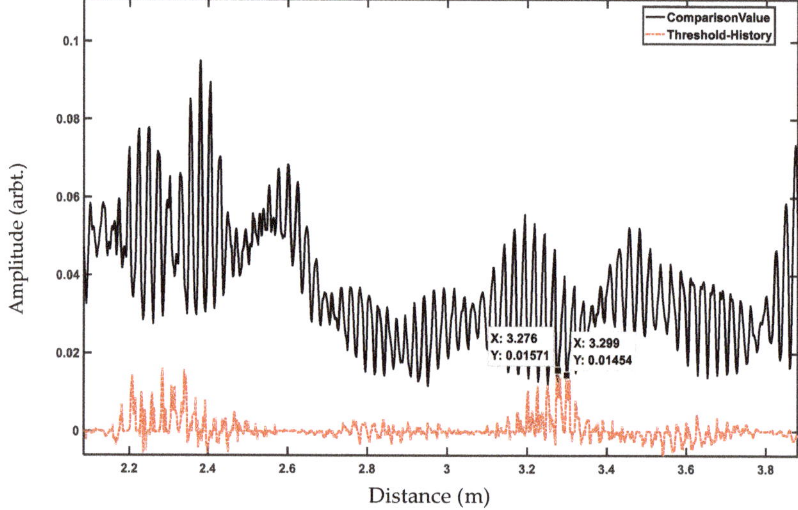

Figure 9. Calculated threshold using method three on Test 5 (red) against the second version comparisonValue (black). In the case of method one, the threshold would be a constant line instead of changing based on the iteration.

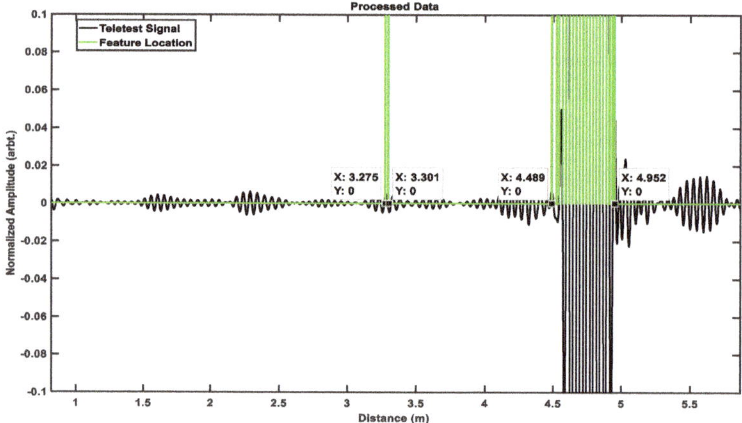

Figure 10. Method three result (green signal) using the average value against the temporal signal (black) for testing a 2% CSA defect.

5.2. First Set of Tests

The detection of Defect 1 is the focus in the first set of tests (test numbers 1–6). In these test cases, Defect 2 does not exist, and therefore, the region between 4–4.5 is considered noise. Furthermore, the coherent flexural noises existing in the Defect-1 region are less than those in the second Defect-2 region, with 34 kHz, 30 kHz, and 37 kHz having the least amounts, respectively. As can be seen from Table 3, the defects in these tests have a low SNR and are almost buried in temporal noise, which makes them difficult to be detected in the temporal domain.

5.2.1. Method One

This method was only successful for tests 1, 2, and 4. Figure 11 shows the limits of the threshold values. Observing the trends with regards to the frequency in tests 1–3, the minimum values are increasing, while the maximum values are decreasing up until the point where the minimum will result in an outlier in Test 3. The same trend is observable for the minimum values in tests 4–6, where, although the maximum values could not be detected, the minimum values are increasing with regard to the frequency. With regard to the defect CSA (and SNR), the algorithm can detect the 2% CSA defect using a 30-kHz excitation frequency (Test 4). Nonetheless, when the frequency is increased to 34 kHz in Test 5, although the defect's SNR is increased by 1.68 dB, this method resulted in a false alarm, even at the minimum limit. Thus, it can be concluded that this algorithm is more sensitive to the frequency and local amplitude of the defect region rather than its temporal SNR and the overall energy of the noises in the region. Since higher energy is associated with lower frequencies, lower frequencies tests would achieve better results using this method for Defect 1.

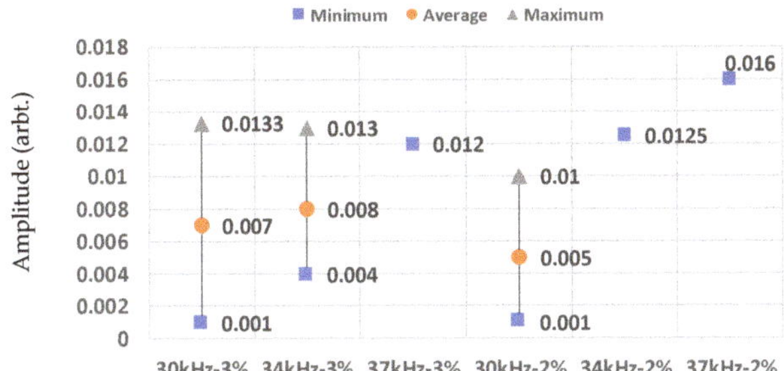

Figure 11. Minimum, average, and maximum values of method one of the first set of tests.

5.2.2. Method Two

This method is effective in all of the tests except for Test 6. With regard to Figure 12, which shows the limits achieved using this method, the relation between the detection limits and frequency is no longer linear. The minimum is increasing with the frequency until it reaches a maximum peak on 34 kHz; then, it starts to decrease when moving toward 37 kHz. The same trend is observable for the detection length and safe zone margins. This means that this algorithm is a function of both the SNR of the noise and the defect, as well as the frequency where the optimum frequency would be around 34 kHz. Nonetheless, it should also be pointed out that when using a lower frequency, the same happens as in the previous method: the minimum value for detection is closer to zero, which means that defects would be easier to detect. Regarding the second thresholding version, the increase in the minimum limits is not as much as those in the maximum, which therefore increases the safe zone margins in all of the tests. This is clearly observable in tests 1 and 2, where the minimum limits remained unchanged. The average and maximum detection lengths of the defect remain unchanged. Therefore, using the second threshold version provides a greater safe zone margin.

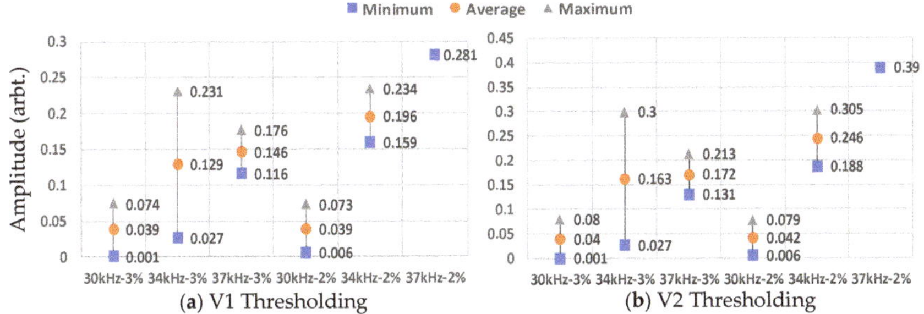

Figure 12. Limits achieved using method two on the first set of tests.

In comparison to method one, this method has a greater range of operating frequencies and provides a larger safety margin for threshold selection. The optimum excitation frequency for this method is 34 kHz for the first set of tests.

5.2.3. Method Three

The average values are calculated automatically by taking the mean of the percentage of active sensors in each iteration; thus, they are fixed in both thresholding versions. The same as method two, this method can detect the defects without any false alarms in tests 1–5. However, only tests with an excitation frequency of 34 kHz can use the average value, since using it for other excitation frequencies would result in detecting outliers. The limits of this method are shown in Figure 13. The tests where the average value is between the minimum and maximum are when the automatic detection method is successful. Furthermore, the averages values achieved using all of the different frequencies are between 0.602–0.554, and tend to decrease slightly when the frequency is increased.

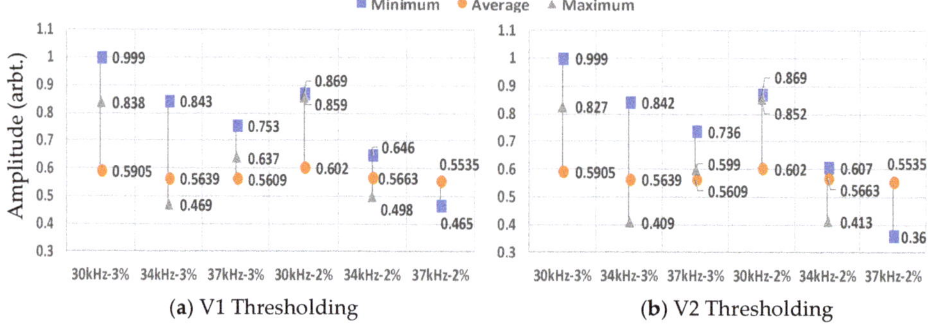

(a) V1 Thresholding   (b) V2 Thresholding

Figure 13. Limits achieved using method three in the first set of tests.

To conclude, this method can automatically detect defects of up to 2% CSA, which are buried in the noise level using a 34-kHz excitation frequency. Furthermore, with regard to the detection length, it can also be said that 34 kHz is the optimum excitation frequency in the first set of tests.

5.2.4. Comparison

Figure 14 shows the amount of increase in the minimum and maximum values achieved using the first and second version of the thresholding. The minimums tend to increase less than the maximum values by at least a factor of two.

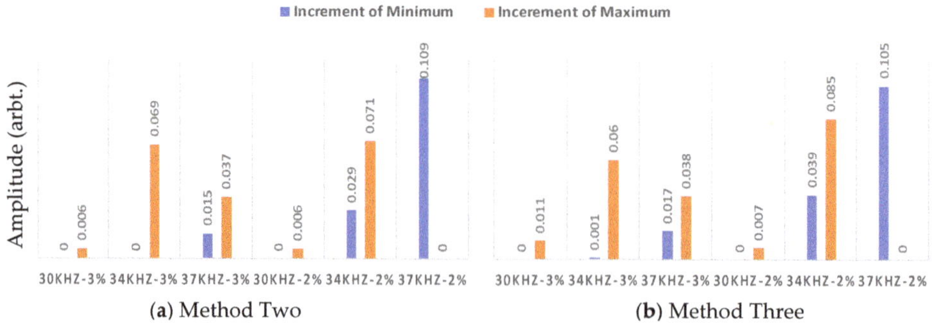

(a) Method Two   (b) Method Three

Figure 14. The increments achieved using the second version of thresholding in the first set of tests.

The generated *safe zone* margins for all of the methods are shown in Figure 15. Except for Test 4, method three tends to have a higher *safe zone* margin than method two, and the best margin is achieved using second thresholding version for method three. These results are especially important on tests 2 and 5, where the optimum excitation frequency of both methods is used. The second version of thresholding always produces a greater safety margin.

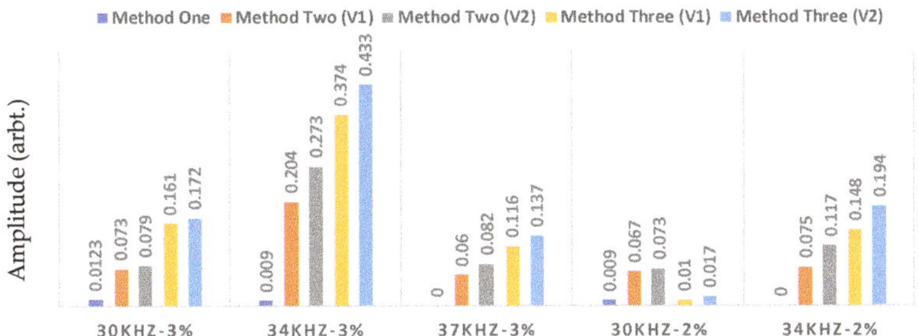

Figure 15. Bar graphs of the Safe zone margins of all methods in the first set of tests.

Figure 16 shows the length of detection achieved by average values with different algorithms. Cases where outliers are detected are not reported. As can be seen, in the optimum frequency, both of the defect lengths achieved by method two and method three are the same. In Test 5, in the second version of thresholding, the length has decreased, since the average value is fixed, and the defect size is smaller; however, in Test 2, the same detection length is achieved using both versions, since the defect is wider. Method three is the best out of these three mentioned ones, since it produces a larger *safe zone* margin, and can detect the defects automatically using the optimum excitation frequency of 34 kHz for the first set of tests.

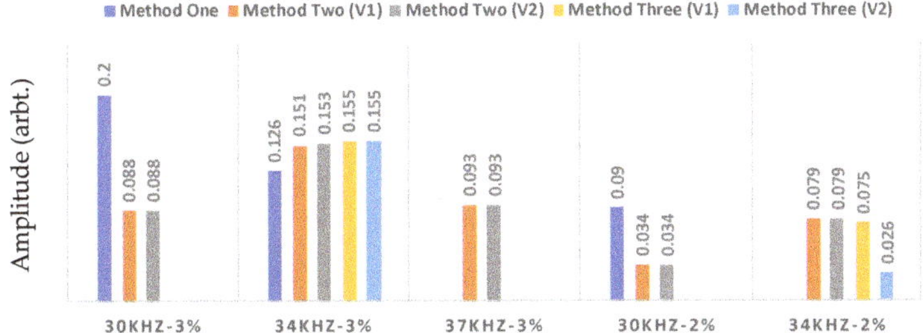

Figure 16. Bar graphs of the detection length of the defect using the average value of all of the methods in the first set of tests.

5.3. Second Set of Tests

In the second set of trials, the focus is on detecting Defect 2. The defect signal is expected to be received between 4–4.5 m. Since the baseline of the defect signal is highly polluted by the coherent flexural noise, higher detection limits are required for detecting the defects, even when the CSA size of the defect is increased. This aim of these tests is to observe the effect of spatial noise on the algorithm and demonstrate its capability in detecting defects in such scenarios. Since excitation frequencies of 37 kHz, 34 kHz, and 30 kHz have the lowest spatial noise in the defect region, respectively, it is expected that the algorithms perform better using the respective mentioned frequencies.

5.3.1. Method One

Figure 17 shows the limits required from method one to detect Defect 2. The only excitation frequency that allowed the detection of defects without any outliers was 37 kHz, while 34 kHz resulted in the detection of defects with at least three outliers in both cases, and in 30 kHz, the defect could not be detected. With 37 kHz as the excitation frequency, it can be seen that when the defect energy decreases, the threshold is increased to 0.009. As this value is close to the maximum allowable threshold (0.01), no smaller defects could be detected even using this frequency. This is unlike the first set of tests, where it was easier to detect the defect using lower frequencies, since Defect 2 is located in a region with higher energies of flexural noises. Consider the case of the 34-kHz frequency, as reported in Figure 11: the maximum limit of detection of Defect 1 without any outliers is approximately 0.013. However, in the case of Defect 2, the minimum level is 0.019 for the 5% CSA defect, which is higher than the previously reported limit. This outcome indicates the detection of outliers in the test. The same can also be said regarding 37 kHz, where previously, the maximum limit was approximately 0.012, and now, the minimum required for the detection of a 4% CSA defect is 0.01, which is below the level where outliers will be detected. It can be concluded that method one is highly affected by the existing energy of flexurals in the same iteration.

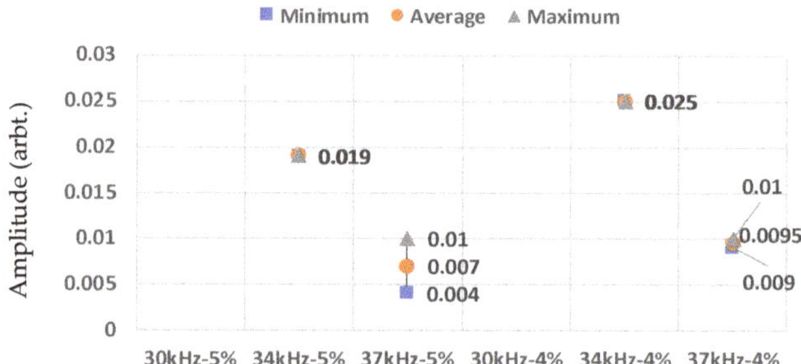

Figure 17. Minimum, average, and maximum values of method one of the second test cases.

5.3.2. Method Two

The results of method two are illustrated in Figure 18. Unlike the first set of tests, where 34 kHz was the optimum frequency of excitation, in these tests, 37 kHz achieves better results. It is not possible to detect Defect 2 without any outliers using the 30-kHz frequency, since the required minimum threshold level, 0.125, is higher than the maximums that are reported in Figure 13. Nonetheless, using the two other frequencies, Defect 2 can be detected. From the margins, it is evident that 37 kHz achieves the highest *safe zone* margin required for detection, where both the minimum and maximum thresholds that are required are respectively lower and higher than the ones required for other frequencies. Furthermore, when the defect size decreases, the limits are not affected significantly, which suggests that when using 37 kHz, smaller defect sizes than 4% CSA can be detected. In fact, in this region, defects with CSAs as small as 2% can only be detected using this frequency using both method two and method three.

Compared to method one, although the limits have been increased in the case of Defect 2, it can be seen that it is not affected as much, since method two is capable of detecting defects using 34 kHz without any outliers and 30 kHz with outliers. Nonetheless, this method also performs best in the absence of flexurals in the same location as the defect signal.

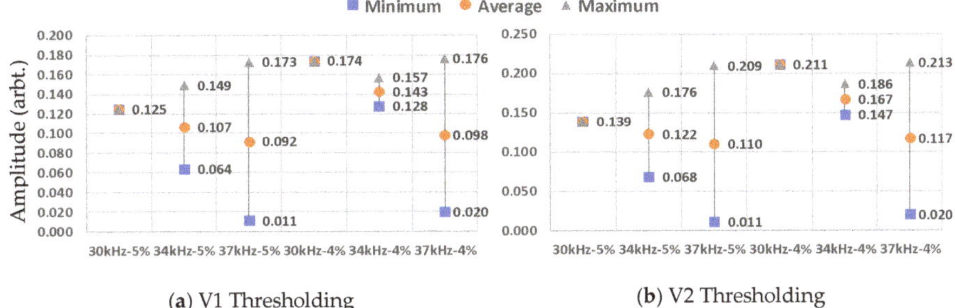

(**a**) V1 Thresholding

(**b**) V2 Thresholding

Figure 18. Limits achieved using method two on the second set of tests.

5.3.3. Method Three

Figure 19 shows the results achieved using method three. These are the same as before in the tests where the average detection values are in between the minimum and maximum thresholds; namely, the tests using 37 kHz as the excitation frequency are the tests where the defects can be automatically detected without any outliers when using this method. Same as the other methods, using 30 kHz for detection of Defect 2 is only possible when outliers are also reported. For the case of 34 kHz, when using the average value, it is not possible to detect Defect 2 without any outliers. However, it can be seen when using the second version of the thresholding value that the minimum detection value is significantly closer to the average value, which clearly illustrates the advantage of this version.

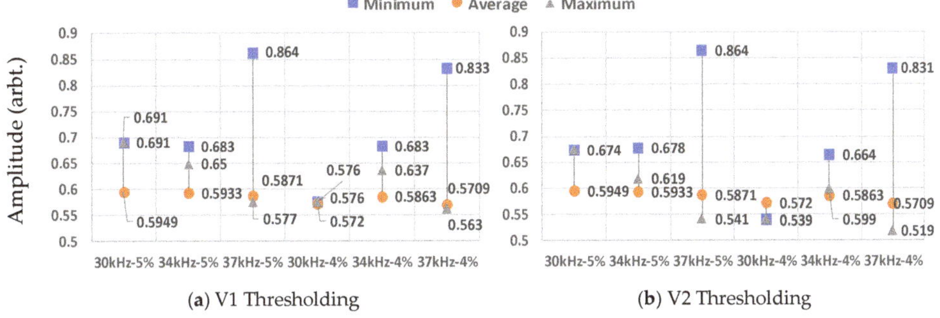

(**a**) V1 Thresholding

(**b**) V2 Thresholding

Figure 19. Limits achieved using method three on the second set of tests.

5.3.4. Comparison

Figure 20 shows the number of increments observed in the limits when V2 is used. It is the same as in the first test case: the minimums tend to increase less than the maximums. In tests using 30 kHz, since outliers are detected, the same amount of increment is reported.

The generated *safe zone* margins for the second set of tests are demonstrated in Figure 21. As expected, V2 always provides a greater *safe zone* margin. This is especially important in the case of method three, since generally, the minimums that are required for detection are well above the average value, while the maximums are close. The benefit of using V2 is clearly illustrated in Figure 19, where upon testing with the 34-kHz frequency, the maximum detection value of V2 in both tests is much closer to the calculated average value.

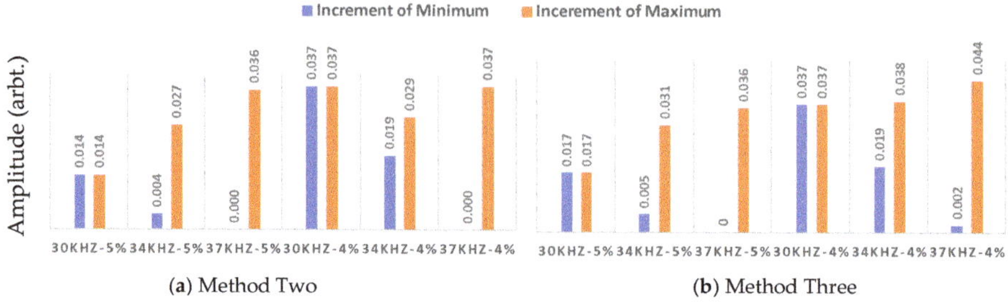

Figure 20. The increments achieved using the second version of thresholding in the second set of tests.

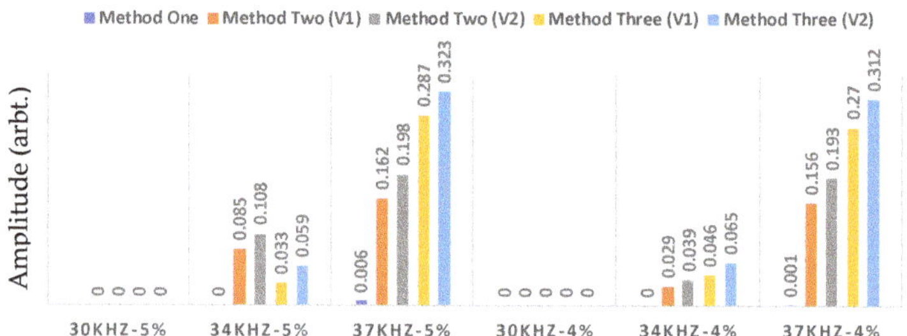

Figure 21. Bar graphs of the Safe zone margins of all methods in the second set of tests.

The average detection lengths achieved using all of the methods are shown in Figure 22 using the optimum frequency (37 kHz) for detecting Defect 2. Both methods two and three achieve the same detection length for 5% CSA; nonetheless, when the defect size decreases, method three tends to have a higher detection length. In the tests with a 34-kHz excitation frequency, method two achieved a higher detection length, since the average value is not set mathematically. Using the optimum frequency, it can be concluded that method three is the best method, since it generated the most detection length with the highest *safe zone* margin.

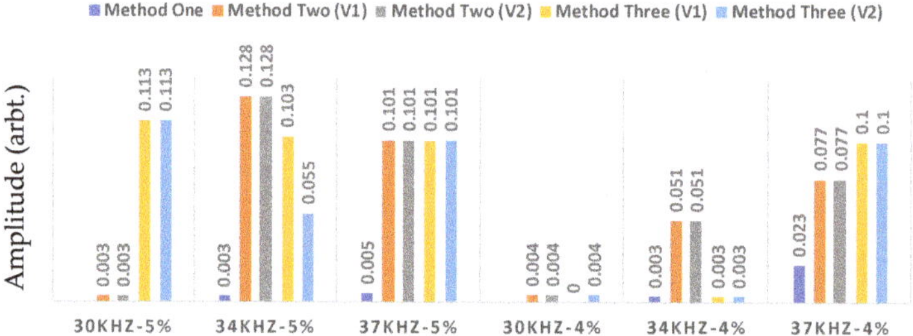

Figure 22. Bar graphs of detection length of the defect using the average value of all of the methods in the second set of tests.

6. Conclusions

In this paper, a novel statistical algorithm was proposed that uses the full capability of the tool-set array of conventional guided waves inspection devices to identify defect signals that have been polluted by coherent noise. Laboratory trials confirmed the effectiveness of this algorithm in detecting defects with low/negative SNRs and its dependence on the frequency.

Three different methods and two different versions were created for generating the results, in which the thresholding scheme of each was different. In method one, the threshold is a static value set by the inspector. In method two, the assigned threshold to each iteration is a percentage taken from the summation of the amplitude of all of the transducers signals with the same as the time-domain signal generated by the normal propagation routine. In method three, the percentage value is determined by determining the number of transducers with the same phase over the total number of transducers, subtracted by an offset. Two versions of method two and method three are also developed, wherein the second version, the summation of signals with the same sign as the temporal signal was subtracted from the comparisonValue, before determining the final assigned threshold of each iteration. The following conclusions can be drawn with regards to the tests done:

1. **Method Comparison:** Method one tended to produce the worst results, as it solely depended on a static value set by the user without considering the characteristics of the signals received in each iteration. Methods two and three can achieve a similar result in most cases; however, in method three, the threshold value can be set automatically, which provides a great benefit in one-off inspection scenarios.

2. **Version Comparison:** Version 2 always produced a wider *safe zone* margin while achieving approximately the same defect detection length using the average values.

3. **Excitation Frequency:** In the first set of tests, by lowering the frequency, a greater *safe zone* margin and a lower threshold value is required for method one. However, in the second set of tests, the exact opposite occurs. Therefore, method one is highly affected by the regional flexural waves, and high torsional energies are required for detection. However, both of the other methods have an optimum excitation frequency of 34 kHz for the first set of tests, and 37 kHz for the second set of tests. This suggests that all of the methods are more affected by the existing spatial noises caused by the flexural waves in the same iteration rather than the temporal noise of the neighborhood. The second set of tests also clearly illustrates the advantages of method two and method three in detecting defects in locations where a higher level of coherent noise exists.

4. **Safe Margin Zones:** Method three (V2) always achieved the highest *safe zone* margin, and is therefore the safest method to use.

5. With regard to these results, method three (V2) is the best method since the threshold can be determined automatically, and it achieves the highest *safe zone* margin. The tests prove that this method can detect defects using both 34 kHz and 37 kHz, depending on the existing coherent noise in the defect location. Furthermore, the algorithm is sensitive to the existing coherent noises in the baseline signal rather than the temporal noise in the tests.

This newly developed algorithm can be used to help the inspectors as an additional tool, allowing them an easier interpretation of the results. This is the main reason why all of the developed methods are mentioned. It also enables the possibility of developing narrowband transducers, as opposed to the currently used wideband transducers, which have a more focused transfer function and stronger excitation power.

In this paper, we have defined a new pathway for solving the problem of defect identification in the UGW inspection of pipelines. In this pathway, we use the information of both space and time to automatically detect defect signals rather than trying to minimize the coherent noise by spatial averaging techniques and interpreting the results based on their properties in a time domain. Thus, the main message of this paper was to illustrate the effectiveness of such approaches, which are feasibly applicable to the current systems of guided-wave testing.

Author Contributions: Conceptualization, H.N.M.; Data curation, H.N.M.; Formal analysis, H.N.M.; Funding acquisition, H.N.M.; Investigation, H.N.M.; Methodology, H.N.M.; Project administration, H.N.M.; Resources, H.N.M.; Software, H.N.M.; Supervision, H.N.M., K.Y. and A.K.N.; Validation, H.N.M.; Visualization, H.N.M., K.Y. and A.K.N.; Writing—original draft, H.N.M.; Writing—review & editing, H.N.M. and A.K.N.

Funding: This research study was made possible by the sponsorship and support of Brunel University London and National Structural Integrity Research Centre (NSIRC).

Conflicts of Interest: The authors declare no conflicts of interest.

References

1. Ostachowicz, W.M. *Guided Waves in Structures for SHM: The Time-Domain Spectral Element Method*; Wiley: Hoboken, NJ, USA, 2012.
2. Lowe, M.J.S.; Cawley, P. *Long Range Guided Wave Inspection Usage—Current Commercial Capabilities and Research Directions*; Imperial College: London, UK, 2016; Available online: http://www3.imperial.ac.uk/pls/portallive/docs/1/55745699.PDF (accessed on 19 October 2018).
3. Lowe, M.J.S.; Alleyne, D.N.; Cawley, P. Defect detection in pipes using guided waves. *Ultrasonics* **1998**, *36*, 147–154. [CrossRef]
4. Thornicroft, K. *Ultrasonic Guided Wave Testing of Pipelines using a Broadband Excitation*; Brunel University London: London, UK, 2015.
5. Pavlakovic, B. Signal Processing Arrangment. U.S. Patent 2006/0203086 A1. 14 September 2006.
6. ASTM Standard E2775–16. *Standard Practice for Guided Wave Testing of above Ground Steel Pipeword Using Piezoelectric Effect Transduce*; ASTM International: West Conshohocken, PA, USA, 2017.
7. Wilcox, P.D. A rapid signal processing technique to remove the effect of dispersion from guided wave signals. *IEEE Trans. Ultrason. Ferroelectr. Freq. Control* **2003**, *50*, 419–427. [CrossRef] [PubMed]
8. Zeng, L.; Lin, J.; Lei, Y.; Xie, H. Waveform design for high-resolution damage detection using lamb waves. *IEEE Trans. Ultrason. Ferroelectr. Freq. Control* **2013**, *60*, 1025–1029. [CrossRef] [PubMed]
9. Yücel, M.K.; Fateri, S.; Legg, M.; Wilkinson, A.; Kappatos, V.; Selcuk, C.; Gan, T.H. Coded Waveform Excitation for High-Resolution Ultrasonic Guided Wave Response. *IEEE Trans. Ind. Inform.* **2016**, *12*, 257–266. [CrossRef]
10. Yücel, M.K.; Fateri, S.; Legg, M.; Wilkinson, A.; Kappatos, V.; Selcuk, C.; Gan, T.H. Pulse-compression based iterative time-of-flight extraction of dispersed Ultrasonic Guided Waves. In Proceedings of the 2015 IEEE 13th International Conference on Industrial Informatics (INDIN), Cambridge, UK, 22–24 July 2015; pp. 809–815.
11. Lin, J.; Hua, J.; Zeng, L.; Luo, Z. Excitation Waveform Design for Lamb Wave Pulse Compression. *IEEE Trans. Ultrason. Ferroelectr. Freq. Control* **2016**, *63*, 165–177. [CrossRef] [PubMed]
12. Malo, S.; Fateri, S.; Livadas, M.; Mares, C.; Gan, T. Wave Mode Discrimination of Coded Ultrasonic Guided Waves using Two-Dimensional Compressed Pulse Analysis. *IEEE Trans. Ultrason. Ferroelectr. Freq. Control* **2017**, *64*, 1–10. [CrossRef] [PubMed]
13. Song, X.; Ta, D.; Wang, W. A base-sequence-modulated golay code improves the excitation and measurement of ultrasonic guided waves in long bones. *IEEE Trans. Ultrason. Ferroelectr. Freq. Control* **2012**, *59*, 2580–2583. [PubMed]
14. Pedram, S.K.; Haig, A.; Lowe, P.S.; Thornicroft, K.; Gan, L.; Mudge, P. Split-spectrum signal processing for reduction of the effect of dispersive wave modes in long-range ultrasonic testing. *Phys. Procedia* **2015**, *70*, 388–392. [CrossRef]
15. Pedram, S.K.; Fateri, S.; Gan, L.; Haig, A.; Thornicroft, K. Split-spectrum processing technique for SNR enhancement of ultrasonic guided wave. *Ultrasonics* **2018**, *83*, 48–59. [CrossRef] [PubMed]
16. Fateri, S.; Boulgouris, N.V.; Wilkinson, A.; Balachandran, W.; Gan, T.H. Frequency-sweep examination for wave mode identification in multimodal ultrasonic guided wave signal. *IEEE Trans. Ultrason. Ferroelectr. Freq. Control* **2014**, *61*, 1515–1524. [CrossRef] [PubMed]
17. Ng, C.-T. On the selection of advanced signal processing techniques for guided wave damage identification using a statistical approach. *Eng. Struct.* **2014**, *67*, 50–60. [CrossRef]
18. Clarke, T.; Simonetti, F.; Cawley, P. Guided wave health monitoring of complex structures by sparse array systems: Influence of temperature changes on performance. *J. Sound Vib.* **2010**, *329*, 2306–2322. [CrossRef]
19. Lu, Y.; Michaels, J.E. A methodology for structural health monitoring with diffuse ultrasonic waves in the presence of temperature variations. *Ultrasonics* **2005**, *43*, 717–731. [CrossRef] [PubMed]
20. Chang, Y.; Zi, Y.; Zhao, J.; Yang, Z.; He, W.; Sun, H. An adaptive sparse deconvolution method for distinguishing the overlapping echoes of ultrasonic guided waves for pipeline crack inspection. *Meas. Sci. Technol.* **2017**, *28*, 035002. [CrossRef]
21. Davies, J.; Cawley, P. The application of synthetic focusing for imaging crack-like defects in pipelines using guided waves. *IEEE Trans. Ultrason. Ferroelectr. Freq. Control* **2009**, *56*, 759–770. [CrossRef] [PubMed]

22. Roberts, A.; Pardini, R.; Diaz, A. High frequency guided wave virtual array SAFT. *Rev. Quant. Nondestruct. Eval.* **2003**, *22*, 205–212.

23. Luo, W.; Rose, J.L.; van Velsor, J.K.; Mu, J. Phased-array focusing with longitudinal guided waves in a viscoelastic coated hollow cylinder. *AIP Conf. Proc.* **2006**, *820*, 869–876.

24. Davies, J. *Inspection of Pipes Using Low Frequency Focused Guided Waves*; Imperial College London: London, UK, 2008.

25. Davies, J.; Simonetti, F.; Lowe, M.; Cawley, P. Review of Synthetically Focussed Guided Wave Imaging Techniques with Application to Defect Sizing. In Proceedings of the European Conference on Nondestructive Testing 2006, Berlin, Germany, 25 September 2006; pp. 1–9.

26. Sicard, R.; Chahbaz, A.; Goyette, J. Guided lamb waves and L-SAFT processing technique for enhanced detection and imaging of corrosion defects in plates with small depth-to-wavelength ratio. *IEEE Trans. Ultrason. Ferroelectr. Freq. Control* **2004**, *51*, 1287–1297. [CrossRef] [PubMed]

27. Fu, H.; Wu, B.; He, C.; Wang, W. A synthetic digital signal processing method of ultrasonic guided wave pipeline inspection. In Proceedings of the International Conference on Image Analysis and Signal Processing, Hubei, China, 21–23 October 2011; pp. 444–446.

28. Meitzler, A.H. Mode Coupling Occurring in the Propagation of Elastic Pulses in Wires. *J. Acoust. Soc. Am.* **1961**, *33*, 435–445. [CrossRef]

29. Rose, J.L. *Ultrasonic Guided Waves in Solid Media*; Cambridge University Press: Cambridge, UK, 2014.

30. Mahal, H.N.; Mudge, P.; Nandi, A.K. Comparison of coded excitations in the presence of variable transducer transfer functions in ultrasonic guided wave testing of pipelines. In Proceedings of the 9th European Workshop on Structural Health Monitoring, Manchester, UK, 10–13 July 2018.

31. Lowe, M.J.S.; Alleyne, D.N.; Cawley, P. The Mode Conversion of a Guided Wave by a Part- Circumferential Notch in a Pipe. *J. Appl. Mech.* **1998**, *65*, 649–656. [CrossRef]

32. Nurmalia. *Mode Conversion of Torsional Guided Waves for Pipe Inspection: An Electromagnetic Acoustic Transducer Technique*; Osaka University: Osaka, Japan, 2013.

33. Teletestndt, Long Range Guided Wave Testing with Teletest Focus+. 2018. Available online: https://www.teletestndt.com/ (accessed on 19 October 2018).

Article

Improved Defect Detection Using Adaptive Leaky NLMS Filter in Guided-Wave Testing of Pipelines

Houman Nakhli Mahal [1,2,*], Kai Yang [3] and Asoke K. Nandi [1]

[1] Department of Electronic and Computer Engineering, Brunel University London, Uxbridge UB8 3PH, UK; asoke.nandi@brunel.ac.uk

[2] NSIRC, Granta Park, Cambridge CB21 6AL, UK

[3] TWI, Granta Park, Cambridge CB21 6AL, UK; kai.yang@twi.co.uk

* Correspondence: houman.nakhlimahal@brunel.ac.uk

Received: 23 November 2018; Accepted: 8 January 2019; Published: 15 January 2019

Abstract: Ultrasonic guided wave (UGW) testing of pipelines allows long range assessments of pipe integrity from a single point of inspection. This technology uses a number of arrays of transducers, linearly placed apart from each other to generate a single axisymmetric wave mode. The general propagation routine of the device results in a single time domain signal, which is then used by the inspectors to detect the axisymmetric wave for any defect location. Nonetheless, due to inherited characteristics of the UGW and non-ideal testing conditions, non-axisymmetric (flexural) waves will be transmitted and received in the tests. This adds to the complexity of results' interpretation. In this paper, we implement an adaptive leaky normalized least mean square (NLMS) filter for reducing the effect of non-axisymmetric waves and enhancement of axisymmetric waves. In this approach, no modification in the device hardware is required. This method is validated using the synthesized signal generated by a finite element model (FEM) and real test data gathered from laboratory trials. In laboratory trials, six different sizes of defects with cross-sectional area (CSA) material loss of 8% to 3% (steps of 1%) were tested. To find the optimum frequency, several excitation frequencies in the region of 30–50 kHz (steps of 2 kHz) were used. Furthermore, two sets of parameters were used for the adaptive filter wherein the first set of tests the optimum parameters were set to the FEM test case and, in the second set of tests, the data from the pipe with 4% CSA defect was used. The results demonstrated the capability of this algorithm for enhancing a defect's signal-to-noise ratio (SNR).

Keywords: adaptive filtering; leaky normalized mean square; ultrasonic guided waves; pipeline inspection; SNR enhancement; signal processing

---

1. Introduction

Ultrasonic guided wave (UGW) allows long-range non-destructive testing (NDT) of pipes from a single point of inspection. The device utilizes a test-tool consisting of a number of transducer arrays that are installed on the pipe. One of the main problems of UGWs is that they are inherently multimodal and dispersive. In general inspection of pipelines, pure axisymmetric modes (torsional and longitudinal), in their non-dispersive regions, are used to decrease the complexity in the interpretation of the results. The general propagation routine of the device generates a single time-domain signal by summing a phase-delayed arrangement of the waveforms received from each transducer [1,2]. The defects are then detected by the inspectors, based on signal-to-noise ratio (SNR), where an anomaly is reported if the signal envelope is higher than the regional noise level [3].

The main advantage of using a transducer array is the excitation of axisymmetric wave and cancellation of the flexural waves, which are mainly considered as coherent noise in the inspection. Nonetheless, due to the imperfect testing condition, perfect cancellation of flexural waves is not achievable. Various approaches have been established in the literature for improving the SNR in the presence of coherent noise. One of the methods is pulse-compression [4–7]. In this technique, a modulated waveform (as opposed to a sine wave) is excited, wherein the reception side results are correlated with the known sequence. The locations with the highest correlation illustrate the same possible location of the defects. However, the performance and the design of an optimum excitation sequence can be dependent on the testing conditions, such as the transfer functions of the transducer [6]. On the other hand, filtering approaches such as split-spectrum processing have also been implemented. In 2018, Pedram et al. [8] utilized split-spectrum processing to enhance the SNR of tests. While significant improvements can be achieved using SSP, the algorithm depends on many different parameters which can be changed on a case by case basis. Furthermore, wavelet de-noising has been investigated by Mallet [9]. In this approach, after a defining a mother wavelet, the signal is decomposed into multiple levels of discrete wavelet transforms where a thresholding technique is applied to remove low-amplitude signals. The undesirable effect of decreasing the noise with this method is the possibility of removing the signal of interest, as the non-random noise components (flexural waves) can have similar amplitudes to those received from axisymmetric waves.

All of the aforementioned techniques can be considered as passive noise cancellation, where a function must be defined beforehand by the user and results are processed based on that function. Furthermore, most of the methods in the literature consider the signals achieved after the device's propagation routine as the main input signals of their implemented techniques while the device inherently produces a multidimensional signal. Therefore, in this paper, we propose to use adaptive filtering which is an active noise cancellation technique. As opposed to passive techniques, active noise cancellation allows the function to update itself to the noise characteristic of each iteration [10].

In the field of signal processing, filtering is one of the most common processes [11]. When the characteristics of noise are known, static filters can be designed that remove the non-coherent unwanted noise of tests (passive noise cancellation). As an example, when the expected bandwidth of the received signal is between 30 and 60 kHz, all other frequencies can be filtered. However, in most real-life scenarios, designing an optimum static filter is not possible since: (1) the characteristic of the noise is not known beforehand and (2) the noise can be iteratively changing. On the other hand, by providing the right multidimensional signal to adaptive filters, it is possible to remove such noise from the tests.

Adaptive filters have been applied in a variety of industries such as communications radar, ultrasonic and medical. The basic concept of adaptive noise cancellers was introduced by Widrow et al. in 1975 [12]. Their work developed the already existing Wiener filter model in order to utilize the residual signal for adaptive update of the filter weights. Many of the derived algorithms, as well as applications, were explained in his work, including canceling periodic interference in electrocardiography and speech signals. Using the same principles, Rajesh et al. [13] proposed an adaptive noise-canceling method for the removal of background noise where the closest sensor to the abdomen of a pregnant patient is used as the reference signal, with the maternal signal captured by a sensor in a different position. By using this configuration, they were able to extract the wanted Fetal signal which was obscured by the signal received from the maternal heart rate. In 2003 [14], Hernandez implemented the adaptive linear enhancer (ALE) algorithm to increase the SNR by using the delayed version of the same signal as the input to the adaptive filter. Ramli et al. [15] reviewed different adaption algorithms, such as the least mean square (LMS), normalized-LMS (NLMS), leaky-LMS and recursive least square (RLS), where it was reported all of these approaches are capable of enhancing the SNR. In communications and speech analysis, one of the common uses of an adaptive noise canceller (ANC) is for echo cancellation [16,17], which can be generated by hybrid transformers while transmitting data, or from the multipath mitigation of the signals to the microphone. In radar technology [18,19], adaptive filters are widely used for multipath cancellation, clutter rejection, and spatial noise reduction by using adaptive beamforming. In ultrasonic testing, Zhu et al. [20] used adaptive filters in order to increase the SNR in ultrasonic non-destructive testing (NDT) of highly scattered materials. In 2010, Monroe et al. [21], tested this configuration for clutter rejection and detection of small targets in ultrasonic backscattered signals. Both researchers reported that adaptive filtering is a robust method of filtering the random ultrasonic grain noise created by the structure. It was reported that if the ultrasonic beam is moved less than a beam width away from the position, the echo from a structural feature is closely related while the grain noise varies. The results were evaluated experimentally using NLMS where an improvement of 5–10 dB was observed with fixed parameters.

In guided-wave testing, the noise is time varying. While the excited wave mode is non-dispersive and axisymmetric, the flexural waves change in time. Therefore, when observing the signal from different points of the pipe the axisymmetric waves will have similar characteristics, while the flexural waves will be variable. In this paper, instead of removing the noise from the single time-domain signal received from the propagation routine, a novel method is proposed to utilize adaptive filtering using the signals received from separate rings to remove the flexural wave modes in the tests. The leaky NLMS method is used for updating the filter weights in each iteration. The NLMS method reduces the amplification of gradient noise and also provides a faster rate of convergence than the LMS algorithm [10,22]; therefore, it is more suited for guided-wave applications with time-varying noise. Furthermore, the introduction of a leakage factor in the update function for the filter weights limits the divergence of the filter weights. Leakage methods were initially introduced to decrease performance degradation caused by limited-precision hardware [23]. The process is equivalent to the introduction of a white noise sequence to the input signal [22]. In the context of this paper, the main reason for applying a leakage factor was to allow faster adaption of the filter weights to the existing noise of each iteration. The performance analysis of leaky NLMS algorithms have been previously done by Sayed et al. [24] and the stability of the algorithm was investigated by Bismor et al. [25]. In mathematical formulation, introducing the leakage factor would only reduce the effect of past filter weights. Therefore, if the same principles of the NLMS algorithm are used, the algorithm will remain stable and bounded. This methodology was initially developed using synthesized data generated by finite element modelling (FEM) of the test system. It was then verified through experimental trials on real pipe data with a defect of 3–8% cross-sectional area (CSA) size, and enhancement of the SNR in comparison to the device's general routine is demonstrated.

In Section 2, the background of guided waves in pipelines and the nature of noise is explained. Section 3 includes the description of the device's general propagation routine and the proposed adaptive filtering strategy. In the results section, for better illustration of the adaptive filtering process, each stage of the process applied to the FEM signal is shown, followed by the results achieved from the experimental tests. It is demonstrated that although the algorithm cannot increase the SNR for small defects, but it will also not decrease it in compared to general propagation routine.

2. Background Information

Guided waves are multimodal [26]. Depending on the excitation frequency, multiple wave modes can be generated at the point of excitation. These modes are categorized based on their displacement patterns (mode shapes) within the structure. The main categories of guided waves in pipes are axisymmetric torsional and longitudinal waves and non-axisymmetric flexural waves. The popular nomenclature used for them is in the format of $X(n,m)$, where X can be replaced by letters L for longitudinal, T for torsional, and F for flexural waves; n shows the harmonic variations of displacement and stress around the circumference; and m represents the order of existence of the wave mode [27]. In an ideal scenario, a single pure family of axisymmetric waves in their non-dispersive region should be excited for general inspection. Nonetheless, this is not possible due to the variations of excitation and the reception transfer function of transducers in real-life scenarios. Furthermore, the interaction of wave modes with the pipe also leads to the generation of different wave modes, which is commonly known as a wave mode conversion phenomenon [26,28,29].

Figure 1 shows the spatial arrival of the signals from different segments of the pipe. These signals are created using the developed model in [30], where a three ring excitation system is used. Each ring contains 32 source points with variable transfer functions. In the figures, blue and red lines show the received amplitude from the transducers of two separate rings while the black dotted line serves as the reference. Any point outside the circle illustrates positive amplitude and the points inside show the negative amplitude of the corresponding source point. The following points can be concluded by comparing the illustrated polar plots:

- The energy of waves received from axisymmetric features (e.g., pipe end) are overwhelmed by the T(0,1) with little variance caused by the existing coherent noise of flexural waves.
- Defect signals have a mixture of both flexural and torsional wave modes; nonetheless, if the cross-sectional area (CSA) of the defect is wide, the T(0,1) will have a greater effect than the flexurals. This is the main reason why adding all the transducers tends to increase the overall energy of T(0,1) while reducing the effect of flexurals.
- Noise regions are mostly consistent with flexurals with almost no axisymmetric wave being detected.

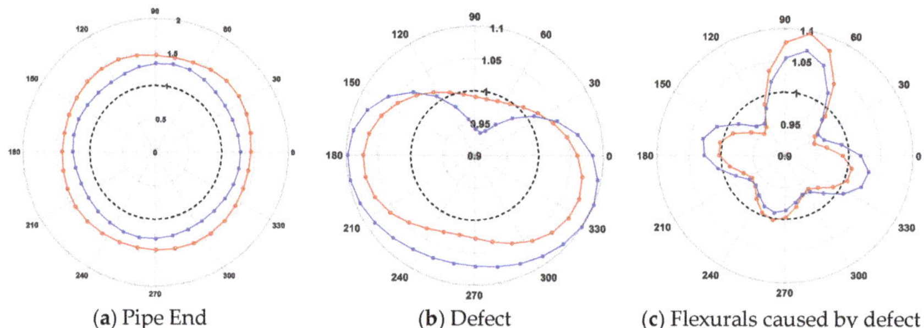

(**a**) Pipe End   (**b**) Defect   (**c**) Flexurals caused by defect

Figure 1. Example of received FEM signals from the array from different sections of a pipe [31].

Furthermore, guided waves exhibit dispersive propagation. Dispersion causes the energy of a signal to spread out in space and time as it propagates [32]. Figure 2a shows an example of a dispersion curve calculated from an 8-inch schedule 40 steel pipe using RAPID software [33]. Wilcox et al. introduced a method to use dispersion curves in order to both simulate [34] and remove [32] the effect of dispersion. By applying the developed equation by Wilcox et al. and the shown dispersion curves in Figure 2a, a simulated propagation of F(4,2) wave mode is created and shown in Figure 2b where the black and red lines show the same wave mode after 1 m and 2 m of propagation, respectively. As can be seen, for a dispersive waveform the temporal resolution and the energy of wave packet is decreased. Therefore, in general inspection, non-dispersive regions of wave modes are excited to allow ease of inspection. In this research, T(0,1) is used as the excitation wave mode since it is non-dispersive across its whole frequency range [35].

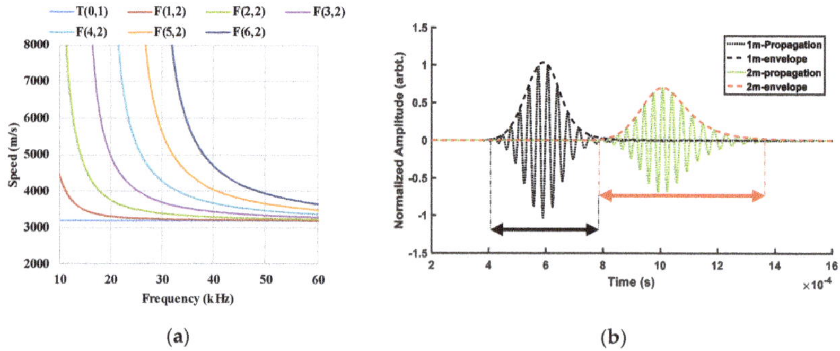

(**a**)   (**b**)

Figure 2. (**a**) Example dispersion curve of T(0,1) wave mode in an 8-inch schedule 40 steel pipe. (**b**) The effect of dispersion on a simulated flexural wave, two propagation distances [31].

Three main signals are expected to be received in guided-wave tests which are shown in Figure 3. The first signal is the superposition of T(0,1) and flexurals where it is the true defect location. The second is the flexurals created due to wave mode conversion [36]; as these signals exhibit the same characteristics of excitation waveform, they can lead to the detection of false alarms. The third category is the non-coherent white noise, which is not correlated to the excitation sequence. This category is the main one which can be removed using generalized filtering techniques, as their characteristic is completely different to the excitation sequence.

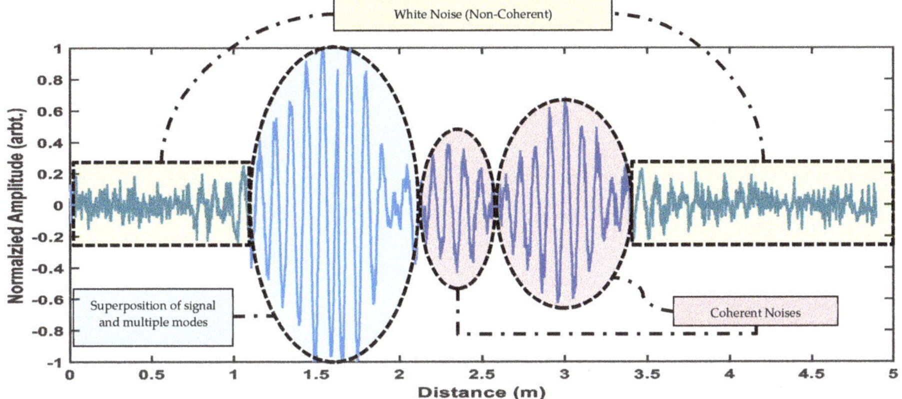

Figure 3. Categories of the signals in the guided-wave inspection [31].

3. Methodology

As explained previously, guided waves are multimodal. To allow inspection of pipelines using guided waves, a single wave mode must be excited in only one direction to allow detection of the anomaly's location. Excitation and detection of flexural waves are difficult and require complex processing algorithms due to their circumferential variation. Therefore, the practice in the general inspection is to excite an axisymmetric wave. For doing so, the toolset of guided-wave testing devices consists of arrays of linearly placed transducers (rings) across the circumference of the pipe. This leads to excitation of a quasi-axisymmetric wave where, in ideal scenarios, the corresponding flexurals of the excitation wave mode will cancel each other out [1,37,38]. If only one array is used, a bidirectional wave mode is generated since there is no directional control in the normal piezo-electric transducer. The most common method of achieving a unidirectional transmission is to use multiple arrays of transducers. By considering the dispersion curves of the generated wave mode and the placement spacing between the rings, waves can be manipulated in such a way to enforce the forward going propagation and suppress the backward leakage. Hence, the forward propagations of each ring are overlapped while the signals received from backward propagation direction are inversely overlapped [1,2,6,39].

On the reception side, signals from the backward direction will still be received due to (1) the impure/non-ideal excitation and (2) the forward propagation signals travel paths after the toolset. Once the device is in monitoring mode, no excitation takes place, meaning if an echo is received from the device, it will continue to travel past the test tool; if any anomalies exist in that direction, a new echo will be generated which will travel in the same way as the forward testing direction. At this point two scenarios will occur:

1. In the first scenario, the signal from a feature in the backward test direction is detected by the rings. The energy of this signal can be reduced by using the same phase-delaying algorithm in the excitation sequence.

2. The second scenario is when the same echo is past the tool. This signal will have the same characteristics as forward propagation and thus cannot be canceled. These are typically known as mirror signals.

The phase-delaying algorithm is once again applied in the general routine of reception to reduce the effect of signals generated from the first scenario. Afterward, the average of all transducer signals is generated, which will reduce the energy of flexural waves. The signal of the rings can also be strongly different which either suggests the existence of no defect or a small defect which is not generating a strong-enough signal to be detected. In the normal approach, as simple summation is used, this will lead to the generation of a residual signal which might be mistaken as a defect in the inspection. The flowchart for this general reception routine, which is the common practice of industry, is shown in Figure 4.

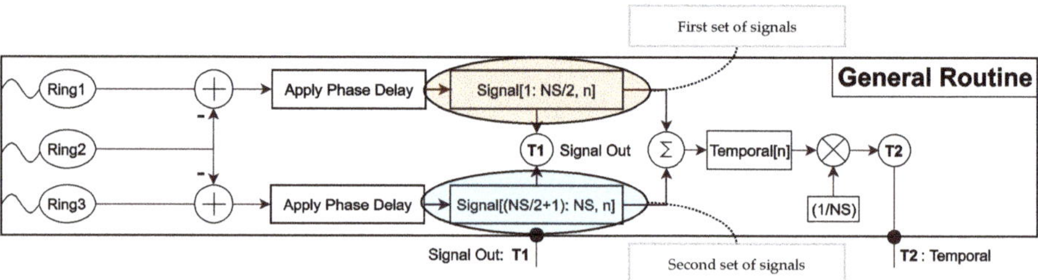

Figure 4. Flowchart of general propagation routine currently used in guided-wave devices.

Adaptive Filtering

The main challenge in the inspection of guided waves is to decrease the energy of coherent noise in the test since non-coherent noise is generally removed by bandpass filtering and averaging. Dispersion characteristics of guided wave modes vary with regards to the structural features, frequency, and wave mode. However, the non-dispersive torsional mode is typically constant with an approximate speed of 3260 m/s. By shifting the rings to overlay the T(0,1) signals together, the wave modes which travel at different speeds will be less correlated with those that travel at constant speed. Furthermore, as shown in Figure 1, the flexural waves are highly variable in the space while the torsional mode has a higher correlation.

In this case, the existing torsional mode is highly correlated while the noise and flexural modes will be less correlated in time. Therefore, instead of the summation of the signals, adaptive filters can be utilized to enhance the signal resolution. The output signal of an finite impulse response (FIR) filter in time-domain can be achieved by the convolution operation using [10,40–42]:

$$y(n) = \sum_{k=0}^{p} w_k(n) \times (n - k) \tag{1}$$

where p is the filter length, n is the iteration number, $w_k(n)$ is the filter coefficients with the index of k, and x(n) is the input signal. In the following equations, both $\mathbf{x(n)}$ and $\mathbf{w(n)}$ are vectors of size p (represented by bold characters). The resultant error signal would be:

$$e(n) = d(n) - y(n) \tag{2}$$

where d(n) represents the required reference (desired) signal; hence, in order to use this method, at least two signals are required. In adaptive filtering, the goal is to minimize the error signal by setting the filter weight. Therefore, in each iteration the filter weights should be updated:

$$\mathbf{w(n+1)} = \mathbf{w(n)} + \Delta\mathbf{w(n)} \tag{3}$$

where $\Delta w(n)$ is the correction step applied to previous filter coefficients to update them for the next iteration; the key component of the adaptive filter is to define $\Delta w(n)$ in a way that it decreases the mean square error over time. One of the robust methods is the normalized least mean square [20,41], which tries to minimize the least mean square error by setting the coefficients according to a variable step size. The formula is as follows:

$$\mathbf{w(n+1)} \;=\; \mathbf{w(n)} \;+\; \beta \frac{\mathbf{x(n)}}{\epsilon + \left|\mathbf{x(n)}\right|^2} \, e(n) \tag{4}$$

where β is a normalized step size with $0 < \beta < 2$, ϵ is some small positive number to bypass the problem when $\left|x(n)\right|^2$ becomes too small. The noise in guided waves is highly variable due to the existence of time-varying multiple modes. To make sure the filter is not over-fitted to the characteristics of previous iterations, a leakage factor, α, is added to reduce the effect of previous iterations [42]:

$$\mathbf{w(n+1)} \;=\; \alpha \mathbf{w(n)} \;+\; \beta \frac{\mathbf{x(n)}}{\epsilon + \left|\mathbf{x(n)}\right|^2} \, e(n) \tag{5}$$

The flow chart of this algorithm is shown in Figure 5. In this setup, instead of adding the signals received from the rings of transducers, they are selected as the desired response $d(n)$ and the input signal $x(n)$. Every other signal is updated based on the characteristics of each iteration and $y(n)$ is the output of the filter. The selected reference and input signals are the two sets generated in the general propagation routine, as marked in Figure 4. Applying this phase delay and summation of arrays within each ring, two sets of signals are created where both have the following components:

- S(t)   Signals of interest/anomalies: Highly correlated and in phase
- N(t)   Flexural and other noise: That are not correlated to S(t). Furthermore, noise sources of two signals are correlated to each other.

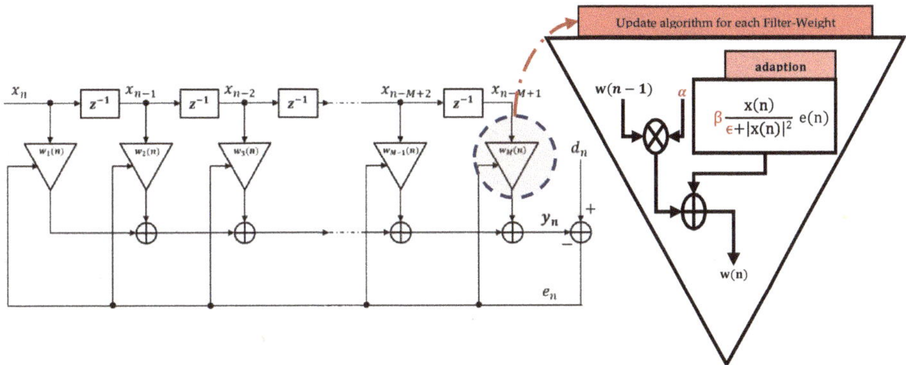

Figure 5. Adaptive linear prediction algorithm for noise cancellation where the red marked parameters are fixed by the user.

This satisfies the need for a reference signal in using adaptive filtering. To cancel the backward leakage, this algorithm is done with the switching of the input and reference signals, where the final result is the summation of the two generated outputs of the adaptive filters.

4. Test Setup

The algorithm was initially developed on a synthesized data generated by FEM using ABAQUS software [30,43]. It was then validated in laboratory trials on real pipes using a Teletest Device [44].

4.1. Finite Element Model

The simulated pipe is based on the characteristics of an 8-inch schedule 40 steel pipe with a length of 6 m, outer diameter of 219.1 mm, thickness of 8.18 mm, Young's Modulus of 210 GPa, density of 7850 kg/m^3 and Poisson's Ratio of 0.3. The excitation frequency in the test was 30 kHz. In order to capture the smallest wavelength in the operating frequency, at least 8 elements are required in the axial direction [45,46]. Therefore, the global seed size was set as 2.5 mm and the time increment was set as 3.11×10^{-7} s to provide a satisfactory accuracy in generation and reception of all the wave modes within the frequency band. The modeled test tool included 3 rings of 32 linearly spaced source points, located 1 m away from the back end of the pipe. Each source point of the rings is numbered clockwise starting from 90°. The schematic of this setup is illustrated in Figure 6.

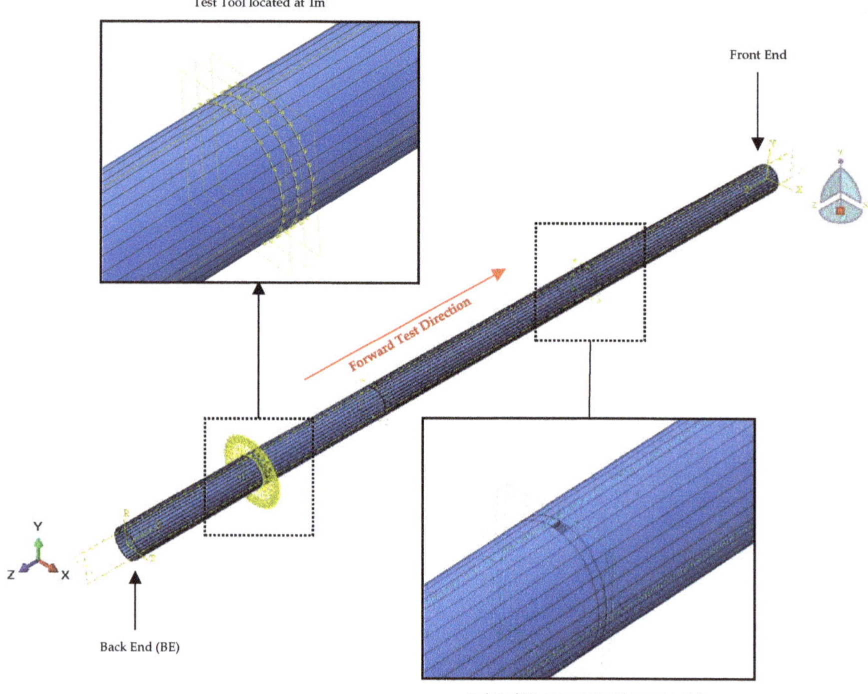

Figure 6. Schematic of the FEM with test tool and defect located at 1 m and 4 m, respectively, from the back end [30].

This type of FEM synthesis of guided wave signals in pipes was developed in previous studies [37, 46–50]. The generated models have also been validated by confirming the accuracy of the received signals from both the device [50], and 3D laser vibrometer [46,47]. Nonetheless, this type of FEM generally produces a limited amount of noise due to the perfect condition of the model. The main aim of this paper and the developed model is to investigate the performance of the algorithm in the presence of coherent noise. Therefore, in order to generate flexural noise in the test, different transfer functions were applied to the excitation sequence of each source point. Since the minimum satisfactory number of elements is used, the transfer functions will not affect the integrity of the model and all the existing wave modes in the test will be captured. Apart from these transfer functions, all other limits and parameters were in line with the verified models in the previous studies.

In 2013, Engineer [51] extracted the broadband frequency response of transducers with regards to various coupling forces using Teletest piezo-electric transducers. The reported frequency response was used as a reference and variability in terms of amplitude and phase was introduced to generate a variable transfer function for each transducer. An example is shown in Figure 7.

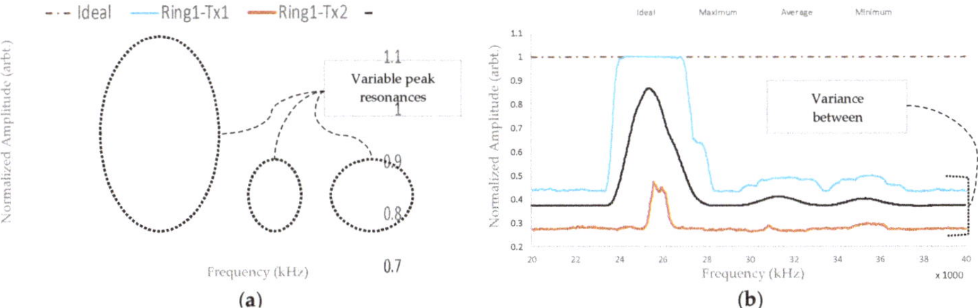

Figure 7. Example of designed transfer functions wherein (**a**) three different source points are compared and (**b**) the maximum, average, and minimum amplitudes across all of the applied transfer functions used in an individual test are shown.

The transfer functions are multiplied by the ideal excitation's frequency response, which is a 10 cycle 30 kHz Hann-windowed sine wave, and then inverted back into the time domain. The generated time-domain signals are used as input files for each of the source points. Additionally, the signals of the second array are phase delayed and inverted with regards to the 30 mm ring spacing cancellation algorithm [1,6]. The overall system works in the pulse-echo mode and the signals were received from the same ideal point sources. The signal generation, as well as the post-processing, were done using MATLAB-R2016a [52]. In order to increase the coherent noise of testing conditions, instead of receiving the signals from all source points, only source point numbers 1, 7, 14 and 26 of each ring have been used for reception (marked in red in Figure 8).

Figure 8. The used reception points of each ring in the FEM.

The signal routes of the FEM test are shown in Figure 9. As can be seen, the first directly received signal from the forward test direction is the defect signal which is received at 3 m, followed by the front-end signal at 5 m. However, as the transfer function and reception points are not ideal, a backward leakage signal will exist, propagating in the forward direction. This leakage would be received at 1 m but due to the backward cancellation algorithm, it will not be observable. However, the reflection of this signal with the defect signal would be received from the forward direction, leading to a false alarm at 4 m. In these tests, since both signals at 3 and 4 m have the same characteristics, they are both considered defect signals even though the second signal is a false detection.

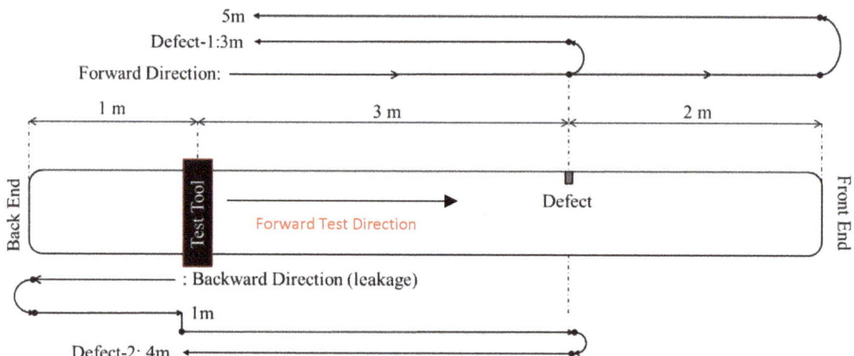

Figure 9. Torsional mode reception route in the FEM test cases.

4.2. Experimental Trials

In the laboratory trials, a Teletest MK4 device [44] was used in order to inspect an 8-inch schedule 40 steel pipe with a length of 6 m. The test tool included a collar of 3 ring transducers with 30 mm ring spacing, located 1.5 m away from the back end of the pipe. Each ring included 24 linearly spaced thickness-shear (d15) piezoelectric transducers. The system provided both excitation and reception of signals in pulse-echo mode with the operating frequency of 20 to 100 kHz. The device can capture the transducers' signals with a sampling frequency of 1 MHz. On the reception side, an internal bandpass filter was applied to filter any frequencies outside the main bandwidth of the excitation bandwidth. The device outputs the time-domain signal received from each ring, as well as the processed signal with the general propagation routine. In this paper, the signals from the rings are extracted in order to apply the algorithm.

The defects in these trials are located 4.5 m from back end of the pipe (3 m away from the test tool). The defects were introduced by removing the pipe's surface material with a saw. The defect sizes were varying from 3% to 8% CSA with steps of 1% CSA (total of 6 defect sizes). The signal routes of the laboratory trials are shown in Figure 10. As opposed to the FEM test setup, the Defect-2 signal, which was generated by reflection of backward leakage signal from the defect, is now overlapped with the front-end signal and is no longer detectable.

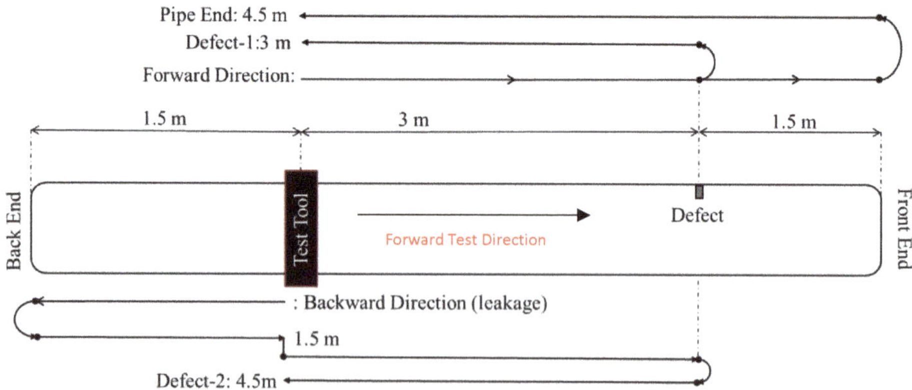

Figure 10. Torsional mode reception route for laboratory trials.

5. Results

In order to better understand this process, an example test case is initially shown where each stage of the algorithm is explained thoroughly. Then the experimental results are demonstrated in the second part of this section where the SNRs are calculated with regards to defects with different CSA sizes and excitation frequencies. In both cases the SNRs are calculated based on the following formula:

$$\text{SNR} = 20 \times \log 10 \left(\frac{\sqrt{\frac{1}{N} \Sigma_1^N \text{Signal}_n}}{\sqrt{\frac{1}{M} \Sigma_1^M \text{Noise}_m}} \right) \tag{6}$$

where Signal is considered the window where the energy of the signal is received (between 3.0 m and 3.5 m depending on the frequency) and everything else is considered noise. In both tests, excitation sequences are 10-cycle Hann-windowed sine waves.

5.1. Example Test Case

This technique is initially developed and validated on synthesized data generated by FEM. The used excitation sequence for generating this data set is 30 kHz and variable transfer functions were used for each of the source points. Figure 11 shows the two sets of signals achieved after applying the phase delays of general propagation routine (blue and red). Four regions are of main interest in this figure, which are (a) backward leakage, (b) flexural noise, (c) Defect-1 and (d) Defect-2. Since both Defect-1 and Defect-2 signals are received from the forward testing direction, they have similar characteristics and thus both should be enhanced by this algorithm. As can be seen, noise region (b) is significantly less correlated than the defect signals (c and d). Furthermore, although the amplitudes of two sets of signals received in the defect regions are not the same, their phases and polarity are in perfect correlation. Therefore, while in the propagation routine their characteristic might be similar but, in this space, the defect and flexural signals are distinguishable.

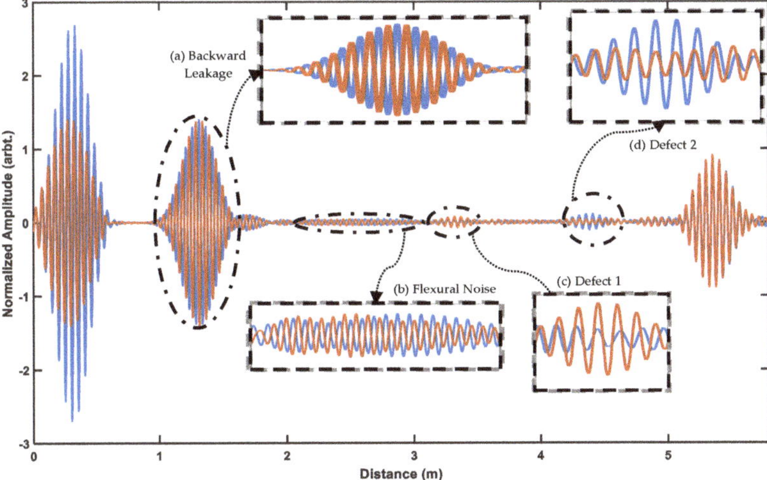

Figure 11. FEM signals (30 kHz) where two defect regions and the flexural noises are marked where blue and red lines show the first and second sets of the signals respectively.

5.2. Cancellation of Backward Leakage

Figure 11a shows the two sets of signals received from the backward testing direction. As can be seen, these signals are inversely overlapped with little differences in amplitude. This is due to the general propagation routine where the received signals from rings are delayed and inverted depending on the ring spacing.

After applying the filtering, the phase of the outcome does not change, therefore by switching the inputs to the adaptive filter, two sets of results will be generated in which the signals from the forward direction are overlapped and the ones from the backward direction are inversely overlapped. This is illustrated in Figure 12, where blue and red signals show the two sets of results achieved from adaptive filtering by switching the inputs. This operation not only enhances the defect signals received from the forward direction (Figure 12c) but it also decreases both the backward cancellation (Figure 12a) and flexural noise regions (Figure 12b) significantly.

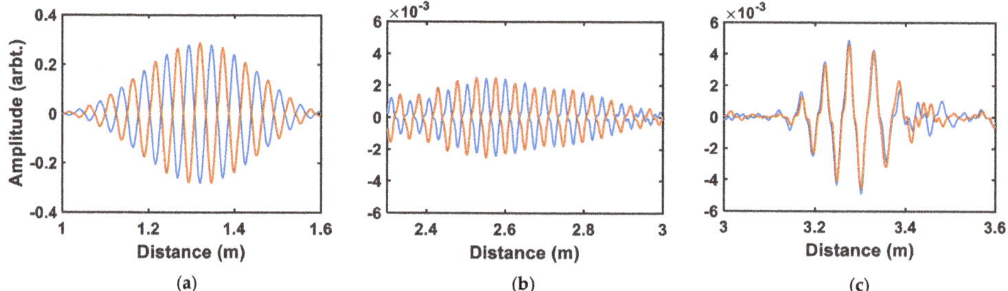

Figure 12. Two sets of results after swapping the inputs to adaptive filter where (**a**) shows the signals received from the backward direction, (**b**) shows the flexural noise, and (**c**) shows the defect signal.

5.3. Adaption of Filter Weights

The used filter order in the tests is five, which means only five past samples are required to process the results (Appendix A). Since the number of available samples is limited, observation of the filter's frequency response is not possible. Therefore, in Figure 13b–f, the changes of the first filter value are shown while Figure 13a shows their corresponding results in the time-domain. The red signals illustrate the results of the leaky-NLMS while the black dotted lines represent the normal NLMS algorithms. While normal NLMS depends more on past values and is not fast to adapt to the current iteration, leaky NLMS makes this adaption faster by reducing the filter weights' dependence on their past values. After each adaption to a big sequence such as backward leakage or defect, the magnitude of the filter weight tends to move toward zero; in other words, the weights forget the characteristics of past iterations significantly faster, which allows them a quicker adaption to the characteristic of the current iteration. Furthermore, as illustrated by blue dotted signals, in the region of defects, a high peak is generated in the magnitude of each filter order set by the leaky-NLMS method. Although other techniques and updated algorithms exist, this fast adaption algorithm, where the filter can forget the characteristics of the previous wave, is the main reason why leaky-NLMS is more effective in the guided-wave inspection of pipelines.

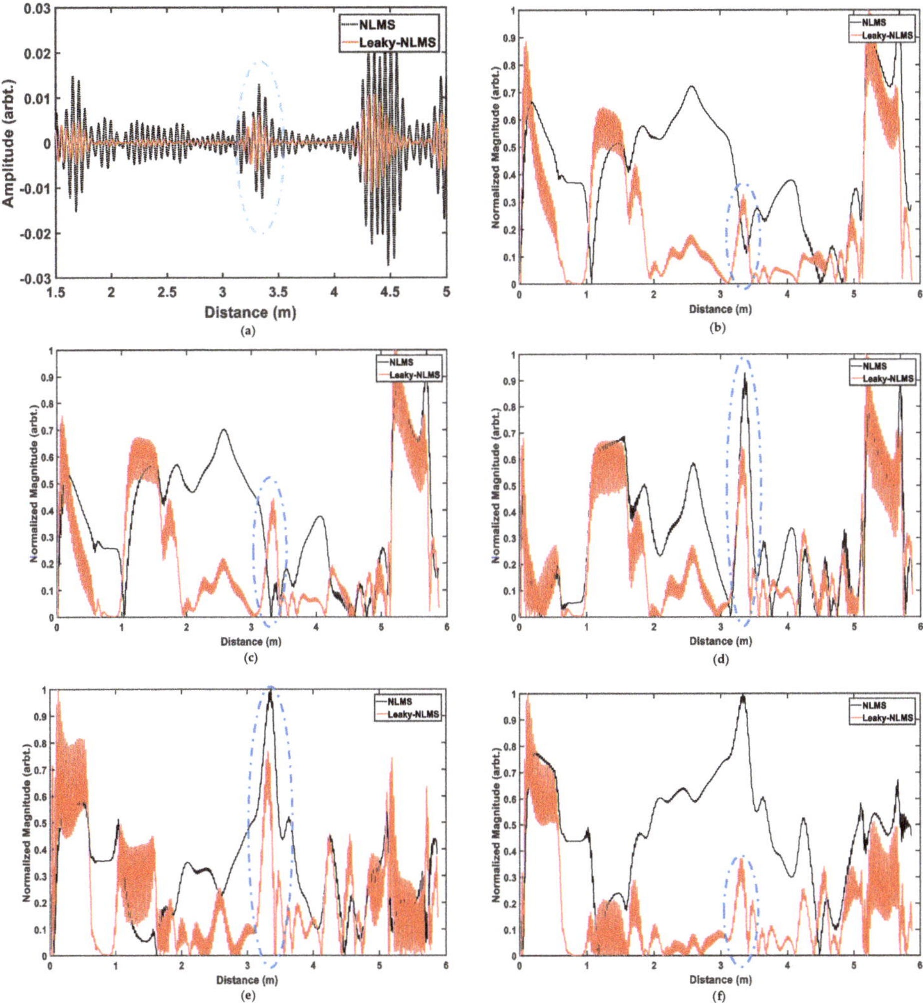

Figure 13. Comparison of the result achieved from NLMS (black) vs leaky LMS (orange) where (**a**) shows the time-domain results and (**b**–**f**) show the normalized magnitudes of filter order 1 to 5, respectively.

5.4. Results Comparison

The performance of the algorithm was assessed in experimental trials by detecting defects of CSA sizes between (and including) 3% and 8%. It is expected that the algorithm would be more effective at lower frequencies since they are more dispersive; furthermore, the excitation power of lower frequencies is higher than those of higher frequencies, both due to the duration of the signals as well as the systems transfer function. Therefore, frequencies between 30 to 50 kHz with a step of 2 kHz were tested. The SNR achieved from the general propagation routine is demonstrated in Appendix B. The peak SNR is always achieved using 40 kHz excitation frequency where the SNR decreases by moving toward either side of that frequency. Furthermore, the SNRs achieved from each frequency are reduced almost linearly when the defect size is decreased; thus, the noise energy is almost constant while the defect energy is decreasing.

5.4.1. Parameter Selection

The results were generated from two sets of parameters (given in Appendix C Table A3):

- Model Parameters: The optimum parameters are achieved by using a brute force search algorithm to find the parameters which give the maximum SNR in the FEM test. The main goal of this test was to assess the response of the algorithm by fixing the parameters to an optimum solution created by the FEM.
- Experimental Parameters: The brute force search algorithm was performed on the 4% CSA sample to find parameters that result in most enhancements of SNR for each frequency. The goal was to find the best testing frequency where the least variations in the enhancements are observed and assess whether fixed parameters can enhance the SNR of defects with lower CSA size (3% test case).

5.4.2. Model-Parameters

The results achieved from model parameters are shown in Figure 14 where (a) shows the achieved SNR values and (b) shows the amount of enhancement. The following conclusions can be extracted from these tests:

- In defects above 5% CSA, the algorithm enhances the results using all frequencies. For defects with CSA size of lower than 4%, the algorithm can enhance the SNR of most frequencies except for 30, 32, 34, and 38 kHz.
- Except from the case of 3% CSA error, both maximum SNR and maximum enhancements are achieved using 34 and 38 kHz in the tests. In the 3% CSA test, the maximum enhancement is for 30 kHz; nonetheless, the final SNRs of 34 and 38 kHz are greater.
- The variations in the case of 30 kHz are less in comparison to all other cases, as shown in Figure 14b.

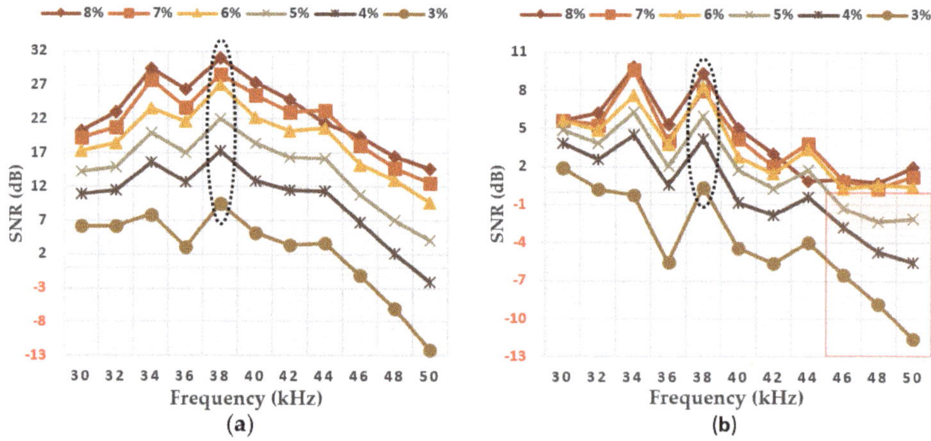

Figure 14. Results achieved using fixed model parameters for all frequencies where (**a**) shows the achieved SNR and (**b**) shows the amount of improvement in comparison to the general propagation routine.

5.4.3. Experimental Parameters

The results of experimental parameters are illustrated in Figure 15. The following conclusions can be drawn from these tests:

- For defects with higher CSA error size (above 5%), lower frequencies tend to achieve higher SNR after processing where the defect signal was enhanced by at least a factor of 2. The greatest

- enhancement is achieved from 34 kHz and the maximum SNR is achieved from 36 kHz for these signals.
- Although higher frequencies will result in lower SNR, their enhancements are approximately the same for different sizes of the defect. Nonetheless, their final SNR is almost always lower than all other frequencies, especially in the cases of frequencies above 46 kHz.
- For 3% and 4% CSA defects, the greatest enhancement and gain are achieved using the 38 kHz frequency. Furthermore, as can be seen in Figure 15b, this frequency is affected less with regards to the CSA loss of the defect.

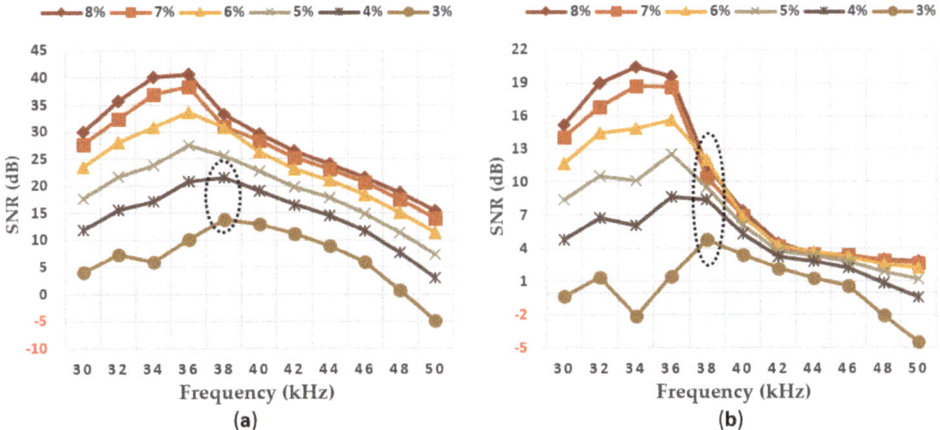

Figure 15. Results achieved using experimental parameters, which are variable for each frequency, where (**a**) shows the achieved SNR and (**b**) shows the amount of improvement compared to the general propagation routine.

5.4.4. Discussion

Figure 16 shows the difference of enhancements between the results of model parameters and experimental parameters. These results show that although temporal SNR of the signals is important, the spatial characteristic of the noise also plays an important role in the amount of enhancement. As an example, while 40 kHz has the highest SNR from the general routine, it does not perform as well as 38 kHz for the algorithm. In most cases of lower CSA defects, the SNR improvements are achieved not by enhancing the axisymmetric signal strength, but by removing the flexural noise. Except for excitation frequencies of 30, 34, and 40 kHz in the 3% CSA defect, all other cases were enhanced when the experimental parameters were used.

Since signals with SNRs higher than 30 dB are more easily detected, it is better to use a frequency which provides a stable enhancement for both low and higher CSA sizes rather than setting parameters which can greatly increase the higher CSA defects while decreasing the SNR of smaller defects. An excitation frequency of 38 kHz provides approximately the same enhancement using both model parameters and experimental parameters, which shows that this frequency is less affected by the parameters in comparison to the others. Furthermore, it can also be seen from the results of model parameters that 38 kHz always achieves the maximum gain even though the parameters are optimized for 30 kHz excitation frequency in the model.

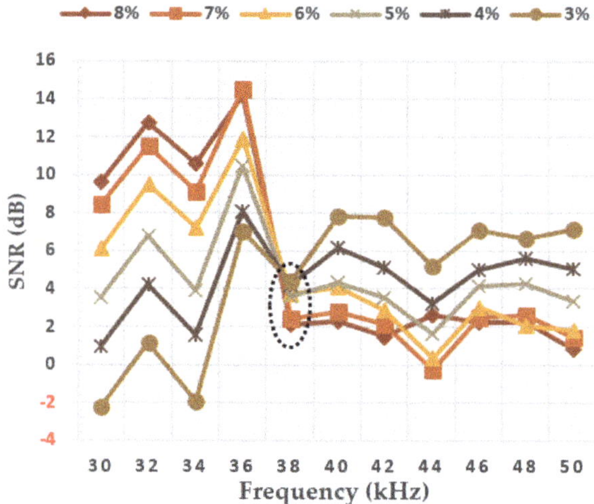

Figure 16. Amount of improvement achieved by experimental parameters as opposed to model parameters.

All frequencies above 40 kHz fail to provide any improvements for defects with CSA sizes smaller than 4% using model parameters. Nonetheless, even the enhancements achieved with experimental parameters in those frequencies are much less than all other frequencies, which suggests that this algorithm tends to perform better at lower frequencies. This is expected since flexurals with lower frequencies are highly dispersive, which adds to the spatial noise of the tests.

It should be born in mind that, even though in some frequencies the mathematically calculated SNR might be less than the one achieved from the general propagation routine, the high-order flexurals are always canceled, which leads to easier interpretation of results by the inspectors. Figure 17 shows the results of the 30 kHz excitation sequence for defects with different CSAs. As can be seen, highly variable noises in the region between 2.5 and 3 m are canceled but those caused by low order flexurals at 2 to 2.5 m are enhanced. While in defects with higher CSA sizes most improvement is achieved due to the amplification of the defect signal, in lower CSA defects (especially in the case of 3% CSA), the noise cancellation is the main reason for enhancement. The enhancement of these low order flexurals is the main reason why the algorithm fails to improve the SNR in some of the testing frequencies. With regards to the dispersion curves (Figure 2a), low order flexurals become non-dispersive on frequencies above 30 kHz with their speed moving toward that of T(0,1), 3260 m/s. Since the phase delays are applied with respect to T(0,1) wave speed, if the flexural waves have the same speed they will be overlapped in the input and desired responses. This is the main reason why such signals might not be canceled and, in fact, in some scenarios, they might even be amplified. Furthermore, considering the 34 kHz filtered signal shown in Figure 18b, the noise envelope is clearly dispersed (2–2.5 m). Figure 19 also shows the results achieved from 38 kHz excitation frequency. In these figures, the input and reference signals for the noise and defect regions are shown as blue and red. The following conclusions can be extracted by observing the difference between a defect and noise signals:

1. If any low order flexurals or noise are overlapping in the input and reference signals, they will not be removed and might be amplified.
2. The amplification of the signals is also dependent on the amplitude of the input signals to the algorithm, e.g., the defect signal in case Figure 18 does not observe the same enhancement of that in case Figure 19 since the blue signal in case Figure 19 is much stronger.
3. Although some flexural signals will remain in the resultant signal, there is an option of further post-processing to enhance the SNR. Such is the case of the noise region located at 2–2.5 m in case Figure 18, where it is clearly dispersed at the end of its window, and the case of 1.5–2 m in

Figure 19, which results from backward leakage. Therefore, by considering other characteristics in the signal, the results can be post-processed to distinguish different modes and further increase the SNR of the test.

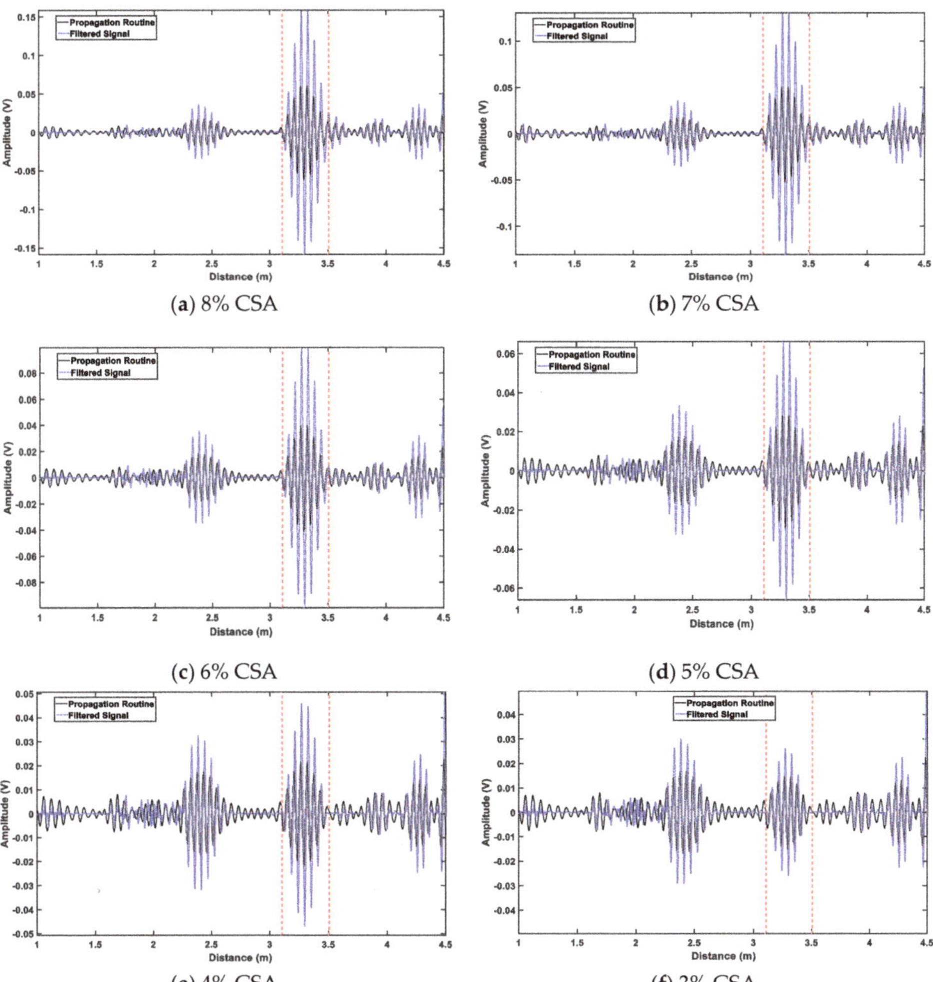

(**a**) 8% CSA

(**b**) 7% CSA

(**c**) 6% CSA

(**d**) 5% CSA

(**e**) 4% CSA

(**f**) 3% CSA

Figure 17. Example of filtered signals using 30 kHz excitation and model parameters where defect size is varying.

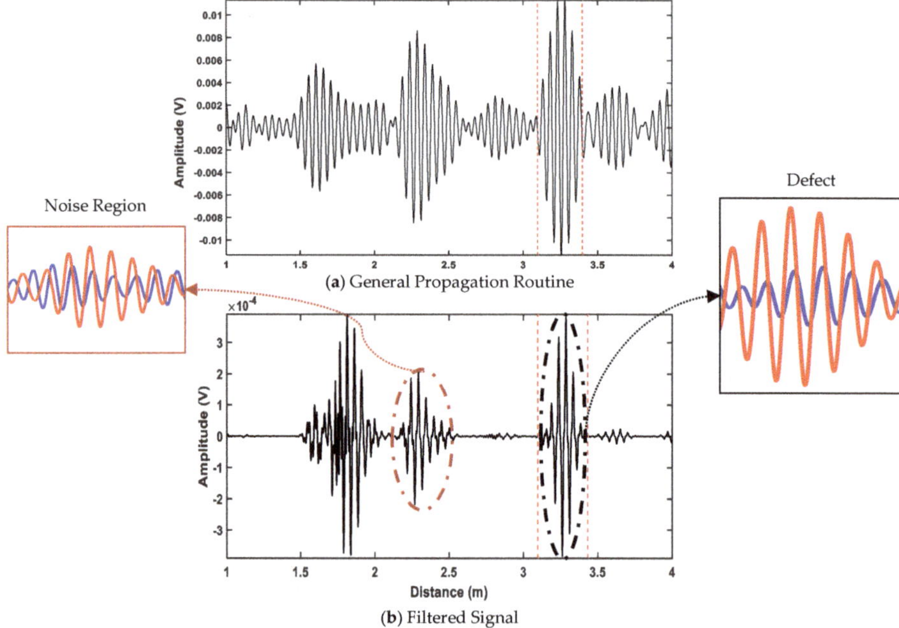

Figure 18. The result of (**a**) general propagation routine vs (**b**) filtered signal with their corresponding inputs (blue and red lines) of noise (left side) and defect (right side) with the excitation frequency of 34 kHz gathered from an experimental pipe with defect size of 3% CSA.

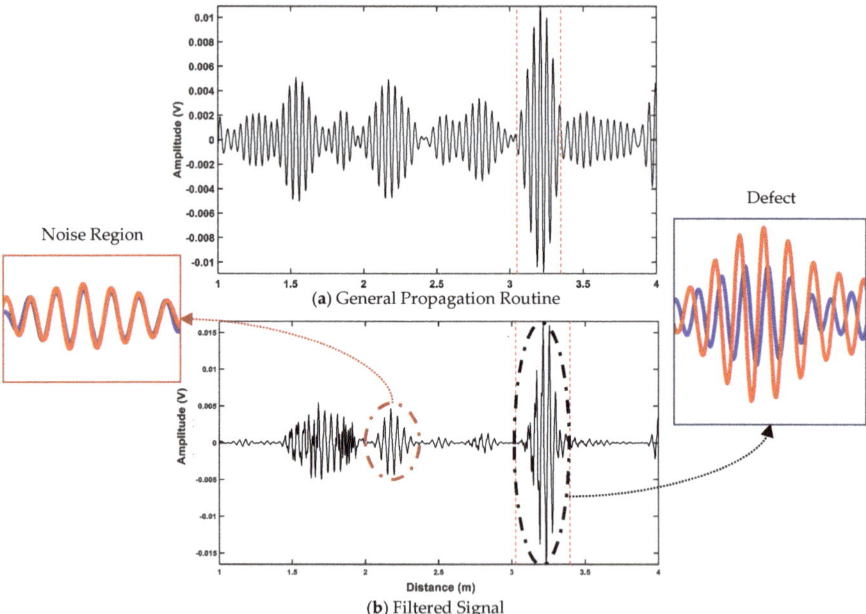

Figure 19. The result of (**a**) general propagation routine vs (**b**) filtered signal with their corresponding inputs (blue and red lines) of noise (left side) and defect (right side) with the excitation frequency of 38 kHz gathered from an experimental pipe with defect size of 3% CSA.

6. Conclusions

In this paper, an adaptive filtering approach is utilized to enhance the SNR of the torsional waves generated from defects in guided wave inspection of pipelines. For doing so, as opposed to using a general propagation routine of guided-wave testing devices, the two sets of signals received from the test tool were used as inputs to the adaptive filter, which resulted in a single output in the time domain. Inherently, in these two sets of signals, the torsional waves are overlapping while the flexurals are varying; this was the basis on which the adaptive filter was able to enhance the SNR of torsional waves. Since flexural waves are variable both in space and time, the used updated algorithm for filter weights was leaky NLMS, which allowed faster adaption to the highly variable flexural noise and enabled a faster reset of the filter weights.

The algorithm was initially developed and validated using the synthesized signal generated from an FEM test case. Afterward, laboratory trail validation took place with two sets of filter parameters: model parameters, where the parameters were set to find the highest SNR for the FEM test case with excitation frequency of 30 kHz; and experimental parameters, where the optimum parameters for each frequency were generated with regards to the real pipe data with 4% CSA defect. In most test cases, all the frequencies resulted in enhancement of the SNR. Nonetheless, the optimum testing frequency was found to be 38 kHz since it achieved both the maximum and greatest enhancement of SNR in tests with defects. Also, this excitation frequency was less affected by the change of parameter as limited enhancement was observed between the results achieved by using model parameters and experimental parameters.

Comparing the results from model parameters and experimental parameters, it is evident that setting the parameters with regards to each individual frequency enhances the achieved SNR. Nonetheless, optimum selection of parameters is a trade-off between the maximum achieved gain and the stable enhancement of the SNR of smaller defects. This was assessed using experimental parameters; when the optimum parameters for 4% CSA defect were used for the case of 3% defect, not all frequencies could improve the SNR. The model parameters test proved that the parameters can be chosen using an FEM model, but it will not result in the maximum gain.

The amount of SNR achieved in each test case depends not only on the enhancements of the torsional wave but also the cancellation of the flexural waves. It must be borne in mind that since the physical arrangement of the rings is fixed, in some cases, especially for lower order flexural waves that have closer wave speeds to torsional waves, noise will not be canceled. However, the higher order flexural noise will be canceled in all the tests and the achieved results can be further processed for torsional wave mode detection.

Author Contributions: Conceptualization, H.N.M. and A.K.N.; methodology, H.N.M. and A.K.N.; software, H.N.M.; validation, H.N.M.; formal analysis, H.N.M.; investigation, H.N.M.; resources, H.N.M.; data curation, H.N.M.; writing—original draft preparation, H.N.M.; writing, reviewing and editing, H.N.M., K.Y., and A.K.N.; visualization, H.N.M. and A.K.N.; supervision, K.Y. and A.K.N.

Funding: This publication was made possible by the sponsorship and support of Brunel University London and National Structural Integrity Research Centre (NSIRC).

Conflicts of Interest: The authors declare no conflicts of interest.

Appendix A

The filter order was selected based on the FEM data. A brute force search was applied to maximize the SNR of Defect-1, calculated by Equation (6). Therefore, the best case of each weight is considered. The generated results are shown in Table A1. The SNR of the unfiltered signal is 5.3 dB. Therefore, this signal is more difficult to detect as it is almost buried in the noise level. As can be seen, by increasing the filter order, the SNR is reduced. When the size of filter order grows above 8, the SNRs of filtered and unfiltered signals are approximately the same. This is because by growing the number of samples, the differences between a torsional and flexural wave become limited as they are in the same frequency band. However, in smaller steps, the changes in flexural waves are more time-varying. Therefore, from these results, 8 can be the considered as the safest number for filter order where no change will be applied to the signal and 2 can be considered as the best filter order for maximizing the SNR.

On the other hand, the best filter order can be dependent on other factors, such as frequency and the existing flexurals of the iteration. Since the main aim of this paper is to test the workability of the adaption algorithm to the guided-wave scenario, the filter order is fixed to 5. This number is a compromise between the safest choice, 8, and the best achieved results, 2, based on FEM with 30 kHz excitation sequence. Furthermore, a filter order of 5 can be considered as a good compromise, as a drop of only 1 dB is observed from the maximum achieved SNR, which is 8.31 dB.

Table A1. Maximum SNR achieved for each filter order on FEM test signal with 30 kHz signal.

Filter Order (#)	2	3	4	5	6	7	8	9	10
Maximum SNR (dB)	8.31	8.27	7.76	7.32	6.78	6.07	5.33	5.07	5.02

Appendix B

The SNR of the defect signal achieved from the general propagation routine of the device is demonstrated in Table A2 and Figure A1.

Table A2. SNR achieved from the general propagation routine (in dB).

CSA	Frequency (kHz)										
(%)	30	32	34	36	38	40	42	44	46	48	50
8	14.68	16.81	19.64	21.04	21.73	22.28	21.84	20.47	18.33	15.70	12.63
7	13.55	15.50	18.19	19.69	20.64	21.39	20.97	19.48	17.20	14.53	11.38
6	11.70	13.57	15.89	17.88	18.78	19.33	18.75	17.27	15.06	12.40	9.24
5	9.29	11.09	13.71	15.00	16.05	16.74	16.05	14.40	12.07	9.38	6.17
4	7.13	8.84	11.12	12.15	13.13	13.76	13.23	11.73	9.50	6.84	3.50
3	4.29	5.97	8.15	8.62	9.08	9.51	9.00	7.60	5.43	2.82	−0.51

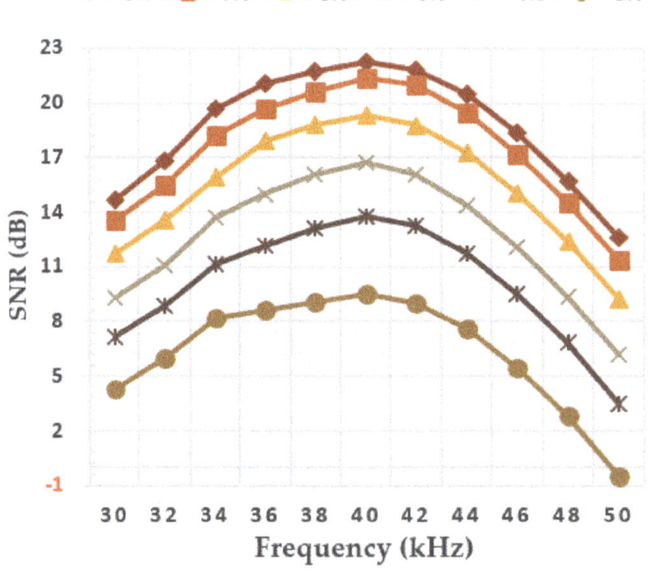

Figure A1. SNR of tests using general propagation routine.

Appendix C

Table A3 shows the parameters used in each test with regards to the frequency.

Table A3. Used parameters in the tests.

Frequency (kHz)	Step Size	Leakage	Compensation
30 (Model)	0.01	0.96	0.01
30	0.01	0.61	0.21
32	0.01	0.46	0.21
34	0.01	0.46	0.26
36	0.01	0.56	0.06
38	0.16	0.81	0.01
40	0.86	0.76	0.01
42	1.41	0.91	0.01
44	1.21	0.96	0.01
46	1.46	0.96	0.01
48	1.46	0.96	0.01
50	1.46	0.96	0.01

References

1. Lowe, M.J.S.; Alleyne, D.N.; Cawley, P. Defect detection in pipes using guided waves. *Ultrasonics* **1998**, *36*, 147–154. [CrossRef]
2. Pavlakovic, B. Signal Processing Arrangment. US Patent 2006/0203086 A1, 14 Septenber 2006.
3. ASTM Standard E2775-16. *Standard Practice for Guided Wave Testing of above Ground steel Pipeword Using Piezoelectric Effect Transducer*; ASTM: West Conshohocken, PA, USA, 2017.
4. Yücel, M.K.; Fateri, S.; Legg, M.; Wilkinson, A.; Kappatos, V.; Selcuk, C.; Gan, T.H. Coded Waveform Excitation for High-Resolution Ultrasonic Guided Wave Response. *IEEE Trans. Ind. Inform.* **2016**, *12*, 257–266. [CrossRef]
5. Fateri, S.; Boulgouris, N.V.; Wilkinson, A.; Balachandran, W.; Gan, T.H. Frequency-sweep examination for wave mode identification in multimodal ultrasonic guided wave signal. *IEEE Trans. Ultrason. Ferroelectr. Freq. Control* **2014**, *61*, 1515–1524. [CrossRef]
6. Thornicroft, K. Ultrasonic Guided Wave Testing of Pipelines Using a Broadband Excitation. Ph.D Thesis, Brunel University London, London, UK, 2015.
7. Malo, S.; Fateri, S.; Livadas, M.; Mares, C.; Gan, T. Wave Mode Discrimination of Coded Ultrasonic Guided Waves using Two-Dimensional Compressed Pulse Analysis. *IEEE Trans. Ultrason. Ferroelectr. Freq. Control* **2017**, *64*, 1092–1101. [CrossRef] [PubMed]
8. Pedram, S.K.; Fateri, S.; Gan, L.; Haig, A.; Thornicroft, K. Split-spectrum processing technique for SNR enhancement of ultrasonic guided wave. *Ultrasonics* **2018**, *83*, 48–59. [CrossRef]
9. Mallet, R. Signal Processing for Non-Destructive Testing. Ph.D Thesis, Brunel Unviersity, London, UK, 2007.
10. Hayes, M.H. *Statistical Digital Signal Processing and Modeling*; John Wiley & Sons, Inc.: New York, NY, USA, 1997.
11. Bellanger, M.G. *Adaptive Digital Filters, Revised and Expanded*, 2nd ed.; Marcel Dekker, Inc.: New York, NY, USA, 2001.
12. Widrow, B.; Glover, J.R.; McCool, J.M.; Kaunitz, J.; Williams, C.S.; Hearn, R.H.; Zeidler, J.R.; Dong, J.E.; Goodlin, R.C. Adaptive Noise Cancelling: Principles and Applications. *Proc. IEEE* **1975**, *63*, 1692–1716. [CrossRef]
13. Rajesh, P.; Umamaheswari, K.; Kumar, V.N. A Novel Approach of Fetal ECG Extraction Using Adaptive Filtering. *Int. J. Inf. Sci. Intell. Syst.* **2014**, *3*, 55–70.
14. Hernandez, W. Improving the response of a wheel speed sensor using an adaptive line enhancer. *Measurement* **2003**, *33*, 229–240. [CrossRef]
15. Ramli, R.M.; Noor, A.O.A.; Samad, S.A. A Review of Adaptive Line Enhancers for Noise Cancellation. *Aust. J. Basic Appl. Sci.* **2012**, *6*, 337–352.

16. Murano, K.; Unagami, S.; Amano, F. Echo Cancellation and Applications. *IEEE Commun. Mag.* **1990**, *28*, 49–55. [CrossRef]
17. Messerrschmi, D. Echo Cancellation in Speech and Data Transmission. *IEEE J. Sel. Areas Commun.* **1984**, *2*, 283–297. [CrossRef]
18. Bürger, W. Space-Time Adaptive Processing: Fundamentals. *Adv. Radar Signal Data Process.* **2006**, *6*, 1–14.
19. Schwark, C.; Cristallini, D. Advanced multipath clutter cancellation in OFDM-based passive radar systems. In Proceedings of the 2016 IEEE Radar Conference (RadarConf 2016), Philadelphia, PA, USA, 2–6 May 2016.
20. Zhu, Y.; Weight, J.P. Ultrasonic nondestructive evaluation of highly scattering materials using adaptive filtering and detection. *IEEE Trans. Ultrason. Ferroelectr. Freq. Control.* **1994**, *41*, 26–33.
21. Monroe, D.; Ahn, I.S.; Lu, Y. Adaptive filtering and target detection for ultrasonic backscattered signal. In Proceedings of the 2010 IEEE International Conference on Electro/Information Technology, Normal, IL, USA, 20–22 May 2010.
22. Haykin, S. *Adaptive Filter Theory*, 5th ed.; Pearson: Harlow, UK, 2014.
23. Cioffi, J. Limited-Precision Effects in Adaptive Filtering [invited paper, special issue on adaptive filtering. *IEEE Trans. Circuits Syst.* **1987**, *34*, 821–833. [CrossRef]
24. Sayed, A.H.; Al-Naffouri, T.Y. Mean-square analysis of normalized leaky adaptive filters. In Proceedings of the 2001 IEEE International Conference on Acoustics, Speech, and Signal Processing (Cat. No.01CH37221), Salt Lake City, UT, USA, 7–11 May 2001; Volume 6, pp. 3873–3876.
25. Bismor, D.; Pawelczyk, M. Stability Conditions for the Leaky LMS Algorithm Based on Control Theory Analysis. *Arch. Acoust.* **2016**, *41*, 731–739. [CrossRef]
26. Lowe, M.J.S.; Cawley, P. *Long Range Guided Wave Inspection Usage—Current Commercial Capabilities and Research Directions*; Imperial College London: London, UK, 2006; Available online: http://www3.imperial.ac.uk/pls/portallive/docs/1/55745699.PDF (accessed on 20 November 2018).
27. Meitzler, A.H. Mode Coupling Occurring in the Propagation of Elastic Pulses in Wires. *J. Acoust. Soc. Am.* **1961**, *33*, 435–445. [CrossRef]
28. Catton, P. *Long Range Ultrasonic Guided Waves for Pipelines Inspection*; Brunel University: London, UK, 2009.
29. Nurmalia. *Mode Conversion of torsional Guided Waves for Pipe Inspection: An electromagnetic Acoustic Transducer Technique*; Osaka University: Osaka Prefecture, Japan, 2013.
30. Nakhli Mahal, N.; Mudge, P.; Nandi, A.K. Comparison of coded excitations in the presence of variable transducer transfer functions in ultrasonic guided wave testing of pipelines. In Proceedings of the 9th European Workshop on Structural Health Monitoring, Manchester, UK, 10–13 July 2018.
31. Nakhli Mahal, N.; Muidge, P.; Nandi, A.K. Noise removal using adaptive filtering for ultrasonic guided wave testing of pipelines. In Proceedings of the 57th Annual British Conference on Non-Destructive Testing, Nottingham, UK, 10–12 September 2018; pp. 1–9.
32. Wilcox, P.D. A rapid signal processing technique to remove the effect of dispersion from guided wave signals. *IEEE Trans. Ultrason. Ferroelectr. Freq. Control* **2003**, *50*, 419–427. [CrossRef]
33. Sanderson, R. A closed form solution method for rapid calculation of guided wave dispersion curves for pipes. *Wave Motion* **2015**, *53*, 40–50. [CrossRef]
34. Wilcox, P.; Lowe, M.; Cawley, P. The effect of dispersion on long-range inspection using ultrasonic guided waves. *NDT E Int.* **2001**, *34*, 1–9. [CrossRef]
35. Guan, R.; Lu, Y.; Duan, W.; Wang, X. Guided waves for damage identification in pipeline structures: A review. *Struct. Control Health Monit.* **2017**, *24*, e2007. [CrossRef]
36. Lowe, M.J.S.; Alleyne, D.N.; Cawley, P. The Mode Conversion of a Guided Wave by a Part-Circumferential Notch in a Pipe. *J. Appl. Mech.* **1998**, *65*, 649–656. [CrossRef]
37. Miao, H.; Huan, Q.; Wang, Q.; Li, F. Excitation and reception of single torsional wave T(0,1) mode in pipes using face-shear d24 piezoelectric ring array. *Smart Mater. Struct.* **2017**, *26*, 025021. [CrossRef]
38. Rose, J.L. *Ultrasonic Guided Waves in Solid Media*; Cambridge University Press: New York, NY, USA, 2014.
39. Alleyne, D.N.; Pavlakovic, B.; Lowe, M.J.S.; Cawley, P. Rapid, Long Range Inspection of Chemical Plant Pipework Using Guided Waves. *Key Eng. Mater.* **2004**, *270–273*, 434–441. [CrossRef]
40. Douglas, S.C. Performance Comparison of Two Implementations of the Leaky LMS Adaptive Filter. *IEEE Trans. Signal Process.* **1997**, *45*, 2125–2129. [CrossRef]
41. Cartes, D.A.; Ray, L.R.; Collier, R.D. Experimental evaluation of leaky least-mean-square algorithms for active noise reduction in communication headsets. *J. Acoust. Soc. Am.* **2002**, *111*, 1758–1771. [CrossRef] [PubMed]

42. Kuo, S.M.; Lee, B.H.; Tian, W. *Real-Time Digital Signal Processing: Implementations and Applications*; John Wiley & Sons: Chichester, UK, 2006.

43. Ssanalysis. Abaqus/Explicit. Available online: http://www.ssanalysis.co.uk/ (accessed on 20 Nov 2018).

44. Teletestndt. Long Range Guided Wave Testing with Teletest Focus+. 2018. Available online: https://www.teletestndt.com/ (accessed on 08 January 2019).

45. Alleyne, D.N.; Lowe, M.J.S.; Cawley, P. The Reflection of Guided Waves from Circumferential Notches in Pipes. *J. Appl. Mech.* **1998**, *65*, 635–641. [CrossRef]

46. Lowe, P.S.; Sanderson, R.M.; Boulgouris, N.V.; Haig, A.G. Inspection of Cylindrical Structures Using the First Longitudinal Guided Wave Mode in Isolation for Higher Flaw Sensitivity. *IEEE Sens. J.* **2016**, *16*, 706–714. [CrossRef]

47. Niu, X.; Duan, W.; Chen, H.; Marques, H.R. Excitation and propagation of torsional T (0, 1) mode for guided wave testing of pipeline integrity Excitation and propagation of torsional T (0, 1) mode for guided wave testing of pipeline integrity. *Measurement* **2018**, *131*, 341–348. [CrossRef]

48. Duan, W.; Niu, X.; Gan, T.; Kanfound, J.; Chen, H.-P. A Numerical Study on the Excitation of Guided Waves. *Metals* **2017**, *7*, 552. [CrossRef]

49. Fateri, S.; Lowe, P.S.; Engineer, B.; Boulgouris, N.V. Investigation of ultrasonic guided waves interacting with piezoelectric transducers. *IEEE Sens. J.* **2015**, *15*, 4319–4328. [CrossRef]

50. Alleyne, D.N.; Cawley, P. The excitation of Lamb waves in pipes using dry-coupled piezoelectric transducers. *J. Nondestruct. Eval.* **1996**, *15*, 11–20. [CrossRef]

51. Engineer, B.A. The Mechanical and Resonant Behaviour of a Dry Coupled Thickness-Shear PZT Transducer Used for Guided Wave Testing in Pipe Line. Ph.D Thesis, Brunel University, London, UK, 2013.

52. Mathworks. Matlab. Available online: https://uk.mathworks.com/products/matlab.html (accessed on 18 April 2018).

Article

Defect Detection using Power Spectrum of Torsional Waves in Guided-Wave Inspection of Pipelines

Houman Nakhli Mahal [1,2,*], **Kai Yang** [3] **and Asoke K. Nandi** [1]

[1] Department of Electronic and Computer Engineering, Brunel University London, Uxbridge UB8 3PH, UK; asoke.nandi@brunel.ac.uk
[2] NSIRC, Granta Park, Cambridge CB21 6AL, UK
[3] TWI, Granta Park, Cambridge CB21 6AL, UK; kai.yang@twi.co.uk
* Correspondence: houman.nakhlimahal@brunel.ac.uk

Received: 24 January 2019; Accepted: 1 April 2019; Published: 6 April 2019

Abstract: Ultrasonic Guided-wave (UGW) testing of pipelines allows long-range assessment of pipe integrity from a single point of inspection. This technology uses a number of arrays of transducers separated by a distance from each other to generate a single axisymmetric (torsional) wave mode. The location of anomalies in the pipe is determined by inspectors using the received signal. Guided-waves are multimodal and dispersive. In practical tests, nonaxisymmetric waves are also received due to the nonideal testing conditions, such as presence of variable transfer function of transducers. These waves are considered as the main source of noise in the guided-wave inspection of pipelines. In this paper, we propose a method to exploit the differences in the power spectrum of the torsional wave and flexural waves, in order to detect the torsional wave, leading to the defect location. The method is based on a sliding moving window, where in each iteration the signals are normalised and their power spectra are calculated. Each power spectrum is compared with the previously known spectrum of excitation sequence. Five binary conditions are defined; all of these need to be met in order for a window to be marked as defect signal. This method is validated using a synthesised test case generated by a Finite Element Model (FEM) as well as real test data gathered from laboratory trials. In laboratory trials, three different pipes with defects sizes of 4%, 3% and 2% cross-sectional area (CSA) material loss were evaluated. In order to find the optimum frequency, the varying excitation frequency of 30 to 50 kHz (in steps of 2 kHz) were used. The results demonstrate the capability of this algorithm in detecting torsional waves with low signal-to-noise ratio (SNR) without requiring any change in the excitation sequence. This can help inspectors by validating the frequency response of the received sequence and give more confidence in the detection of defects in guided-wave testing of pipelines.

Keywords: signal processing; defect detection; torsional wave; power spectrum; sliding window; pipeline inspection; ultrasonic guided-waves (UGWs)

---

1. Introduction

Pipelines are the main means of transferring oil and gas. They are usually installed in hostile locations and must be inspected to avoid failures that would be harmful to the environment. Ultrasonic guided-wave (UGW) enables long-range inspection of pipelines from one single test point. For example, the UGW technique in pulse-echo mode can enable inspections of up to 50 m (in normal testing conditions); hence, it is cost efficient in the inspection of large structures. The received signals are inspected in time-domain where inspectors spot anomalies in the data based on their local signal-to-noise ratio (SNR). In any region where the envelope of the signal has higher energy than the regional noise level, an anomaly is reported which can indicate a defect or a feature of the pipe [1–3]. UGWs are multimodal: for each excitation frequency, multiple wave modes are transmitted and received. In the inspection, the goal is to achieve a pure axisymmetric wave, but due to the imperfect testing conditions [4,5], nonaxisymmetric waves will also be received. This is the main source of the coherent noise in the inspection. To allow ease of inspection, many researchers have applied digital signal processing methods that use the differences in these wave modes to either detect the axisymmetric wave or filter the nonaxisymmetric waves to increase the SNR of the received anomalies. One of the widely used techniques in the signal processing of guided-waves is the work of Wilcox et al. [6], which uses previously calculated dispersion curves in order to compensate for the effect of dispersion for the wave mode of interest in a certain propagation distance. In 2013, Zeng et al. [7] used the basis of the dispersion compensation technique and introduced a novel method to design waveforms in order to precompensate for a certain propagation distance.

Pulse compression is another approach that has been previously investigated in the literature. Instead of exciting a narrowband sine wave sequence, pulse compression uses a known coded sequence and processes the received signal by applying match-filtering in order to detect the known sequence. One of the initial attempts of testing pulse compression in the guided-wave was done by Rodriguez et al. [8,9], where chirp sequences were excited using air-coupled piezoelectric transducer in order to generate Lamb waves in aluminum plates. Higher SNR and peak values for the signals of interests were achieved in the experiments; however, the effect of dispersion was not considered, and therefore a decrease in signal amplitude was expected. In 2010, they published another paper using the same system where the phase modulation based on Golay codes were used where the enhancement of 21 dB in SNR using a 16-bit Golay code was achieved compared to the conventional pulse transmission. Mehmet et al. [10,11], introduced an iterative dispersion compensation for removing the dispersion effect of guided-wave propagation. Most recently, Malo et al. [12] developed a two-dimensional compressed pulse analysis in order to enhance the achieved SNR of each wave mode. In the pulse compression approaches, the results are always a trade-off between the spatial resolution and SNR, as having a result with higher propagation energy and δ-like correlation means that the signal duration must be increased. Furthermore, the transfer function of the excitation system can affect the accuracy of the coded waveforms. These two factors indicate that the initial waveform design and the accuracy of inspection using this method depend on the testing conditions.

On the other hand, other approaches have also been reported in the literature where narrowband sine waves were used as excitation sequence. Kamran et al. investigated split-spectrum processing [13–15], where the time-domain signal is decomposed into multiple signals in different frequency bands, and then recombined in time-domain in order to remove the coherent noise. In their most recent work [15], it was reported that using the optimum filter parameters and polarity thresholding method for recombination, the SNR of defects with sizes as small as 2% cross-sectional area (CSA) could be improved significantly. Nonetheless, the technique depends on various parameters which can depend on the pipes' characteristics. Another recently developed method is the spectral subtraction, investigated by Duan et al. [16]. In this method, the noise signature is calculated using a small section of the retrieved signal where no real pipe feature exists. Afterwards, this signal is subtracted from the total signal using a sliding window where a significant reduction of coherent noise level could be achieved. Nonetheless, achieving noise signature in the practical inspection of pipelines is difficult as the location of defects and even pipe features might be unknown.

Guided-waves are dispersive and multimodal [17]. Depending on the excitation frequency, multiple wave modes can be generated at the point of excitation. Guided-wave modes are categorised based on their displacement patterns (mode shapes) within the structure. Three main families of waves exist in pipes, which are longitudinal, torsional and flexural. Longitudinal and torsional waves are axisymmetric waves, while flexural waves are nonaxisymmetric. The popular nomenclature used for them are in the format of $X(n,m)$, where X can be replaced by the letters L for longitudinal, T for torsional and F for flexural waves; n shows the harmonic variations of displacement and stress around the circumference and m represents the order of existence of the wave mode [18]. In general inspection, the practice is to generate a single pure axisymmetric wave which tends to ease the interpretation of results. However, variations in the transducers' transfer function and their placement will cause the flexural waves to be received; furthermore, wave mode conversion also causes flexural waves to be received [5,17–19].

Dispersion causes the energy of a signal to spread out in space and time as it propagates [6]. Figure 1a shows an example of a dispersion curve calculated from 8-inch schedule 40 steel pipe using RAPID software [21]. Wilcox et al. introduced a method to use dispersion curves in order to both simulate [22] and remove [6] the effect of dispersion. Using the developed formula, an example of a dispersive wave is shown in Figure 1b, which is based on the dispersion curves of F (4,2). Unlike T (0,1), which is nondispersive across its whole frequency range [23], flexural waves are dispersive. Therefore, for ease of inspection, the torsional wave which is both axisymmetric and nondispersive is used.

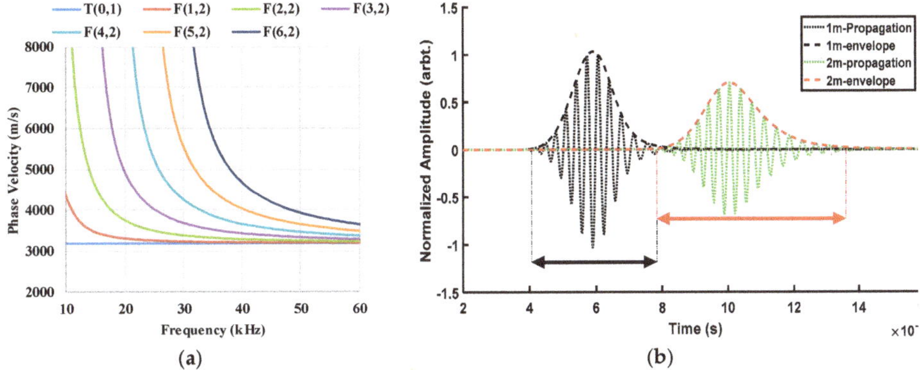

Figure 1. (**a**) Example dispersion curve of T(0,1) wave mode in an 8″ schedule 40 steel pipe. (**b**) The effect of dispersion on a simulated flexural wave for two propagation distances [20].

The common concept in all the aforementioned signal processing methods is the usage of dispersion and multimodality of guided-waves in order to increase the SNR. Afterwards, the signals are inspected in time-domain to report on the location of anomalies. It has already been demonstrated in the literature that the spectral domain can be processed to remove the noise. Looking at the problem with another perspective, instead of noise removal, the spectral domain can also be used to detect the signal of interests (defect) automatically. The excitation sequence is not dispersive and axisymmetric; hence, no significant changes in the frequency response of the received signal should be inspected. Therefore, a sliding window is used where the power spectrum of each iteration is compared with the one achieved from the excitation sequence.

The primary motivation is to enable automated inspection of pipelines without the need for any inspectors. However, at this stage, the developed algorithm can be used as a tool, along with the conventional detection methods, to provide more certainty in interoperation of the results and reduce the number of outliers called due to the coherent noise. It is expected that experience of this algorithm will help develop automated inspection of pipelines. It should be noted that the experiments were performed with shorter pipe length. Nevertheless, the detectability of the defects is mostly dependent on the characteristics of the received signal. The algorithm can easily detect a perfect axisymmetric feature such as a weld in long-range. Nonetheless, for defects which are generally smaller in size and asymmetric, the hypothesis is that the feature can be detected if the characteristic of the torsional wave is still detectable in the received signal. It should also be noted, that as the inspection pipe length increases the energy of the flexural waves reduce due to their dispersion in time and the effect of attenuation on them. Therefore, false alarms using this method would be expected to be less in long-range inspections.

The paper is organised as follows. Section 2 explains the proposed method for detection of defects using torsional waves. Section 3 demonstrates the results of the method on both synthesised and real experimental data and, finally, Section 4 concludes the paper.

2. Methodology

In ultrasonic testing, the excitation sequence is known, and the inspectors generally look for signals with the same characteristics in the time domain to find their signals of interest, such as a defect in materials. In UGW, one of the common excitation sequences is a 10-cycle Hann windowed signal with different frequency, depending on the testing conditions. However, these signals are typically polluted by coherent noise which has similar temporal characteristics to those of defect signals.

It was mentioned beforehand that in the guided-wave, torsional waves are nondispersive and axisymmetric. The flexurals, which are the main source of coherent noise in the tests, are dispersive and nonaxisymmetric. In the power spectrum of an inspected pure torsional wave, the same characteristics of the spectrum of the excitation sequence must be observed; this is not the case for flexural waves due to their temporal and spatial variances. It should also be mentioned that asymmetric features (which is the case for most defects) generally reflect both torsional and flexural waves; where their corresponding reflection coefficients depends on the geometry of the feature. However, in this research, we focus our detection based on the detection of the torsional waves. The torsional wave is always reflected, and it is received symmetrically around the pipe circumference regardless of the features' geometry or the wave propagation distance. Furthermore, in some cases, the flexural waves are considered as the coherent noise received due to mode conversion from the known features, leading to detection of false alarms in the tests. Hence, the focus is in detection of torsional wave in order to find the location of features.

In this paper, we propose a condition-based comparison of the power spectrum achieved from a moving window of the received signal and the spectrum of the normal excitation sequence. Keeping in mind that, since both flexural and torsional waves have the same main bandwidth (BW), correlating them would not be sufficient for distinguishing between them. Nonetheless, using this proposed method the torsional wave is identifiable. Another advantage of this method is that it is relying on the excitation sequence, which is known and set manually by the inspectors.

This algorithm consists of three main aspects: (1) initialisation, which initialises the excitation sequence and extract the required features for comparison; (2) main loop, that uses the advancing window and carries out the pre- and postprocessing of the conditions; and (3) conditions, which constitute the main processing, where the spectrum of each iteration is compared with the one achieved from excitation sequence.

Since the signals are processed digitally, and due to the limited resolution in this domain, in the following section all formulae and definitions are presented based on the sample number for the time-domain and bin number for the power spectrum. In doing so, better performance, in terms of speed is achieved as no unnecessary interpolation, is needed. The sampling frequency (Fs) used in the tests is 1 MHz which is a fairly common sampling frequency of guided-wave inspection devices.

2.1. Initialisation

Initially, the excitation sequence is created where the inspector inputs the centre frequency and the number of cycles. Then, the *"windowSize"* is calculated based on the number of samples required for the excitation sequence. This sequence is already normalised by its maximum value. The normalisation bounds the signal amplitudes between 1 and −1 by dividing each sample by the maximum amplitude within the signal. Since 10 cycles are typically used for inspection, the duration of the excitation sequence is small; therefore, this signal is zero padded by a factor of 4 to achieve better resolution in the spectral density. Zero padding is the operation of adding extra zeroes to the end of the sequence in order to increase the resolution of the power spectrum [24]. The power spectrum is then evaluated by applying the fast Fourier transform [24], where the magnitudes of the spectrum are saved in *"signalRef"* and the corresponding frequency for each bin is saved in *"frqList"*.

The first extracted feature from the spectrum is the maximum magnitude and its corresponding bin number which are stored in *"maxRef"* and *"idFC"*, respectively. This should be the bin representing the centre frequency of excitation as set before by the inspector. Afterwards, the lowest and highest frequencies of the 10 dB BW of this spectrum are calculated. The bin numbers of the lowest and highest frequencies within this BW are saved in *"teFLID"* and *"teFHID"*. If the previously calculated *"idFC"* is not in the centre of this spectrum, these numbers are expanded away from the centre frequency so both the lower half and the upper half of the spectrum have the same amount of bin numbers. Then, the greater magnitude achieved from these (border) bins is stored in *"teMax"*. The flowchart of this function is shown in Figure 2 and a summary of the extracted variables are shown in Table 1. All values except *"windowSize"* are needed for the conditions function.

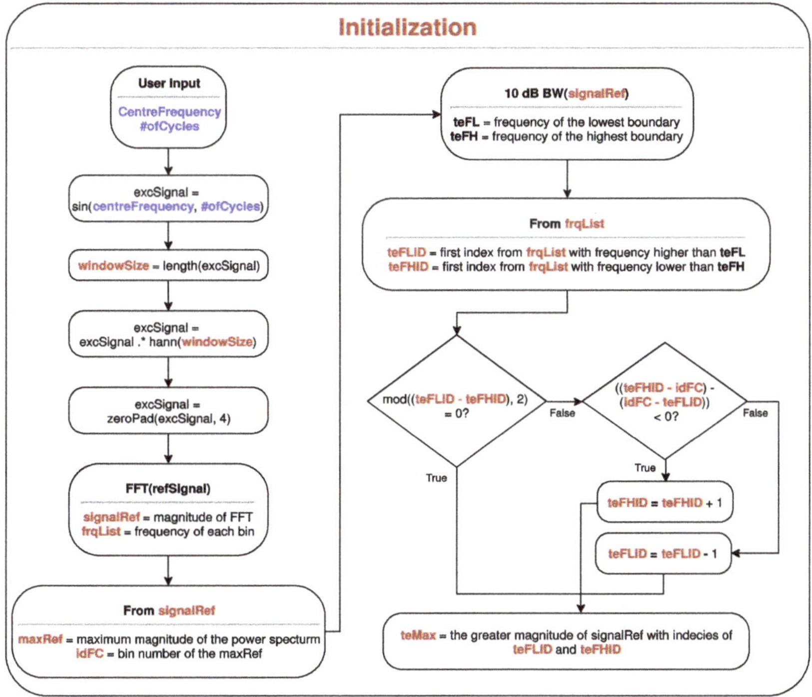

Figure 2. Flowchart of the initialisation.

Table 1. Description of variables extracted from the Initialisation.

Variable Name	Description	Conditions
windowSize	Length of the moving window	-
signalRef	The power spectrum of the normalised excitation sequence	C2, C3, C4
frqList	List of the corresponding frequency for each bin number	C0
maxRef	Maximum magnitude achieved from signalRef	C4
idFC	Bin number of maxRef	C0
teFLID	Bin number of the lowest frequency within 10 dB bandwidth	C0, C1, C3
teFHID	Bin number of the highest frequency within 10 dB bandwidth	C0, C1, C3
teMax	The greater magnitude between teFLID and teFHID from signalRef	C3

2.2. Conditions

In this function, all previously extracted characteristics are used in order to compare the spectrum of the moving window with the excitation sequence. The focus of this paper is on the similarity of the spectrum rather than the energy of the signals; thus, all signals are normalised before calculating the spectrum. Five conditions (features) exist in total. Since defect signals are affected by the flexural noises, it is expected that the received signals from smaller defects would not result in a perfectly matching spectrum. Hence, in each condition, safety variables are defined in order not to over-fit the conditions to the reference. These variables are set by the inspectors to provide a safety tolerance for detecting weak signals. It should be borne in mind that these tolerances have typically small values and can be fixed. Furthermore, they are set only once at the start of the algorithm and does not change iteratively.

2.2.1. Condition Zero (C0)

The centre frequency of the windowed spectrum must be near the centre frequency of the excitation sequence, defined by *"idFC"*. In case of small defect signals, where less energy is reflected, they are more affected by the coherent noise which will cause a slight shift in the centre frequency. Therefore, a slight shift in the frequency is acceptable which is defined by variable *"maxFCallow"*. If such a shift is detected, the corresponding bin numbers of *"idFC"*, *"teFLID"* and *"teFHID"*, which are required for the other conditions, will be shifted towards the new centre frequency of the spectrum. If this shift of centre frequency is more than maximum allowable shift, it means that the current iteration is significantly different to the excitation sequence and all other conditions will automatically fail. The maximum allowable shift, *"maxFCAllow"* is set to 3 kHz, which is a small margin in comparison to the total BW of the signal.

2.2.2. Condition One (C1)

In the Hann windowed sine waves, the lower half of the BW tends to increase in magnitude when moving towards the centre frequency then it starts to decrease. In this condition, it is checked that there is no discontinuity in this trend; in other words that no local maxima exist. In doing so, the 10 dB BW is of the main focus as the outside edges are typically affected by the flexural noises due to their lesser magnitudes. The reason for this is that in guided-wave inspection, (Hann) windowed, narrowband excitation sequences are used. Therefore, the main energy spread is within the main BW of the signal while a limited amount will be assigned to the borders of BW. The flexurals are dispersive, depending on their dispersion curves, their spectra tend to shift towards other frequencies, but the nondispersive torsional wave will not observe this change. Therefore, the energy of flexural waves will be stronger outside the main BW. Choosing 10 dB BW is a good compromise since more than 90% of the bins is covered using this value and only a few bins on the sides are neglected. Using the bin number of lowest frequency and the highest frequency in the 10 dB BW, *"teFLID"* and *"teFHID"*, it is checked whether the magnitude of each respective bin is increasing when moving toward the centre frequency. Nonetheless, a variable can also be defined as *"diffVal"* to allow small differences to be neglected. Currently, no tolerance is needed, and it is kept as zero.

2.2.3. Condition Two (C2)

In the excitation sequence, the strongest magnitudes are detected in the region of the centre frequency. Since they have the strongest magnitude, they will be less affected by the noise in the region. In this condition, it is confirmed that the neighbourhood of the centre frequency has relatively the same magnitude. The number of neighbourhoods in each side of the centre frequency is defined by *"cBins"*, and the maximum allowable difference is defined based on a percentage of current iteration maximum magnitude defined by *"maxPerc"*. In this paper, the number of the neighbourhood is set as 2 and the maximum allowable difference is set as 20% of the maximum magnitude of each iteration.

2.2.4. Condition Three (C3)

The window size is set as the length of the excitation sequence. Due to this limited duration of the signal, in the iteration's power spectrum, there should not be any magnitudes greater than the 10 dB BW boundary, *"teMax"*, of the excitation sequence other than the frequencies within these boundaries. Nonetheless, in the immediate bins next to the boundaries, *"teFLID"* and *"teFHID"*, the magnitudes are more affected by the coherent noise and can be greater than the *"teMax"*. Therefore, *"sBins"* is defined as the allowable number of immediate bins, which will be neglected during this comparison. This value is set to 1.

2.2.5. Condition Four (C4)

Each iteration window is normalised before its power spectrum is calculated. As the window size is limited to the excitation sequence, in the power spectrum of the iteration, a loss of magnitude can be expected due to the coherent noise in the signals. However, the maximum magnitude of the iteration cannot be greater than the one achieved from the excitation sequence, *"maxRef"*. Since normalisation is taking place, a small safety margin is defined as *"sMargin"*, to compensate for the small variances between the magnitudes defined. This value is set to 1, which is much less than the maximum magnitudes achieved from excitation sequences.

2.2.6. Final Results

Iterations are marked as the signal of interest (one) if all the aforementioned conditions are met and will be marked as noise (zero) if any of them are failed. Furthermore, the maximum correlation between the 10 dB BW of the power spectrum of the excitation sequence and current iteration is calculated. The cross-correlation of two sequences $x[n]$ and $y[n]$ is defined as [24]

$$r_{xy}[l] = \sum_1^N y[n+l]x[n] \tag{1}$$

where l is the (time) shift (or) lag, n is the sample number of each sequence and N is the number of samples in the sequences. The correlation operation measures the degree to which two sequences are similar. As it can be seen in the formula, the output is given in a vector of time shifts. Therefore, by considering the maximum amplitude of $r_{xy}[l]$, the maximum similarity between two sequences is measured. Nonetheless, since in C0, the peaks are shifted, no time shifting is required to achieve the maximum value and the results will become the summation of multiplication of each corresponding bin. Both the cross-correlation value and the binary condition results (*"Total"*) will be returned to the main loop where the final *"detection"* results will be generated. The flowchart of this function is shown in Figure 3.

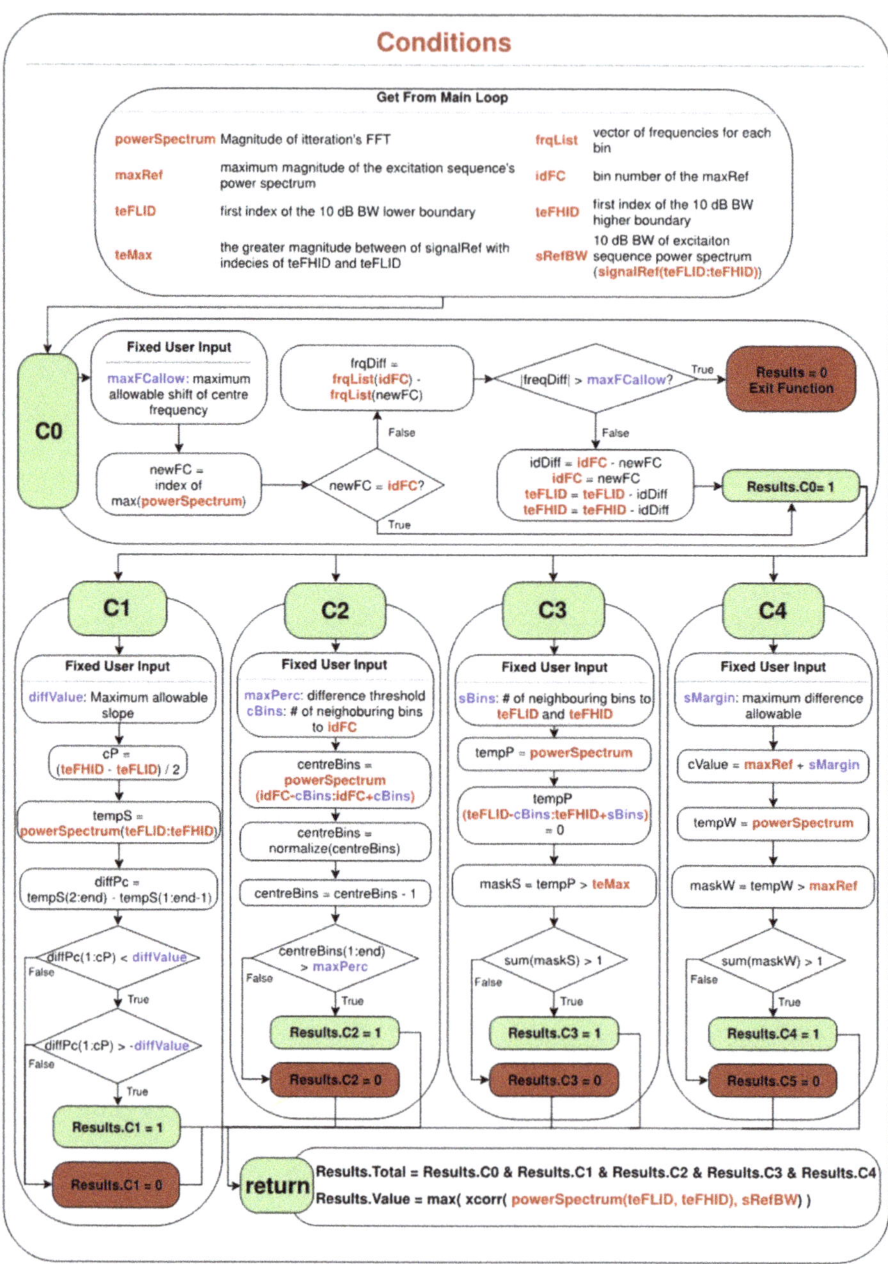

Figure 3. Flowchart of conditions.

2.3. Main Loop

The initialisation is done before the main loop in order to extract the required features. Afterwards, the start of the iteration is delayed by the length of the window size. In practice, it means that the system can be linear and time invariant, and can be implemented in real-time as it only requires past samples of the signals. In each iteration, a temporary signal, *"WindowedSignal"*, is created which holds the past samples of the signals with the same length as the excitation sequence. Since the power spectra are needed to be compared, the signals are first normalised and zero padded by a factor of 4. The power spectrum of this windowed signal is calculated and passed to the conditions function. After processing, the "*Results*" object is returned where it contains the detection variable "*Total*" and the correlation value of the spectrums as "*Value*". If the signal of interest is detected, e.g., "*Total*" is equal to one, then the current iteration represents a similar power spectrum to the one of excitation sequence.

One approach would be to mark the current iteration number as one which would mean that the current iteration is representing the signal of interest; however, this is not the best representation as the current iteration might not be the best representative of the windowed signal characteristics. In this paper, the index of the maximum amplitude in each windowed signal is detected. This index appears to be a better representation since the region with more energy has a stronger influence on the power spectrum. Since the excitation sequence is Hann windowed, the centre of the signal will hold the highest amount of signals' energy. This as in this location, the concentration of signal energy is higher, stronger flexural features are required to disturb reduce/change the characteristics of this region. Therefore, the centre instead of adding the correlation value to the sides of the window (current iteration number), the value is cumulatively added to the location of the highest amplitude within the window, which represents the centre of the Hann windowed sequence in a torsional wave. This value is stored in "detection" vector. Therefore, the correlation value is added to the location where the signal represents the strongest amplitude in the "*detection*" vector. The reason why they are added rather than replaced is that since it is an iterative process, different windows can have the same index as the maximum; by doing a cumulative sum for each index, the certainty in defect detection is increased and the amplitudes assigned for the outliers and noises are decreased. The main loop finishes when the end of the signal is reached, at which point the "*detection*" vector is normalised and is plotted against the original signal to show the defect locations and their normalised correlation values. The flow chart of this function is shown in Figure 4.

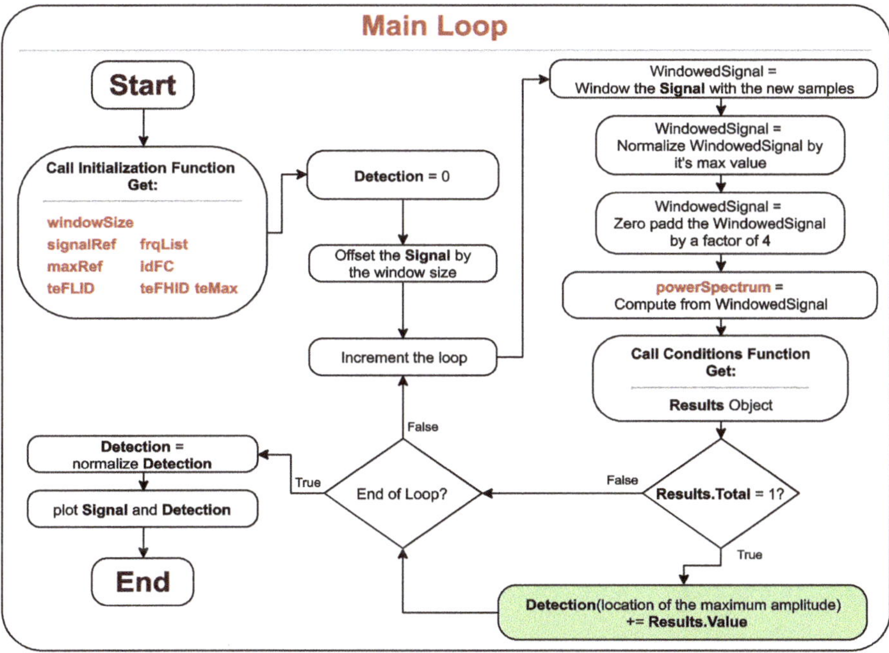

Figure 4. Flowchart of the main loop.

3. Results

This method was initially developed on the FEM test case. As explained previously, the FEM model considers both spatial and temporal characteristics of guided-waves in order to simulate a coloured noise with similar characteristics of the noise in the real inspection. For a better understanding of the methodology, each condition is firstly explained using this FEM signal. To validate this method, the results on experimental pipes are shown. SNRs reported in this section as follows

$$\text{SNR} = 20 * \log 10 \left(\frac{\sqrt{\frac{1}{N} \sum_1^N \text{Signal}[n]}}{\sqrt{\frac{1}{M} \sum_1^M \text{Noise}[m]}} \right) \qquad (2)$$

where "*Signal*" is considered as the samples where the defect (torsional wave) is expected to be received and every other sample is considered as Noise. In all tests, the excitation sequences are of 10-cycle Hann windowed sine waves, which are commonly used in the field inspections.

3.1. FEM Test Case

The generated signal in the FEM test case has the excitation frequency of 30 kHz, as shown in Figure 5. This setup of the model was previously explained in [25]. Modelling of guided-wave signals using FEM have been widely used and validated in the literature [2,26–28]. However, the major difference in this model is that the excitation sequences have variable transfer functions and fewer reception points are used; this tends to add flexural noise in data. In this figure, for better illustration of the signal, the received signal from (forward) the pipe end and the starting dead zone are removed. The defect signal is expected to be received at 3–3.5 m. Since highly variable transfer functions have been used, the backward cancellation algorithm is not able to cancel the backward going signal perfectly. This reversed wave reflected from the pipe end is received at 1–1.5 m. Due to this backward leakage, another signal of interest, which is a torsional wave, is expected to be received at approximately 4–4.5 m. This signal is generated when the echo of the backward leakage and the defect signal (in the forward testing direction) is received at the test tool. All other regions are the noise caused by the flexural waves.

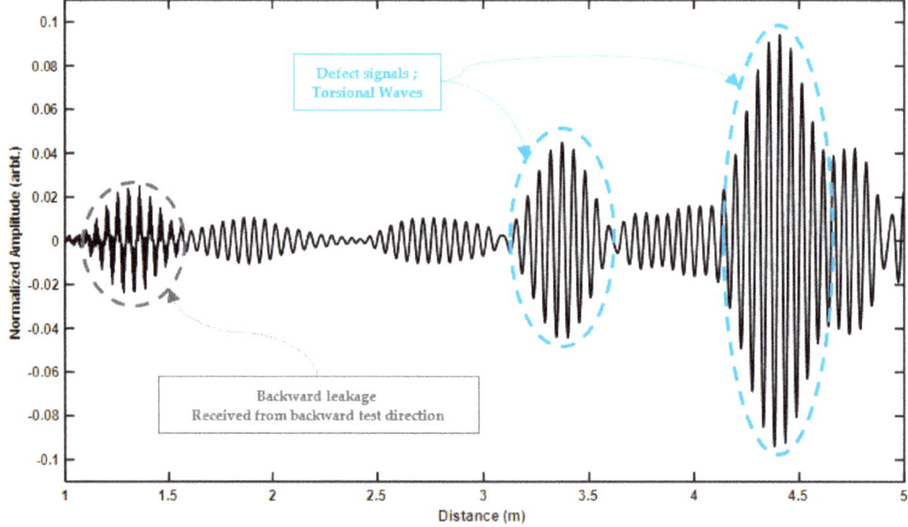

Figure 5. Synthesised data generated from finite element model (FEM) test case.

3.1.1. Condition Zero

The starting condition is that the maximum magnitude of each iteration's power spectrum must belong to the approximate region of the centre frequency. This approximation is set by the inspector ("*maxFCAllow*"). It is a necessary factor for the detection of smaller defects since the spectrum will be more affected by the flexural waves, which leads to a shift in centre frequency. An example of an outlier based on this condition is shown in Figure 6, where the blue box shows an example allowable region for frequency shift, and the grey circle shows the bin with the maximum magnitude in this iteration. Since it is outside the allowable region, the outcome of this condition will be zero, which will automatically force all other conditions to zero. The used value for the maximum allowable shift is 3 kHz which is a small margin in comparison to the overall BW of the excitation sequence. In the case of FEM signal, since the flexural wave does not affect the defect significantly, similar results can be achieved even when this value is set to zero.

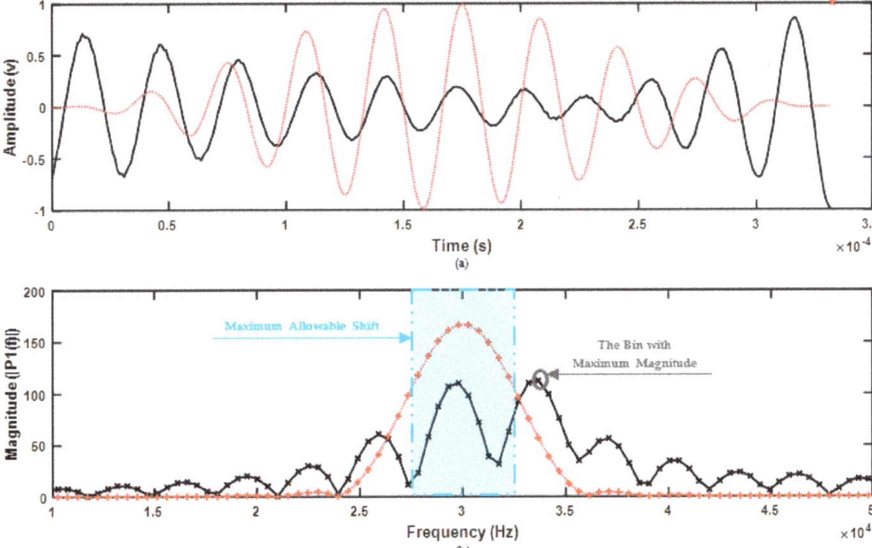

Figure 6. Example of an outlier case detected using Condition Zero (C0), where (**a**) shows the time domain of the iteration window and (**b**) is its respective power spectrum. The red lines (dotted, +) show the references achieved from excitation sequence and the black lines (x, solid) show the results from each iteration.

3.1.2. Condition One

In simple terms, the magnitudes of each bin in the 10 dB BW of the iterations' spectra must increase when moving toward the centre frequency. Due to dispersive nature and multimodality of guided-waves, this condition is not true for many iterations as a combination of waves will be received with different centre frequencies. Nonetheless, this condition must be true in the case of the torsional wave as it will have the same characteristic of the excitation waveform. Figure 7 shows an example of an outlier detected based on this condition. As can be seen, in the region between 26.5 to 27.5 kHz, the magnitude of three frequencies are actually decreasing, and a local minima is created. This suggests that the iteration is actually two separate wave modes with two different centre frequencies.

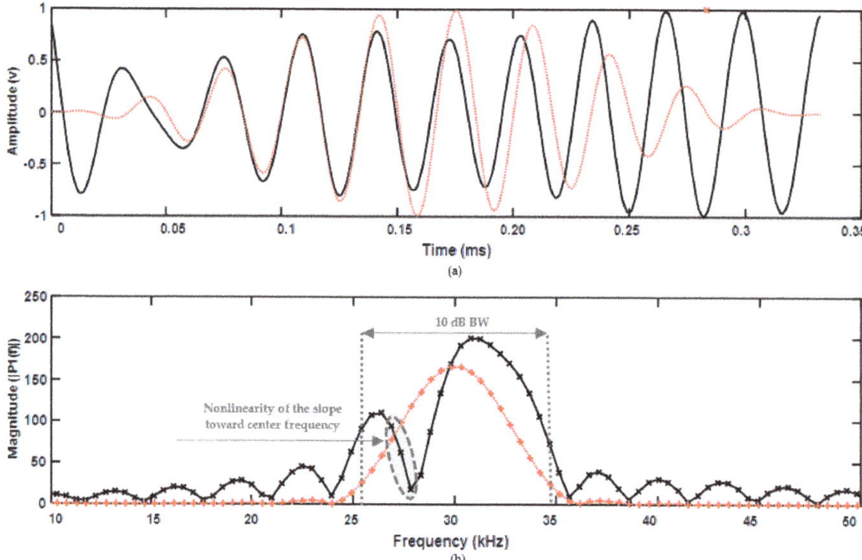

Figure 7. Example of an outlier case detected using Condition One (C1), where (**a**) shows the time domain of the iteration window and (**b**) is its respective power spectrum. The red lines show (dotted, +) the references achieved from excitation sequence and the black lines (solid, x) show the results from each iteration.

In this condition, in order to avoid marginal errors due to the limited number of samples in the window, a safety variable ("*diffValue*"), so the condition holds true if a slight decrease in the magnitude of each consecutive bin, is detected. In these tests, this value is kept at zero which means strictly positive increase of the magnitudes must be observed when moving toward centre frequency. The reason is, in case of guided-waves testing, if the flexural waves are strong enough that the frequency characteristics of the excitation waveform are changed, the signal must not be marked as a defect; since, in effect, the iteration is representing a strong existnace of a flexural wave.

3.1.3. Condition Two

The strongest magnitudes are in the neighbourhood area of the centre frequency; hence, they are less affected by the flexural waves. This condition verifies that the magnitudes from centre frequencies are not below a certain limit. Figure 8 shows an example of an outlier detected from this condition. Two limits must be set for this condition: (1) the number of neighbouring in "*cBins*" and (2) the percentage of allowable drop with regards to the maximum magnitude detected in the iteration ("*maxPerc*"). In these tests, "*cBins*" is set as two (of each side of centre frequency), which means a total of four bins is checked and thus limits this validation to the main central bins of the spectrum. In the reference signal, it can be seen that at the border of this region, an approximate 10% drop is observed. Since in practical testing, the magnitudes will be more affected by the flexural waves and testing condition, the allowable drop is set to 20% of the maximum magnitude of each iteration.

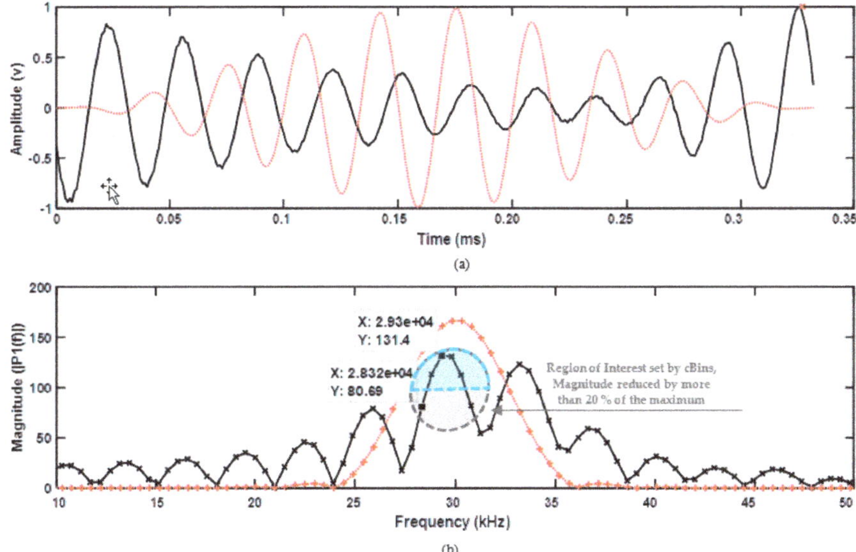

Figure 8. Example of an outlier case detected using Condition Two (C2), where (**a**) shows the time domain of the iteration window and (**b**) is its respective power spectrum. The red lines (dotted, +) show the references achieved from excitation sequence and the black lines (solid, x) show the results from each iteration.

3.1.4. Condition Three

In the third condition, the magnitudes of frequencies outside the 10 dB BW of the excitation sequence are checked. Since the window size is limited to the length of the excitation sequence, no other strong peaks in other frequencies should be detected. This can be verified by confirming that no other magnitudes are above the lowest magnitude within 10 dB BW. Figure 9 shows an example for an outlier detected using this condition. The blue boxes illustrate the expected regions while the grey boxes illustrate the bins with higher magnitudes than the thresholds. It should be noted that this threshold is defined by the reference signal rather than the excitation sequence. Since the boundaries are affected more by flexural waves, a safe margin can be provided by expanding the bins away from the centre frequency ("*sBins*"). In these tests, this value is defined as one, which means only one bin is neglected from each side of the spectrum and all other bins must have magnitudes less than the minimum magnitude detected in the 10 dB BW.

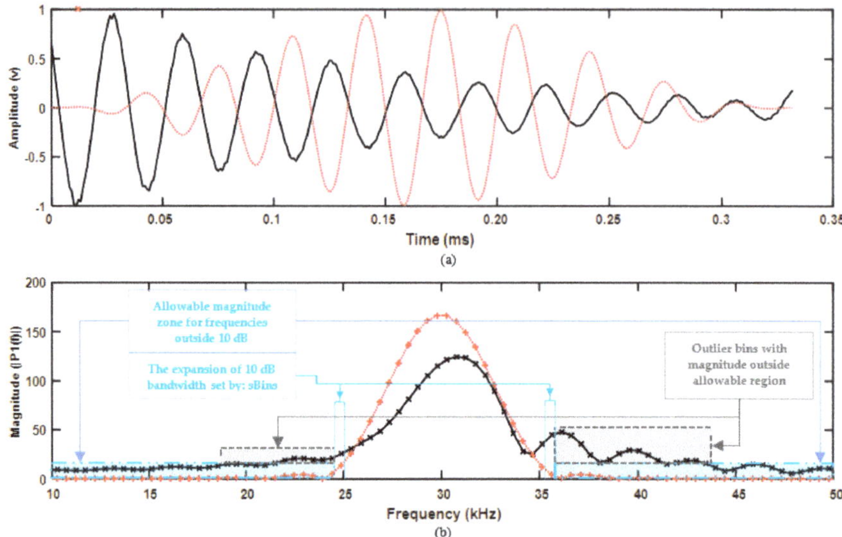

Figure 9. Example of an outlier case detected using Condition Three (C3), where (**a**) shows the time domain of the iteration window and (**b**) is its respective power spectrum. The red lines (dotted, +) show the references achieved from excitation sequence and the black lines (solid, x) show the results from each iteration.

3.1.5. Condition Four

In this condition, it is verified that the magnitude achieved in the power spectrum of iteration is not more than the reference. The reference signal is in the perfect form of the excitation when it is normalised and not affected by any noise, while any the windowed iterations are subject to noise; hence, the maximum magnitude of the iteration cannot be more than the reference. Nonetheless, due to the finite resolution of the windows, a safety value is defined as "*sMargin*" to allow small differences between the two to be neglected.

Keeping in mind that in the case of FEM signal, since source points were not placed linearly and variable transfer functions were used, the characteristics of the generated signals can be different. This is clearly illustrated in Figure 10, where the received has more than 10 cycles; hence, the maximum magnitude in the power spectrum of the iteration would be increased. The shown windowed signal is approximately located around the 4-metre region (second torsional wave), where the results of impact from backward leakage and the front defect are received from the forward direction. As can be seen, the condition will not hold in this case; however, when setting the threshold as 10% of the maximum magnitude (15) the result of the first detection is increased and this wave will also be detected. Changing this value to 15% increases the detection amplitude of the second signal and the pipe end even more, but the first defect signal remains constant. At 20%, the maximum values are achieved (while the overall detection result of the defect becomes less due to the increment of the associated value to the pipe end), and anything above does not change the generated results. Nonetheless, for the purpose of experimental tests, this value is only set as one which is less even than 1% of the maximum magnitude of the reference. This ensures that only calculation errors due to the limitation of the window size are neglected.

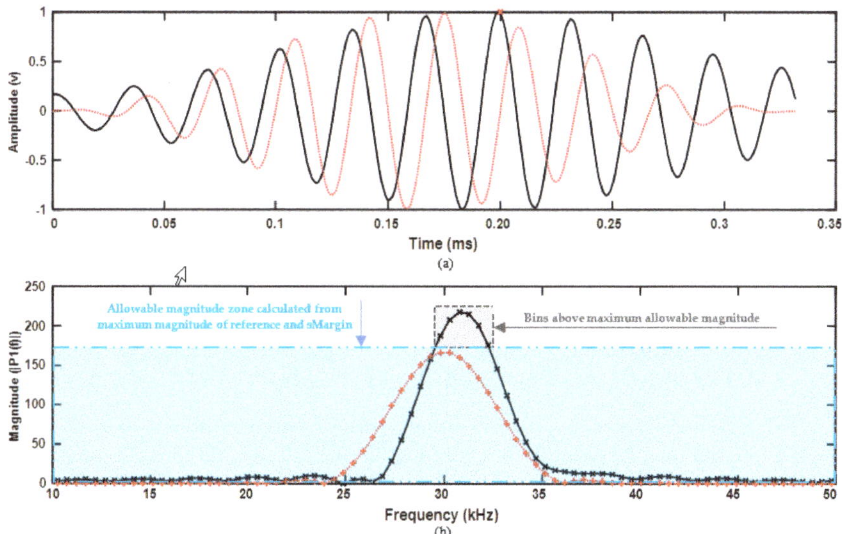

Figure 10. Example of an outlier case detected using Condition Four (C4), where (**a**) shows the time domain of the iteration window and (**b**) is its respective power spectrum. The red lines (dotted, +) show the references achieved from excitation sequence and the black lines (solid, x) show the results from each iteration.

3.1.6. Results Total

The correlation results of the reference and the iteration's power spectrum is added to the location of maximum voltage in each window. Furthermore, as the correlation result is typically a large number, the final result is normalised by the maximum value, which is typically the pipe end. This is illustrated in Figure 11 where it can be seen that the value of 0.1 is associated exactly to the centre of the defect signal. As illustrated, knowing only the excitation waveform, the approximate location of the defect is clearly detectable using this method.

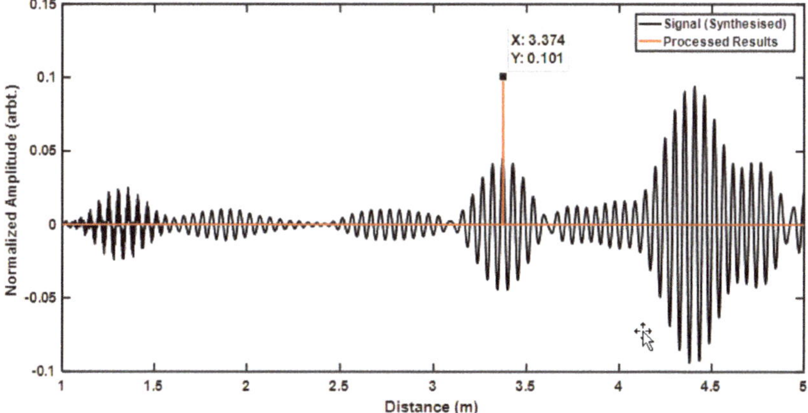

Figure 11. The final result (red line) overlaid on the time-domain signal (black line) from the FEM Case. The defect size is 3% cross-sectional area (CSA) and the excitation frequency is 30 kHz.

The binary detection results of each condition for each iteration, as well as the total results, is demonstrated in Figure 12. In this figure, it is clearly demonstrated outliers exist in the results of each individual condition, but by combining all of them, only the iterations associated with the defect and pipe end are being detected. Furthermore, high correlation results are achieved from all iterations, irrespective of it being noise or defect; which is the main reason why doing a cross-correlation is not enough for detecting the defect location in the case of guided-wave testing. However, even with the existence of the coloured-noise, by checking the mentioned condition, the detection becomes possible.

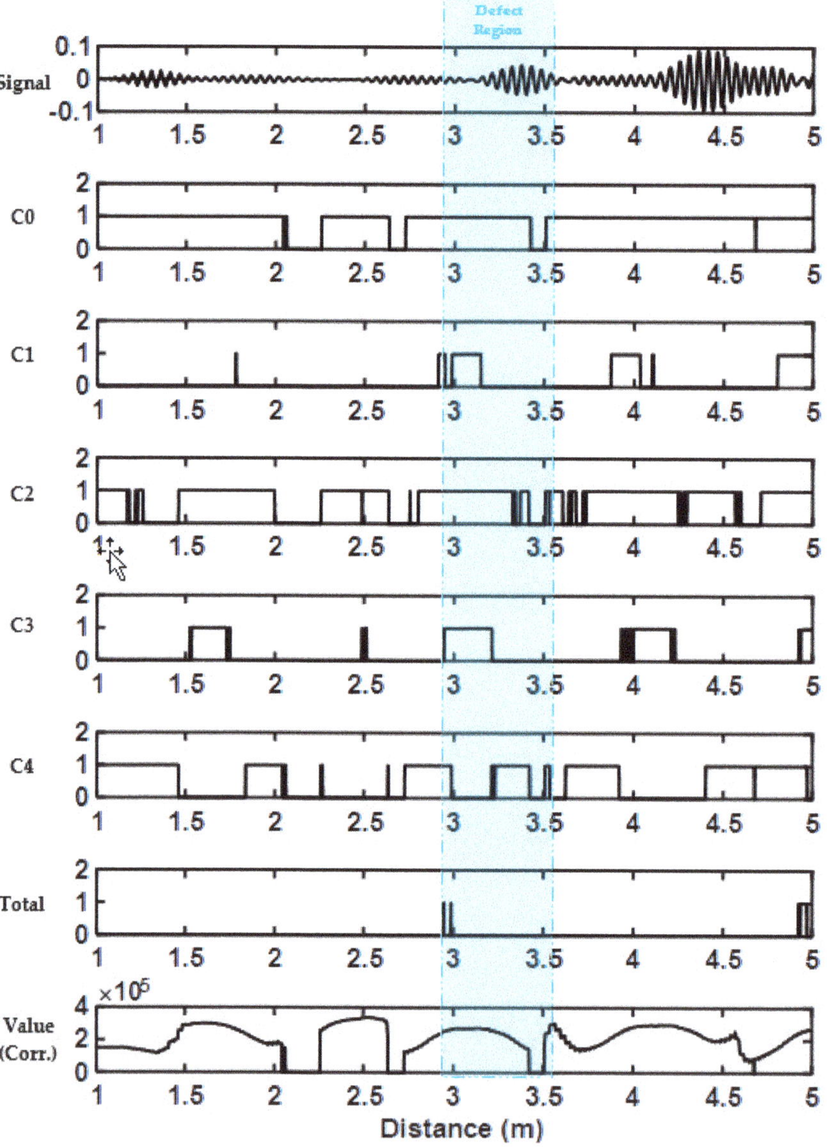

Figure 12. The generated results from the FEM test case.

In the same format, Figures 13 and 14 show the results achieved from the 3% CSA defect from the experimental results. The same aforementioned conclusions can be drawn with regard to Figure 13. However, it can be seen that in the experimental results since the transfer functions are more linear, the outcome of condition 4 (C4) is true in most iterations. The results are shown in Figure 14, where the defect location is clearly marked using this algorithm. In overall, conditions one and three are the most effective ones which filter out most of the outliers from the tests.

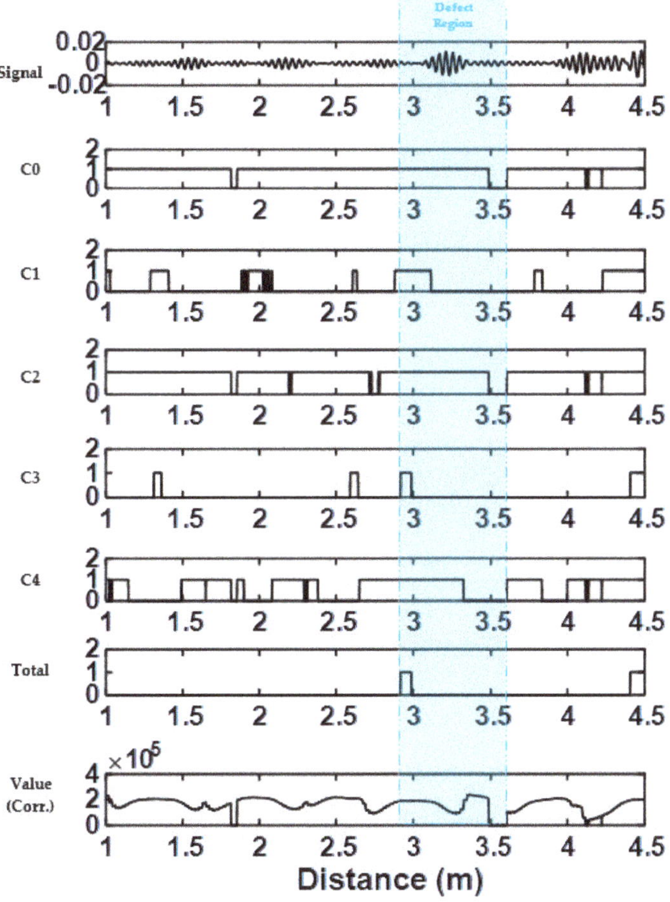

Figure 13. The generated results from the experimental pipe with a defect of 3% CSA and testing frequency of 38 kHz.

These five conditions have not been optimised. Nevertheless, they were all designed based on the physical phenomena of guided-waves propagation within pipes. As can be seen in both the results of FEA and experimental tests, only in C3 and C4 could the defects be detected. However, having more relevant conditions, especially for practical inspection of pipes, may increase the certainty of the generated results. Nonetheless, the conditions are not necessarily required to be in the same Fourier Domain as stated in this algorithm, and the technique can be combined with other approaches, such as Wiener filtering, in order to improve its performance.

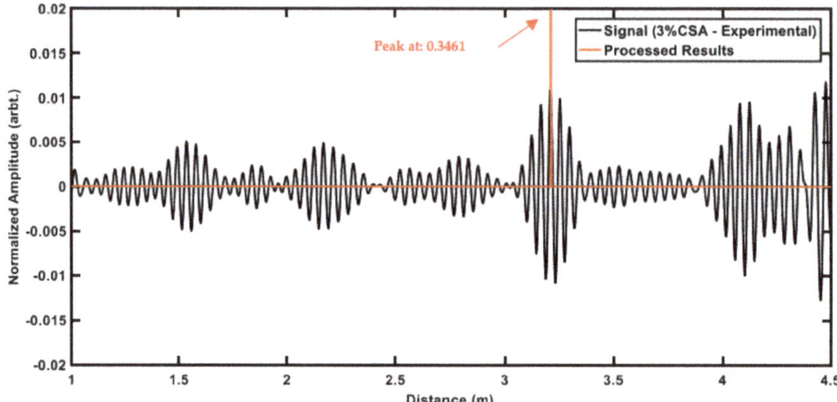

Figure 14. The final result (red line) overlaid on the time-domain signal (black line) from experimental test case with the excitation frequency of 38 kHz and defect size of 3% CSA.

3.2. Experimental Results

A Teletest MK4 device [29], manufactured by Teletest branch of Eddify technology family in Cambridge (UK), was used in order to inspect an eight-inch schedule 40 steel pipe with a total length of six metres. The test tool included three rings of 24 linearly spaced thickness-shear (d15) transducers with 30-mm spacing, which are located 1.5 m away from the back end of the pipe. The device is capable of operating in pulse-echo mode with the operating frequency of 20 to 100 kHz, and maximum sampling frequency of 1 MHz. On the reception side, the individual sections received from each transducer is postprocessed in order to reduce the effect of echoes received from backward testing direction and flexural waves [5,17,30]. The schematic of the signal reception routes with this setup is shown in Figure 15.

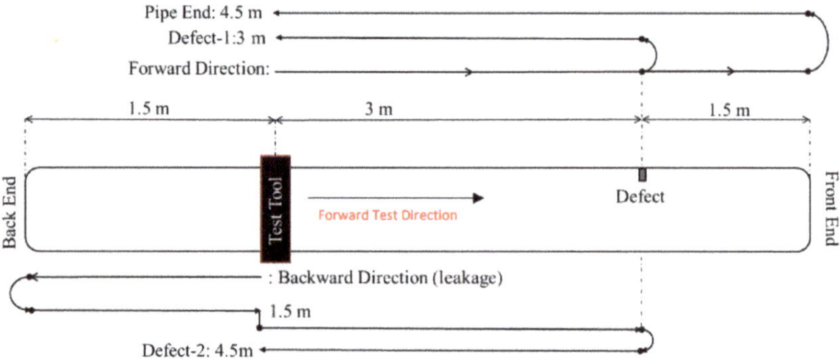

Figure 15. Torsional mode reception route for laboratory trials.

The algorithm was assessed using data gathered from real pipes with three different defect size of 2%, 3% and 4% CSA. The defects were introduced using a saw-cut by thinning the wall thickness of the pipe's circumference. The schematics of the defects and their approximate sizes are shown respectively in Figure 16 and Table 2.

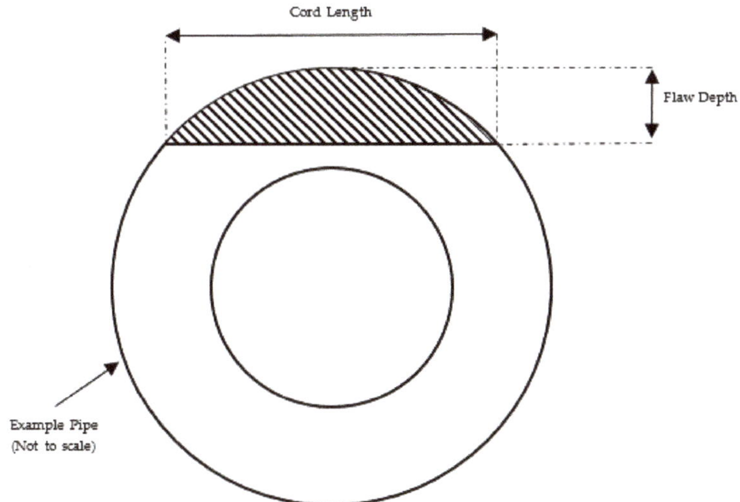

Figure 16. Schematic of the saw-cut defects introduced in the wall of the pipe (not to scale).

Table 2. Defect specifications.

CSA (%)	Flaw Depth (mm)	Cord Length (mm)	Arc Length (mm)
2	3.1	51.75	52.24
3	4.1	59.37	60.12
4	5	65.43	66.48

Depending on the system specification and test specimen, the maximum excitation power achieved by using frequencies in the range of 34 to 42 kHz. As can be seen in Figure 1a, most of the created flexurals from conversion of torsional wave is highly dispersive in the lower frequency region. Since dispersion affects the power spectrum of the received signal, it is expected that this algorithm works better with lower frequencies. However, the final choice of excitation frequency is a trade-off between the capability of the system to excite the required waveform and having highly dispersive flexural waves. Furthermore, it should be noted that the reflected wave amplitude at different frequencies is generally dependent on the flaw geometry, and this could effect the detection capability at different frequencies. In order to illustrate this dependency, 10 different frequencies have been used varying from 30 to 50 kHz with steps of 2 kHz. The achieved SNR from each signal of experimental tests are demonstrated in Figure 17 and Table 3. As can be seen, 40 kHz achieved the best SNR in the tests; with a neglectable difference in the case of 2% CSA defect signal in comparison to other frequency (34 and 42 kHz). Nonetheless, all these defects are either close to the coherent noise level, which are generated by the unwanted flexural waves or are buried within it. Therefore, detecting these defects using normal inspection procedures without calling any outliers are challenging.

Table 3. Achieved SNR from defect signal in each test case (in dB).

CSA (%)	Frequency (kHz)										
	30	32	34	36	38	40	42	44	46	48	50
4	7.13	8.84	11.12	12.15	13.13	13.76	13.23	11.73	9.50	6.84	3.50
3	4.29	5.97	8.15	8.62	9.08	9.51	9.00	7.60	5.43	2.82	−0.51
2	1.46	2.67	3.15	2.29	2.00	2.74	3.09	2.67	0.89	−1.43	−4.44

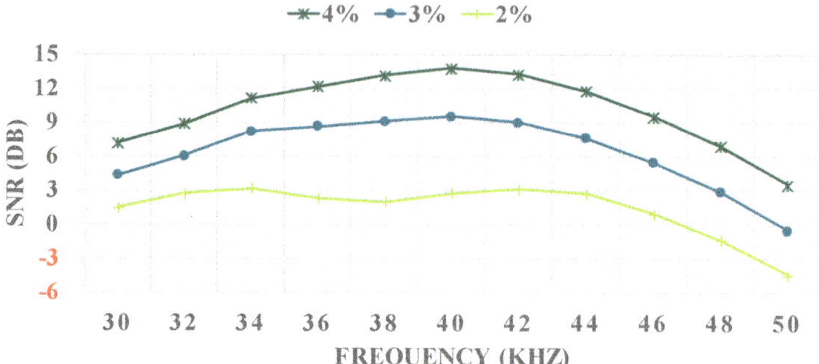

Figure 17. Signal-to-noise ratio (SNR) of the defect signal in each experimental test case.

Figure 18 shows the processed results from each of these test cases where (a) shows the amplitude of detection and (b) shows the amplitude of outlier in the test. Outliers in these tests are considered as any detection corresponding to regions where the signal from neither defect or pipe features is expected to be received. In this figure, each line represents a defect with different CSA size and the value zero represents no detection. Furthermore, it can be seen that the algorithm works best in frequency range of 34–42 kHz. With these frequencies, defects with sizes above 3% can be detected with minor or no outliers existing in the results. In lower frequencies, although the amplitude of outliers is smaller than most test cases, defects with sizes smaller than 4% CSA cannot be detected. On the other hand, frequencies above 42 kHz generally result in the detection of outliers with higher amplitudes, where when using 48 and 50 kHz, no defect can be detected. The main reason is the transfer function of the system, where it is not capable of generating the expected power spectrum for the excitation waveform. As an example, consider the received signal from 50 kHz test case on 4% CSA pipe, is shown in Figure 19. The pipe end signal is a pure reflection of the excitation sequence where limited changes should be observed due to the coherent noise. The retrieved signal from the pipe end observes a 5 kHz shift in centre frequency. Furthermore, the generated sequence can also have a greater number of cycles which will result in higher maximum energy in an iteration's power spectrum in comparison to the reference. Therefore, because of these significant changes which are caused by generation of the sequences, it is not recommended to use higher than 42 kHz excitation frequencies with this algorithm.

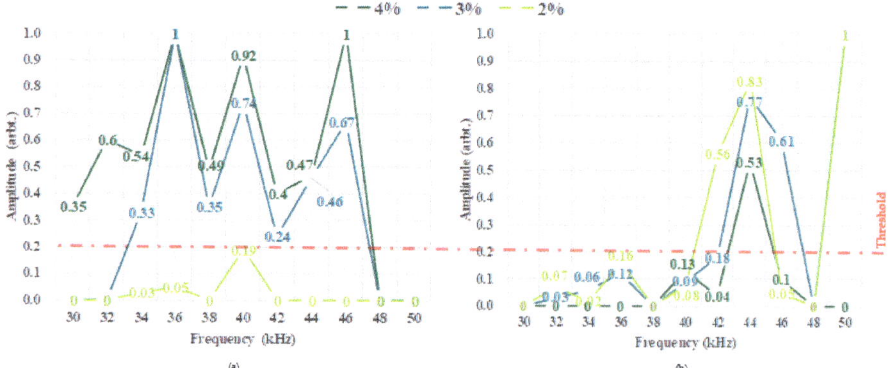

Figure 18. Results achieved using the algorithm where (a) shows the detection amplitude of defect signal and (b) shows the detection amplitude of the outlier. Each line represents a defect with different CSA size. The red dotted line represents the amplitude threshold for filtering the outliers.

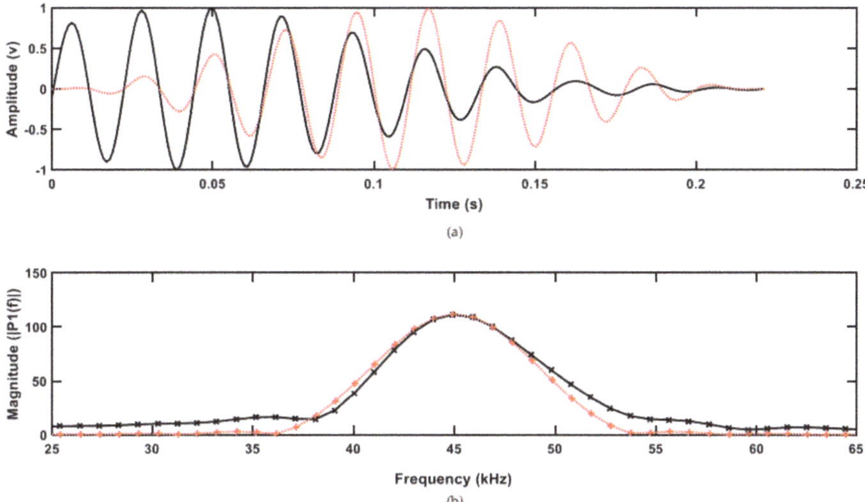

Figure 19. (**a**) The signal received from pipe end using 50 kHz and (**b**) its corresponding power spectrum. The red lines (dotted, +) show the references achieved from excitation sequence and the black lines (solid, x) show the results from each iteration.

The only frequencies capable of detecting 2% CSA defect are 34, 36 and 40 kHz. With regards to Figure 20, which demonstrates the defect to outlier detection ratio, the detection amplitude achieved from defect from 34 and 40 kHz is higher by a factor of 1.5 and 2.38 than the outliers', while, in the case of 36 kHz, this factor is only 0.36 which means the amplitude for defect detection is smaller. Nonetheless, for defects with higher CSA size, the detected defect to outlier amplitude ratio from tested frequencies in the range of 34 to 42 kHz is almost always higher than a factor of 2. This suggests that by setting an amplitude threshold, defects above 3% CSA size can be detected without any outliers. This threshold is shown in Figure 18 by the red dotted line which in this case is set as 0.2. It should also be borne in mind that doing so, would remove the detection of 2% CSA defect, especially in the case of 40 kHz since even though the values are smaller than this threshold, but the amplitude of defect detection (0.19) is approximately twice than those of outlier (0.08). The only two frequencies which do not detect any outliers in any of the tests are 30 and 38 kHz. Nonetheless, using 38 kHz, 2% CSA defect and using 30 KHz, both 2% and 3% CSA defects are not detected.

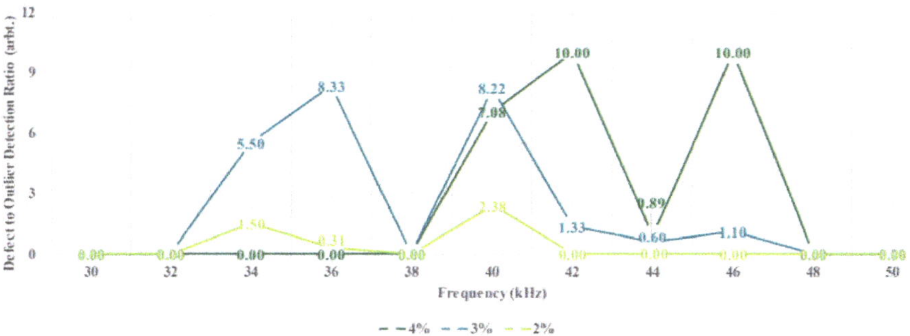

Figure 20. The ratio of detection amplitude of defect to outliers. In cases where the defect is not detected, the ratio is set as zero.

4. Conclusions

In this paper, a novel method is proposed to detect the location of defects using the known excitation power spectra of the torsional waves in guided-wave inspection of pipelines. For doing so, as opposed to traditional inspection using time-domain signal, the spectral domain of the signal is compared with the characteristics of the known excitation sequence. The method works by applying a moving window to the received signal from the inspection, where in each iteration the signal is normalised and its corresponding power spectrum is generated. Since both defect signal and noise are within the same frequency band, by only measuring the correlation between reference and iterations power spectrum, the defect signal will not be distinguishable. Therefore, before measuring the correlation, a total of five conditions were checked to verify that a similar power spectrum to the reference was achieved:

- Centre frequency shift must be small.
- Considering the 10 dB bandwidth of the iteration's power spectrum, the magnitude of each respective frequency bin must be increased when moving toward the centre frequency.
- The achieved magnitude from the neighbouring bins to the centre frequency must be closely related.
- Frequencies outside the 10 dB bandwidth must have less magnitude than the minimum one achieved from within the 10 dB bandwidth.
- The maximum magnitude achieved must be less than the one from the reference.

For each condition, a safety margin is set by the user to neglect minor differences due to the digitisation of the signals and the limitation of the available sample in each window. Nonetheless, before applying the algorithm on guided-wave data from pipes, these limits can be fixed as they are set just to provide a tolerance. On the other hand, this method can also be tested in traditional ultrasonic testing, where there might be a need of tweaking these parameters.

The algorithm was initially developed and validated on synthesised data from FEM. Afterwards, experimental data form real pipes with defect CSA sizes of 2, 3 and 4%, were used to validate its capability and also its limitations. It was illustrated that the choice of excitation frequency is a trade-off between the ability of the system to generate the given excitation sequence and having highly dispersive flexural waves (which are considered as the main source of the noise in these tests). As such, it was observed that frequencies in the range of 34 to 40 kHz are optimal, and can detect defects reliably with larger CSA sizes than 3% with no or very few outliers. In the case of 2% CSA defect, only frequencies of 34 and 40 kHz provided robust detection, where the ratio of the defect to outlier amplitudes is higher by a factor of 1.5. On the other hand, 38 kHz is the only frequency within this range where no outliers were detected in any of the cases. With regards to the test results, due to the significant changes in the excitation sequences, it is not recommended to use this algorithm with frequencies higher than 42 kHz. Since this method only requires a time-domain signal and the characteristics of the excitation sequence, it can easily be combined with other methods to further enhance the reliability of defect detection in guided-wave testing of pipelines.

Author Contributions: Conceptualisation, H.N.M.; Data Curation, H.N.M.; Formal Analysis, H.N.M.; Funding Acquisition, A.K.N.; Investigation, H.N.M.; Methodology, H.N.M.; Project Administration, A.K.N. and H.N.M.; Resources, H.N.M.; Software, H.N.M.; Supervision, K.Y. and A.K.N.; Validation, H.N.M.; Visualisation, H.N.M., K.Y. and A.K.N.; Writing—Original Draft, H.N.M.; Writing—Review & Editing, H.N.M., K.Y. and A.K.N.

Funding: This research study was made possible by the sponsorship and support of Brunel University London and the National Structural Integrity Research Centre (NSIRC).

Conflicts of Interest: The authors declare no conflicts of interest.

References

1. Lowe, M.J.S.; Alleyne, D.N.; Cawley, P. Defect detection in pipes using guided waves. *Ultrasonics* **1998**, *36*, 147–154. [CrossRef]

2.   Ostachowicz, W.; Kudela, P.; Krawczuk, M.; Zak, A. *Guided Waves in Structures for SHM: The Time-Domain Spectral Element Method*; Wiley: Hoboken, NJ, USA, 2012; ISBN 0470979836.

3.   ASTM Standard E2775-16. *Standard Practice for Guided Wave Testing of Above Ground Steel Pipeword Using Piezoelectric Effect Transducer*; ASTM: West Conshohocken, PA, USA, 2017.

4.   Nakhli Mahal, H.; Mudge, P.; Nandi, A.K. Comparison of coded excitations in the presence of variable transducer transfer functions in ultrasonic guided wave testing of pipelines. In Proceedings of the 9th European Workshop on Structural Health Monitoring, Manchester, UK, 10–13 July 2018.

5.   Nakhli Mahal, H.; Yang, K.; Nandi, A. Detection of Defects Using Spatial Variances of Guided-Wave Modes in Testing of Pipes. *Appl. Sci.* **2018**, *8*, 2378. [CrossRef]

6.   Wilcox, P.D. A rapid signal processing technique to remove the effect of dispersion from guided wave signals. *IEEE Trans. Ultrason. Ferroelectr. Freq. Control* **2003**, *50*, 419–427. [CrossRef] [PubMed]

7.   Zeng, L.; Lin, J.; Lei, Y.; Xie, H. Waveform design for high-resolution damage detection using lamb waves. *IEEE Trans. Ultrason. Ferroelectr. Freq. Control* **2013**, *60*, 1025–1029. [CrossRef] [PubMed]

8.   Garcia-Rodriguez, M. Lamb Wave generation with an air-coupled piezoelectric array using square chirp excitation. In Proceedings of the International Congress on Acoustics, Madrid, Spain, 2–7 September 2007; pp. 2–7.

9.   Garcia-Rodriguez, M.; Yaez, Y.; Garcia-Hernandez, M.J.; Salazar, J.; Turo, A.; Chavez, J.A. Application of Golay codes to improve the dynamic range in ultrasonic Lamb waves air-coupled systems. *NDT E Int.* **2010**, *43*, 677–686. [CrossRef]

10.   Yücel, M.K.; Fateri, S.; Legg, M.; Wilkinson, A.; Kappatos, V.; Selcuk, C.; Gan, T.H. Pulse-compression based iterative time-of-flight extraction of dispersed Ultrasonic Guided Waves. In Proceedings of the 2015 IEEE 13th International Conference on Industrial Informatics (INDIN), Cambridge, UK, 22–24 July 2015; pp. 809–815.

11.   Yücel, M.K.; Fateri, S.; Legg, M.; Wilkinson, A.; Kappatos, V.; Selcuk, C.; Gan, T.H. Coded Waveform Excitation for High-Resolution Ultrasonic Guided Wave Response. *IEEE Trans. Ind. Inform.* **2016**, *12*, 257–266. [CrossRef]

12.   Malo, S.; Fateri, S.; Livadas, M.; Mares, C.; Gan, T. Wave Mode Discrimination of Coded Ultrasonic Guided Waves using Two-Dimensional Compressed Pulse Analysis. *IEEE Trans. Ultrason. Ferroelectr. Freq. Control* **2017**, *64*, 1092–1101. [CrossRef] [PubMed]

13.   Pedram, S.K.; Fateri, S.; Gan, L.; Haig, A.; Thornicroft, K. Split-spectrum processing technique for SNR enhancement of ultrasonic guided wave. *Ultrasonics* **2018**, *83*, 48–59. [CrossRef] [PubMed]

14.   Pedram, S.K.; Haig, A.; Lowe, P.S.; Thornicroft, K.; Gan, L.; Mudge, P. Split-spectrum signal processing for reduction of the effect of dispersive wave modes in long-range ultrasonic testing. *Phys. Procedia* **2015**, *70*, 388–392. [CrossRef]

15.   Pedram, S.K.; Mudge, P.; Gan, T.-H. Enhancement of ultrasonic guided wave signals using a split-spectrum processing method. *Appl. Sci.* **2018**, *8*, 1815. [CrossRef]

16.   Duan, W.; Kanfoud, J.; Deere, M.; Mudge, P.; Gan, T.-H. Spectral subtraction and enhancement for torsional waves propagating in coated pipes. *NDT E Int.* **2018**, *100*, 55–63. [CrossRef]

17.   Lowe, M.J.S.; Cawley, P. *Long Range Guided Wave Inspection Usage—Current Commercial Capabilities and Research Directions*; Imperial College London: London, UK, 2006. Available online: http://www3.imperial.ac.uk/pls/portallive/docs/1/55745699.PDF (accessed on 20 November 2018).

18.   Catton, P. *Long Range Ultrasonic Guided Waves for Pipelines Inspection*; Brunel University: Uxbridge, UK, 2009.

19.   Nurmalia. *Mode Conversion of Torsional Guided Waves for Pipe Inspection: An Electromagnetic Acoustic Transducer Technique*; Osaka University: Osaka, Japan, 2013.

20.   Sanderson, R. A closed form solution method for rapid calculation of guided wave dispersion curves for pipes. *Wave Motion* **2015**, *53*, 40–50. [CrossRef]

21.   Wilcox, P.; Lowe, M.; Cawley, P. The effect of dispersion on long-range inspection using ultrasonic guided waves. *NDT E Int.* **2001**, *34*, 1–9. [CrossRef]

22.   Guan, R.; Lu, Y.; Duan, W.; Wang, X. Guided waves for damage identification in pipeline structures: A review. *Struct. Control Health Monit.* **2017**, 1–17. [CrossRef]

23.   Nakhli Mahal, H.; Mudge, P.; Nandi, A.K. Noise removal using adaptive filtering for ultrasonic guided wave testing of pipelines. In Proceedings of the 57th Annual British Conference on Non-Destructive Testing, Nottingham, UK, 10–12 September 2018; pp. 1–9.

24. Proakis, J.G.; Manolakis, D.G. *Digital Signal Processing*, 4th ed.; Pearson: London, UK, 2007.
25. Nakhli Mahal, H.; Yang, K.; Nandi, A.K. Improved Defect Detection Using Adaptive Leaky NLMS Filter in Guided-Wave Testing of Pipelines. *Appl. Sci.* **2019**, *9*, 294. [CrossRef]
26. Fateri, S.; Lowe, P.S.; Engineer, B.; Boulgouris, N.V. Investigation of ultrasonic guided waves interacting with piezoelectric transducers. *IEEE Sens. J.* **2015**, *15*, 4319–4328. [CrossRef]
27. Gresil, M.; Giurgiutiu, V.; Shen, Y.; Poddar, B. Guidelines for Using the Finite Element Method for Modeling Guided Lamb Wave Propagation in SHM Processes. In Proceedings of the 6th European Workshop on Structural Health Monitoring, Dresden, Germany, 3–6 July 2012; pp. 1–8.
28. Miao, H.; Huan, Q.; Wang, Q.; Li, F. Excitation and reception of single torsional wave T(0,1) mode in pipes using face-shear d24 piezoelectric ring array. *Smart Mater. Struct.* **2017**, *26*, 1–9. [CrossRef]
29. Teletestndt. Long Range Guided Wave Testing with Teletest Focus+. 2018. Available online: https://www.teletestndt.com/ (accessed on 14 January 2019).
30. Dürager, C.; Boller, C.; Cornish, A. Damage feature extraction from measured Lamb wave signals using a model-based approach. In Proceedings of the 8th European Workshop on SHM (EWSHM), Bilbao, Spain, 5–8 July 2016; Volume 4, pp. 5–8.

Article

A Modified Leakage Localization Method Using Multilayer Perceptron Neural Networks in a Pressurized Gas Pipe

Qi Wu [1] and Chang-Myung Lee [1,*]

Department of Mechanical and Automotive Engineering, University of Ulsan, 93 Daehak-ro, Nam-Gu, Ulsan 44610, Korea; wuqidake@gmail.com
* Correspondence: cmlee@ulsan.ac.kr

Received: 15 March 2019; Accepted: 8 May 2019; Published: 13 May 2019

Featured Application: To improve the leakage location accuracy in a pressurized gas pipe, this paper proposes a modified leakage location method using multilayer perceptron neural networks based on the acoustic emission signal.

Abstract: Leak detection and location in a gas distribution network are significant issues. The acoustic emission (AE) technique can be used to locate a pipeline leak. The time delay between two sensor signals can be determined by the cross-correlation function (CCF), which is a measure of the similarity of two signals as a function of the time delay between them. Due to the energy attenuation, dispersion effect and reverberation of the leakage-induced signals in the pipelines, the CCF location method performs poorly. To improve the leakage location accuracy, this paper proposes a modified leakage location method based on the AE signal, and combines the modified generalized cross-correlation location method and the attenuation-based location method using multilayer perceptron neural networks (MLPNN). In addition, the wave speed was estimated more accurately by the peak frequency of the leakage-induced AE signal in combination with the group speed dispersive curve of the fundamental flexural mode. To verify the reliability of the proposed location method, many tests were performed over a range of leak-sensor distances. The location results show that compared to using the CCF location method, the MLPNN locator reduces the average of the relative location errors by 14%, therefore, this proposed method is better than the CCF method for locating a gas pipe leak.

Keywords: acoustic emission; nondestructive testing; leakage location; fault diagnosis

---

1. Introduction

Pipelines are used to transport natural gas, water, oil products and other easy-flowing products, and pipeline leakage is a common phenomenon in the continuous transportation process due to the corrosion of pipe walls [1], third-party interference [2], aging of the pipes [3], etc. Leakage may lead to environmental pollution, energy loss, injuries, and even fatalities. Therefore, leakage location is one of the paramount concerns of pipeline operators and researchers.

Pipeline leakage detection methods include direct and indirect methods [4]. Direct methods detect the leaked product outside the pipeline using fiber optic sensing, moisture measurements, gas tracing, vapor sensing, etc. [5]; however, these methods suffer from low sensitivies and expensive sensors. Indirect methods detect the leakage by measuring the pipeline gas pressure, flow, temperature, etc. using sensors mounted inside the pipeline; however, many pressurized vessels in industry are inaccessible, which necessitates nondestructive testing. Acoustic emission (AE) [6] is a nondestructive testing technique that provides the ability to rapidly locate small-scale leaks [7]; it has thus now become a common leak location method. AE signals can be generated by friction, cavitation, impact, or leakage in a high-pressure pipe. In this study, the AE signals in the pipe wall are detected by AE sensors mounted on the pipe's exterior surface.

Leakage location commonly relies on attenuation-based methods or time-of-flight-based methods [8]. The former methods are based on the reduction in the AE signal strength as the propagation distance increases. The prerequisite for the latter methods is that the wave speed is a known constant; however, the wave speed varies as a function of frequency due to the group speed dispersion nature of gas-leakage-induced guided waves [9]. To determine the speed of the AE signal, Li et al. [10] derived the wavenumber formulae to predict the wave speeds, and thereby halved the location error compared to that using the theoretical wave speed. Davoodi and Mostafapour [11] derived the motion equation of the pipe for the simply supported boundary condition by the standard form of Donnell's nonlinear cylindrical shell theory. Nishino et al. [12] generated wide-band cylindrical waves in aluminum pipes by the laser-ultrasonic method; the experimental results showed good agreement between the theoretical and experimental group speed dispersion curves. In this study, the group speed of the leakage-induced signal was estimated by its peak frequency in combination with the group speed dispersive curve of the fundamental flexural mode.

When leakage-induced signals transmit along the pipe wall, signal distortion arises due to energy attenuation, dispersion effect [13] and reverberation. The cross-correlation function (CCF) can determine the time delay based on the similarity of two signals. However, the degree of signal distortion affects the similarity of two signal waveforms measured by two AE sensors, and this interference degrades the performance of the CCF method. To improve the leakage location accuracy, Davoodi and Mostafapour [14,15] located a leak on a high-pressure gas pipe by wavelet transform, filtering technique, and CCF. Yu et al. [16] located the leakage by dual-tree complex wavelet transform and singular value decomposition. Brennan et al. [17] estimated the time delay between two channels by the time and frequency domain methods, with almost identical test results being obtained. Sun et al. [18] proposed a leakage aperture recognition and location method using the root mean square entropy of local mean deposition and Wigner-Ville time-frequency analysis. Guo et al. [19] proposed an adaptive noise cancellation method using the empirical mode decomposition, the test result showed that the time delay estimation of the leakage-induced signal is more accurate. Yang et al. [20] estimated the characteristics related to the two propagation channels by blind system identification. The travel duration of the leakage source signals from the leak point to either sensor was extracted from the identified propagation channel. This paper proposes a modified generalized cross-correlation (GCC) location function, which can compensate for the weakening effect of the different propagation paths on the leakage-induced signals. After the multilayer perceptron neural networks (MLPNN) is trained by the distances between the impact points and the sensors and the energy ratios of two impulse response signals measured by two sensors, it can obtain the relationship between the signal energy ratios and the vibration excitation location, and thus can locate a pipeline leak. This location method can be used to locate a leak in a pipe made of isotropic materials, such as metallic, plastic, etc., whose elastic modulus, Poisson's ratio and density of which should be constant. In this paper, the impulse response signals are generated by the collision between an impulse hammer and the pipe.

Although the proposed method is applicable for pipes with variable operating pressure, this study only deals with the location of a single leak source when the gas pressure inside the pipe is constant. The rest of this paper is organized as follows. Section 2 introduces the modified GCC location method and wave speed estimation based on modal analysis of gas-leakage-induced guided waves. Section 3 introduces the principle of the leak location method based on the AE signal energy attenuation and the MLPNN. Section 4 presents the experimental setup, data collection and the signal processing for leakage location. Section 5 analyzes the experiment results. Section 6 presents the study conclusions.

2. The Principle of Leak Location Method Based on the Generalized Cross-Correlation Function

2.1. The Modified Generalized Cross-Correlation Location Method

Figure 1 presents a pressurized piping system: l is the distance between the two sensors, and l_1 and l_2 are the distances between the leak point and sensors 1 and 2, respectively.

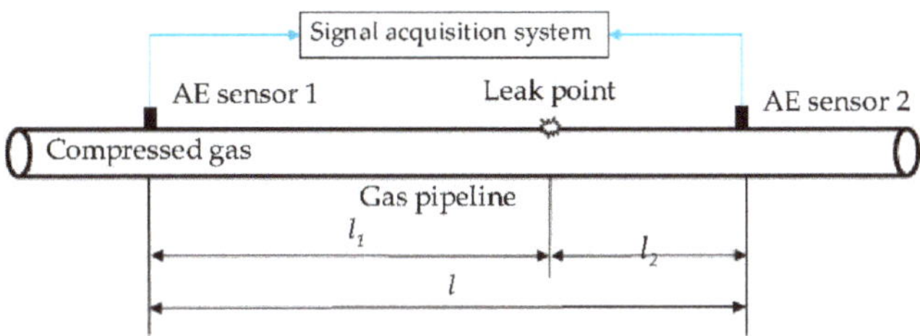

Figure 1. Schematic of the pressurized piping system.

If there is a leak between the two AE sensors, the two measured signals $x_1(t)$ and $x_2(t)$ can be mathematically modeled [21] as

$$x_1(t) = h_1(t) \otimes s(t) + n_1(t), x_2(t) = h_2(t) \otimes s(t+d) + n_2(t), \tag{1}$$

where $s(t)$ is the leakage-induced signal measured at the leak location, the parameter t is the discrete-time counter, and d denotes the time delay between two measured signals. $h_1(t)$ and $h_2(t)$ are the discrete-time impulse response functions between the leak point and each sensor, respectively, and they provide the relationships between an impulse signal and each sensor signal. $n_1(t)$ and $n_2(t)$ are the background noises picked up by the two sensors, $\otimes$ is the discrete-time convolution operator. $s(t)$, $n_1(t)$ and $n_2(t)$ are assumed as stationary random signals. The noises measured by the two sensors are assumed to be mutually unrelated and unrelated with the leakage-induced signal, so the CCF between two measured signals with noise and without noise are equivalent, and can be expressed as follow:

$$R_{s1s2}(\tau) = R_{x1x2}(\tau), \tag{2}$$

where $R_{s1s2}(\tau)$ and $R_{x1x2}(\tau)$ are the CCF without noise and with noise, respectively, in terms of the time delay τ. Since the CCF is sensitive to the non-leak vibration noises [21], the time delay is commonly estimated by the GCC function [22], which can be written as:

$$R_{x1x2}(\tau) = F^{-1}\{\Psi_g(\omega)S_{x1x2}(\omega)\}, \tag{3}$$

where $F^{-1}\{\cdot\}$ denotes the inverse discrete Fourier transform, ω denotes the frequency, and $\Psi_g(\omega)$ is the frequency weighting function of the GCC function. $S_{x1x2}(\omega)$ is the cross-spectral density between two measured signals $x_1(t)$ and $x_2(t)$ [23], and can be written as:

$$S_{x1x2}(\omega) = \frac{1}{2\pi} \lim_{T\to\infty} E\left[\frac{X_1^*(\omega)X_2(\omega)}{T}\right] = S_{ss}(\omega)|H_1^*(\omega)H_2(\omega)|e^{i\omega d}, \tag{4}$$

where $E[\cdot]$ denotes the expectation operator, $X_1(\omega)$ and $X_2(\omega)$ are the N-point discrete Fourier transforms (DFT) of $x_1(t)$ and $x_2(t)$, respectively, $H_1(\omega)$ and $H_2(\omega)$ are the DFT of $h_1(t)$ and $h_2(t)$, respectively, $S_{ss}(\omega)$ is the auto-spectral density of $s(t)$, and $(\cdot)^*$ means complex conjugation. Since multiplication in the frequency domain corresponds to convolution in the time domain [24], the GCC function between two measured signals in terms of the time delay τ can be derived by substituting Equation (4) into Equation (3), and expressed as:

$$R_{x1x2}(\tau) = F^{-1}\left\{S_{ss}(\omega)\Psi_g(\omega)\left|H_1^*(\omega)H_2(\omega)\right|e^{i\omega d}\right\} = R_{ll}(\tau) \otimes_g (\tau) \otimes \delta(\tau + d), \tag{5}$$

where $R_{ll}(\tau) = F^{-1}\{S_{ss}(\omega)\}$ is the autocorrelation function of $s(t)$, and $\Psi_g(\tau) = F^{-1}\{\Psi_g(\omega)|H_1{}^*(\omega)H_2(\omega)|\}$, $\delta(\tau + d) = F^{-1}\{e^{i\omega d}\}$. The time delay between two measured signals is equal to the quotient of the offset of the GCC maximum peak and the sampling frequency. $\delta(\tau + d)$ is a function equal to zero everywhere except for $\tau = -d$, and the Dirac Delta function's integral over the entire real line is equal to one. If $\Psi_g(\omega) = 1$, the Dirac delta function $\delta(\tau + d)$ delayed by d will be broadened by $S_{ss}(\omega)$, $H_1(\omega)$ and $H_2(\omega)$ [25]. In fact, $S_{ss}(\omega)$ is often unknown as it depends on the leakage aperture, the gas pressure difference between inside and outside the pipe [10], etc. However, $H_1(\omega)$ and $H_2(\omega)$ can be obtained by the impulse response method [26], and can be calculated by using:

$$H_i(\omega) = \frac{Y_i(\omega)}{X(\omega)}, \ i = 1, 2, \tag{6}$$

where the parameter i indicates the sensor serial number, $Y_i(\omega)$ denotes the DFT of the impulse response signal measured by the corresponding sensor, and $X(\omega)$ is the DFT of the impulse signal measured by the impact hammer. In this study, the impulse signal is assumed as the Dirac delta function. Therefore, to compensate for the weakening effect of the pipes on the AE signals, the modified frequency weighting function of the GCC function is written as:

$$\Psi_g(\omega) = \frac{1}{\left|H_1^*(\omega)H_2(\omega)\right|}. \tag{7}$$

The modified GCC function $R_{x1x2^g}(\tau)$ can be derived by substituting Equations (4) and (7) into Equation (3), and expressed as:

$$R_{x1x2}^g(\tau) = F^{-1}\left\{S_{ss}(\omega)e^{i\omega d}\right\} = R_{ll}(\tau) \otimes \delta(\tau + d). \tag{8}$$

Compared with Equation (5), pre-filter $\Psi_g(\omega)$ removes the effects of the propagation paths on the AE signal, $\delta(\tau + d)$ is only broadened by $S_{ss}(\omega)$. Therefore, the modified frequency weighting function can increase the degree of correlation between two measured signals and improve the accuracy of the time delay estimation. The modified frequency weighting function differs from the PHAT processor function [25] in that the frequency weighting function involves pre-whitening of the leakage-induced signal. To reduce the effects of noise on the time delay estimation, only the leakage-induced signal with greater power spectral density should be extracted and processed by the proposed pre-filter.

Figure 2 shows a schematic of the modified GCC location method, which contains N-point DFT, complex conjugate, cross power spectrum, pre-filter and N-point inverse discrete Fourier transform (IDFT).

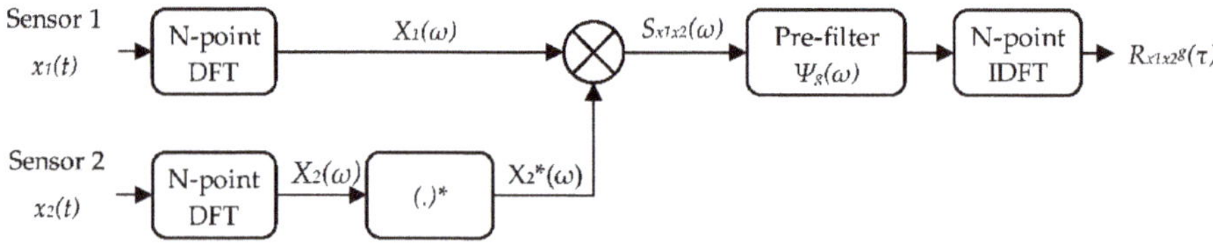

Figure 2. Schematic of the implementation of the modified GCC location method.

The leakage can be located by:

$$l_{est} = \frac{l + v(\omega_0)d_{est}}{2},$$

(9)

where l is the distance between the two sensors, l_{est} is the estimated distance between the leak and one sensor, and d_{est} is the time delay estimation. The frequency-dependent speed $v(\omega_0)$ [9] of AE signals along the pipe wall is discussed in Section 2.2. In this study, the relative location error with respect to the distance between the leak and one sensor is determined as:

$$\delta l = \frac{|l_1 - l_{est}|}{l_1} \times 100\%,$$

(10)

The cross-correlation coefficient (CCC) [27] in terms of the time delay τ can be used to evaluate the correlation of two measured signals and it is given by:

$$\rho(\tau) = \frac{R_{x1x2}(\tau)}{\sqrt{R_{x1x1}(0)R_{x2x2}(0)}},$$

(11)

where $R_{x1x1}(0)$ and $R_{x2x2}(0)$ are the amplitudes in the autocorrelation functions of $x_1(t)$ and $x_2(t)$, respectively, for $\tau = 0$. The larger the maximum peak in CCC is, the higher the degree of the correlation between the two signals is [28].

2.2. Wave Speed Estimation Based on Modal Analysis of Gas-Leakage-Induced Guided Waves

Since the gas solid coupling between the in-pipe gas and the pipe wall can be ignored due to the acoustic impedance mismatch between them, the gas-filled pipes can be approximated as hollow cylinders [9]. The dispersive behaviors of the guided waves in a gas pipe can be analyzed by the guided wave theory of the hollow cylinders [29]. The guided waves in a pipe are grouped into three families: longitudinal mode, torsional mode, and flexural mode. These modes are denoted by $L(0,m)$, $T(0,m)$ and $F(n,m)$, respectively, where n and m represent the circumferential and radial modal parameters, respectively. The theoretical analysis shows that the radial vibration signal collected by AE sensors is dominated by the dispersive fundamental flexural mode $F(1,1)$ [30].

If the material of the pipe wall is changed, the group speed dispersive curve needs to be calculated again based on some parameters before determining the group speed of the leak-induced signal, those parameters are external diameter, thickness, elastic modulus, Poisson's ratio and density of the pipe wall. The material and geometric parameters of the given pipe are shown in Table 1. According to the parameters of the given pipe, the group speed dispersive curves of the flexural modes shown in Figure 3 can be obtained using the guided wave theory of hollow cylinders in MATLAB R2018b. The speed of the AE signal can be online estimated by the peak frequency of the AE signal in combination with the known group speed dispersive curve of the fundamental flexural mode [9].

Table 1. Material and geometric parameters of the given pipe.

a (mm)	b (mm)	μ	ρ (kg/m^3)	G (GPa)	E (GPa)
13.9	16.7	0.35	8000	76	205

a is the inner radius of the pipe, b is the outer radius of the pipe, μ is the Poisson's ratio of the pipe wall material, ρ is the density of the pipe wall material, G is the shear modulus of the pipe wall material, E is the Young's modulus of the pipe wall material.

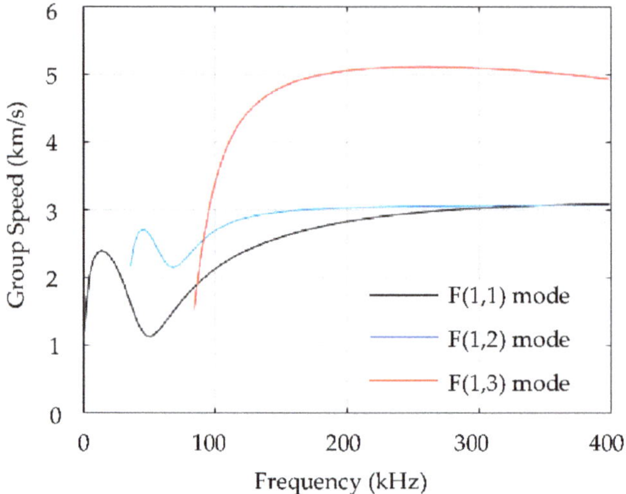

Figure 3. Group speed dispersive curves of the flexural modes in the frequency band 0–400 kHz for the given gas pipe.

3. The Principle of the Leak Location Method Based on the AE Signal Energy Attenuation

3.1. AE Signal Energy Attenuation

The energy of the AE signal reduces as the propagation distance increases due to geometric spreading, structural scattering, and energy absorption [14]. The weakening effect of the pipe on AE signals is similar with the low-pass filter [10], in that the cut-off frequency decreases as the propagation distance increases. When the AE signals propagate in one direction along the pipe wall, the energy attenuation of the AE signal can be described by an exponential decay [31]:

$$E = E_0 e^{-kl_i},$$ (12)

where E is the energy of the sensor signal, E_0 is the AE signal energy at the leak point and k is the attenuation coefficient. The AE signal energy is proportional to the integral of the square of the signal amplitude [32], and can be written as:

$$E = \int_0^{\Delta T} v^2 dt,$$ (13)

where ΔT represents the length of sample time in seconds, and v represents the instantaneous signal amplitude in voltages. If the distance between the vibration source and the sensor is constant, the weakening effect of the pipe on the AE signal will be also constant, irrespective of any leakage-induced signals or impulse response signals. The closer the sensor is to the leak source, the greater the energy of the measured signals. Then the energy ratios of two measured signals will gradually increase or decrease as the distance between the vibration source and one of the two sensors increases. Therefore, a pipeline leakage can be located by the signal energy ratios without knowing the wave speed or the time delay estimation.

3.2. Multilayer Perceptron Neural Networks Classifier

The mathematical model of the biological neural network is defined as the artificial neural network [33]. The relationships between the input and output vectors can be obtained after the artificial neural networks are trained. Neural network can be applied to solve nonlinear problems, MLPNN is a commonly used model. Figure 4 shows the perceptron neural networks with three layers. The input and output layers accept the external input vectors in the learning mode. Every neuron in the input and hidden layers connects to every neuron in the hidden and output layers, respectively. The input layer L_1 is comprised of P variables $x = [x_1, \cdots, x_P]^T$, which are processed by the following equation:

$$a_j = \sum_{i=1}^{P} \omega_{ij}^{(1)} x_i + \omega_0^{(1)}, \ j = 1, \cdots, M, \tag{14}$$

where M is the number of neurons in the hidden layer, and the superscript indicates that the parameter $\omega_{ij}^{(1)}$ is the connection weight between the neuron x_i in the input layer L_1 and the neuron z_j in the hidden layer L_2. The parameter $\omega_0^{(1)}$ represents the bias of the variable a_j, which prevents a_j from becoming zero. Then a_j is transformed by an activation function, which is the logistic sigmoid function in this study, and it is given by:

$$z_j = f(a_j) = \frac{1}{1 + e^{-a_j}}. \tag{15}$$

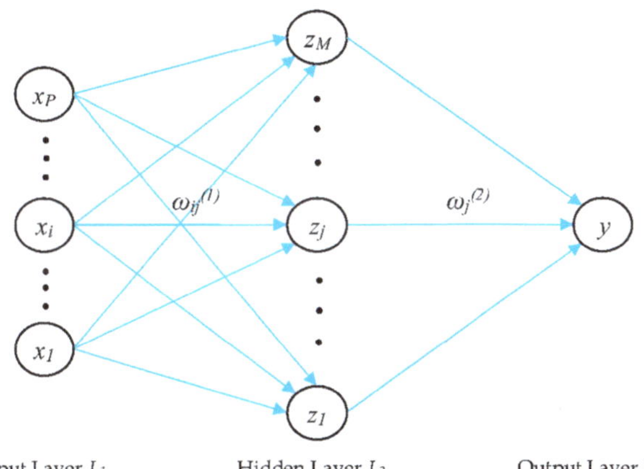

Figure 4. The MLPNN with input, hidden, and output layers.

Similarly, the neuron y in the output layer L_3 receives the variables z_j as shown in the following equation:

$$y = \sum_{j=1}^{M} \omega_j^{(2)} z_j + \omega_0^{(2)}, \tag{16}$$

where $\omega_j^{(2)}$ and $\omega_0^{(2)}$ are the connection weight and the bias in the hidden layer, respectively. For the regression problem, the aim of the training process is to minimize the squared error function as shown in the following equation by the gradient descent method:

$$E(\omega) = \frac{|y(\omega) - d|^2}{2}, \tag{17}$$

where d is the distance between the vibration source and one sensor, ω is the set of all weights and biases, and $y(\omega)$ represents the neuron in the output layer. All the weights and biases need to be updated by training the MLPNN on labeled data until the error function becomes small enough or the maximum iteration is reached; then the MLPNN can be used as a leakage locator.

4. Laboratory Experiments

4.1. Experimental Setup and Data Collection

The experimental setup consists of two parts: a pressurized piping system and a signal acquisition system, as shown in Figure 5. If the pipe gas pressure is too low, the energy of the leak-induced signal will be small and difficult to detect. Therefore, in the pressurized piping system, an air compressor (NH-5, Hanshin product) supplies 7 bar gas pressure to a 15m-long steel pipe (ASTM A 312 TP304) with a pipe wall thickness and external diameter of 2.8 mm and 33.4 mm, respectively. One end of the pipe was blocked, and the other end was connected to the air compressor via a hose. To avoid the air compressor vibration noise affecting the experimental results, the air compressor was placed far away from the piping system. The pipe was supported by a sponge material to reduce energy dissipation of the AE signal on the pipe wall to the ground. The diameter of the leakage orifice in the early stage of a pipeline leakage is usually small. To verify that the proposed method can locate a leak in this stage, a simulated leak orifice with a diameter of 0.3 mm was punched in the pipe wall. The simulated leak source, the AE sensor mounted on the pipe wall, an impulse hammer and the placement of two AE sensors along the pipe are shown in Figures 6–9, respectively. The pipe wall had been cleaned with sandpaper before the AE sensors were mounted with the magnetic hold downs and vacuum grease couplant.

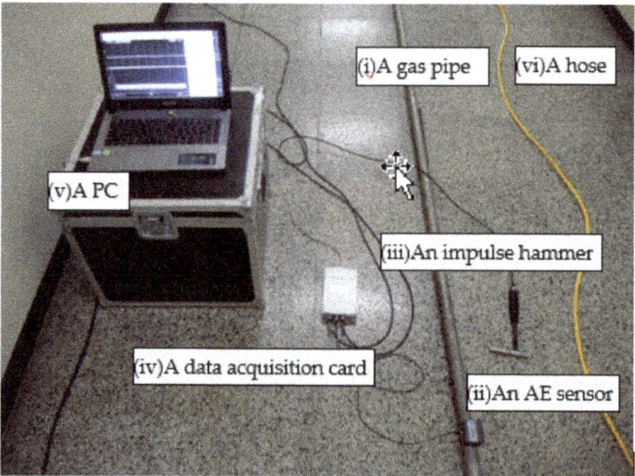

Figure 5. Experiment system diagram: (i) a gas pipe, (ii) two AE sensors, (iii) an impulse hammer, (iv) a data acquisition card, (v) a PC, and (vi) a hose.

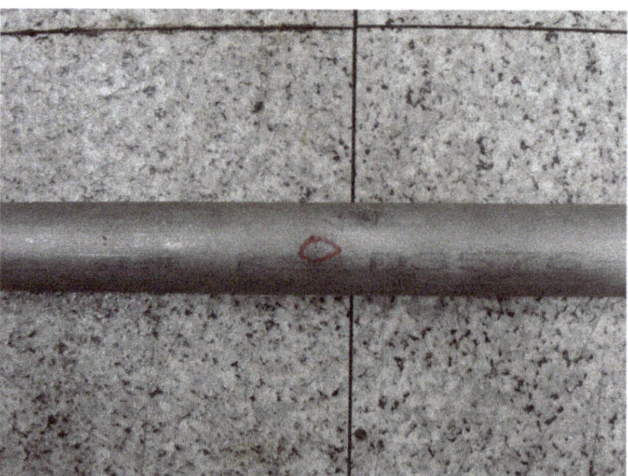

Figure 6. A simulated leak orifice of the pipe.

Figure 7. The AE sensor mounted with the magnetic hold down and vacuum grease couplant.

Figure 8. An impulse hammer.

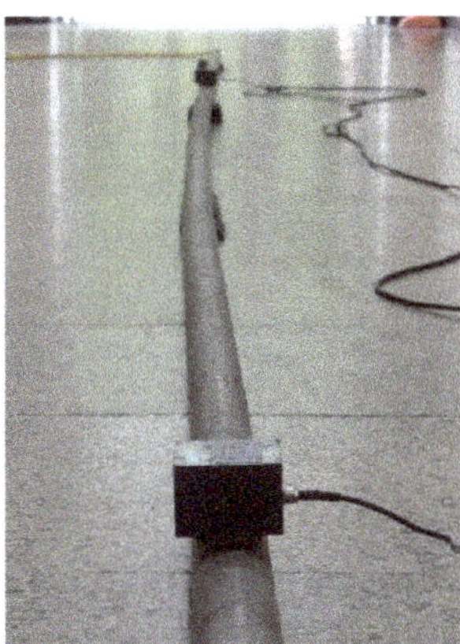

Figure 9. The placement of two AE sensors along the pipe.

The signal acquisition system consists of two R15a sensors with a resonant frequency of 150 kHz, an impulse hammer (086C03), a data acquisition card (USB-5132, NI product) and a personal computer (PC) equipped with NI-SCOPE software. Since the sensitivity of the AE sensors is high in the frequency range between 50 kHz and 400 kHz, data were captured at a sampling rate of 800 kHz in the experiment. Analog to digital conversion was performed by the data acquisition card connected to the PC. Data were recorded into the PC for the signal processing and preprocessed by the NI-SCOPE software, including the preamplification and anti-aliasing filtering.

In the experiment, the training sets for MLPNN include the distances between one sensor and each impact location and the impulse response signals energy ratios between two sensor signals in different frequency bands. The impulse signal was measured by the sensor in the impact hammer, and the impulse response signals were measured by AE sensors when the pipe was tapped by the impact hammer, and the tapping locations were distributed along a straight line between the two AE sensors. To avoid the simulated leak orifice affecting the impulse response signal, the impact test was performed before the simulated leak orifice was punched. The test sets for MLPNN were the energy ratios of leakage-induced signals in different frequency bands. The leakage-induced signals were generated by the friction force between the jet of the leaking gas and the simulated leak orifice.

4.2. Signal Processing for Leakage Location

The leakage location procedure for the MLPNN locator is shown in Figure 10, and the MLPNN locator operates in two different modes:

1. In the learning mode, the training sets of the MLPNN contain the input vectors and the output vectors. The input vectors are the energy ratios of two impulse response signals in corresponding frequency bands. Since the impulse response signals are non-stationary signals, and the wavelet packet analysis is suitable for processing the non-stationary signals, then the impulse response signals were processed by the wavelet packet transform in this study. The output vectors are the distances between one sensor and each impact location. The learning procedure for the MLPNN is as follows:

1. Since the energy of a leakage-induced signal at frequency band from 100 to 150 kHz is higher than at other frequency bands, and three-level decomposition can extract the signal in this frequency range, so all the impulse response signals are decomposed into three layers of wavelet packet coefficients, and the signals are reconstructed in eight frequency bands by wavelet packet coefficients. The Haar wavelet is used in this study.
2. Calculate the energies of the decomposed signals in step 1 by Equation (13); then calculate the energy ratios of two impulse response signals in the corresponding frequency band.
3. Train the MLPNN with the input and output vectors. The input vectors are the energy ratios in step 2, and the output vectors are the distances between one sensor and each impact location, and the epochs in the training method is set to 0.01.

2. In the application mode, the trained MLPNN locator can locate the leakage by using the energy ratios of the leakage-induced signals in the frequency bands with greater power spectral density. The leakage location procedures for the MLPNN locator are described as follows:

1. Measure the leakage-induced signals by using the two AE sensors.
2. Determine the peak frequency of the leakage-induced signals by using the power spectral density.
3. Estimate the group speed of the leakage-induced signals by using the peak frequency in step 2 and the known group speed dispersive curve of the fundamental flexural mode.
4. Calculate the CCC by using the CCF method and the modified GCC method.
5. Compare the maximum peak amplitudes of the two CCCs obtained in step 4. $\rho_{x1x2}{}^c$ and $\rho_{x1x2}{}^g$ are the maximum peak amplitudes of the CCCs obtained by using the CCF method and the modified GCC method, respectively. If $\rho_{x1x2}{}^g$ is greater than $\rho_{x1x2}{}^c$, it implies that compared to the CCC obtained by using CCF method, the CCC obtained by using modified GCC method is more reliable; then perform step 6, otherwise perform step 7.
6. Calculate the distance between the leakage and sensor 1 by Equation (9). The group speed of the leakage-induced signal was determined in step 3, and the time delay was determined by the sampling frequency and the offset of the CCC maximum peak obtained by the modified GCC method in step 4.

7. Decompose the leakage-induced signals into three layers wavelet packet coefficients, reconstruct the signals in eight frequency bands by using the wavelet packet coefficients.
8. Calculate the energy ratios of the decomposed signals in the frequency bands with the greater power spectral density.
9. Locate the leakage by using the energy ratios in step 8 and the trained MLPNN locator.

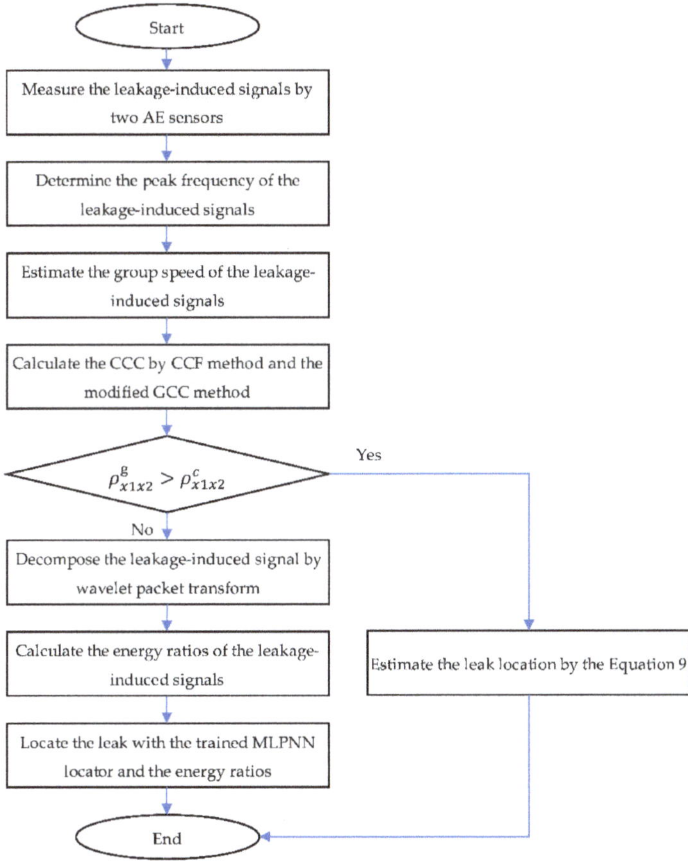

Figure 10. Flowchart of the leakage location method for the MLPNN locator.

5. Results and Discussion

5.1. Characteristics of the Frequency Domain

Figure 11 shows the power spectral densities of the leakage-induced signals collected at the distances of 1 m and 2 m away from the simulated leak source. The center frequency of a leak-induced signal is proportional to the gas jet velocity and inversely proportional to the leakage aperture [10], and is also related to the resonant frequency of the sensor and the propagation distance between the leak and the sensor. As shown in Figure 11, the amplitude of the power spectral density decreases with increasing propagation distance. Since the peak frequency of leakage-induced signals is 124 kHz, the wave speed can be determined to be 2404 m/s by the peak frequency in combination with the known group speed dispersive curve of the fundamental flexural mode.

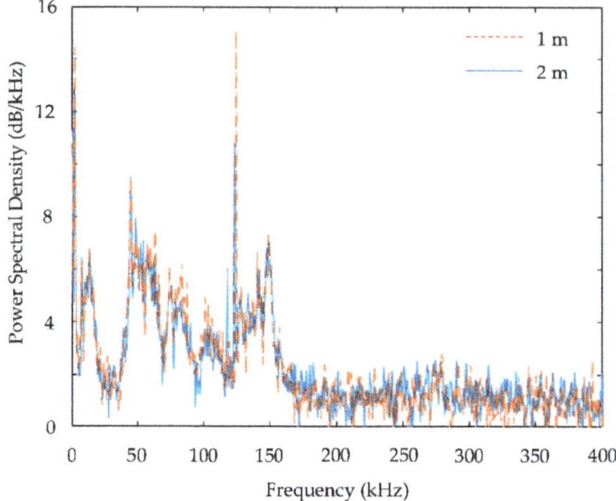

Figure 11. Comparison of the power spectral densities of the leakage-induced signal in the different distances ($L_1 = 1$ m, 2 m) away from the leak source.

5.2. Characteristics of the Signal Energy Ratios

To verify the consistency of weakening effect of the pipe on two kinds of signals, which are the impulse response signals and leakage-induced signals measured by maintaining a distance of 14 m between the two sensors while varying the distance between sensor 1 and the simulated leak orifice. Since the maximum amount of leakage-induced signal energy occurs in frequencies near 124 kHz, the signal-to-noise ratio at this frequency band is higher than at other bands, the impulse response signal in frequencies near 124 kHz was also extracted with wavelet packet transform, and the signal energy ratios between the two sensors were calculated. Figure 12 shows the fitting curves of the signal energy ratios as a function of the propagation distance. The horizontal axis denotes the distance between sensor 1 and the simulated leak orifice. The vertical axis denotes the energy ratios of the leakage-induced signal and the impulse signal measured by the two sensors. The difference between the two fitting curves is small, which implies that the weakening effect of the pipe on the AE signals is proportional to the propagation distance, irrespective of any leakage-induced signals or impulse response signals. Therefore, the signal energy ratio can also be used for locating the vibration excitation source.

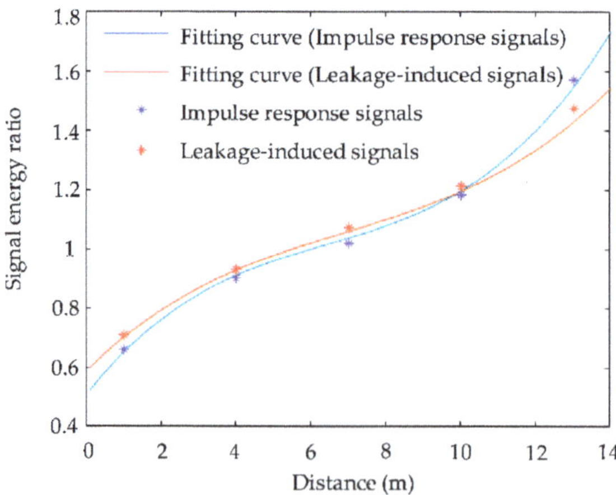

Figure 12. Fitting curves of the signal energy ratios of the AE signals measured by the two sensors.

5.3. Characteristics of the Time Domain.

Figure 13a,b show the impulse response signals collected at distances of 0.4 m and 13 m away from the impact location, respectively. The maximum amplitude of the impulse response signals decreases as the propagation distance increases, which indicates that the energy of the impulse response signal decreases due to energy attenuation. Moreover, the closer the vibration source is to the sensor, the earlier the impulse response signal begins to oscillate.

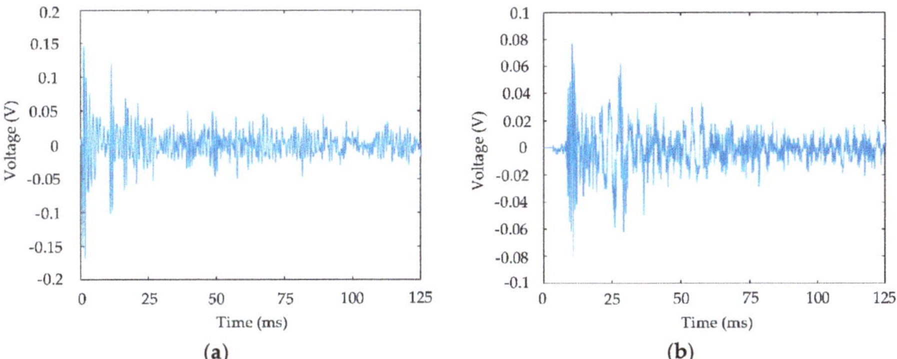

Figure 13. Impulse response signals acquired at different distances from sensor 1: (**a**) 0.4m and (**b**) 13m.

Figure 14a,b show the CCC obtained by the CCF method and the modified GCC method, respectively. The time delay estimation only depends on the shift value of the maximum peak in the cross-correlation coefficient with respect to the central point, and can be calculated as the product of the sampling period and the shift of the cross-correlation function maximum peak [34]. Two maximum peaks can be observed, and the time delay estimations are −0.00020 seconds and −0.00034 seconds, respectively. The distance estimations between the leak and sensor 1 are 1.83 m and 1.99 m, respectively. Since the relative location error in this study is equal to the quotient of the leakage location error and the distance between the leak and sensor 1, the real distance between the simulated leak orifice and sensor 1 is 2.16 m, so the relative location errors are 15% and 8%, respectively. The maximum peak amplitudes of the two CCCs are 0.168 and 0.236, respectively. Hence, the modified pre-filter can improve the accuracy of the time delay estimation and increase the degree of the correlation between the two measured signals.

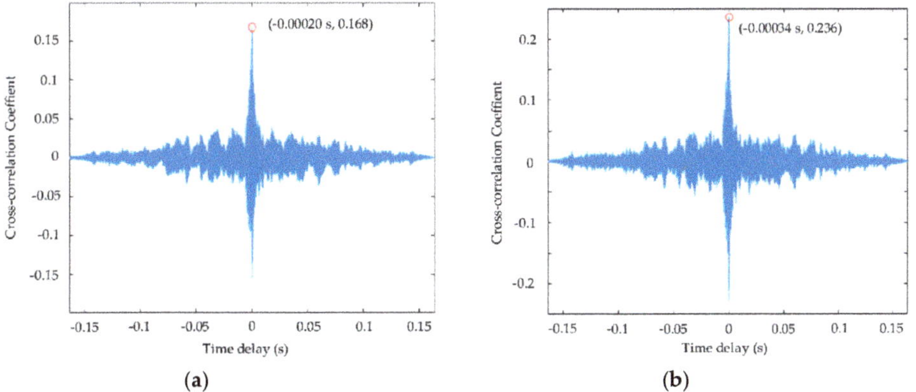

Figure 14. Cross-correlation coefficients (CCCs) obtained by two methods: (**a**) CCF location method, and (**b**) modified GCC location method.

5.4. Leakage Location Analysis

To verify the reliability of the MLPNN locator, the gas leakage was located by the CCF location method and the MLPNN locator under different conditions, which are maintaining a distance of 2.16 m between the simulated leak orifice and sensor 1 and varying the distance (1.08 m, 2.36 m and 3.54 m) between the simulated leak orifice and sensor 1. In addition, the group speed of the leakage-induced signal was determined by the peak frequency of the leakage-induced signal and the group speed dispersive curve of the fundamental flexural mode. Tables 2 and 3 show the location results obtained by the CCF method and the MLPNN locator. The relative location errors in Table 3 show that the relative location errors increase as the distance between the two sensors increases, which implies that some high frequency components of the leakage-induced signal attenuate as the propagation distance increases.

Table 2. The location results obtained by the CCF location method.

L (m)	3.24	3.24	3.24	3.24	4.52	4.52	4.52	4.52	5.70	5.70	5.70	5.70
f (kHz)	124.2	123.8	123.4	125.4	105.5	103.4	107.2	106.7	104.7	104.3	104.7	103.5
v (m/s)	2404	2400	2397	2415	2200	2173	2222	2215	2190	2186	2190	2175
l_1 (m)	2.16	2.16	2.16	2.16	2.16	2.16	2.16	2.16	2.16	2.16	2.16	2.16
$\hat{l}_1$ (m)	1.78	1.71	1.82	1.72	1.70	1.77	1.75	1.79	1.83	1.74	1.79	1.83
Δl_1 (m)	0.38	0.45	0.34	0.44	0.46	0.39	0.41	0.37	0.33	0.42	0.37	0.33
δl_1 (%)	17.6	20.1	15.7	20.4	21.3	18.1	19.0	17.1	15.3	19.4	17.1	15.3

L is the distance between the two AE sensors, f is the peak frequency of the leakage-induced signal, v is the group speed estimation of the leakage-induced signal, l_1 is the real distance between the leak and sensor 1, $\hat{l}_1$ is the distance estimation between the leak and sensor 1, Δl_1 is the leakage location error, δl_1 is the relative location error.

Table 3. The location results obtained by the MLPNN locator.

L (m)	3.24	3.24	3.24	3.24	4.52	4.52	4.52	4.52	5.70	5.70	5.70	5.70
f (kHz)	124.2	123.8	123.4	125.4	105.5	103.4	107.2	106.7	104.7	104.3	104.7	103.5
v (m/s)	2404	2400	2397	2415	2200	2173	2222	2215	2190	2186	2190	2175
l_1 (m)	2.16	2.16	2.16	2.16	2.16	2.16	2.16	2.16	2.16	2.16	2.16	2.16
$\hat{l}_1$ (m)	2.19	2.18	2.20	2.25	2.26	2.09	2.25	2.27	1.99	2.23	2.28	1.98
Δl_1 (m)	0.03	0.02	0.04	0.09	0.10	0.07	0.09	0.11	0.17	0.07	0.12	0.18
δl_1 (%)	1.4	0.9	1.9	4.2	4.6	3.2	4.2	5.1	7.9	3.2	5.6	8.3

Figure 15a,b show the histograms of the relative location errors in Tables 2 and 3, respectively. The vertical axis represents the number of measurements, and the horizontal axis represents the percentage of relative location error. These two tables show the measurement results of twelve groups by using the proposed method and the cross-correlation function, respectively, with twelve measurements in both Figure 15a,b. Despite the small sample size, all the measurement results are typical and can show the location accuracy of the two location methods. The averages of the relative location errors are 18% and 4%, respectively, which indicates that the average of the relative location errors obtained by the MLPNN locator is reduced by 14% compared to that using the CCF location method. Consequently, the MLPNN locator is more suitable for locating a leak in a gas pipe than the CCF location method is. However, the MLPNN locator suffers the drawback of needing to be trained again if one of the following experimental conditions changes: for example, the distance between the two AE sensors, the geometric parameters of the pipe wall, or the material of the pipe wall.

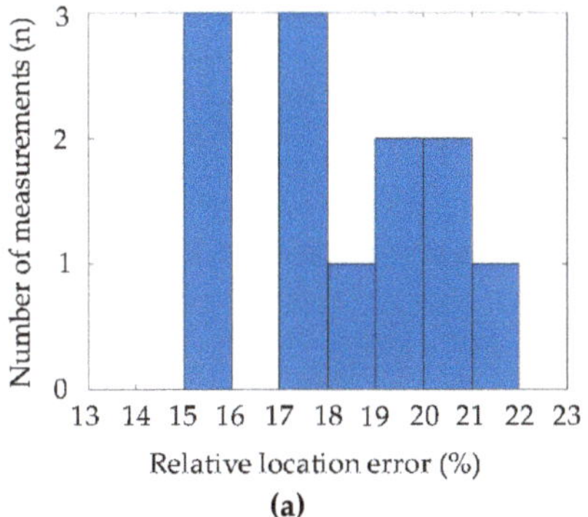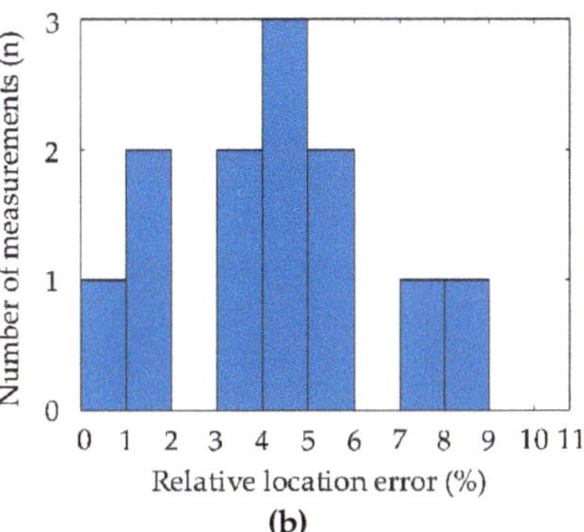

Figure 15. Histograms of the relative location errors obtained by two methods: (**a**) CCF location method and (**b**) MLPNN locator.

6. Conclusions

This paper has proposed a leakage location method using the modified generalized cross-correlation (GCC) location method in combination with the attenuation-based location method using multilayer perceptron neural networks (MLPNN). The GCC location method can compensate for the weakening effect of the different propagation paths on the leakage-induced signals, can increase the degree of the correlation between two measured signals and improve the accuracy of the time delay estimation. Unlike the cross-correlation function (CCF) location method, the MLPNN locator locates a leak not only by the time delay estimation between two measured signals but also by the energy ratios of two leakage-induced signals measured by two sensors in the frequency bands with greater power spectral density. In addition, the wave speed was determined more accurately by using the peak frequency in combination with the group speed dispersive curve of the fundamental flexural mode. The MLPNN locator was experimentally verified by using a pressurized piping system and a signal acquisition system. The average of the relative location errors obtained by the MLPNN locator was reduced by 14% compared to that using the CCF location method. Hence, the MLPNN locator is more suitable for locating a leak in a gas pipe than the CCF location method is.

Data Availability: The data used to support the findings of this study are available from the corresponding author upon request.

Author Contributions: Conceptualization, C.-M.L.; methodology, Q.W. and C.-M.L.; software, Q.W.; validation, Q.W.; formal analysis, Q.W.; investigation, Q.W.; resources, C.-M.L.; data curation, X.X.; writing—original draft preparation, Q.W.; writing—review and editing, Q.W.; visualization, Q.W.; supervision, C.-M.L.; project administration, C.-M.L.; funding acquisition, C.-M.L.

Funding: This research was funded by a 2017 grant from the Russian Science Foundation (Project No. 17-19-01389).

Conflicts of Interest: The authors declare no conflicts of interest.

References

1. Edalati, K.; Rastkhah, N.; Kermani, A.; Seiedi, M.; Movafeghi, A. The use of radiography for thickness measurement and corrosion monitoring in pipes. *Int. J. Press. Vessels Pip.* **2006**, *83*, 736–741. [CrossRef]
2. Liang, W.; Hu, J.; Zhang, L.; Guo, C.; Lin, W. Assessing and classifying risk of pipeline third-party interference based on fault tree and SOM. *Eng. Appl. Artif. Intell.* **2012**, *25*, 594–608. [CrossRef]
3. Vargas-Arista, B.; Hallen, J.M.; Albiter, A. Effect of artificial aging on the microstructure of weldment on API 5L X-52 steel pipe. *Mater. Charact.* **2007**, *58*, 721–729. [CrossRef]
4. Ferraz, I.N.; Garcia, A.C.; Bernardini, F.C. Artificial Neural Networks Ensemble Used for Pipeline Leak Detection Systems. In Proceedings of the 7th International Pipeline Conference IPC 2008, Calgary, AB, Canada, 29 September–3 October 2008; pp. 739–747.

5. Nehorai, A.; Paldi, E.; Porat, B. Detection and Localization of Vapor-Emitting Sources. *IEEE Trans. Signal Process.* **1995**, *43*, 243–253. [CrossRef]

6. Xiao, Q.; Li, J.; Bai, Z.; Sun, J.; Zhou, N.; Zeng, Z.A. Small Leak Detection Method Based on VMD Adaptive De-Noising and Ambiguity Correlation Classification Intended for Natural Gas Pipelines. *Sensors* **2016**, *16*, 2116. [CrossRef] [PubMed]

7. Faerman, V.A.; Cheremnov, A.G.; Avramchuk, V.V.; Luneva, E.E. Prospects of Frequency-Time Correlation Analysis for Detecting Pipeline Leaks by Acoustic Emission Method. *IOP Conf. Ser. Earth Environ. Sci.* **2014**, *21*, 012041. [CrossRef]

8. Rewerts, L.E.; Roberts, R.R.; Clark, M.A. Dispersion compensation in acoustic emission pipeline leak location. In *Review of Progress in Quantitative Nondestructive Evaluation*; Springer: Boston, MA, USA, 1997; Volume 16, pp. 427–434.

9. Li, S.; Wen, Y.; Li, P.; Yang, J.; Dong, X.; Mu, Y. Leak location in gas pipelines using cross-time-frequency spectrum of leakage-induced acoustic vibrations. *J. Sound Vib.* **2014**, *333*, 3889–3903. [CrossRef]

10. Li, S.; Wen, Y.; Li, P.; Yang, J.; Yang, L. Determination of acoustic speed for improving leak detection and location in gas pipelines. *Am. Inst. Phys.* **2014**, *85*, 024901. [CrossRef]

11. Davoodi, S.; Mostafapour, A. Modeling Acoustic Emission Signals Caused by Leakage in Pressurized Gas Pipe. *J. Nondestruct. Eval.* **2013**, *32*, 67–80. [CrossRef]

12. Nishino, H.; Takashina, S.; Uchida, F.; Takemoto, M.; Ono, K. Modal analysis of hollow cylindrical guided waves and applications. *Jpn. J. Appl. Phys.* **2001**, *40*, 364–370. [CrossRef]

13. Grabec, I. Application of correlation techniques for localization of acoustic emission sources. *Ultrasonics* **1978**, *16*, 111–115. [CrossRef]

14. Davoodi, S.; Mostafapour, A. Gas leak locating in steel pipe using wavelet transform and cross-correlation method. *Int. J. Adv. Manuf. Technol.* **2013**, *70*, 1125–1135. [CrossRef]

15. Mostafapour, A.; Davoodi, S. Leakage Locating in Underground High Pressure Gas Pipe by Acoustic Emission Method. *J. Nondestruct. Eval.* **2012**, *32*, 113–123. [CrossRef]

16. Yu, X.C.; Liang, W.; Zhang, L.B.; Jin, H.; Qiu, J.W. Dual-tree complex wavelet transform and SVD based acoustic noise reduction and its application in leak detection for natural gas pipeline. *Mech. Syst. Signal Process.* **2016**, *72–73*, 266–285. [CrossRef]

17. Brennan, M.J.; Gao, Y.; Joseph, P.F. On the relationship between time and frequency domain methods in time delay estimation for leak detection in water distribution pipes. *J. Sound Vib.* **2007**, *304*, 213–223. [CrossRef]

18. Sun, J.D.; Xiao, Q.Y.; Wen, J.T.; Zhang, Y. Natural gas pipeline leak aperture identification and location based on local mean decomposition analysis. *Measurement* **2016**, *79*, 147–157. [CrossRef]

19. Guo, C.; Wen, Y.; Li, P.; Wen, J. Adaptive noise cancellation based on EMD in water-supply pipeline leak detection. *Measurement* **2016**, *79*, 188–197. [CrossRef]

20. Yang, J.; Wen, Y.; Li, P. Leak location using blind system identification in water distribution pipelines. *J. Sound Vib.* **2008**, *310*, 134–148. [CrossRef]

21. Choi, J.; Shin, J.; Song, C.; Han, S.; Park, D.I. Leak Detection and Location of Water Pipes Using Vibration Sensors and Modified ML Prefilter. *Sensors* **2017**, *17*, 2104. [CrossRef]

22. Knapp, C.; Carter, G. The generalized correlation method for estimation of time delay. *IEEE Trans. Accoust. Speech Signal Process.* **1976**, *244*, 320–327. [CrossRef]

23. Gao, Y.; Brennan, M.J.; Joseph, P.F.; Muggleton, J.M.; Hunaidi, O. A model of the correlation function of leak noise in buried plastic pipes. *J. Sound Vib.* **2004**, *277*, 133–148. [CrossRef]

24. Gao, Y.; Brennan, M.J.; Joseph, P.F.; Muggleton, J.M.; Hunaidi, O. On the selection of acoustic/vibration sensors for leak detection in plastic water pipes. *J. Sound Vib.* **2005**, *283*, 927–941. [CrossRef]

25. Gao, Y.; Brennan, M.J.; Joseph, P.F. A comparison of time delay estimators for the detection of leak noise signals in plastic water distribution pipes. *J. Sound Vib.* **2006**, *292*, 552–570. [CrossRef]

26. Hur, J.; Kim, S.; Kim, H. Water hammer analysis that uses the impulse response method for a reservoir-pump pipeline system. *J. Mech. Sci. Technol.* **2017**, *31*, 4833–4840. [CrossRef]

27. Almeida, F.C.; Brennan, M.J.; Joseph, P.F.; Dray, S.; Whitfield, S.; Paschoalini, A.T. Towards an in-situ measurement of wave velocity in buried plastic water distribution pipes for the purposes of leak location. *J. Sound Vib.* **2015**, *359*, 40–55. [CrossRef]

28. Li, S.; Zhang, J.; Yan, D.; Wang, P.; Huang, Q.; Zhao, X.; Cheng, Y.; Zhou, Q.; Xiang, N.; Dong, T. Leak detection and location in gas pipelines by extraction of cross spectrum of single non-dispersive guided wave modes. *J. Loss Prev. Process Ind.* **2016**, *44*, 255–262. [CrossRef]
29. Xiao, Q.; Li, J.; Sun, J.; Feng, H.; Jin, S. Natural-gas pipeline leak location using variational mode decomposition analysis and cross-time–frequency spectrum. *Measurement* **2018**, *124*, 163–172. [CrossRef]
30. Li, S.; Wen, Y.; Li, P.; Yang, J.; Wen, J. Modal analysis of leakage-induced acoustic vibrations in different directions for leak detection and location in fluid-filled pipelines. In Proceedings of the IEEE International Ultrasonics Symposium, Chicago, IL, USA, 3–6 September 2014.
31. Reuben, R.L.; Steel, J.A.; Shehadeh, M. Acoustic Emission Source Location for Steel Pipe and Pipeline Applications: The Role of Arrival Time Estimation. *Proc. Inst. Mech. Eng. Part E J. Process Mech. Eng.* **2006**, *220*, 121–133.
32. Mostafapour, A.; Davoodi, S. Continuous leakage location in noisy environment using modal and wavelet analysis with one AE sensor. *Ultrasonics* **2015**, *62*, 305–311. [CrossRef]
33. Gopi, E.S. *Algorithm Collections for Digital Signal Processing Applications Using Matlab*; Springer: Dordrecht, The Netherlands, 2010.
34. Ionel, R.; Ionel, S.; Bauer, P.; Quint, F. Water leakage monitoring education: cross correlation study via spectral whitening. In Proceedings of the IECON 2014—40th Annual Conference of the IEEE Industrial Electronics Society, Dallas, TX, USA, 29 October–1 November 2014.

 *applied sciences*

Article

Using the Partial Wave Method for Wave Structure Calculation and the Conceptual Interpretation of Elastodynamic Guided Waves

Christopher Hakoda and Cliff J. Lissenden *

Department of Engineering Science and Mechanics, Pennsylvania State University, University Park, PA 16802, USA; cnh137@psu.edu
* Correspondence: Lissenden@psu.edu

Received: 8 May 2018; Accepted: 8 June 2018; Published: 12 June 2018

Featured Application: Shortcut to wave-structure calculation for elastodynamic guided waves and prediction of acoustic leakage behavior.

Abstract: The partial-wave method takes advantage of the Christoffel equation's generality to represent waves within a waveguide. More specifically, the partial-wave method is well known for its usefulness when calculating dispersion curves for multilayered and/or anisotropic plates. That is, it is a vital component of the transfer-matrix method and the global-matrix method, which are used for dispersion curve calculation. The literature suggests that the method is also exceptionally useful for conceptual interpretation, but gives very few examples or instruction on how this can be done. In this paper, we expand on this topic of conceptual interpretation by addressing Rayleigh waves, Stoneley waves, shear horizontal waves, and Lamb waves. We demonstrate that all of these guided waves can be described using the partial-wave method, which establishes a common foundation on which many elastodynamic guided waves can be compared, translated, and interpreted. For Lamb waves specifically, we identify the characteristics of guided wave modes that have not been formally discussed in the literature. Additionally, we use what is demonstrated in the body of the paper to investigate the leaky characteristics of Lamb waves, which eventually leads to finding a correlation between oblique bulk wave propagation in the waveguide and the transmission amplitude ratios found in the literature.

Keywords: partial wave method; slowness curves; lamb wave; stoneley wave; mode sorting; acoustic leakage; rayleigh wave; surface waves; elastodynamics

---

1. Introduction

In this paper, we demonstrate how the partial-wave method, as first introduced by Solie and Auld [1], can be used to gain a new perspective on guided waves. This new perspective adds to our understanding of the structure of guided waves, and provides valuable insight into the leaky characteristics of Lamb waves (which are not to be confused with the study of leaky Lamb waves and their corresponding dispersion curve solutions).

At present, our understanding of guided waves is primarily informed by the derivation of characteristic equations and dispersion curve solutions. The dispersion curve solutions in particular have become significantly more accessible with the widespread use of the semi-analytical finite element (SAFE) method. However, the SAFE method does not provide the same level of understanding that would be possible by analyzing a characteristic equation. The most well-known example of this is that the two-part Lamb wave characteristic equation shows that all of the Lamb waves can be divided into two types: symmetric and anti-symmetric. Hence, the importance of the present work. However, this paper will not cover the calculation of dispersion curves, attenuation, energy loss, or anisotropic materials. Neither will it cover the derivation of characteristic equations for various guided waves.

For the purpose of clarity, in this paper, bulk waves are plane waves that travel through the bulk of a media, and surface waves are plane waves that travel along the surface of a media. It is emphasized that all of the waves are considered to have a planar wavefront.

Sections 2 and 3 will provide a brief history and the fundamentals of the partial-wave method and the Christoffel equation, respectively. These sections give special attention to the diction used and the assumptions made, as they are pertinent to how the interpretation presented in this paper deviates from the literature. Section 4 describes this new method of interpretation. It discusses in detail how the slowness curve solutions of the Christoffel equation can be used to interpret the characteristics of most guided waves, and the Lamb wave solutions in particular. Sections 5 and 6 apply the new method of interpretation to a variety of guided waves, which include Rayleigh waves, Stoneley waves, Lamb waves, and shear horizontal waves.

Section 7 focuses on Lamb waves, and a method developed in Section 6 is used to investigate the contributions of each partial wave to Lamb waves from 50 kHz to 20 MHz for a 1-mm thick aluminum plate. This section also identifies characteristics of Lamb waves that have not been formally discussed in the literature.

Section 8 uses the results from the previous sections to develop an understanding of the leaky characteristics of Lamb waves. Various characteristics of the acoustic leakage from the waveguide can be easily determined, including the angle of propagation and the amplitude of emission based on the interface conditions. A correlation between the amount of acoustic leakage and the refraction qualities of obliquely travelling bulk waves in the guided wave is also identified.

2. The Partial Wave Method

The term 'partial-wave method' first appeared in a paper by Solie and Auld in 1973 [1]; specifically, it was called "the method of partial waves (or transverse resonance)". This method was described as follows: "In this method, the plate wave solutions are constructed from simple exponential-type waves (partial waves), which reflect back and forth between the boundaries of the plates ... " They go on to write, "A basic principle of the partial wave method is that the partial waves are coupled to each other by reflections at the plate boundaries".

In Auld's 1990 monograph [2], he discusses the method further: "[the superposition of partial waves method] can be further simplified by making use of the *transverse resonance principle*, which has been used to great advantage in electromagnetism" [emphasis added by Auld]. That is, what is known colloquially as the partial-wave method can be broken down into two parts. First, there is the superposition of the partial-waves method, which uses the principle of superposition with the Christoffel equation's solutions (i.e., partial waves) [2]. Second, there is the transverse resonance principle, which assumes that the standing wave perpendicular to wave propagation can only exist if the partial waves satisfy the relevant boundary conditions [2,3]. Modern application of the partial-wave method includes the calculation of dispersion curves using the global-matrix method or the transfer-matrix method for anisotropic plates and/or multilayered plates [4,5].

3. The Christoffel Equation and the Use of Slowness Curves

The Christoffel equation is critical to the flexibility and strength of the partial-wave method. The Christoffel equation can be derived by assuming a plane wave (Equation (4)) propagating in a solid elastic media (Equations (1)–(3)).

$$\sigma_{ij,j} = \rho u_{i,tt} \tag{1}$$

$$\sigma_{ij} = C_{ijkl}\epsilon_{kl} \tag{2}$$

$$\epsilon_{ij} = \frac{1}{2}\left(u_{i,j} + u_{j,i}\right) \tag{3}$$

$$u_i(x, y, z, t) = U_i e^{i(-k_j r_j + \omega t)} \tag{4}$$

Here σ_{ij}, ϵ_{ij}, and u_i are the stress, strain, and displacement; ρ and C_{ijkl} are the mass density and elastic stiffness; k_j and ω are the wave vector and angular frequency; $i, j, k, l = x, y, z$, and t is the time variable. When combined, Equations (1)–(4) yield the Christoffel equation:

$$C_{ijkl} k_k U_l k_j - \rho \omega^2 U_i = 0 \tag{5}$$

When Equation (5) is solved, a slowness curve/surface can be calculated and plotted in an inverse phase velocity space.

3.1. Deriving the Expression for the Slowness Curves of an Isotropic Solid

For an isotropic linear elastic solid, Equation (2) can be replaced with:

$$\sigma_{ij} = \lambda \epsilon_{kk} \delta_{ij} + 2\mu \epsilon_{ij} \tag{6}$$

where λ and μ are the Lamé parameters and δ is the Kronecker delta. As a side note, in addition to replacing Equation (2) with Equation (6), if Equation (4) is replaced with:

$$u_i^L(x, t) = U_o \hat{x} e^{i(-k_x x + \omega t)} \tag{7}$$

or:

$$u_i^S(x, t) = U_o \hat{z} e^{i(-k_x x + \omega t)} \tag{8}$$

then, the Christoffel equation can be used to derive the relationship between the Lamé parameters, the longitudinal wave speed, c_L, and the shear wave speed, c_S:

$$\lambda + 2\mu = \rho c_L^2 \tag{9}$$

$$\mu = \rho c_S^2 \tag{10}$$

where $\hat{x}$ and $\hat{z}$ are the unit vectors along the x-axis and the z-axis. Using Equation (6), assuming $k_y = k_z = 0$ (i.e., wave propagation in the x–z plane), and dividing through by ω^2, Equation (5) can be rearranged into matrix form:

$$A u_i = 0 \tag{11}$$

where:

$$A = \begin{bmatrix} (\lambda + 2\mu)c_x^{-2} + \mu c_z^{-2} - \rho & 0 & c_x^{-1} c_z^{-1}(\lambda + \mu) \\ 0 & \mu\left(c_x^{-2} + c_z^{-2}\right) - \rho & 0 \\ c_x^{-1} c_z^{-1}(\lambda + \mu) & 0 & \mu c_x^{-2} + (\lambda + 2\mu)c_z^{-2} - \rho \end{bmatrix} \tag{12}$$

Equations (11) and (12) represent a system of linear homogeneous equations, where $c_x^{-1} = k_x/\omega$ and $c_z^{-1} = k_z/\omega$. The non-trivial solution to Equation (11) only exists if:

$$\det(A) = 0 \tag{13}$$

Expanding Equation (13) based on Equation (12) results in:

$$\left(\rho - (\lambda + 2\mu)\left(c_x^{-2} + c_z^{-2}\right)\right)\left(\rho - \mu\left(c_z^{-2} + c_x^{-2}\right)\right)^2 = 0 \tag{14}$$

The solutions to the characteristic equation (Equation (14)) are:

$$c_x^{-2} + c_z^{-2} = \frac{\rho}{(\lambda + 2\mu)} \tag{15}$$

$$c_x^{-2} + c_z^{-2} = \frac{\rho}{\mu} \tag{16}$$

If Equations (9) and (10) are considered, the right-hand side of Equations (15) and (16) become c_L^{-2} and c_S^{-2}, respectively. These represent the slowness curve solutions to the Christoffel equation of an isotropic material. It should be noted that in Equation (14), the squared term represents the multiplicity in the shear wave's slowness curve solutions.

3.2. Interpretation of Slowness Curve Solutions

From the previous section, it is clear that the solutions to the Christoffel equation do not take into consideration the boundary conditions (BCs) of the problem. That is, the solution only depends on the material properties of the elastic solid in which the plane waves propagate. As a result, in the literature, it is shown that the Christoffel equation can be used to determine the reflection and refraction characteristics of plane waves at oblique incidence to an interface between anisotropic elastic solids [6–8]. Additionally, the energy velocity vector and the bulk wave's skew angle can be calculated [4].

Slowness curves are often plotted in terms of the 'inverse phase velocity'. However, since phase velocity is not a vector, the use of the term 'inverse phase velocity' to describe the slowness vector can be confusing. A different notation that may be easier to conceptually understand is plotting slowness curves on the k_z/ω and k_x/ω axis instead of the c_z^{-1} versus c_x^{-1} axis. Therefore, if a ray is drawn from the origin to a point on the slowness curve, it is understood that its unit vector describes the wave propagation direction in the x–z plane.

Since the focus of the partial-wave method is on guided waves, for the purposes of this paper, it is assumed that the wave is propagating parallel to the boundary/interface, which we take to be in the x-direction. The z-axis is through the thickness of the elastic solid. This coordinate system will be maintained throughout the entirety of the paper, with additional details depending on the problem being considered. Further, continuing to assume that $k_y = 0$, and using the frequency–wavenumber (ω-k_x) pairs obtained from any one of the many methods for calculating dispersion curves (i.e., enforcing the BC and solving the eigenvalue problem), the Christoffel equation becomes a complex valued quadratic eigenvalue problem in three dimensions. The solution, should it exist, will always be composed of six complex valued eigensolutions. These six solutions are the partial waves of the partial-wave method. The partial waves can be paired up into three pairs, and each pair is composed of an upward and a downward propagating wave (i.e., the magnitudes of the eigenvalues are the same, but the signs are opposite) [1,4].

In Auld [2], there is a discussion of regions for an increasing k_x, in which the partial waves have specific characteristics. Also, in Auld [2], it is recognized that at an interface, k_x is the same for all of the partial waves in order to satisfy Snell's law, and that a given guided wave must theoretically have a single k_x value for a given ω. Auld gives an example for an isotropic material, which has two slowness curves and three regions (see Equations (15) and (16)):

- Region 1: $0 < c_p < c_S$ All six partial waves are surface waves (imaginary-valued k_z) (see Figure 1a).
- Region 2: $c_S < c_p < c_L$: Two partial waves are surface waves (imaginary-valued k_z), and four are bulk waves (real-valued k_z) (see Figure 1b).
- Region 3: $c_L < c_p$: All six partial waves are bulk waves (real-valued k_z) (see Figure 1c).

The inverses of the shear and longitudinal group velocities (c_S^{-1} and c_L^{-1}) are also the radii of the two slowness curves, which are circles for an isotropic material. By Crandall's paper [6], it is also understood that outside of these circles, the value for k_Z becomes imaginary, and the magnitude traces a hyperbola.

Auld also gives examples of slowness curve representations for a shear horizontal (SH) mode composed of obliquely travelling bulk waves, and a piezoelectric plate with leaky electromagnetic radiation [2].

This concludes the state-of-the-art section of the paper. How these theories and concepts allow us to reinterpret our understanding of guided waves in a novel way is discussed from here onward.

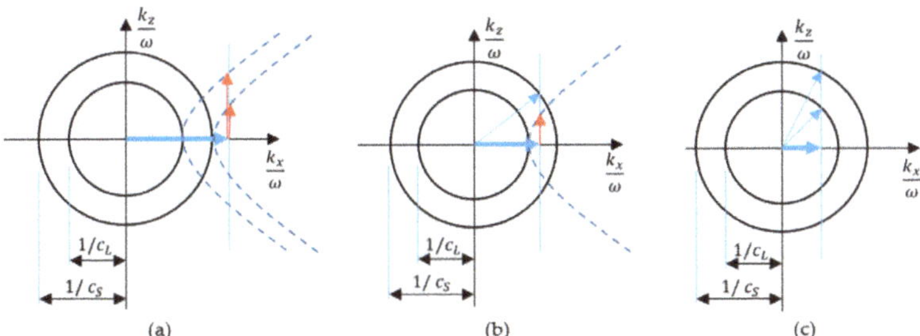

Figure 1. Depiction of the characteristics of (**a**) region 1, (**b**) region 2, and (**c**) region 3 with respect to the slowness curves of an isotropic solid. Slowness curves with solid lines denote real-valued k_z solutions, and dotted lines denote imaginary-valued k_z solutions. This figure is a compilation of elements from the figures in Auld [2] and Crandall [6].

4. Reinterpreting the Dispersion Curves

As mentioned previously, the Christoffel equation is only dependent upon the material properties of the elastic solid in which the plane waves propagate. The BCs are not used until the partial-wave method is applied. Not only that, but the boundary conditions are not limited to the conventional traction-free BCs (e.g., Lamb waves). As a result, the partial-wave method is not limited to any specific type of waveguide. In fact, most types of guided waves should be representable using slowness curves. For some, such as the shear horizontal guided waves, this approach is more cumbersome than the standard derivation. However, its value lies in how it provides a fundamental foundation for all of the guided waves, and a common method for interpreting and comparing each one. When considering reflection/transmission, mode conversion, or multi-mode propagation, this perspective is quite beneficial.

In modern literature, perhaps the most commonly used dispersion curves are of Lamb waves in an aluminum plate in terms of phase velocity versus the frequency–thickness product. Using the three regions depicted in Figure 1, we can divide the phase velocity dispersion curves into three regions, as shown in Figure 2. By doing this, it is possible to view how the dispersion curve solutions match up with the partial waves' characteristics. For example, this is a good way to dispel the potential misconception that guided waves are composed of multiple reflecting bulk waves. That is, the original assumption of plane wave propagation (Equation (4)), which is fundamental to the Christoffel equation, does not exclude the surface wave solution (i.e., imaginary-valued k_z) as k_x increases. It would be more accurate to say that guided waves are a superposition of reflecting bulk waves and surface waves with elliptical particle motion (this will be shown in the later calculations and examples).

In fact, guided waves composed solely of reflecting bulk waves exist only in Region 3 (c), where they are highly dispersive. However, there is one exception to this rule: there exists a range of phase velocities in Region 2 (Figure 1b), where only bulk waves propagate, which will be discussed in a later section.

The A0 mode and Rayleigh waves exist in Region 1 (Figure 1a), where each guided wave can be represented in terms of six surface waves. The low-frequency S0 mode is in Region 2 (b), and is composed of two surface waves and two bulk waves (as will be shown later in the paper), excluding the two shear horizontal partial waves.

The line separating Region 1 (Figure 1a) and Region 2 (Figure 1b) has great significance, since many Lamb wave modes appear nearly non-dispersive near it. That is, the higher-order Lamb waves approach the shear wave speed, and the A0 and S0 modes approach the Rayleigh wave speed for large ω values. The shear horizontal modes also converge to this line for large ω values. The line separating Region 2 (Figure 1b) and Region 3 (Figure 1c) has many higher-order Lamb modes, which appear nearly non-dispersive close to it as well.

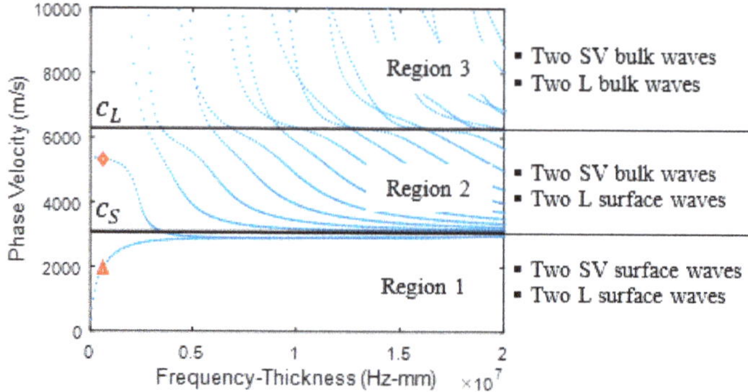

Figure 2. Lamb wave dispersion curves for an aluminum plate divided according to the three regions shown in Figure 1.

The A0 and S0 Lamb wave modes' tendency to approach the Rayleigh wave phase velocity can also be explained using this reinterpretation, since increasing k_x will increase the magnitude of the imaginary-valued k_z (see Figure 1a). Once k_z increases sufficiently, the surface waves decay with the depth from the surface before interacting with the opposite traction-free boundary, thereby imitating a Rayleigh wave. This also explains why quasi-Rayleigh waves [9] have similar wave structures to Rayleigh waves. That is, the quasi-Rayleigh wave and the Rayleigh wave are both composed of partial waves (with slight differences in amplitude and exponential decay).

Many higher-order Lamb modes have wave structures that are sinusoidal through the thickness, but have slightly larger or smaller amplitudes near the traction-free boundaries. In Region 2 (Figure 1b), this could be explained by surface waves affecting the wave structure near the traction-free boundaries, while the reflecting bulk waves create the sinusoidal appearance.

This reinterpretation of guided waves provides for the relatively simple calculation of wave structures based on the superposition of the partial-wave eigensolutions. Based on the preceding concepts and with wave propagation in the x-direction, $c_p = \omega/k_x = c_x$. Solving Equations (15) and (16) for c_z^{-1} results in:

$$c_z^{-1} = \pm\sqrt{c_L^{-2} - c_x^{-2}} \tag{17}$$

and:

$$c_z^{-1} = \pm\sqrt{c_S^{-2} - c_x^{-2}} \tag{18}$$

respectively. Equations (9), (10) and (17) can then be substituted into Equation (11) to get matrix A as a function of c_x^{-1} and c_L^{-1}:

$$A^{(L)} := \begin{bmatrix} (\lambda+\mu)\left(c_L^{-2} - c_x^{-2}\right) & 0 & \pm\sqrt{c_L^{-2} - c_x^{-2}}\,c_x^{-1}(\lambda+\mu) \\ 0 & -(\lambda+\mu)c_L^{-2} & 0 \\ \pm\sqrt{c_L^{-2} - c_x^{-2}}\,c_x^{-1}(\lambda+\mu) & 0 & (\lambda+\mu)\left(-c_x^{-2}\right) \end{bmatrix} \tag{19}$$

And Equations (9), (10) and (18) can be substituted into Equation (11) to get matrix A as a function of and c_x^{-1} and c_S^{-1}:

$$A^{(S)} := \begin{bmatrix} (\lambda+2\mu)c_x^{-2} + \mu\left(c_S^{-2} - c_x^{-2}\right) - \mu c_S^{-2} & 0 & \pm\sqrt{c_S^{-2} - c_x^{-2}}\,c_x^{-1}(\lambda+\mu) \\ 0 & 0 & 0 \\ \pm\sqrt{c_S^{-2} - c_x^{-2}}\,c_x^{-1}(\lambda+\mu) & 0 & \mu c_x^{-2} + (\lambda+2\mu)\left(c_S^{-2} - c_x^{-2}\right) - \mu c_S^{-2} \end{bmatrix} \tag{20}$$

Finding the reduced row echelon form of $A^{(L)}$ and $A^{(S)}$ gives the solution to the eigenvectors, which contain, at most, two free variables (i.e., m and p). Considering $A^{(L)}$ and $A^{(S)}$, and also the "$\pm$" terms in them, which correspond with the sign of k_z, results in four possible eigenvectors (not including the multiplicity). For the positive k_z eigenvalue of the longitudinal wave, the eigenvector is:

$$u_i^{(+L)} = \begin{bmatrix} \dfrac{c_x^{-1}}{\sqrt{c_L^{-2} - c_x^{-2}}}\, m \\ 0 \\ m \end{bmatrix} \tag{21}$$

While for the negative k_z eigenvalue of the longitudinal wave, the eigenvector is:

$$u_i^{(-L)} = \begin{bmatrix} -\dfrac{c_x^{-1}}{\sqrt{c_L^{-2} - c_x^{-2}}}\, m \\ 0 \\ m \end{bmatrix} \tag{22}$$

For the positive k_z eigenvalue of the shear wave, the eigenvector is:

$$u_i^{(+S)} = \begin{bmatrix} \dfrac{-\sqrt{c_S^{-2} - c_x^{-2}}}{c_x^{-1}}\, m \\ p \\ m \end{bmatrix} \tag{23}$$

While for the shear wave's negative k_z eigenvalue, the eigenvector is:

$$u_i^{(-S)} = \begin{bmatrix} \dfrac{\sqrt{c_S^{-2} - c_x^{-2}}}{c_x^{-1}}\, m \\ p \\ m \end{bmatrix} \tag{24}$$

The superscripts L and S correspond to the longitudinal and the shear waves, respectively. The S can be made more specific by using SV and SH to denote the shear vertical and shear horizontal polarizations. Examining Equations (21) and (22), it is clear that for $c_x^{-1} > c_L^{-1}$, the square root becomes imaginary-valued, and for $c_x^{-1} < c_L^{-1}$, the square-root becomes real-valued. These two scenarios correspond with the surface wave and bulk wave solutions, respectively. The same holds true for Equations (23) and (24) with c_S instead of c_L, with $p = 0$. For $c_x^{-1} \gg c_L^{-1}$ or $c_x^{-1} \gg c_S^{-1}$ and positive k_z, the corresponding eigenvector is:

$$u_i^{(+)} = \begin{bmatrix} -im \\ 0 \\ m \end{bmatrix} \tag{25}$$

and for negative k_z it is:

$$u_i^{(-)} = \begin{bmatrix} im \\ 0 \\ m \end{bmatrix} \tag{26}$$

Hence, for very large c_x^{-1} values (relative to their respective wave speeds), the eigenvectors become circularly polarized, regardless of whether it is a shear wave or longitudinal wave. Lastly, for a given k_x and ω, the k_z values of the four partial waves (not including the multiplicity of the shear waves) can be calculated using Equations (15) and (16):

$$k_z^{(\pm L)} = \pm \omega \sqrt{c_L^{-2} - c_x^{-2}} \tag{27}$$

$$k_z^{(\pm S)} = \pm\omega\sqrt{c_S^{-2} - c_x^{-2}} \tag{28}$$

As implied previously, since the solution of the Christoffel equation does not depend on the boundary/interface conditions, a linear combination of the partial waves should be able to represent most elastodynamic guided waves, regardless of the boundary/interface conditions. We demonstrate this in the following sections. The material properties shown in Table 1 will be used. For each partial-wave eigenvector calculated in the following examples, the free variable is set equal to unity; then, the resulting eigenvector is normalized into a unit vector by dividing by its magnitude.

Table 1. Material properties used for the calculations shown in this paper.

	Aluminum [4]	Tungsten [4]	Air [10]
ρ_B, kg/m^3	2700	19,250	1.2
λ_B, GPa	56.98	199.4	-
μ_B, GPa	25.95	158.6	-
c_L, m/s	6350	5180	343
c_S, m/s	3100	2870	-

5. Rayleigh Waves and Generalized Rayleigh Waves

5.1. Rayleigh Waves

The Rayleigh wave, which travels at the surface of a half-space, is a good starting example, since its solution is well-known, and this different approach will mirror the conventional derivation in the literature. The term 'generalized Rayleigh waves' will be used to refer to a class of wave propagation solutions developed following the original Rayleigh wave derivation, as was done by Stoneley [11]. This will be considered later in section V.B.

Assuming an aluminum half-space with the boundary at $z = 0$, an example Rayleigh wave calculation is completed using the previously mentioned methods. The eigensolutions for the six partial waves at a frequency of 3.188×10^6 rad/s and a wavenumber of 1103.4 rad/m are shown in Table 2. The frequency and wavenumber were chosen to match the non-dispersive phase velocity solution to the standard Rayleigh wave problem.

Table 2. Eigenvalues and normalized eigenvectors from the Christoffel equation for aluminum at $\omega = 3.188 \times 10^6$ rad/s and $k_x = 1103.4$ rad/m.

Partial Waves	k_z, rad/m	u_x	u_y	u_z
$(+L)$	982.6i	$-0.747i$	0	0.665
$(-L)$	$-982.6i$	0.747i	0	0.665
$(+SV)$	400.0i	$-0.341i$	0	0.940
$(-SV)$	$-400.0i$	0.341i	0	0.940
$(+SH)$	400.0i	0	1	0
$(-SH)$	$-400.0i$	0	-1	0

There are several notable results in Table 2:

- All of the k_z values are imaginary, since Rayleigh waves exist in Region 1 (Figure 1a);
- $k_z^{(\pm L)}$ is larger in magnitude than $k_z^{(\pm SV)}$ and $k_z^{(\pm SH)}$ because of the hyperbolic shape of the imaginary-valued slowness curves;
- $k_z^{(+SV)}$, $k_z^{(-SV)}$, $k_z^{(+SH)}$, and $k_z^{(-SH)}$ have a multiplicity due to the isotropic material properties;
- The eigenvectors are complex, which suggests elliptical particle motion [9];
- $u_i^{(+SH)}$ and $u_i^{(-SH)}$ are not complex-valued.

As shown in Figure 3, only the exponentially decaying solutions are necessary for the Rayleigh wave solution, because the domain is a half-space.

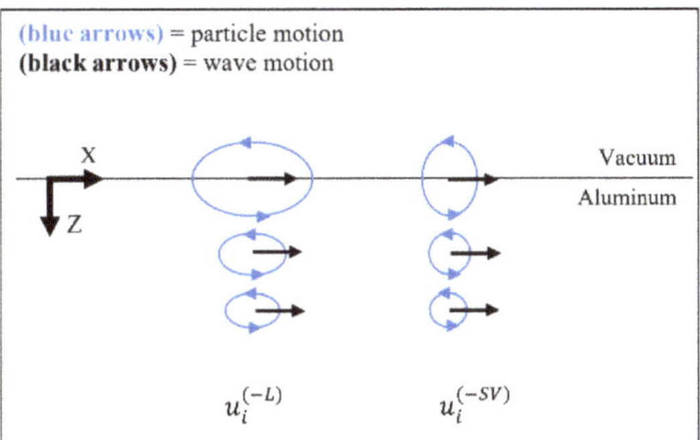

Figure 3. Conceptual depiction of the partial waves, which are necessary for the Rayleigh-wave problem, overlaid on a traction-free boundary of an aluminum half-space.

Although the derivation method is different, closer inspection of the k_z solutions shows that they match the values of q and s given by Viktorov [9]. In the conventional derivation, the Rayleigh wave solution is a superposition of two exponentially decaying terms that are dependent upon the phase velocity and material properties, so this close agreement is expected. A superposition of the two partial waves, when $u_i(x, z, t) = u_i(z)e^{i(-k_x x + \omega t)}$, can be represented as:

$$u_i(z) = B^{(-L)} \left(u_i^{(-L)} e^{-ik_z^{(-L)} z} \right) + B^{(-SV)} \left(u_i^{(-SV)} e^{-ik_z^{(-SV)} z} \right) \tag{29}$$

The amplitudes of each partial wave, $B^{(-L)}$ and $B^{(-SV)}$, are solved for by substituting into one of the BCs:

$$\sigma_{zz} = 0 \text{ at } z = 0, \tag{30}$$

$$\sigma_{xz} = 0 \text{ at } z = 0, \tag{31}$$

and assuming $B^{(-L)} = 1$. This approach is taken in lieu of satisfying both BCs (by setting the determinant of the coefficients matrix equal to zero), since k_x is an approximate value, and the determinant is never exactly zero, so most attempts at getting a reduced row echelon form of the matrix result in the trivial solution of the eigenvector. To avoid this dilemma, we set $B^{(-L)} = 1$, and normalized as necessary. $\sigma_{xz} = 0$ could be used in place of the other BC, but only one is necessary. Solving for $B^{(-SV)}$, it is found that $B^{(-SV)} = -1.154$. Plotting Equation (29) with the calculated values of $B^{(-L)}$ and $B^{(-SV)}$ gives the same wave structure for a Rayleigh wave as that found in Viktorov [9].

5.2. Stoneley Waves

The same approach can be applied to generalized Rayleigh waves. Generalized Rayleigh waves refer to guided waves that demonstrate Rayleigh wave-like characteristics, such as for example, Stoneley waves, Scholte waves, and Love waves [11–13]. For this example, Stoneley waves will be the focus, but a similar approach can be applied to the others as well.

Stoneley waves, which propagate along the interface between two half-spaces, will be analyzed as an example. Stoneley waves typically have a non-dispersive phase velocity that is slightly less than the denser material's bulk shear wave velocity, which means that it is in Region 1 (Figure 1a). A Rayleigh wave-like wave structure should exist in the denser half-space [2]. A tungsten half-space rigidly connected to an aluminum half-space will be used for the example. Using Auld's example (i.e., on pages 104–105 in [2]), the six eigensolutions are calculated for a frequency of 3.188×10^6 rad/s and a wavenumber of 1172.1 rad/m. The frequency and wavenumber are chosen to meet the non-dispersive phase velocity solution to the Stoneley wave problem [2]. The eigensolutions at this frequency and wavenumber are shown in Tables 3 and 4 for tungsten and aluminum, respectively.

Table 3. Eigenvalues and normalized eigenvectors from the Christoffel equation for tungsten at $\omega = 3.188 \times 10^6$ rad/s and $k_x = 1172.1$ rad/m.

Partial Waves	k_z, rad/m	u_x	u_y	u_z
$(+L)$	$997.6i$	$-0.762i$	0	0.648
$(-L)$	$-997.6i$	$0.762i$	0	0.648
$(+SV)$	$374.5i$	$-0.304i$	0	0.953
$(-SV)$	$-374.5i$	$0.304i$	0	0.953
$(+SH)$	$374.5i$	0	1	0
$(-SH)$	$-374.5i$	0	-1	0

Table 4. Eigenvalues and normalized eigenvectors from the Christoffel equation for aluminum at $\omega = 3.188 \times 10^6$ rad/s and $k_x = 1172.1$ rad/m.

Partial Waves	k_z, rad/m	u_x	u_y	u_z
$(+L)$	$1058.1i$	$-0.742i$	0	0.671
$(-L)$	$-1059.1i$	$0.742i$	0	0.671
$(+SV)$	$562.5i$	$-0.433i$	0	0.902
$(-SV)$	$-562.5i$	$0.433i$	0	0.902
$(+SH)$	$562.5i$	0	1	0
$(-SH)$	$-562.5i$	0	-1	0

For both tungsten and aluminum, the Stoneley wave is in Region 1 (Figure 1a). To construct the Stoneley wave displacement solution, the following partial waves are needed: solutions $(-L)$ and $(-SV)$ from aluminum's Christoffel equation, and solutions $(+L)$ and $(+SV)$ from the tungsten's Christoffel equation. Solutions $(-L)$ and $(-SV)$ from the aluminum are considered to be exponentially decaying solutions for $+z$ values, while solutions $(+L)$ and $(+SV)$ from the tungsten are considered to be exponentially decaying solutions for $-z$ values. These are shown in Figure 4 with respect to the interface.

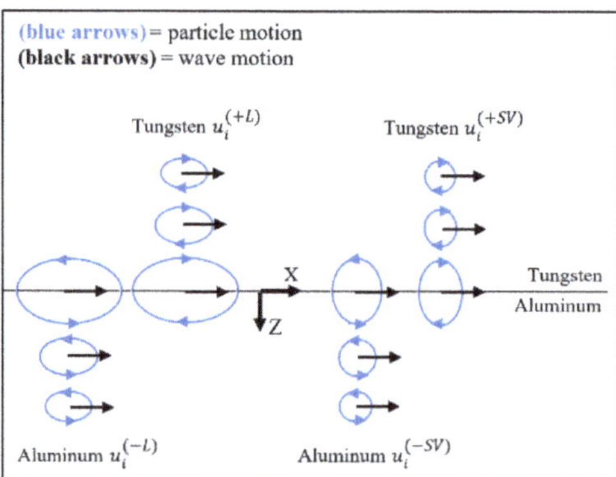

Figure 4. Conceptual depiction of the partial waves from tungsten and aluminum, which are necessary for the Stoneley wave problem, overlaid on the two half-spaces.

The superposition of partial waves for each material, when $u_i(x,z,t) = u_i(z)e^{i(-k_x x + \omega t)}$, is shown in Equations (32) and (33). For tungsten (using values from Table 3):

$$u_i^W(z) = B^{(+L)}\left(u_i^{(+L)}e^{-ik_z^{(+L)}z}\right) + B^{(+SV)}\left(u_i^{(+SV)}e^{-ik_z^{(+SV)}z}\right) \tag{32}$$

while for aluminum (using values from Table 4):

$$u_i^{Al}(z) = B^{(-L)}\left(u_i^{(-L)}e^{-ik_z^{(-L)}z}\right) + B^{(-SV)}\left(u_i^{(-SV)}e^{-ik_z^{(-SV)}z}\right) \tag{33}$$

Since the two half-spaces are assumed to be welded at the interface, the three of the four available interface conditions:

$$\sigma_{zz}^{Al} = \sigma_{zz}^{W} \, \text{at} \, z = 0, \tag{34}$$

$$\sigma_{xz}^{Al} = \sigma_{xz}^{W} \, \text{at} \, z = 0, \tag{35}$$

$$u_{z}^{Al} = u_{z}^{W} \, \text{at} \, z = 0, \tag{36}$$

$$u_{x}^{Al} = u_{x}^{W} \, \text{at} \, z = 0, \tag{37}$$

can be used to solve for the amplitudes $B^{(-L)}$, $B^{(+SV)}$, and $B^{(-SV)}$ (i.e., assuming $B^{(+L)} = 1$).

Using these interface conditions, it was found that $B^{(-L)} = -0.3465$, $B^{(+SV)} = -1.2181$, $B^{(-SV)} = 0.3108$. Equations (32) and (33) can now be plotted to get the Stoneley wave's wave structure, as shown in Figure 5.

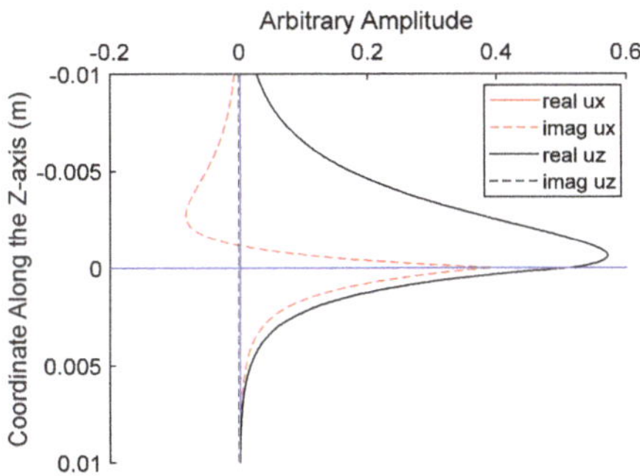

Figure 5. Wave structure of a Stoneley wave calculated using the partial-wave method at an aluminum/tungsten interface. Blue lines are the coordinate axis. $-z$ is in the tungsten half-space, and $+z$ is in the aluminum half-space.

6. Lamb Waves and Shear Horizontal Waves

Consider a 1-mm thick aluminum plate with traction-free BCs, and the same frequency as was analyzed in the previous section. The wavenumbers are calculated based on the Lamb wave dispersion curve solutions at that frequency.

6.1. A0 Lamb Wave Mode

The six eigensolutions at a frequency of 3.188×10^6 rad/s and a wavenumber of 1688.7 rad/m are shown in Table 5. This frequency–wavenumber pair corresponds with the red-triangle symbol plotted in the dispersion-curve plot shown in Figure 2. Similar to the Christoffel equation solutions used for the Rayleigh -wave case, the solutions in Table 5 are typical of Region 1 (Figure 1a), in which the A0 mode exists. Unlike the Rayleigh wave case, there is now a bottom surface, so the exponentially increasing eigensolutions are now necessary, although they will be written as exponentially decreasing from the bottom surface upward as $e^{-ik_z(z-H)}$. Using the coordinate system shown in Figure 6, assignment to the top or bottom of the plate means that the partial waves will be defined using $e^{-ik_z(z)}$ or $e^{-ik_z(z-H)}$, respectively. For simplicity, the rule used in this paper to categorize the partial waves between the top or bottom of the plate is shown in Table 6.

Table 5. Eigenvalues and normalized eigenvectors from the Christoffel equation for aluminum at $\omega = 3.188 \times 10^6$ rad/s and $k_x = 1688.7$ rad/m.

Partial Waves	k_z, rad/m	u_x	u_y	u_z
$(+L)$	$1612.4i$	$-0.723i$	0	0.691
$(-L)$	$-1612.4i$	$0.723i$	0	0.691
$(+SV)$	$1339.5i$	$-0.621i$	0	0.784
$(-SV)$	$-1339.5i$	$0.621i$	0	0.784
$(+SH)$	$1339.5i$	0	1	0
$(-SH)$	$-1339.5i$	0	-1	0

Table 6. Method for assigning partial waves to the top or bottom of the plate using the coordinate system shown in Figure 6 and assuming $e^{-ik_z z}$.

	Top of Plate	Bottom of Plate
Real-valued (bulk waves)	$k_z^{(+L)}$ and $k_z^{(+SV)}$	$k_z^{(-L)}$ and $k_z^{(-SV)}$
Imaginary-valued (surface waves)	$k_z^{(-L)}$ and $k_z^{(-SV)}$	$k_z^{(+L)}$ and $k_z^{(+SV)}$

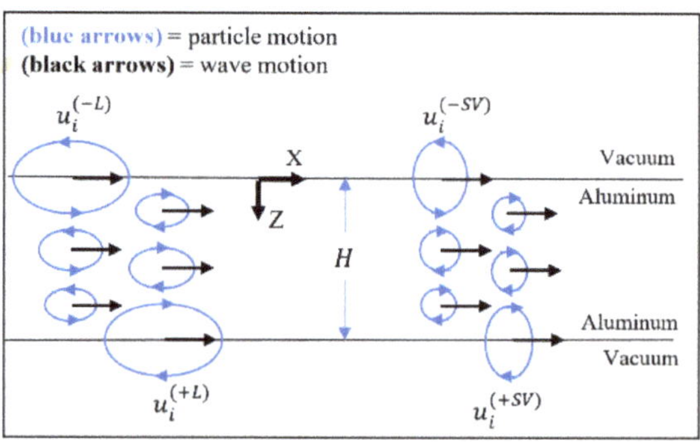

Figure 6. Conceptual depiction of the partial waves, which are necessary for the A0 Lamb wave problem, overlaid on a plate in vacuum.

The superposed displacement profile in the z-direction is constructed by considering the symmetry of the problem:

$$u_i(z) = B^{(+L)}u_i^{(+L)}e^{-ik_z^{(+L)}(z-H)} + B^{(-L)}u_i^{(-L)}e^{-ik_z^{(-L)}z} + B^{(+SV)}u_i^{(+SV)}e^{-ik_z^{(+SV)}(z-H)} +$$
$$B^{(-SV)}u_i^{(-SV)}e^{-ik_z^{(-SV)}z}, \tag{38}$$

where $u_i(x,z,t) = u_i(z)e^{i(-k_x x + \omega t)}$. For partial waves existing on the bottom surface of the plate, it is shown in Equation (38) that the z-position is offset by the thickness of the plate. Just as in the Rayleigh wave problem, $B^{(-L)}$, $B^{(+SV)}$, and $B^{(-SV)}$ can be solved by using three of the four BC available (assuming that $B^{(+L)} = 1$):

$$\sigma_{zz} = 0 \text{ at } z = 0, \tag{39}$$

$$\sigma_{xz} = 0 \text{ at } z = 0, \tag{40}$$

$$\sigma_{zz} = 0 \text{ at } z = H, \tag{41}$$

$$\sigma_{xz} = 0 \text{ at } z = H, \tag{42}$$

It was found that $B^{(-L)} = 1$, and $B^{(+SV)} = B^{(-SV)} = -1.0284$. As before, Equation (38) can now be used to plot the wave-structure of the A0 mode at this frequency.

6.2. S0 Lamb Wave Mode

The six eigensolutions at a frequency of 3.188×10^6 rad/s and a wavenumber of 591.57 rad/m are shown in Table 7, where the shear vertical solutions have real-valued k_z and real-valued eigenvectors. This frequency–wavenumber pair corresponds with the red-diamond symbol plotted in the dispersion-curve plot shown in Figure 2. These solutions correspond with the bulk waves reflecting within the plate waveguide (see Figure 7).

Table 7. Eigenvalues and normalized eigenvectors from the Christoffel equation for aluminum at $\omega = 3.188 \times 10^6$ rad/s and $k_x = 591.57$ rad/m.

Partial Waves	k_z, rad/m	u_x	u_y	u_z
$(+L)$	$312.93i$	$-0.884i$	0	0.468
$(-L)$	$-312.93i$	$0.884i$	0	0.468
$(+SV)$	841.13	-0.818	0	0.575
$(-SV)$	-841.13	0.818	0	0.575
$(+SH)$	841.13	0	1	0
$(-SH)$	-841.13	0	-1	0

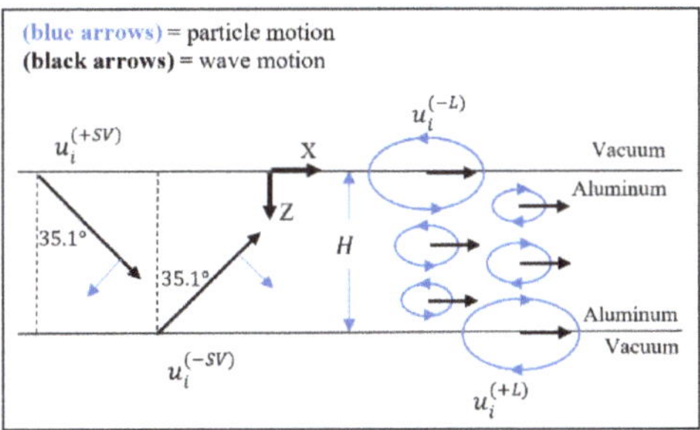

Figure 7. Conceptual depiction of the partial waves, which are necessary for the S0 Lamb wave problem, overlaid on a plate in vacuum. The angle for the propagation direction of the shear vertical partial waves was calculated using $\tan^{-1}(k_x/k_z)$.

The displacement solution is constructed in a manner analogous to the A0 mode, except now there are bulk waves present:

$$u_i(z) = B^{(+L)}u_i^{(+L)}e^{-ik_z^{(+L)}(z-H)} + B^{(-L)}u_i^{(-L)}e^{-ik_z^{(-L)}z} + B^{(-SV)}u_i^{(-SV)}e^{-ik_z^{(-SV)}(z-H)} + \\ B^{(+SV)}u_i^{(+SV)}e^{-ik_z^{(+SV)}z}. \tag{43}$$

$B^{(+L)}$, $B^{(-SV)}$ and $B^{(+SV)}$ are determined by using three of the four BCs (Equations (39)–(42)) (assuming that $B^{(-L)} = 1$), and are found to be $B^{(+L)} = -0.998$, $B^{(-SV)} = -0.2139 + 0.4780i$ and $B^{(+SV)} = 0.2139 - 0.4779i$. As before, Equation (43) can now be used to plot the wave structure of the S0 mode at this frequency. Occasionally, the plotted wave structure will be at a phase that is inconsistent with the conventional wave structure figures. This can be corrected by multiplying all of the displacements by $e^{-i\phi}$, where $\phi = \tan^{-1}\left(\frac{imag(u_x)}{real(u_x)}\right)$. This multiplication is equivalent to letting the wave propagate over some non-zero distance and/or time.

6.3. Shear Horizontal Guided Wave Mode

Since the shear horizontal (SH) guided wave modes are similar to each other, a single discussion of how each behaves is possible. The dispersion relationship found in Rose [4] is:

$$(M\pi)^2 = \left(\frac{\omega h}{c_S}\right)^2 - (kh)^2 \tag{44}$$

M is the guided wave mode number, and $h = H/2$. The phase velocity is equal to the shear wave velocity for $M = 0$, and for large ω values, it approaches the shear wave velocity for all $M > 0$. The shear wave velocity represents the boundary between regions 1 and 2 (Figure 1a,b). That is, $k_z = 0$ and $k_x = \frac{\omega}{c_S}$. The SH partial wave exists as a bulk wave in the plate that is propagating parallel to the traction-free boundaries. When $M > 0$ and ω is not large, then the SH guided wave modes increase in phase velocity and are highly dispersive. With the increasing phase velocity, the incidence angles of the SH bulk waves decrease as the solution of the Christoffel equation traces the shear slowness curve (the slowness curve representation of this dispersive SH guided wave mode is shown in Auld [2]).

7. The Contribution of Various Partial Waves in Each Lamb Wave

7.1. Symmetry and Opposing Phases

Based on the results from the S0 and A0 Lamb wave examples, it would seem as if there is an inherent symmetry to the problem, since the amplitudes of the partial wave pairs often have equal magnitudes or are out of phase by $180°$. This is due to the problem's symmetric geometry. This leads to the following simplification to the displacement solution used to calculate the wave structure:

$$u_i(z) = B^{(L)}\left(q\, u_i^{(\pm L)} e^{-ik_z^{(\pm L)}(z-H)} + u_i^{(\mp L)} e^{-ik_z^{(\mp L)}z}\right)$$
$$+ B^{(SV)}\left(q\, u_i^{(\pm SV)} e^{-ik_z^{(\pm SV)}(z-H)} + u_i^{(\mp SV)} e^{-ik_z^{(\mp SV)}z}\right) \tag{45}$$

where q is either $+1$ or -1. Determining which partial waves should be assigned to the top or bottom of the plate is done in the same way as in the previous examples. As before, it is still implied by the use of the Christoffel equation that $u_i(x,z,t) = u_i(z)e^{i(-k_x x + \omega t)}$. In practice though, a few test calculations reveal that although the amplitudes of the partial-wave pairs are close to equivalent, they are not exactly equivalent. At best, Equation (45) can be used as a general description of the characteristics of many of the Lamb waves, but it is not a mathematically viable solution for our current application. That being said, the general description is invaluable for analyzing the characteristics of Lamb waves.

7.2. Using Symmetry to Summarize the Contribution of Longitudinal and Shear Partial Waves

One of the advantages of the partial-wave method approach is that the contribution of each partial wave can be determined for each Lamb wave mode. That is, if the amplitudes are normalized by dividing by:

$$B_{mag} = \left(|B^{(+L)}|^2 + |B^{(-L)}|^2 + |B^{(+SV)}|^2 + |B^{(-SV)}|^2\right)^{1/2} \tag{46}$$

and since $|B^{(+L)}| \approx |B^{(-L)}|$ due to the symmetry of the problem, a representation of the amount of longitudinal partial wave contribution can be calculated using the average of $B^{(+L)}$ and $B^{(-L)}$:

$$B_{avg}^{(L)} = \frac{|B^{(+L)}| + |B^{(-L)}|}{2B_{mag}} \tag{47}$$

This average is used to account for the approximate nature of the solutions. $B_{avg}^{(L)}$ is representative of the amount of the longitudinal partial waves that make up a given Lamb wave mode, and is overlaid on the dispersion curves as an intensity plot in Figure 8a.

Similarly, since $|B^{(+SV)}| \approx |B^{(-SV)}|$ due to the symmetry of the problem, the contribution of the shear vertical partial waves can also be shown by calculating:

$$B_{avg}^{(SV)} = \frac{|B^{(+SV)}| + |B^{(-SV)}|}{2B_{mag}},\tag{48}$$

and plotting it in Figure 8b.

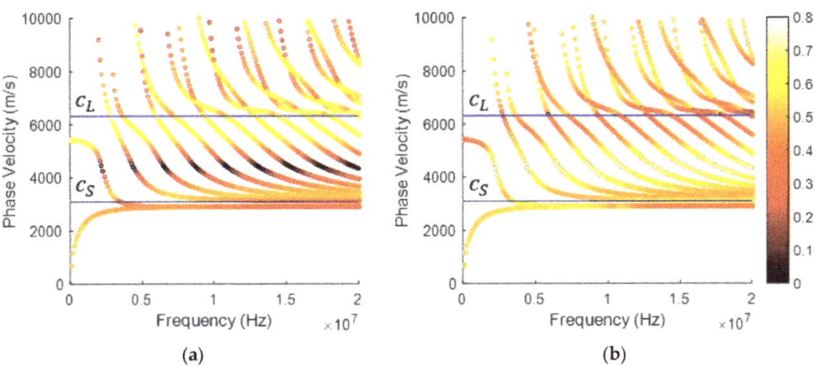

Figure 8. Amplitude of the (**a**) longitudinal (i.e., $B_{avg}^{(L)}$) and (**b**) shear vertical (i.e., $B_{avg}^{(SV)}$) partial waves after normalizing and averaging the magnitude for various points on the Lamb wave dispersion curves for a 1-mm thick aluminum plate.

In Figure 8, it can be seen that in Region 1 (Figure 1a), the apportionment of longitudinal (i.e., $B_{avg}^{(L)}$) and shear vertical (i.e., $B_{avg}^{(SV)}$) partial waves are similar. However, in regions 2 and 3 (Figure 1b,c) however, this is no longer the case. In Region 2 (Figure 1b), the Lamb waves around 4000 m/s are almost entirely made up of bulk shear vertical partial waves, and Lamb waves just below 6000 m/s are made up of mostly surface longitudinal partial waves. Interestingly, Lamb modes have phase and group velocities of $\sqrt{2}c_s$ and are comprised of shear bulk waves propagating at $\pm 45°$ [14]. Region 2 (Figure 1b) also has a clear phase velocity dependence on the amount of each partial wave, but no frequency dependence. In Region 3 (Figure 1c), a certain degree of mode dependence appears, but it does not appear to differentiate between symmetric and anti-symmetric modes.

7.3. Using Opposing Phases to Sort Symmetric and Anti-Symmetric Lamb Wave Modes

Another advantage of using the partial-wave method is that the phase difference between the partial waves at the top and bottom of the plate can be different depending on the type of Lamb wave. That is, in symmetric modes, the top partial wave will be about 180° out-of-phase with the bottom partial wave. Anti-symmetric modes will typically have the top and bottom partial waves in-phase. This can be demonstrated by calculating q from Equation (45):

$$q = \frac{B_{Top}^{(L)}}{B_{Bottom}^{(L)}}\tag{49}$$

which should be equivalent to:

$$q = \frac{B_{Top}^{(SV)}}{B_{Bottom}^{(SV)}} \tag{50}$$

However, since the frequency–wavenumber pairs are approximations, q will vary from -1 to $+1$ for most Lamb waves. The values of q are overlaid on the dispersion curves in Figure 9. It is worth emphasizing that the phase difference between the top and bottom partial waves applies to both the longitudinal and shear vertical partial waves, as shown in Equation (45).

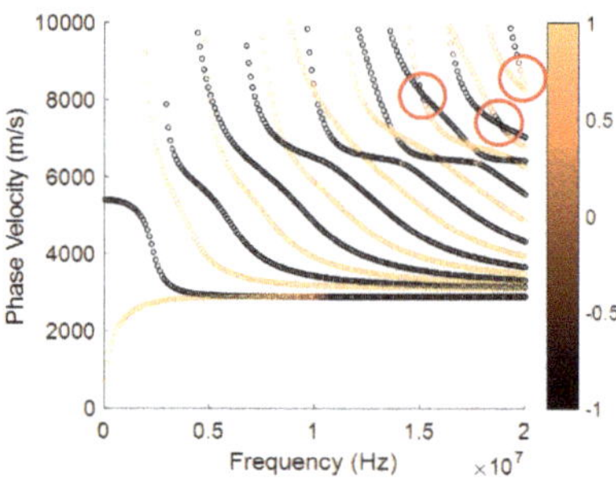

Figure 9. q value which is indicative of the phase difference between the top and bottom partial waves for various points on the Lamb wave dispersion curves for a 1-mm thick aluminum plate. The color axis is constrained to $+1$ and -1.

Figure 9 demonstrates the phase difference characteristics of symmetric and anti-symmetric Lamb waves, but it also shows how those characteristics break down at high frequencies where dispersion curves intersect. At these intersection points, a phase difference (either $+1$ or -1) will dominate the surrounding points of both dispersion curves. The clearest examples of this phenomenon are circled in red in Figure 9.

8. Leaky Characteristics of Lamb Waves

Lamb waves travel through a traction-free plate. Leaky Lamb waves are the waves that propagate through the same waveguide when the traction-free boundaries are replaced with fluid half-spaces. The theoretical literature on leaky Lamb waves focuses on the calculation of dispersion curves and determining how these curves differ from the conventional Lamb wave dispersion curves [15,16]. The focus of this section is not this, but rather the leaky characteristics of Lamb waves, assuming there is no change in the dispersion curves. That is, the dispersion curves of a plate surrounded by a vacuum will be used as an approximation for the dispersion curves of a plate surrounded by air. This is a universal standard practice. The goal will be to use the partial-wave method to determine the characteristics of the acoustic leakage into the fluid from the Lamb wave travelling in the plate waveguide.

Before continuing, it should be stated that for fluids, the Eulerian form of the balance of linear momentum for acoustics is:

$$\rho \left[\frac{\partial}{\partial t} v_i + v_j \left(\frac{\partial}{\partial x_j} v_i \right) \right] = -\frac{\partial}{\partial x_i} P \tag{51}$$

This complicates the velocity–pressure relationship of the acoustic fluid. To simplify this, it is assumed that the displacement is time-harmonic, which makes the particle velocity:

$$v_i = i\omega u_i. \tag{52}$$

It is also assumed that the displacement is spatially harmonic and longitudinal in polarization such that:

$$u_i(x,y,z,t) = U_o \hat{k}_i e^{-ik_j x_j + i\omega t}, \tag{53}$$

where $\hat{k}_j$ is the unit vector denoting the direction of wave propagation. Using Equations (52) and (53) in Equation (51) results in:

$$i\rho|v_i|(\omega - |k_i||v_i|) = i|k_i|P. \tag{54}$$

The Eulerian form can approximate the Lagrangian form with these assumptions if:

$$\omega \gg |k_i|\omega U_o. \tag{55}$$

Equation (55) can be rearranged to:

$$\omega \ll \frac{c_p}{U_o}. \tag{56}$$

If it is assumed that the phase velocity is around 10^3 m/s and that displacements are about 10^{-6} m, then for frequencies much less than 10^9 rad/s, the Eulerian form should approximate the Lagrangian form. As a result, the simpler pressure–velocity relationship:

$$\rho c_L i\omega U_o = P, \tag{57}$$

which was derived from the Lagrangian form of the balance of linear momentum for fluids assuming a time-harmonic displacement, can be used for the following example.

As an acoustic wave in a fluid, the slowness curve can be plotted as a circle with a radius equal to $1/c_L^{air}$. Plotted with respect to aluminum's slowness curve, the order of magnitude difference in radii makes all but the very dispersive part of the A0 mode, at very low frequencies, emit acoustic bulk waves (see Figure 10). At these low frequencies, the A0 mode is highly dispersive, and has a phase velocity lower than the acoustic wave speed in air.

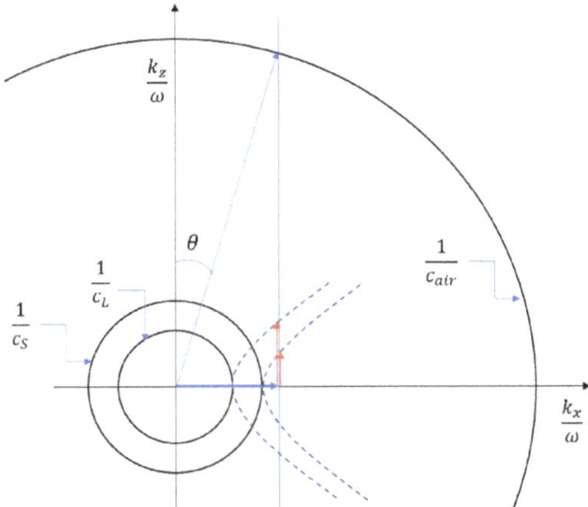

Figure 10. Depiction of the aluminum slowness curves compared with the slowness curve for air. Slowness curves with solid lines denote real-valued k_z solutions, and dotted lines denote imaginary-valued k_z solutions (figure not to scale).

The comparison of slowness curves from various media, which are shown in Figure 10, is similar to an example in Auld [2], but that example discussed the electromagnetic emission from a piezoelectric waveguide. The approach relies on Snell's law, dictating that the wavenumber (i.e., k_x)

should be equal at the interface between media. However, if Snell's law were to be used directly, complex-valued angles (i.e., θ) would need to be used.

Based on this slowness curve representation, it can be concluded that the acoustic leakage will be an obliquely travelling bulk wave. The angle of incidence of the acoustic leakage would be between $0°$ and $20°$ for most Lamb waves. If these findings are applied to air-coupled transducers, it would suggest that there is an optimum angle of reception for the air-coupled transducers.

The partial waves for air can be represented as an obliquely traveling longitudinal wave:

$$u_i^{air}(x, z, t) = A_{air}\hat{k}_i e^{-ik_j x_j + i\omega t}, \tag{58}$$

and the partial waves of a Lamb wave in an aluminum plate (see Equations (38) and (43)) can be used with five of the six available interface conditions:

$$P^{air} = -\sigma_{zz}^{Al} \text{at} z = 0, \tag{59}$$

$$u_z^{air} = u_z^{Al} \text{at} z = 0, \tag{60}$$

$$\sigma_{xz}^{Al} = 0 \text{at} z = 0, \tag{61}$$

$$P^{air} = -\sigma_{zz}^{Al} \text{at} z = H, \tag{62}$$

$$u_z^{air} = u_z^{Al} \text{at} z = H, \tag{63}$$

$$\sigma_{xz}^{Al} = 0 \text{at } z = H. \tag{64}$$

where an expression for P^{air} can be found by using Equations (57) and (58) to find that:

$$P^{air}(x, z, t) = iB^{air}\rho c_L \omega e^{-ik_j x_j + i\omega t} \tag{65}$$

Although stress components are expressly used in the boundary conditions detailed in this paper, it is understood that they describe traction continuity. Equations (59)–(64) represent the traction and displacement continuity conditions indicative of a fluid–solid interface. We assume that $B^{(+L)} = 1$, $B^{(-L)}$, $B^{(+SV)}$, $B^{(-SV)}$, $B^{(+air)}$, and $B^{(-air)}$ can be calculated as before by using the continuity conditions in Equations (59)–(64). $B^{(+air)}$ and $B^{(-air)}$ in this case would correspond with the amplitudes of the longitudinal waves at the top and bottom of the plate. All of the amplitudes were normalized by dividing each by:

$$B_{mag}^{air} = \left(|B^{(+L)}|^2 + |B^{(-L)}|^2 + |B^{(+SV)}|^2 + |B^{(-SV)}|^2 + |B^{(+air)}|^2 + |B^{(-air)}|^2\right)^{1/2} \tag{66}$$

Equations (38) and (58) can then be used to plot the wave structure, and verify that the interface conditions are satisfied.

Since it is now possible to calculate the amplitude of the acoustic leakage, every point of the Lamb wave dispersion curve can be input to determine the ideal guided wave mode for maximum (or minimum) acoustic leakage. Figure 11 shows sample calculations for a 1-mm aluminum plate. The color axis of Figure 11 is the average amplitude of acoustic leakage from Lamb waves:

$$B_{avg}^{(air)} = \frac{|B^{(+air)}| + |B^{(-air)}|}{2B_{mag}^{air}} \tag{67}$$

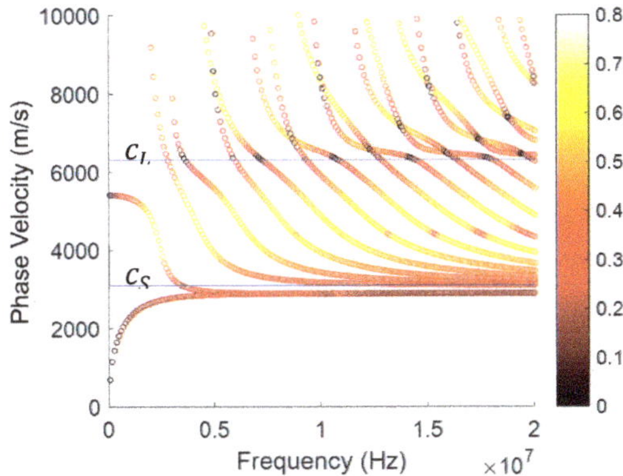

Figure 11. Amplitude of the longitudinal partial waves in the air half-space after normalizing and averaging the magnitude for various points on the Lamb wave dispersion curves for a 1-mm thick aluminum plate. The Lamb wave dispersion curves were calculated assuming traction-free boundary conditions (BC).

Assuming there is a direct correlation between the strength of acoustic leakage and the amplitude of the acoustic wave coupled to the Lamb wave, Figure 11 suggests several things:

- the A0 mode can have a lower amplitude of acoustic leakage than the S0 mode (when normalized by the amplitude of the wave structure);
- the amount of acoustic leakage is not mode-dependent;
- comparisons to Figure 8 suggest a potential relationship between the dominant partial waves present and the amplitude of the acoustic leakage;
- when $c_p = c_L$, the leakage of the symmetric modes shown in Figure 11 agrees with the previous analysis by Rose [4](Section 8.7.1), which attributes the lack of leakage to the lack of out-of-plane displacement at that surface.

Perhaps most interesting is that the maximum acoustic leakage is concentrated in Region 2 (Figure 1b) of the dispersion curves, while Region 1 (Figure 1a) has noticeably less acoustic leakage. A possible explanation for why this would be is that the guided wave modes in Region 1 (Figure 1a) are composed of surface partial waves, which are inherently coupled to the interface.

In Region 2 (b), bulk shear vertical partial waves propagate through the thickness of the plate. Incident upon the aluminum–air interface, this can be modeled as an oblique incidence problem to determine the reflection and transmission characteristics, as shown in Figure 12 where θ_i is the incident angle of a shear vertical bulk wave.

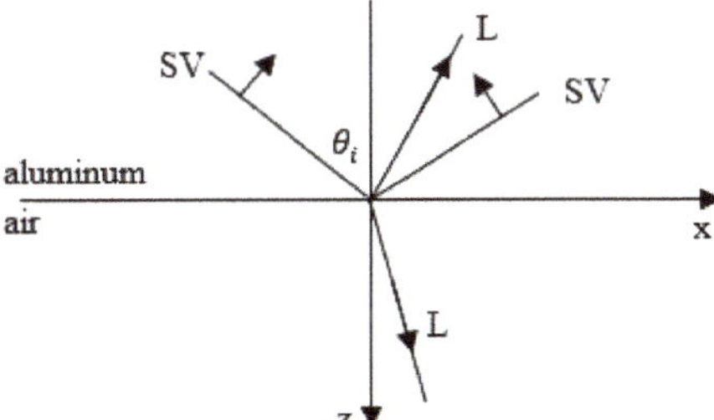

Figure 12. Depiction of the oblique plane wave problem being considered.

Displacement terms for each bulk wave are constructed according to particle polarization and wave propagation direction. The same BCs that are detailed in Equations (59)−(64) are applied to this problem. Along with Snell's law, these BCs are used to calculate the displacement amplitude ratios as a function of incidence angle, θ_i, as shown in Figure 13:

- R_L is the particle displacement amplitude ratio between the reflected longitudinal wave in the aluminum and the incident shear wave in the aluminum.
- R_S is the particle displacement amplitude ratio between the reflected shear wave in the aluminum and the incident shear wave in the aluminum.
- T_L is the particle displacement amplitude ratio between the transmitted longitudinal wave in the air and the incident shear wave in the aluminum.

It is noted that in place of $\sin\theta$, the ratio of wavenumbers were used, e.g., $\sin\theta_{Al}^{(L)} = k_x^{(L)}/|k_i^{(L)}|$.

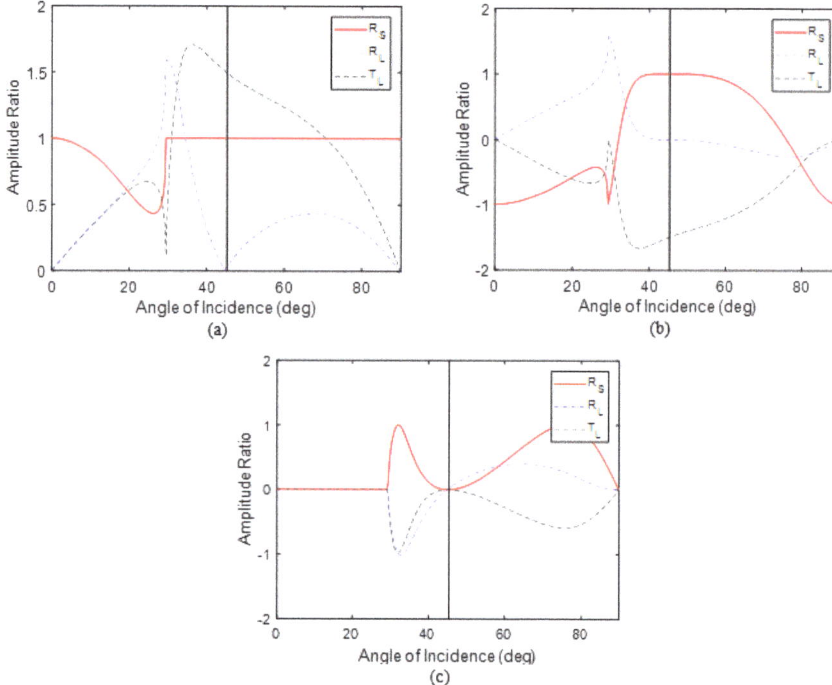

Figure 13. (**a**) The absolute value, (**b**) real part, and (**c**) imaginary part of the amplitude ratios for an incident shear vertical bulk wave at an aluminum–air interface.

By setting up Snell's law to be an equivalence between a bulk wave and an arbitrary wave in the same plate, the relation is:

$$c_S^{-1}\sin\theta_i = \left(\frac{|k_i|}{\omega}\right)\sin\theta. \tag{68}$$

Considering that θ is defined with respect to the z-axis, Snell's law becomes:

$$c_S^{-1}\sin\theta_i = k_x/\omega = c_p. \tag{69}$$

The black vertical line in Figure 13 corresponds with an incidence angle of about 45.45°, which translates to a phase velocity of 4350 m/s by Snell's law. For reference, a phase velocity at the shear wave speed, 3100 m/s, corresponds with an incidence angle of 90°, and a phase velocity at the longitudinal wave speed, 6350 m/s, corresponds with an incidence angle of 29°. The angles in the vicinity of 45.45° are associated with very little mode conversion to longitudinal modes, but most importantly to a large transmission of acoustic longitudinal waves to the air half-space. The behavior of the waves within the plate is consistent with what is Region 2 of Figure 8b.

Region 3 (c) would correspond to smaller angles of incidence as the phase velocity increases. For angles of incidence that are less than the first critical angle (i.e., $k_x/\omega < 1/c_L$), the mode conversion to longitudinal waves is significant with progressively less transmission of acoustic leakage as the angle of incidence decreases. At the larger phase velocities in Region 3 (Figure 1c) (i.e., $c_p > c_L$), the longitudinal bulk waves traveling through the plate would then begin to contribute significantly to any acoustic leakage.

In many studies of leaky Lamb waves, the fluid medium in which the wave leaks is water; thus, it is worth mentioning that the dispersion curves do not change significantly when water coupling is included [15,16]. As a result, many of the conclusions made in this example calculation pertaining to the form of the acoustic leakage can be found to agree, to some degree, with the findings in Hayashi and Inoue's paper [15]. Specifically, the findings that the oblique propagation of the longitudinal acoustic wave in the water and the evanescent propagation of it at a phase velocity less than the longitudinal wave speed in water agree. We reiterate that Hayashi and Inoue [15] and Gravenkamp et al. [16] focus on the study of the dispersion curves of leaky Lamb waves, which is different from what was done in this example calculation.

9. Conclusions

This paper exploits the Christoffel equation, slowness curves, and the partial-wave method to establish a foundation on which guided waves can be analyzed, compared, and explained. The approach was used to calculate the wave structures of many types of guided waves, once the frequency–wavenumber pairs were known. It is, of course, possible to calculate the wave structures using the well-known methods for dispersion curve calculation, but those methods can often be calculation-intensive. There are analytical expressions, but those leave little room for conceptual interpretation. Thus, the approach detailed in this paper is seen as a relatively easy, general, and informative method for calculating wave structures of specific modes of interest. Indicative of that are the Lamb wave characteristics observed in the paper, which relate to the phase difference between the top and bottom partial waves, and the amount by which each partial wave contributes to each Lamb wave. It was shown, among other things, that in Region 2 (Figure 1b), there is a phase velocity range in which the Lamb waves consist of partial shear vertical bulk waves traveling at ±45°. These are known as Lame waves, and they exhibit the most leakage into the fluid half-spaces surrounding the plate. Likewise, there is a similar band where all of the modes are dominated by longitudinal surface waves. It was also shown that the symmetric and anti-symmetric modes can be characterized by the phase difference between the partial waves associated with the top surface and those associated with the bottom surface.

The example calculation of the leaky characteristics of Lamb waves is applicable to air-coupled transducers. The optimum angle calculation is merely Snell's law, but the calculation of the amplitude of acoustic leakage provides us with a potentially simple methodology to compare the amount of acoustic leakage from various Lamb waves. Also, the potential relationship between the acoustic leakage from Lamb waves and the well-known oblique incidence problems could help to explain some of the acoustic leakage phenomenon using familiar and relatively simple theories. Beyond Lamb waves, a similar approach can be used for Rayleigh waves.

For other guided waves such as the Lamb-like waves in a plate on a half-space, the method detailed in this paper should still apply. That is, there will be six partial waves (i.e., four in the plate and two in the half-space) and six BCs (i.e., two traction-free at the top of the plate, and two traction continuity and two displacement continuity at the interface).

Lastly, the partial-wave method is not at all restricted to isotropic materials; it can also be used with anisotropic materials. If a similar approach is attempted for anisotropic materials, the immediately discernable differences would be that there are four regions instead of three in the dispersion curves, and three slowness curves instead of two. Some symmetry in the eigensolutions of the Christoffel equation may also disappear as a result of the anisotropy. That being said, the general approach is still expected to hold true, although further analysis is needed to confirm this.

Author Contributions: C.H. conducted the modeling and drafted the manuscript. C.L. advised C.H. and helped edit the manuscript.

Funding: This research received no external funding.

Conflicts of Interest: The authors declare no conflicts of interest.

References

1. Solie, L.; Auld, B. Elastic waves in free anisotropic plates. *J. Acoust. Soc. Am.* **1973**, *54*, 50–65. [CrossRef]
2. Auld, B. *Acoustic Fields and Waves in Solids*; Robert, E., Ed.; Krieger Publishing Company: Malabar, FL, USA, 1990.
3. Rozzi, T.; Mongiardo, M. *Open Electromagnetic Waveguides*; The Institution Engineering and Technology: London, UK, 1997.
4. Rose, J.L. *Ultrasonic Waves in Solid Media*; Cambridge University Press: New York, NY, USA, 1999.
5. Lowe, M. Matrix Techniques for Modeling Ultrasonic Waves in Multilayered Media. *IEEE Trans. Ultrason. Ferroelectr. Freq. Control.* **1995**, *42*, 525–542. [CrossRef]
6. Crandall, S.H. On the Use of Slowness Diagrams to Represent Wave Reflections. *J. Acoust. Soc. Am.* **1970**, *47*, 1338–1342. [CrossRef]
7. Henneke, E.G. Reflection-Refraction of a Stress Wave at a Plane Boundary between Anisotropic Media. *J. Acoust. Soc. Am.* **1972**, *51*, 210–217. [CrossRef]
8. Achenbach, J. *Wave Propagation in Elastic Solids*; North Holland Elsevier: New York, NY, USA, 1973.
9. Viktorov, I. *Rayleigh and Lamb Waves: Physical Theory and Applications*; Springer Science+Business Media: New York, NY, USA, 1967.
10. Blackstock, D. *Fundamentals of Physical Acoustics*; John Wiley and Sons, Inc.: New York, NY, USA, 2000.
11. Stoneley, R. Elastic Waves at the Surface of Separation of Two Solids. *R. Soc. Proc.* **1924**, *106*, 416–428. [CrossRef]
12. Scholte, J. The Range of Existence of Rayleigh and Stoneley Waves. *Geophys. J. Int.* **1947**, *5*, 120–126. [CrossRef]
13. Scholte, J. On the Stoneley-wave equation. I. *Proc. Acadamy Sci. Amst. (KNAW)* **1942**, *45*, 20–25.
14. Matsuda, N.; Biwa, S. Phase and group velocity matching for cumulative harmonic generation in Lamb waves. *J. Appl. Phys.* **2011**, *109*, 094903. [CrossRef]
15. Hayashi, T.; Inoue, D. Calculation of leaky Lamb waves with a semi-analytical finite element method. *Ultrasonics* **2014**, *54*, 1460–1469. [CrossRef] [PubMed]
16. Gravenkamp, H.; Birk, C.; Song, C. Numerical modeling of elastic waveguides coupled to infinite fluid media using exact boundary conditions. *Comput. Struct.* **2014**, *141*, 36–45. [CrossRef]

Article

Lamb Wave Local Wavenumber Approach for Characterizing Flat Bottom Defects in an Isotropic Thin Plate

Guopeng Fan [1], Haiyan Zhang [1,*], Hui Zhang [1], Wenfa Zhu [1,2] and Xiaodong Chai [2]

[1] School of Communication and Information Engineering, Shanghai University, Shanghai 200444, China; phdfanry@shu.edu.cn (G.F.); zh6154@126.com (H.Z.); zhuwenfa1986@163.com (W.Z.)
[2] School of Urban Railway Transportation, Shanghai University of Engineering Science, Shanghai 201620, China; xdchai@sues.edu.cn
* Correspondence: hyzh@shu.edu.cn; Tel.: +86-021-6613-7262

Received: 28 June 2018; Accepted: 24 July 2018; Published: 10 September 2018

Abstract: This paper aims to use the Lamb wave local wavenumber approach to characterize flat bottom defects (including circular flat bottom holes and a rectangular groove) in an isotropic thin plate. An air-coupled transducer (ACT) with a special incidence angle is used to actuate the fundamental anti-symmetric mode (A0). A laser Doppler vibrometer (LDV) is employed to measure the out-of-plane velocity over a target area. These signals are processed by the wavenumber domain filtering technique in order to remove any modes other than the A0 mode. The filtered signals are transformed back into the time-space domain. The space-frequency-wavenumber spectrum is then obtained by using three-dimensional fast Fourier transform (3D FFT) and a short space transform, which can retain the spatial information and reduce the magnitude of side lobes in the wavenumber domain. The average wavenumber is calculated, as a real signal usually contains a certain bandwidth instead of the singular frequency component. Both simulation results and experimental results demonstrate that the average wavenumber can be used not only to identify shape, location, and size of the damage, but also quantify the depth of the damage. In addition, the direction of an inclined rectangular groove is obtained by calculating the image moments under grayscale. This hybrid and non-contact system based on the local wavenumber approach can be provided with a high resolution.

Keywords: Lamb wave; local wavenumber; air-coupled transducer; wavenumber domain filtering; hybrid and non-contact system

---

1. Introduction

Lamb waves have shown great potential for structural health monitoring (SHM) in plate-like structures [1]. Their attractive features include sensitivity to a variety of damage types and the capability of traveling relatively long distances [2]. Damage imaging methods based on Lamb waves have been widely studied by many researchers, such as delay-and-sum (DAS) imaging [3,4], time reversal focusing imaging [5,6], diffraction tomography imaging [7], and ultrasonic phased array [8].

Lamb waves are dispersive and multimodal [9]. Moreover, the propagating Lamb waves may include incident, reflected, and converted waves when they encounter a sudden thickness variation [10,11], such as circular flat bottom holes and rectangular grooves. Obviously, various wave modes make the interpretation of Lamb waves very difficult [12].

In previous literature, the mode identification and separation of Lamb waves have been widely investigated in the time domain, time-frequency domain, and frequency-wavenumber domain [13,14]. Short-time Fourier transform (STFT) with a sliding window can make the variant distribution of the energy spectrum act as a function of time, which is capable of identifying Lamb wave mode [15].

Frequency-wavenumber analysis based on two-dimensional fast Fourier transform (2D FFT) is a useful approach to distinguish multiple modes. Moreover, the individual Lamb mode can be separated by a 2D band-pass filter [16]. It should be noted that the spatial information is also lost [17].

Compared with the above methods, the local wavenumber approach combined with a spatial window is viewed as an effective way to obtain the space-frequency-wavenumber spectrum. Furthermore, the spatial information can be reserved and the local features of the material can be clearly characterized, including, for instance, the thickness distribution of the material.

Many researchers have tried to utilize the above techniques for SHM. Yu et al. adopted a short space 2D Fourier transform to obtain the frequency-wavenumber spectrum at various spatial locations. Consequently, a through-thickness crack of an aluminum plate was discriminated by analyzing the space-frequency-wavenumber spectrum [18], but the size of the crack was not quantified. In another study, Yu et al. proved that spatial wavenumber imaging method was able to characterize the location and length of a crack. However, multiple cracks and cracks with various orientations were not researched in their work [19]. Rogge et al. utilized the local wavenumber analysis to characterize the impact damage in composite laminates [20]. The location of a preset delamination was clearly characterized, while its depth was not quantified.

Rao et al. reconstructed the thickness distribution of a liquid loaded plate by using an ultrasonic guided wave tomography method based on full waveform inversion (FWI) [21,22]. It was noted that the remaining wall thickness was reconstructed with high accuracy and resolution. Chronopoulos et al. adopted an inverse wave and finite element approach to recover the thickness and density, as well as all independent mechanical characteristics of layered composite structures [23]. However, one drawback of this method is that it is not efficient, as the inversion process is complicated and time-consuming. Therefore, a rapid method for SHM is desired.

Air-coupled ultrasonic inspection is a promising non-contact method for the rapid, non-contact inspection of materials [24,25]. The main advantages of this technique are the absence of any contact and the ability to generate a relatively pure Lamb wave mode when an appropriate incidence angle is adopted. The laser Doppler vibrometer (LDV) based on the Doppler effect has a high spatial resolution in obtaining the propagating Lamb waves [26]. Harb and Yuan accomplished the target of actuating the pure A0 mode (anti-symmetric mode) in an isotropic aluminum plate by employing an air-coupled transducer (ACT) and an LDV [27].

A hybrid non-contact system composed of an ACT and an LDV is presented in this paper. The local wavenumber approach is adopted to acquire the average wavenumber distribution of an isotropic thin plate, which contains circular flat bottom holes or a rectangular groove. The main goal is not only to extract defect features in location, size, and shape, but also quantify the depth of the damage by combining the wavenumber with the phase velocity dispersion curve. The direction of a rectangular groove can be attained by calculating the image moments under grayscale. Several factors which can affect test quality are analyzed in the discussion section. Moreover, the solutions are presented when the proposed approach is applied for an unknown damage.

This paper is organized with seven sections, including this introduction. Section 2 illustrates the theory of the local wavenumber approach. Section 3 presents the results of tests performed on numerically simulated examples. Experimental results are explicated in Section 4. Section 5 concludes the discussion. Section 6 provides recommendation for future research. Section 7 summarizes the main work.

2. Theory for the Local Wavenumber Approach

2.1. Sound Field Transmission and Reception

The A0 mode is the appropriate Lamb wave mode for characterizing circular flat bottom holes and a rectangular groove in an isotropic thin plate, as the phase velocity of the A0 mode is sensitive to the thickness-frequency product. First of all, a special incidence angle can be easily calculated based on Snell's law. Then, the fundamental anti-symmetric mode (A0) can be actuated by an ACT with this special incidence angle. A scanning grid is predefined before using a non-contact LDV to measure the velocities of the desired mode. Obviously, the velocity $U(x, y, t)$ is a function of both time and space. Figure 1 shows the theoretical dispersion curves obtained by a commercial software package called Disperse (version 2.0.20a, Imperial College London, London, UK) [28]. In this paper, the incidence angle is set at 12.75°, which is suitable for an ACT to actuate the A0 mode in an aluminum plate with the frequency-thickness product of 0.3 MHz·mm. Figure 2 shows a schematic diagram, which is used for describing the process of sound field transmission and reception.

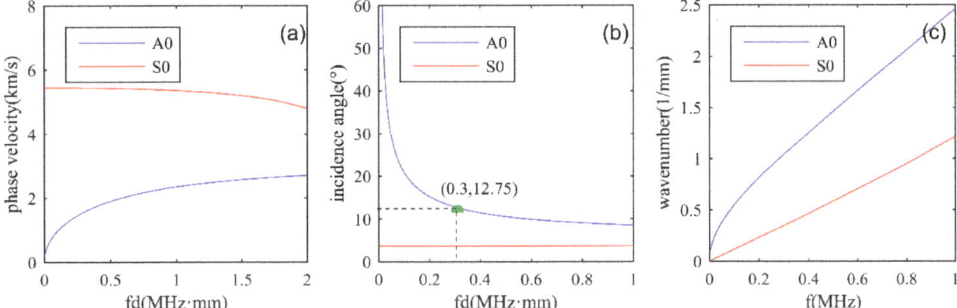

Figure 1. Dispersion curves of an aluminum plate. (**a**) Phase velocity; (**b**) incidence angle; (**c**) wavenumber.

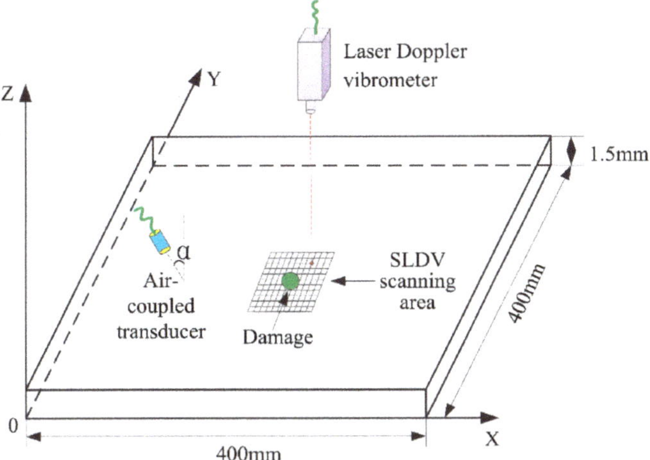

Figure 2. Schematic diagram of Lamb wave transmission and reception.

2.2. Identification and Separation of Lamb Wave Modes in the Wavenumber Domain

Only the fundamental anti-symmetric mode (A0) can be actuated by ACT with the special incidence angle; however, mode conversion still exists due to the interaction between the Lamb wave and the flat bottom defects. The received Lamb wave signals usually comprise the incident, reflected, and converted wave modes. By using three-dimensional fast Fourier transform (3D FFT) [29], the full wavefield data can be transformed into the wavenumber domain, where various wave modes can be distinguished.

$$U(k_x, k_y, f) = F_{3D}(U(x, y, t)) \tag{1}$$

The entire wavefield signals are filtered by the wavenumber domain filter for the purpose of removing any modes other than the A0 mode.

$$\widetilde{U}(k_x, k_y, f) = W_k(k_x, k_y, f)U(k_x, k_y, f) \tag{2}$$

$$W_k(k_x, k_y, f) = \begin{cases} 0 & |k| \leq k_1 \\ 1 & k_1 < |k| < k_2 \\ 0 & |k| \geq k_2 \end{cases} \tag{3}$$

$$k = \begin{bmatrix} k_x \\ k_y \end{bmatrix} \tag{4}$$

These filtered signals in the wavenumber domain should be transformed back into the time domain by using three-dimensional inverse fast Fourier transform (3D IFFT). $\widetilde{U}(x, y, t)$ is the filtered wavefield.

$$\widetilde{U}(x, y, t) = F_{3D}^{-1}(\widetilde{U}(k_x, k_y, f)) \tag{5}$$

2.3. Acquisition of the Average Wavenumber

A spatial window is used for reducing the magnitude of the side lobes. It should be noted that the spatial window size is at least twice the wavelength of the A0 mode in this paper. The centered coordinate of this spatial window is located at (x_m, y_n). Under this circumstance, the process of obtaining the average wavenumber is clearly illustrated as below.

First, a short space transform is conducted in order to retain the spatial information and reduce the magnitude of side lobes in the wavenumber domain. Specifically, the wavefield data is multiplied by a spatial window to produce the windowed dataset $\widetilde{U}_{mn}(x, y, t)$. The spatial window is non-zero for only a short period in space. The Hanning window W_{xy} is usually adopted [30].

$$\widetilde{U}_{mn}(x, y, t) = \widetilde{U}(x, y, t)W_{xy} \tag{6}$$

$$W_{xy} = \begin{cases} 0.5\left[1 + \cos(2\pi \frac{r}{D_x})\right] & \text{if } r \leq D_x/2 \\ 0 & \text{otherwise} \end{cases} \tag{7}$$

where r and D_x can be expressed as:

$$r = \sqrt{(x - x_m)^2 + (y - y_n)^2} \tag{8}$$

$$D_x \geq 2\lambda \tag{9}$$

Here, λ denotes the wavelength of the Lamb wave. D_x is the window length. These signals processed with the spatial window are transformed into the spatial-frequency-wavenumber domain by using 3D FFT.

$$\widetilde{U}_{mn}(k_x, k_y, f) = F_{3D}[\widetilde{U}_{mn}(x, y, t)] \tag{10}$$

The weight of $\|k\|$ must be considered before calculating the local wavenumber k_{mn}, as presented in the following equation:

$$k_{mn}(f) = \frac{\sum\limits_{k}\left[\left|\widetilde{U}_{mn}(k_x, k_y, f)\right|^2 \|k\|\right]}{\sum\limits_{k}\left|\widetilde{U}_{mn}(k_x, k_y, f)\right|^2} \tag{11}$$

where $\|k\|$ is the Euclidean norm of the 2D wavenumber range [31]. The average wavenumber $\overline{k_{mn}}$ over the selected frequency band is calculated because a real signal usually contains a certain bandwidth instead of the singular frequency component. The average wavenumber of each point can be obtained by changing the centered coordinate of this spatial window and repeating the above steps. The flow chart of the local wavenumber approach is shown in Figure 3.

$$\overline{k_{mm}} = \frac{1}{N}\sum_{i=1}^{N} k_{mn}(f_i) \tag{12}$$

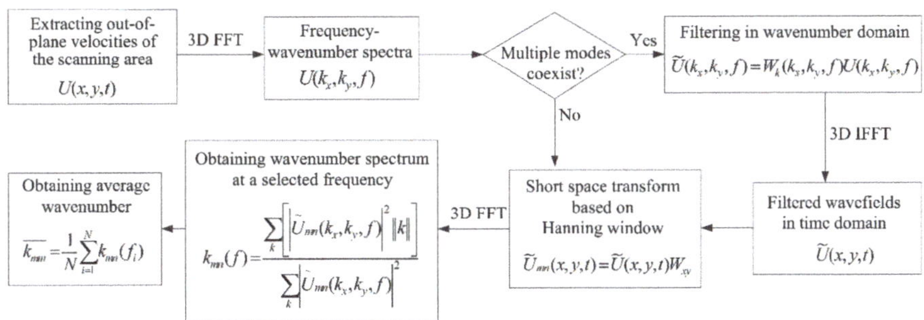

Figure 3. Flow chart of the local wavenumber approach.

2.4. Depth Reconstruction

The average wavenumber $\overline{k_{mn}}$ at each point can be obtained by the local wavenumber approach. Hence, the phase velocity $C_p(x_m, y_n)$ can be calculated as below:

$$C_p(x_m, y_n) = \frac{2\pi f}{\overline{k_{mn}}} \tag{13}$$

The relationship between the phase velocity and the plate thickness can be gained from the phase velocity dispersion curve shown in Figure 1a. Consequently, the plate thickness $d(x_m, y_n)$ of the target area can be reconstructed according to this relationship. Then, the depth distribution can be written as:

$$h(x_m, y_n) = D - d(x_m, y_n) \tag{14}$$

where D is the thickness of the pristine plate. In order to evaluate the depth, the maximum depth error E_{depth} is defined as below:

$$E_{depth} = \frac{|h_{\max} - h_0|}{h_0} \tag{15}$$

Here, $h_{\max}$ denotes the maximum value of the reconstructed depth and h_0 is the actual value.

3. Finite Element Simulation

3.1. Finite Element Model

A commercial software package called PZFlex was used to create the finite element model, which is shown in Figure 4. PZFlex is the time domain finite element method (FEM). The complex acoustic wave propagation in a three-dimensional model can be simulated by using this method [32]. The upper and bottom surfaces of the model are free, while the remaining four sides are absorbing. The size of the 6061-T6 aluminum plate is 400 mm × 400 mm × 1.5 mm, and the other material properties are listed in Table 1. The incident point of the ACT is (100, 200), and its incidence angle is set at 12.75°, which is suitable to actuate the A0 mode on aluminum with the frequency-thickness product of 0.3 MHz·mm. The damage located at the middle of the plate is a circular flat bottom hole with a diameter of 10 mm and a depth of 1 mm. This damage is surrounded by a rectangular scanning area (60 mm × 60 mm), which is divided into 121 × 121 points with a spatial interval of 0.5 mm. The spatial interval should not be more than half of the wavelength according to the sampling theorem. In our test, the phase velocity of the A0 mode is about 1551 m/s, and the central frequency of the received signal is 200 kHz. Therefore, the wavelength of the A0 mode is 7.775 mm (λ = 1511/200 = 7.775 mm). The sampling interval is only 0.5 mm (λ/15). Apparently, the sampling interval satisfies the sampling theorem and can provide a good spatial resolution. The out-of-plane velocities over the scanning area are recorded by the LDV.

In this paper, a five-cycle sinusoidal tone burst modulated by a Hanning window with a central frequency of 200 kHz is applied as the excitation signal, which is shown in Figure 5. Eleven specimens listed in Table 2 are used to evaluate the performance of the local wavenumber approach from different aspects.

Case 1 and Case 2 are used to evaluate the performance of the proposed method when the circular flat bottom hole changes in depth.

Case 1 and Case 3 are used to evaluate the performance of the proposed method when the circular flat bottom hole changes in radius.

Case 4 to Case 7 are used to study the impacts caused by the excitation source position and the distance among adjacent defects. The excitation source in Case 4 and Case 5 is located at (100, 200), while it is located at (200, 100) in Case 6 and Case 7. The gap between the two flat bottom holes is 20 mm in Case 4 and Case 6, while it is 10 mm in Case 5 and Case 7.

Case 8 and Case 9 are used to investigate whether the impact which the big hole exerts on the small one is connected with their positions relative to the excitation source when they are in the same line.

Case 10 and Case 11 are used to identify the direction of the rectangular groove.

Table 1. Properties of the 6061-T6 aluminum plate.

Property	Value
Density	2700 (kg/m^3)
Young's modulus	68.9 (GPa)
Poissons'ratio	0.33

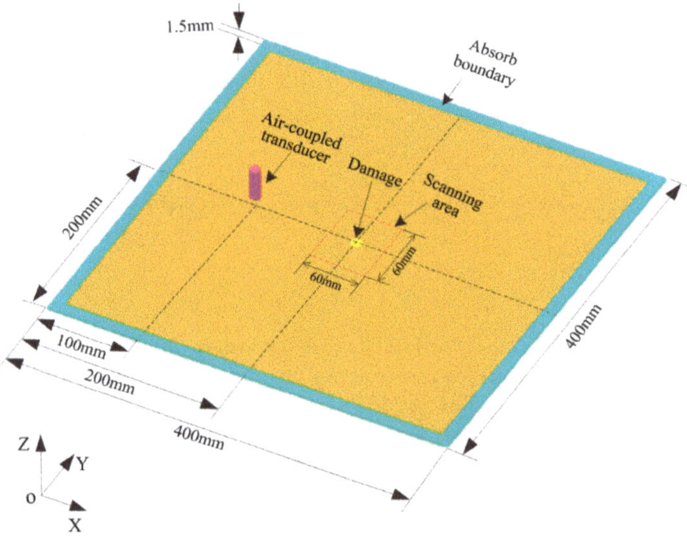

Figure 4. Finite element model.

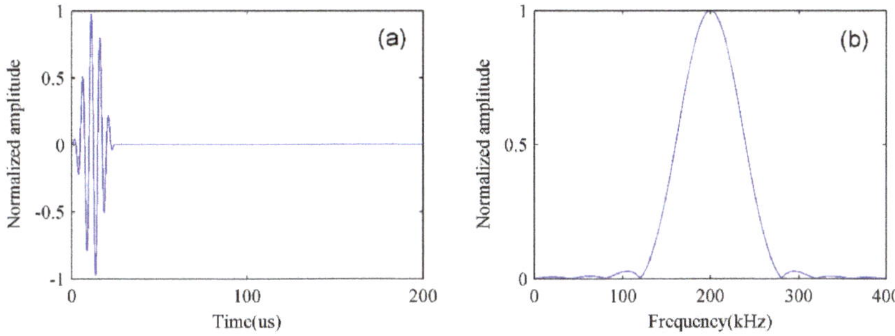

Figure 5. The excitation signal in the (**a**) time-domain; (**b**) frequency-domain.

Table 2. Specimen parameters.

Specimen	Damage Type	Excitation Source Position (mm)	Central Position of Damage (mm)	Radius (mm)	Depth (mm)	Inclined Degree (°)	Width (mm)	Length (mm)
Case 1	One flat bottom hole	(100, 200)	(200, 200)	5	1	/	/	/
Case 2	One flat bottom hole	(100, 200)	(200, 200)	5	0.5	/	/	/
Case 3	One flat bottom hole	(100, 200)	(200, 200)	3.5	1	/	/	/
Case 4	Two flat bottom holes	(100, 200)	No. 1: (200, 215)	5	1	/	/	/
			No. 2: (200, 185)	5	1	/	/	/
Case 5	Two flat bottom holes	(100, 200)	No. 1: (200, 210)	5	1	/	/	/
			No. 2: (200, 190)	5	1	/	/	/
Case 6	Two flat bottom holes	(200, 100)	No. 1: (200, 215)	5	1	/	/	/
			No. 2: (200, 185)	5	1	/	/	/
Case 7	Two flat bottom holes	(200, 100)	No. 1: (200, 210)	5	1	/	/	/
			No. 2: (200, 190)	5	1	/	/	/
Case 8	Two flat bottom holes	(200, 100)	No. 1: (200, 210)	5	1	/	/	/
			No. 2: (200, 190)	3.5	1	/	/	/
Case 9	Two flat bottom holes	(200, 100)	No. 1: (200, 210)	3.5	1	/	/	/
			No. 2: (200, 190)	5	1	/	/	/
Case 10	One groove	(100, 200)	(200, 200)	/	1	90	4	24
Case 11	One groove	(100, 200)	(200, 200)	/	1	135	4	24

3.2. FEM Results

3.2.1. Mode Separation

The interactions between Lamb waves and the damage have been analyzed by many researchers. Bhuiyan et al. adopted the wave damage interaction coefficient (WDIC) to investigate the scattering of Lamb waves and demonstrated that the damage in the plate acted as a non-axisymmetric secondary source [33]. In another report, Bhuiyan et al. described the 3D interaction (frequency, incident direction, and azimuth direction) of Lamb waves with the damage [34]. Testoni et al. adopted a wavenumber filter to filter out the main wavefield generated by the acoustic source. Thus, the wave scattered by the delamination can be clearly observed in the filtered signal [35]. In this paper, Case 1 is taken as an example to describe the interaction between Lamb waves and the damage; moreover, the mode separation is introduced in detail.

The snapshots of the Lamb waves are shown in Figure 6. It can be observed that only the A0 mode can be actuated by an ACT with a special incidence angle. However, the S0 mode will be generated when the A0 mode encounters a flat bottom defect. Following this, the incident A0, reflected A0, transmitted A0, and S0 mode will coexist for a long time. It is worth noting that the amplitude of the S0 mode is very low.

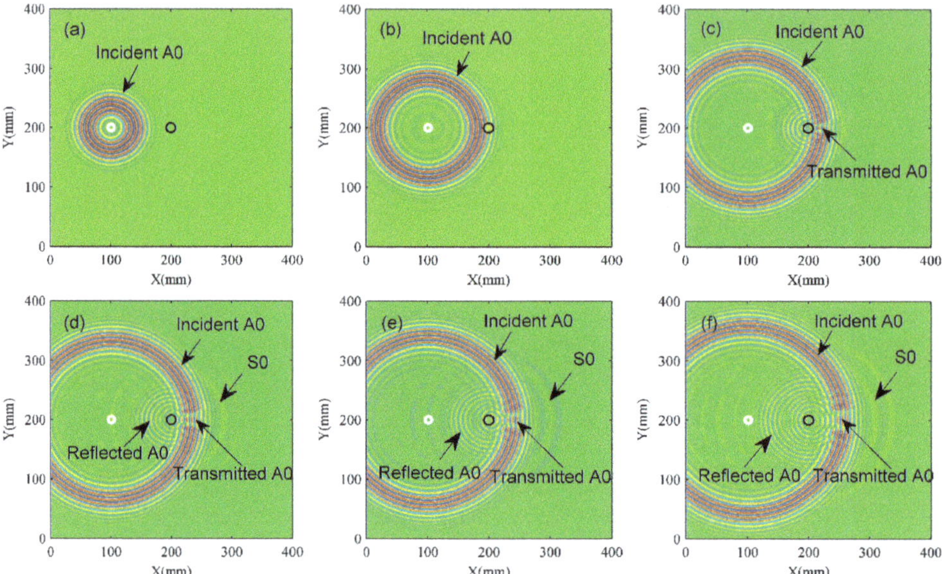

Figure 6. Snapshots of Lamb waves. (**a**) 74.18 us; (**b**) 91.31 us; (**c**) 105.89 us; (**d**) 111.60 us; (**e**) 114.14 us; (**f**) 120.48 us. The white circle denotes the incident point of the air-coupled transducer (ACT). The black circle represents the actual location of a defect.

The Lamb wave propagation was analyzed in detail. The next section will be focused on mode separation. The frequency-thickness product is 0.3 MHz·mm at the undamaged area. Under this situation, the theoretical wavenumbers of the A0 mode and S0 mode are 0.8102 rad/mm and 0.2312 rad/mm, respectively. The frequency-thickness product is only 0.1 MHz·mm at the damage. Hence, the theoretical wavenumbers of the A0 mode and S0 mode are 1.3082 rad/mm and 0.2311 rad/mm, respectively. Apparently, the wavenumber of the A0 mode changes more obviously than that of the S0 mode when the thickness changes. Figure 7 shows the mode separation by using 3D FFT and the wavenumber domain filtering technique. The wavenumber spectrum is obtained at a frequency of 200 kHz. The images are given in dB scale so that the wavenumber spectrum can be clearly visualized. The wavenumber spectrum for the pristine plate is shown in Figure 7a. The wavenumber spectrum for the damaged plate is shown in Figure 7b. The high-pass filter in wavenumber domain is shown in Figure 7c. The filtered wavenumber spectrum is shown in Figure 7d. By comparing the wavenumber spectrum in Figure 7a,b, we can confirm the fact that the A0 mode has interacted with the defect. In addition, the S0 mode generated by mode conversion is very weak, as the amplitude of the S0 mode is much lower than that of the A0 mode in the time domain.

Generally, a band-pass filter should be applied to reserve the A0 mode according to the theory of the local wavenumber approach. However, the wavenumber distribution of the A0 mode is wide in our test, while the wavenumber distribution of the S0 mode is very narrow. Moreover, the intensity of the S0 mode is weak. Therefore, it is easy to design a high-pass filter rather than a band-pass filter. The high-pass filter should ensure that the wavenumbers of the A0 mode are included in the pass band.

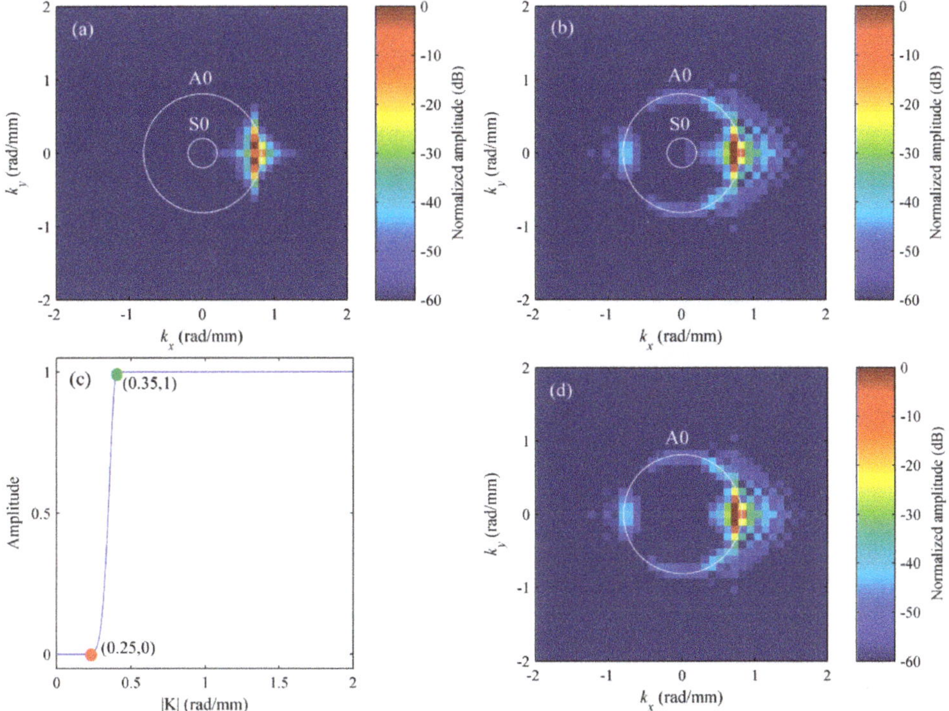

Figure 7. Mode separation based on the spatial-wavenumber filtering technique. (**a**) Wavenumber spectrum for the pristine plate; (**b**) wavenumber spectrum for the damaged plate; (**c**) high-pass filter in the wavenumber domain; (**d**) filtered wavenumber spectrum. White circles are theoretical wavenumber curves for the S0 and A0 modes.

3.2.2. FEM Results of a Circular Flat Bottom Hole

In this paper, the wavelength of the A0 mode is 7.775 mm. Therefore, the window radius ($D_x/2$) should be larger than 7.775 mm. It is assumed that the window radius is 8 mm. The average wavenumber can be obtained using Equation (12). Figure 8 shows the acquisition of the average wavenumber for Case 1. The frequency interval is 15 kHz. It can be seen that the average wavenumber contributes to improve the test quality at the scanning boundary by comparing Figure 8d,h. Moreover, lower image noise is seen in the image when the average wavenumber is adopted.

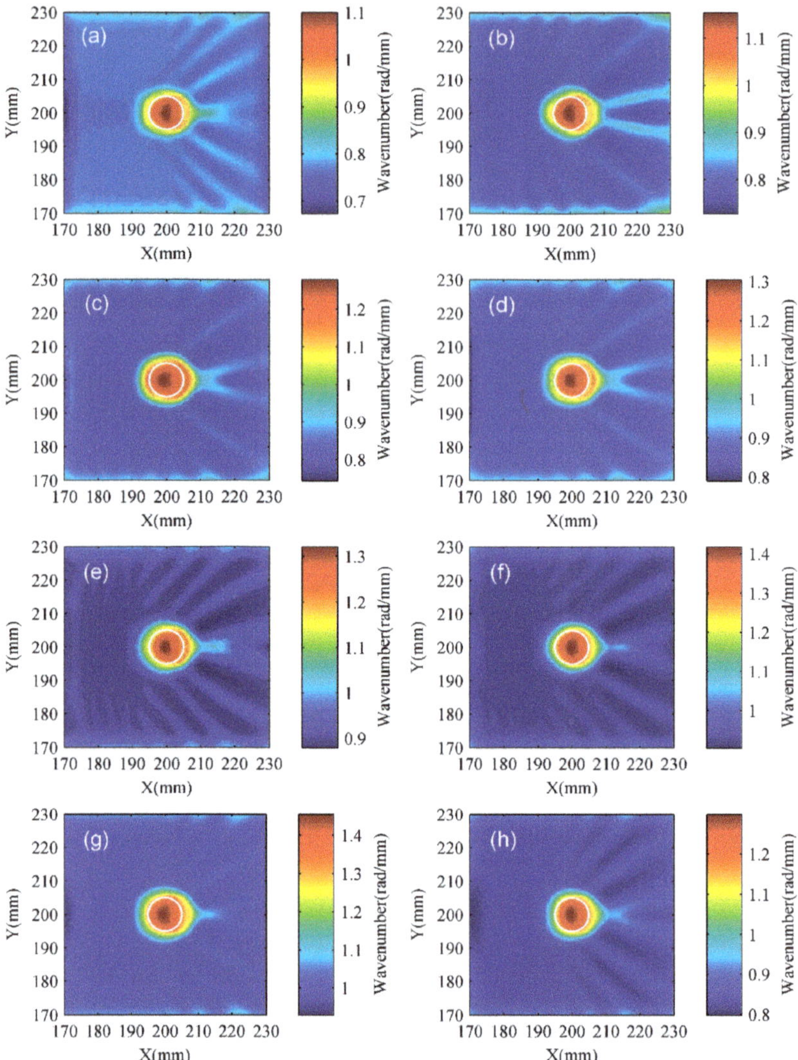

Figure 8. The acquisition of average wavenumber for Case 1. (**a**) Wavenumber distribution obtained at f = 155 kHz; (**b**) wavenumber distribution obtained at f = 170 kHz; (**c**) wavenumber distribution obtained at f = 185 kHz; (**d**) wavenumber distribution obtained at f = 200 kHz; (**e**) wavenumber distribution obtained at f = 215 kHz; (**f**) wavenumber distribution obtained at f = 230 kHz; (**g**) wavenumber distribution obtained at f = 245 kHz; (**h**) average wavenumber distribution.

In fact, the best test result is not always obtained under the condition in which the window radius is equal to the wavelength. It is meaningful to analyze the test results under various window radii in order to find the optimum size. The test results shown in Figure 9 indicate that the damage region characterized by the proposed approach will expand as the window radius increases. Meanwhile, the maximum wavenumber attained under the various radii will decrease as the window radius increases, which is shown in Figure 10. It is only the window radius of 8 mm that can make the identified damage region match well with its actual size. Therefore, 8 mm is the optimum size. All of the test results were obtained under this condition.

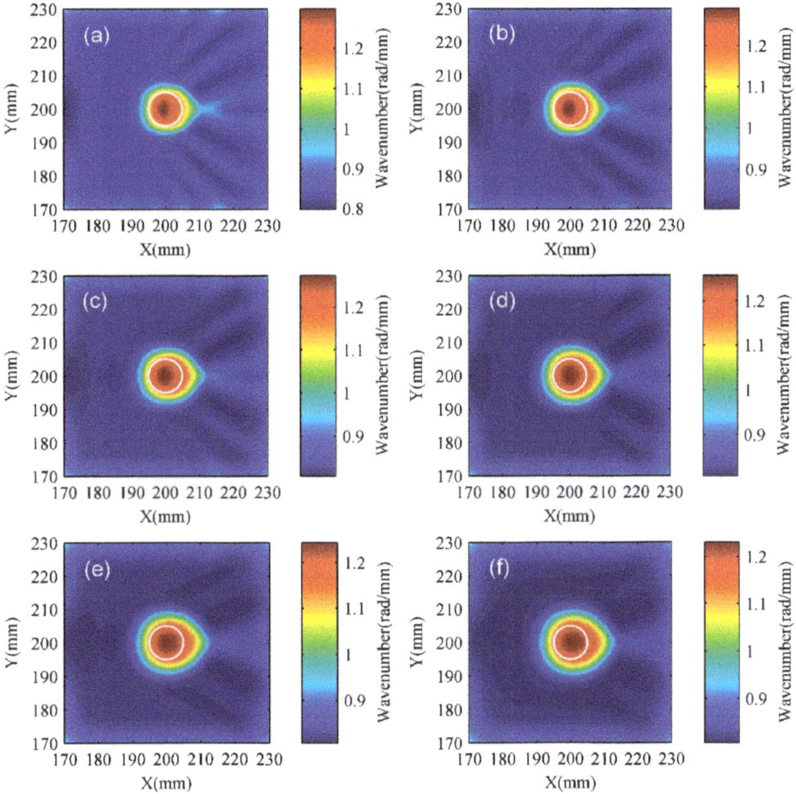

Figure 9. The test results obtained under various window radii. (**a**) 8 mm; (**b**) 9 mm; (**c**) 10 mm; (**d**) 11 mm; (**e**) 12 mm; (**f**) 13 mm. The white circle represents the actual location of the damage. The central frequency of the excitation signal is 200 kHz.

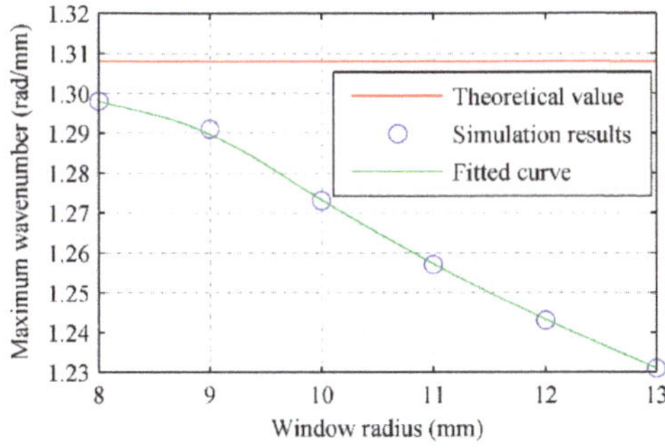

Figure 10. The maximum wavenumber obtained under various window radii. The central frequency of the excitation signal is 200 kHz.

Case 1 and Case 2 only contain a circular flat bottom hole with the same radius and different depths. Case 1 and Case 3 only contain a circular flat bottom hole with different radii and the same depth. The specimen parameters are listed in Table 2. The simulation results of Case 1 to Case 3 are shown in Figure 11. The size and shape of the defect can be clearly visualized. The depth estimations are shown in Table 3. The depth error of the deep hole is 1.13%, while it reaches 3.26% for the shallow hole. Apparently, the depth estimation of the deep hole is more accurate than that of the shallow one. The depth error of the big hole is 1.13%, while it reaches 8.17% for the small one. Hence, the depth estimation of the big hole is more accurate than that of the small one.

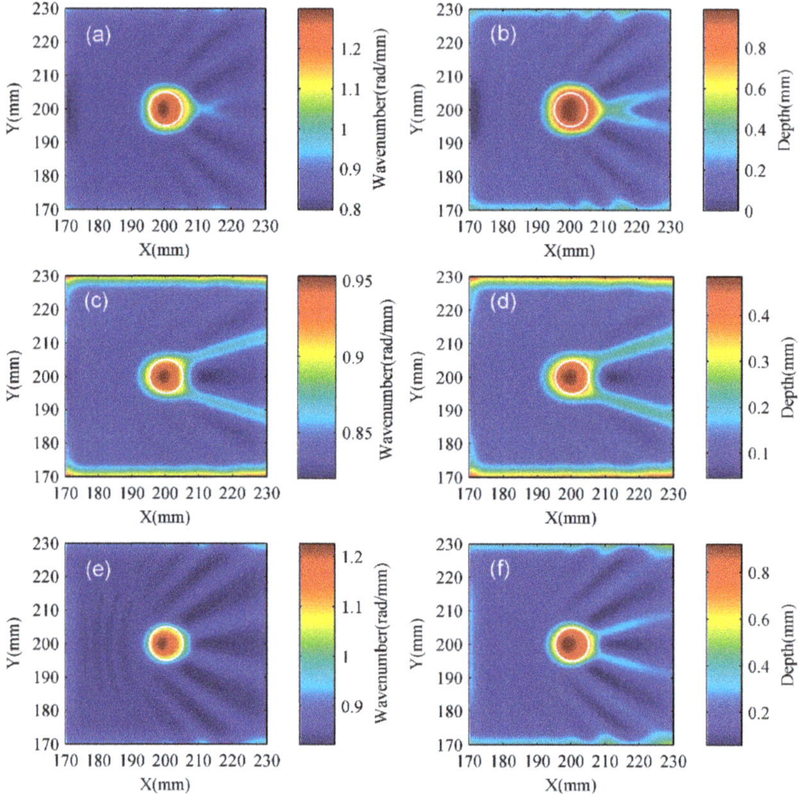

Figure 11. Simulation results of Case 1 to Case 3. (**a**) Wavenumber distribution of Case 1; (**b**) depth distribution of Case 1; (**c**) wavenumber distribution of Case 2; (**d**) depth distribution of Case 2; (**e**) wavenumber distribution of Case 3; (**f**) depth distribution of Case 3. The white circle represents the actual location of the damage. The actual defect depth of Case 1 and Case 3 is 1 mm, while the actual defect depth of Case 2 is 0.5 mm.

Table 3. The depth evaluations of Case 1 to Case 3.

Specimen	Actual Depth (mm)	Reconstructed Depth (mm)	Estimation Error of Depth
Case 1	1	0.9887	1.13%
Case 2	0.5	0.4837	3.26%
Case 3	1	0.9183	8.17%

3.2.3. FEM Results of Two Circular Flat Bottom Holes with the Same Radius and Depth

Case 4 to Case 7 contain two circular flat bottom holes with the same radius and depth, but the excitation source location and the gap between two defects are different. The specimen parameters are listed in Table 2. The simulation results of Case 4 to Case 7 are shown in Figure 12. It is observed that the shape and location of the defects are recognizable. The two holes exhibit a symmetrical distribution when the excitation source is located at (100, 200). In this condition, the better result appears under a big gap between the defects. The two holes and the excitation source are in the same line when the excitation source is located at (200, 100). In this condition, it can be noticed that the hole far away from the excitation source seems smaller than the near one. Moreover, the closer the defects are, the more obvious the effect is. Table 4 shows that the depth errors of Case 4 to Case 7 are all within 2%.

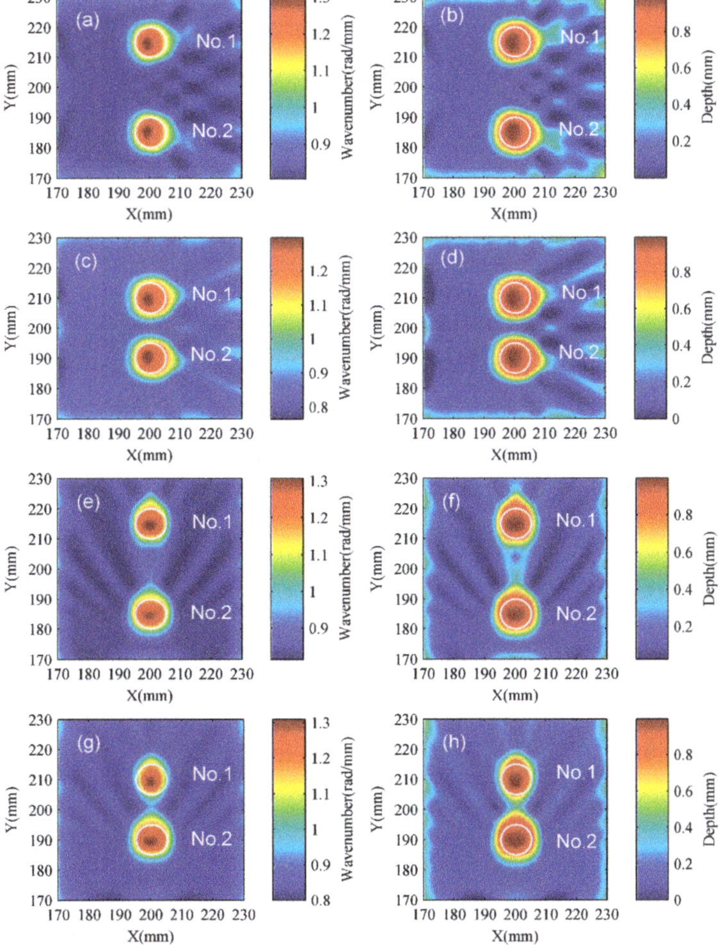

Figure 12. Simulation results of Case 4 to Case 7. (**a**) Wavenumber distribution of Case 4; (**b**) depth distribution of Case 4; (**c**) wavenumber distribution of Case 5; (**d**) depth distribution of Case 5; (**e**) wavenumber distribution of Case 6; (**f**) depth distribution of Case 6; (**g**) wavenumber distribution of Case 7; (**h**) depth distribution of Case 7. The white circle represents the actual location of the damage. The actual depth of the defect is 1 mm.

Table 4. The depth evaluations of Case 4 to Case 7.

Specimen		Actual Depth (mm)	Reconstructed Depth (mm)	Estimation Error of Depth
Case 4	No. 1	1	0.9917	0.83%
	No. 2	1	0.9917	0.83%
Case 5	No. 1	1	0.9887	1.13%
	No. 2	1	0.9887	1.13%
Case 6	No. 1	1	0.9946	0.54%
	No. 2	1	0.9976	0.24%
Case 7	No. 1	1	0.9887	1.13%
	No. 2	1	0.9976	0.24%

3.2.4. FEM Results of Two Circular Flat Bottom Holes with Different Radius and the Same Depth

Two circular flat bottom holes with different radii and the same depth are used in Case 8 and Case 9. It is important to note that these defects and the excitation source are in the same line, which means that the defects have the different positions relative to the excitation source. The specimen parameters are listed in Table 2. Simulation results are shown in Figure 13. It can be seen that the shape and location of two holes can be identified, which demonstrates that the local wavenumber approach has a high resolution even in distinguishing adjacent defects. Table 5 shows the depth evaluations of Case 8 and Case 9. The depth error of the big hole is within 1%, while the depth error of the small one is in the range of 4% to 9%. Thus, the depth estimation of the big hole is more accurate than that of the small one. Besides, the relative position is also taken into consideration. The depth error of the small hole is 8.17% when it is in front of the big one. Apparently, this level is equivalent to that of Case 3. However, the depth error of the small hole is reduced to 4.76% when it is behind the big one. This indicates that the impact which the big hole exerts on the small one is related to their positions when they and the excitation source are in the same line.

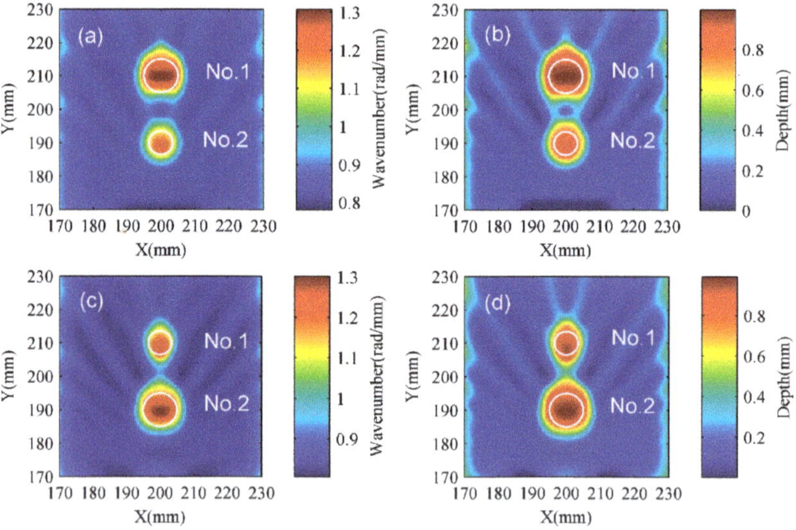

Figure 13. Simulation results of Case 8 and Case 9. (**a**) Wavenumber distribution of Case 8; (**b**) depth distribution of Case 8; (**c**) wavenumber distribution of Case 9; (**d**) depth distribution of Case 9. The white circle represents the actual location of the damage. The actual depth of the defect is 1 mm.

Table 5. The depth evaluations of Case 8 and Case 9.

Specimen		Actual Depth (mm)	Reconstructed Depth (mm)	Estimation Error of Depth
Case 8	No. 1	1	0.9976	0.24%
	No. 2	1	0.9183	8.17%
Case 9	No. 1	1	0.9524	4.76%
	No. 2	1	0.9917	0.83%

3.2.5. FEM Results of an Inclined Rectangular Groove

Case 10 is a rectangular groove with an inclined angle of 90°. Case 11 is a rectangular groove with an inclined angle of 135°. The specimen parameters are listed in Table 2. The simulation results of Case 10 and Case 11 are shown in Figure 14. The damage shape can be clearly visualized. It should be noted that the length of this defect is a little shorter than the actual value, while its width is obviously larger than the actual value. In addition, the orientation of the rectangular groove can be easily obtained and the computational process is introduced as below.

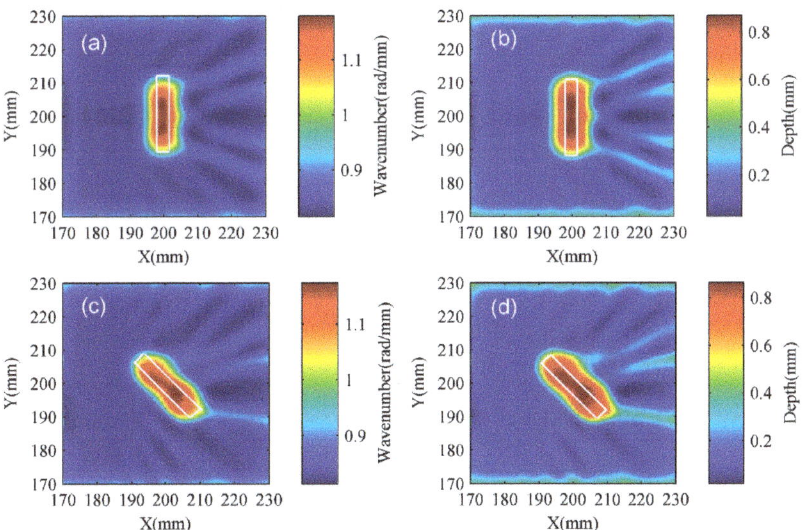

Figure 14. Simulation results of Case 10 and Case 11. (**a**) Wavenumber distribution of Case 10; (**b**) depth distribution of Case 10; (**c**) wavenumber distribution of Case 11; (**d**) depth distribution of Case 11. The white rectangle represents the actual location of the damage. The actual depth of the defect is 1 mm.

In image processing, computer vision, and related fields, an image moment is a certain particular weighted average (moment) of the image pixels' intensities. Image moments and moment invariants are useful to describe global features in image processing [36]. Image moments are calculated under grayscale in order to distinguish the direction of the rectangular groove. $G(x, y)$ denotes the pixel intensity of the gray value, where x, y are the row and column of the pixel matrix, respectively. The gray image must be processed by a threshold as we focus attention on the area where the rectangular groove is likely to exist rather than the whole area. $I(x, y)$ is the pixel intensity obtained by setting a threshold at the level of -3 dB. The zero moment is shown in Equation (16). The first moment is shown in Equation (17). The second moment is shown in Equation (18). These image moments can be used to extract image features [37].

$$M_{00} = \sum_x \sum_y I(x, y) \tag{16}$$

$$\begin{cases} M_{10} = \sum_x \sum_y x \times I(x,y) \\ M_{01} = \sum_x \sum_y y \times I(x,y) \end{cases} \tag{17}$$

$$\begin{cases} M_{20} = \sum_x \sum_y x^2 \times I(x,y) \\ M_{02} = \sum_x \sum_y y^2 \times I(x,y) \\ M_{11} = \sum_x \sum_y x \times y \times I(x,y) \end{cases} \tag{18}$$

M_{00}, M_{10}, and M_{01} are used to obtain the centroid coordinates of the feature image [38]. The calculation formulas are written as below.

$$\begin{cases} \bar{x} = \dfrac{M_{10}}{M_{00}} \\ \bar{y} = \dfrac{M_{01}}{M_{00}} \end{cases} \tag{19}$$

The spindle orientation of the feature image can be obtained by calculating an angle, which is shown in the following equation.

$$\Theta = \frac{\arctan(2\mu'_{11}/(\mu'_{20} - \mu'_{02}))}{2} \tag{20}$$

where μ'_{11}, μ'_{20}, and μ'_{02} can be given by the following equations:

$$\begin{cases} \mu'_{20} = \dfrac{M_{20}}{M_{00}} - \bar{x}^2 \\ \mu'_{02} = \dfrac{M_{02}}{M_{00}} - \bar{y}^2 \\ \mu'_{11} = \dfrac{M_{11}}{M_{00}} - \bar{x} \times \bar{y} \end{cases} \tag{21}$$

Figure 15 shows the orientation of an inclined rectangular groove. The depth and angle evaluations are shown in Table 6. It is easy to find that the angle errors of Case 10 and Case 11 only reach 0.00% and 0.12%, respectively, while the depth errors of Case 10 and Case 11 rise to 12.95% and 13.60%, respectively. One reasonable explanation is clarified in the discussion section.

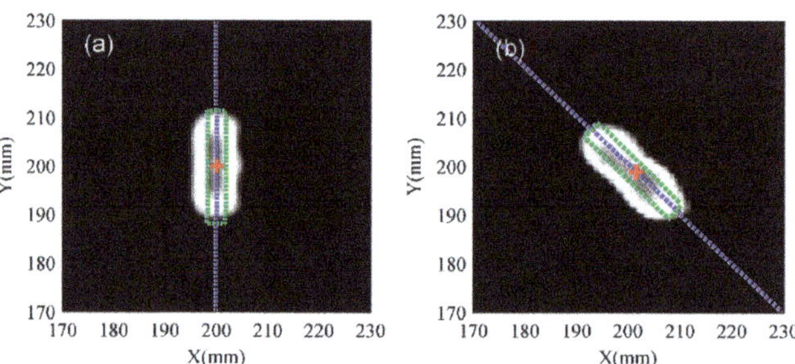

Figure 15. The direction identification of an inclined rectangular groove based on the numerical result. (a) 90°; (b) 135°. The red plus (+) indicates the centroid location. The blue dashed line is the spindle. The green dash line is the boundary of the inclined rectangular groove. The background image is a grayscale image which is processed by a threshold at the level of −3 dB.

Table 6. The depth and angle evaluations of Case 10 and Case 11.

Specimen	Actual Depth (mm)	Reconstructed Depth (mm)	Estimation Error of Depth	Actual Incidence Angle (°)	Reconstructed Angle (°)	Estimation Error of Angle
Case 10	1	0.8705	12.95%	90	90.00	0.00%
Case 11	1	0.8640	13.60%	135	134.84	0.12%

4. Experimental Verification

4.1. Experimental Setup

The experimental system is composed of an ACT, an LDV, an ultrasonic power device, a signal amplifier, a digital phosphor oscilloscope, and a data acquisition interface. The ACT is a piezoelectric ceramic-based transducer with a 200 kHz center frequency, and a bandwidth of $\pm 25\%$ of the center frequency. The incidence angle is set at $12.75°$, which is suitable for the ACT to actuate the A0 mode. The LDV is employed to measure the out-of-plane velocity on the surface of the plate. The sampling frequency is set at 500 MHz. The signal should be executed at a suitable time when the waveform displayed on the oscilloscope tends to be stable. Each measurement is repeated 10 times in order to reduce the measurement error. The received signals need be amplified by a wideband amplifier for the purpose of improving the signal-to-noise ratio. The gain of the signal amplifier is 60 dB. Moreover, it can be adjusted according the central frequency of the excitation signal. The data acquisition subsystem provides detailed information of the recording signals, such as the sample interval, trigger point, vertical scale, horizontal scale, etc. The experimental system is shown in Figure 16. The experimental conditions are in agreement with the simulation conditions.

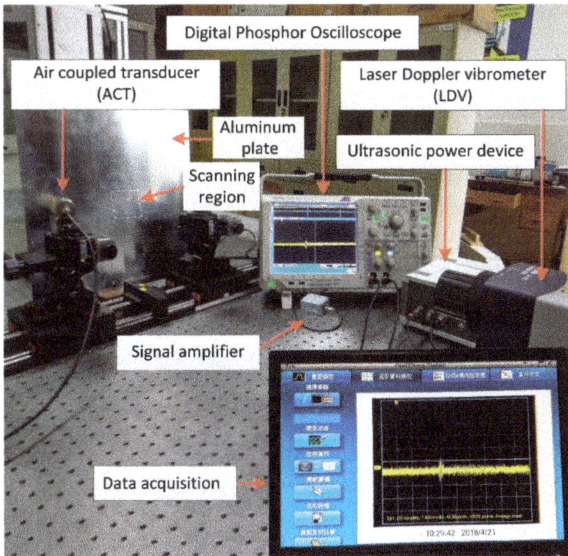

Figure 16. Schematic of the experimental setup.

4.2. Experimental Results

For the purpose of removing the noise, the received signals are filtered by a Chebyshev band-pass filter [39] with a bandwidth of 140 kHz to 260 kHz. Figure 17 shows the experimental signal in the time domain and frequency domain. Figures 18–22 are the experimental results.

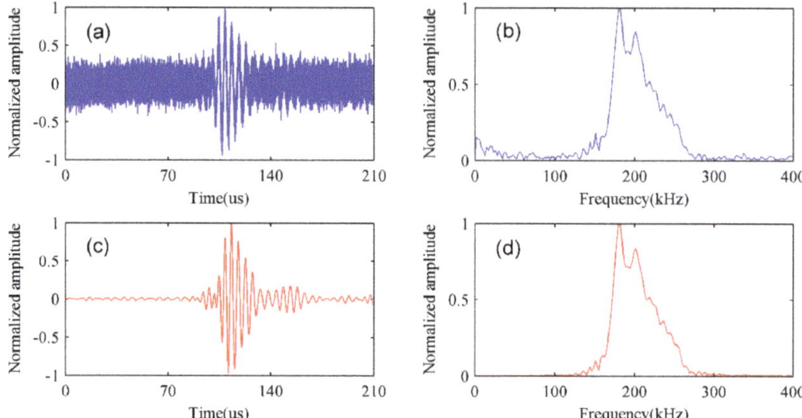

Figure 17. Band-pass filtering of an experimental signal. (**a**) Original signal; (**b**) spectrum of the original signal; (**c**) filtered signal; (**d**) spectrum of the filtered signal.

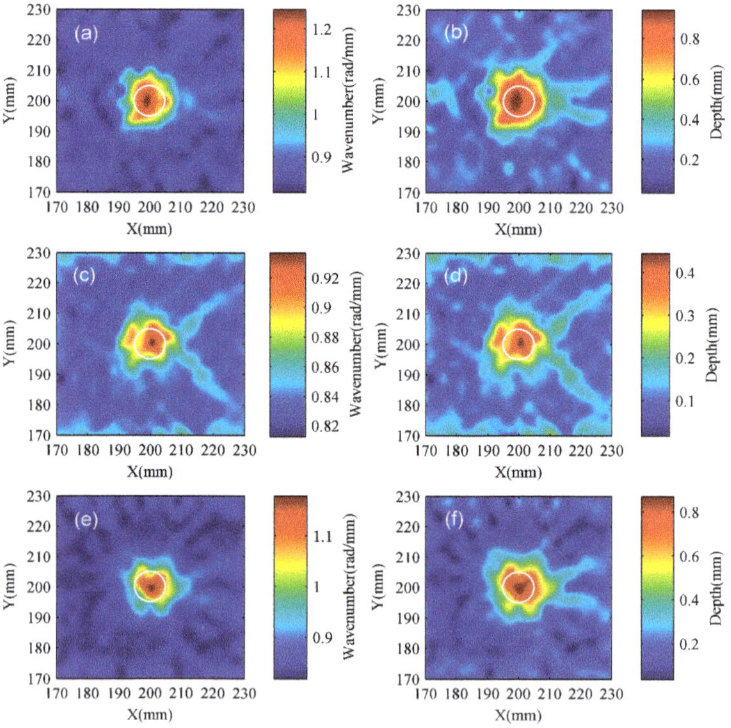

Figure 18. Experimental results of Case 1 to Case 3. (**a**) Wavenumber distribution of Case 1; (**b**) depth distribution of Case 1; (**c**) wavenumber distribution of Case 2; (**b**) depth distribution of Case 2; (**d**) wavenumber distribution of Case 3; (**e**) depth distribution of Case 3. The white circle represents the actual location of the damage. The actual defect depth of Case 1 and Case 3 is 1 mm, while the actual defect depth of Case 2 is 0.5 mm.

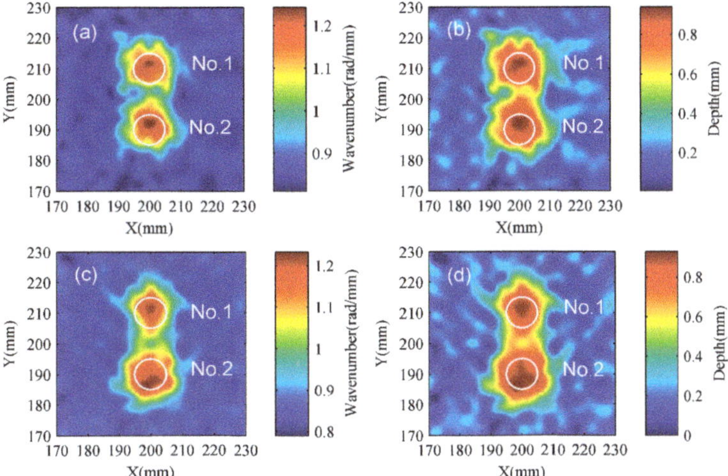

Figure 19. Experimental results of Case 5 and Case 7. (**a**) Wavenumber distribution of Case 5; (**b**) depth distribution of Case 5; (**c**) wavenumber distribution of Case 7; (**d**) depth distribution of Case 7. The white circle represents the actual location of the damage. The actual depth of the defect is 1 mm.

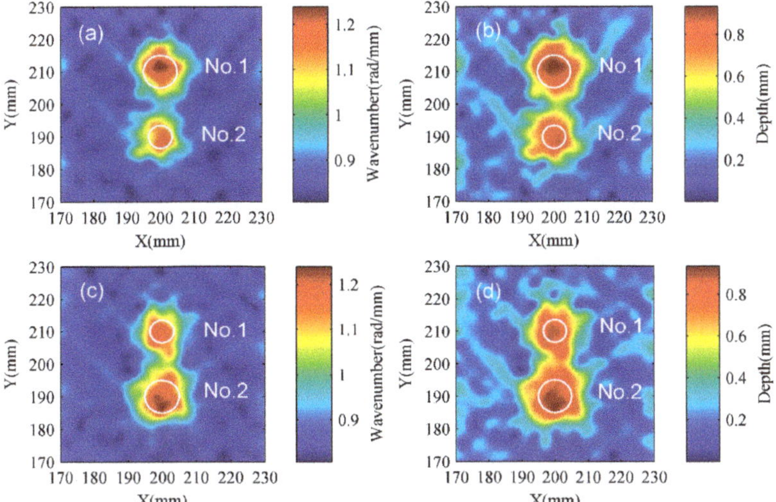

Figure 20. Experimental results of Case 8 and Case 9. (**a**) Wavenumber distribution of Case 8; (**b**) depth distribution of Case 8; (**c**) wavenumber distribution of Case 9; (**d**) depth distribution of Case 9. The white circle represents the actual location of the damage. The actual depth of the defect is 1 mm.

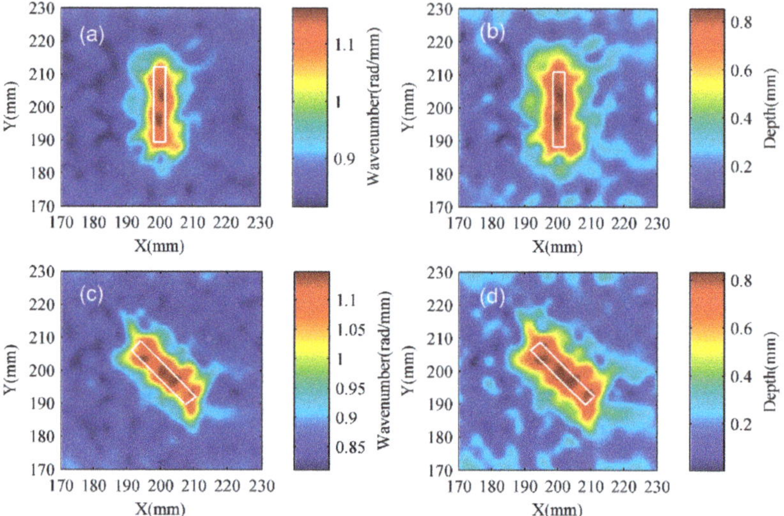

Figure 21. Experimental results of Case 10 and Case 11. (**a**) Wavenumber distribution of Case 10; (**b**) depth distribution of Case 10; (**c**) wavenumber distribution of Case 11; (**d**) depth distribution of Case 11. The white rectangle represents the actual location of the damage. The actual depth of the defect is 1 mm.

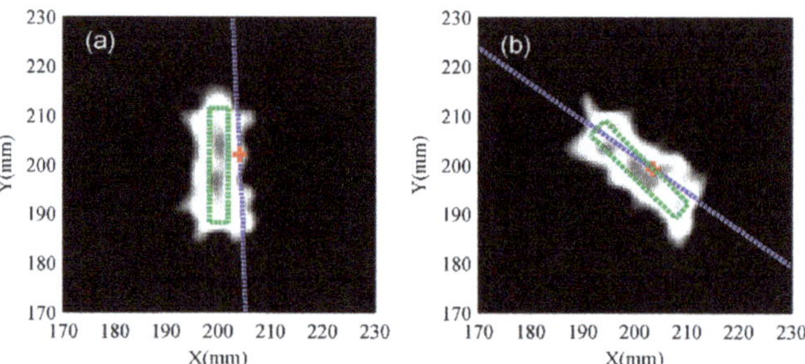

Figure 22. The direction identification of an inclined rectangular groove based on the experimental result. (**a**) 90°; (**b**) 135°. The red plus (+) indicates the centroid location. The blue dashed line is the spindle. The green dash line is boundary of the inclined rectangular groove. The background image is a grayscale image which is processed by a threshold at the level of −3 dB.

The experimental results have good agreement with the simulation results. Figure 18 shows that the depth estimation of the deep hole is more accurate than that of the shallow one. In addition, the depth estimation of the big hole is more accurate than that of the small one. Figure 19 reveals that the circular flat bottom hole far away from the excitation source seems smaller than the near one when the excitation source and two identical defects are in the same line. Figure 20 shows that the proposed approach is capable of identifying the adjacent defects with different sizes. Meanwhile, it indicates that the impact that the big hole exerts on the small one is connected to their positions relative to the excitation source when they are in the same line. Figure 21 shows that the width of the rectangular groove is larger than its actual value. Figure 22 shows the direction identification for an inclined rectangular groove. Tables 7 and 8 indicate that the evaluation results still need to be improved due to the existence of multi-reflection between the adjacent defects and the spatial window effect, which will be analyzed in the discussion section.

Table 7. The depth evaluations of Case 1 to Case 3, Case 5, Case 7, Case 8, and Case 9.

Specimen		Actual Depth (mm)	Reconstructed Depth (mm)	Estimation Error of Depth
Case 1		1	0.9432	5.68%
Case 2		0.5	0.4490	10.20%
Case 3		1	0.8673	13.27%
Case 5	No. 1	1	0.9308	6.92%
	No. 2	1	0.9401	5.99%
Case 7	No. 1	1	0.9152	8.48%
	No. 2	1	0.9277	7.23%
Case 8	No. 1	1	0.9339	6.61%
	No. 2	1	0.8636	13.64%
Case 9	No. 1	1	0.8866	11.34%
	No. 2	1	0.9246	7.54%

Table 8. The depth and angle evaluations of Case 10 and Case 11.

Specimen	Actual Depth (mm)	Reconstructed Depth (mm)	Estimation Error of Depth	Actual Incidence Angle (°)	Reconstructed Angle (°)	Estimation Error of Angle
Case 10	1	0.8510	14.90%	90	92.37	2.63%
Case 11	1	0.8345	16.55%	135	143.46	6.27%

5. Discussion

In this paper, the size of a spatial window is 16 mm, while the diameter of a circular flat bottom hole is no more than 10 mm. Hence, this spatial window will not only contain the propagating waves located at the damage, but also contain those located at the undamaged area. According to the dispersion curves, it should be noted that the wavenumber of the A0 mode will decrease with the increase in the frequency thickness product. More specifically, the wavenumber located at the damage is larger than that located at the undamaged area. Moreover, the deeper the circular flat bottom hole is, the more prominent the difference is. Therefore, this window effect cannot be ignored.

(1) The depth estimation of the deep hole is more accurate than that of the shallow one, as revealed by comparing the results of Case 1 and Case 2.

The frequency-thickness product of the shallow hole is much larger than that of the deep hole, which means that the wavenumber of the former is smaller than that of the latter. Additionally, there is no significant difference between the shallow hole and the undamaged area in terms of the wavenumber. According to Equation (11), it is easy to understand that the shallow hole will be easily affected by the window effect relative to the deep hole.

(2) The depth estimation for the big hole is more accurate than that of the small one. This conclusion was drawn by comparing the results of Case 1 and Case 3.

One reasonable explanation for this is that the local wavenumber is affected by the difference in size between the defect and the spatial window. The small diameter of a circular flat bottom hole means that the spatial window will contain more propagating waves located at the undamaged area where the wavenumbers are small. According to Equation (11), it is easy to understand that the local wavenumber will decrease as the lower wavenumbers are present in the spatial window. The smaller the diameter of a circular flat bottom hole is, the smaller the local wavenumber and the reconstructed depth will be.

(3) The results of Case 4 and Case 5 indicate that the better test result appears under the big gap between defects when two circular flat bottom holes are symmetrical to the excitation source.

Apparently, multi-reflection between two defects is low when they are far apart from each other. Furthermore, they are affected at the same level as they are symmetrical to the excitation source.

(4) The circular flat bottom hole far away from the excitation source seems smaller than the near one when the excitation source and two identical defects are in the same line. Moreover, the closer the two defects are, the more obvious the effect is. This conclusion was obtained by comparing the results of Case 6 and Case 7.

In order to determine the explanation for the above phenomenon, several numerical simulations were conducted under different conditions, which are listed in Table 9. The wavenumber distributions of the spatial window are shown in Figure 23. There is only one defect existing in Figure 23a–c, while there are two defects coexisting in Figure 23d–f. The wavenumber distributions of Figure 23a,d are obtained under the spatial window whose center is at (200, 210). Great differences between Figure 23a,d can be found. The wavenumber distributions of Figure 23b,e are obtained under the spatial window whose center is at (200, 215). There are some differences between Figure 23b,e. The wavenumber distributions of Figure 23c,f are obtained under the spatial window whose center is at (200, 220). Only a few differences between Figure 23c,f can be found. The fundamental reason for this is that the defect near to the excitation source can change the incident wave in direction and amplitude, which dominates the distribution of the wavenumber. Furthermore, it can be clearly found that the impact which the near defect exerts on the far one is gradually reduced with the increase of the distance between them when the excitation source and two identical defects are in the same line.

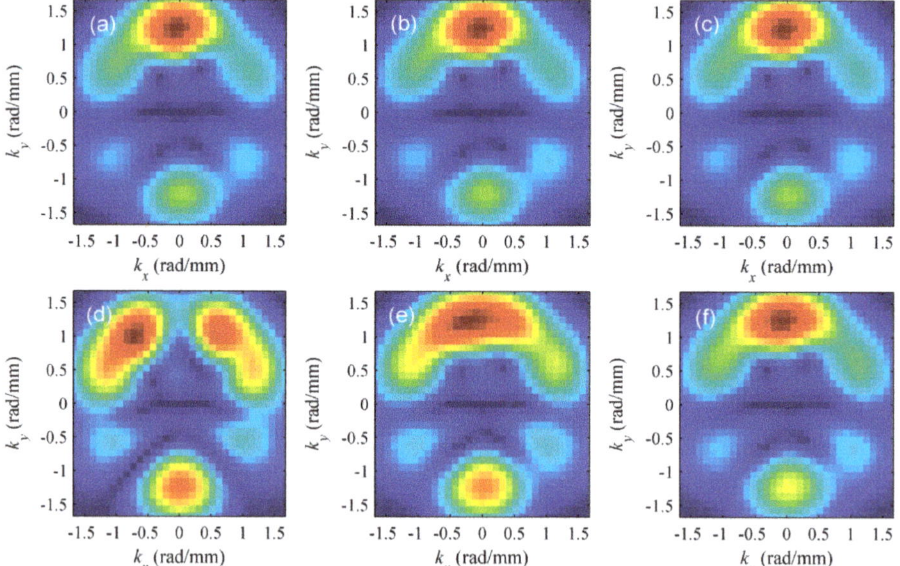

Figure 23. The wavenumber distributions of the spatial window when different conditions are taken into consideration. (**a**) Condition 1; (**b**) condition 2; (**c**) condition 3; (**d**) condition 4; (**e**) condition 5; (**f**) condition 6. The conditions are listed in Table 9.

Table 9. The conditions of Figure 23.

Condition	Damage Type	Excitation Source Position (mm)	Central Position of Damage (mm)	Damage Radius (mm)	Damage Depth (mm)	Central Position of Spatial Window (mm)	Spatial Window Radius (mm)
Condition 1	One flat bottom hole	(200, 100)	(200, 210)	5	1	(200, 210)	8
Condition 2	One flat bottom hole	(200, 100)	(200, 215)	5	1	(200, 215)	8
Condition 3	One flat bottom hole	(200, 100)	(200, 220)	5	1	(200, 220)	8
Condition 4	Two flat bottom holes	(200, 100)	No. 1: (200, 210) No. 2: (200, 190)	5 5	1 1	(200, 210)	8
Condition 5	Two flat bottom holes	(200, 100)	No. 1: (200, 215) No. 2: (200, 185)	5 5	1 1	(200, 215)	8
Condition 6	Two flat bottom holes	(200, 100)	No. 1: (200, 220) No. 2: (200, 180)	5 5	1 1	(200, 220)	8

(5) The results of Case 8 and Case 9 indicate that the impact which the big hole exerts on the small one is related to their positions when the excitation source and multiple defects are in the same line.

The direction and amplitude of the incident wave can be changed obviously when the big hole is in front of the small one. Hence, the wavenumber distributions located at the small hole also change. The impact that the big hole exerts on the small one comes mainly from the reflection wave when the big hole is behind the small one. However, it is the incident wave that plays a crucial role in changing the wavenumber distributions, because the incident wave has a larger amplitude than the reflection wave. Therefore, the test result for the small hole will only change a little when the big hole is behind it.

(6) The depth errors for Case 10 and Case 11 are more than 10%.

The length of the rectangular groove is larger than that of the spatial window. However, the width of the former is still far smaller than that of the latter. Hence, the local wavenumber is seriously affected by the window effect.

(7) Several measures are adopted to improve the test quality when the proposed approach is applied for an unknown damage.

In practice, both the damage size and its position are usually unknown. A large amount of data will be acquired if the entire structure is inspected. Thus, a preliminary step is desired. More specifically, it is easy to acquire the approximate location of a defect by using reliable methods, such as DAS or the compact phased array (CPA) method [40]. Then a small scanning region where the defect is likely to exist can be obtained. Under the circumstance, the local wavenumber approach can be conducted without consuming much time. The spatial window size is at least twice the wavelength of the excited mode. It should be noted that the best test result is not always obtained under the minimum size of the spatial window. In fact, the best test result is obtained under the optimum size due to the fact that the optimum size contributes to diminishing the spatial window effect. The optimum size is also fixed when the central frequency of the excitation signal and the excited mode are fixed. The optimum size can be attained by adjusting the size of the spatial window until the maximum wavenumber of the wavenumber map is acquired. Therefore, patience is important during the period of seeking the optimum size. The process of improving the test quality is shown as below.

The excitation signal with a higher central frequency is adopted as the wavelength of the excited mode is small. Under this condition, the minimum size of the spatial window may be smaller than the optimum size. The wavelength of the A0 mode is about 5 mm when the central frequency of the excitation signal is 400 kHz. Hence, the minimum radius of the spatial window is 5 mm. The radius will be increased at an interval of 1 mm. The test results obtained under various window radii are shown in Figure 24. The maximum wavenumber obtained under various window radii is shown in Figure 25.

Figure 24 shows that the identified damage region will expand with the increase of the window radius. It was found that the test results are poor when the radius is small. The identified damage region is almost equal to the actual size under a special window radius (8 mm). Figure 25 shows that the maximum wavenumber increases with the size of the spatial window and gradually approaches to the theoretical value at the early stage. The maximum wavenumber is very close to the theoretical value under a special window radius (8 mm). However, the maximum wavenumber will decrease when the size of the spatial window continues to be increased.

It is worth that the wavelength of the excited mode is large when the central frequency of the excitation signal is low. Hence, it may lead to the minimum size of the spatial window being larger than the optimum size. Under this circumstance, a good test result is obtained at the minimum size of a spatial window. The test results are shown in Figures 9 and 10.

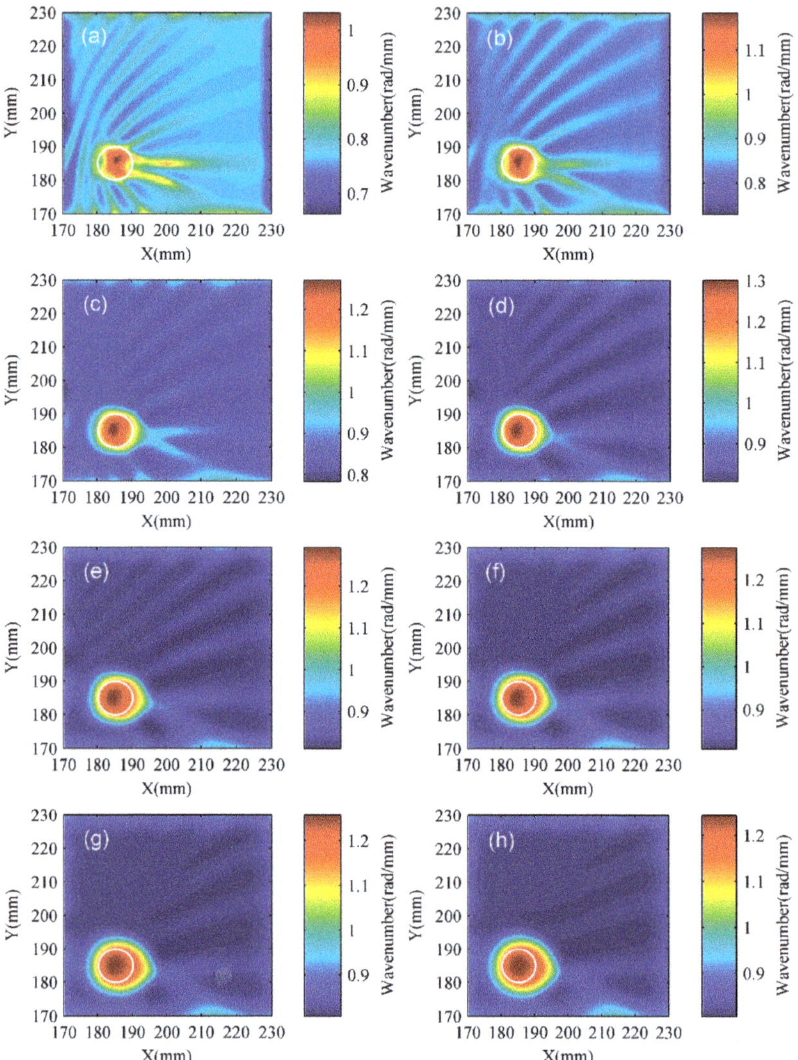

Figure 24. The test results obtained under various window radii. (**a**) 5 mm; (**b**) 6 mm; (**c**) 7 mm; (**d**) 8 mm; (**e**) 9 mm; (**f**) 10 mm; (**g**) 11 mm; (**h**) 12 mm. The white circle represents the actual location of the damage. The central frequency of the excitation signal is 400 kHz. A circular flat bottom hole is located in (185, 185).

(8) The limitation of the proposed approach should be taken into consideration when it is applied for characterizing defects existing in anisotropic materials.

The delamination of composite laminated plates was taken as an example to illustrate the proposed approach. The interactions between Lamb waves and the delamination are complex. It is difficult to identify and isolate a specific mode for the reconstruction analysis. Therefore, the average wavenumber is less accurate. In addition, the position of the delamination cannot be quantified precisely.

In summary, the test quality can be affected by the spatial window effect and the reflection between multiple defects. It is necessary to avoid making the excitation source and multiple defects in the same line in order to reduce the interference between adjacent defects. The optimum size of a spatial window need be sought based on the test results. Increasing the central frequency of the excitation signal is a way to decrease the wavelength of the excited mode and helps to improve the test quality. However, the high frequency can easily produce complex modes. Therefore, it is important to maintain a balance between increasing the central frequency of the excitation signal and reducing the mode complexity.

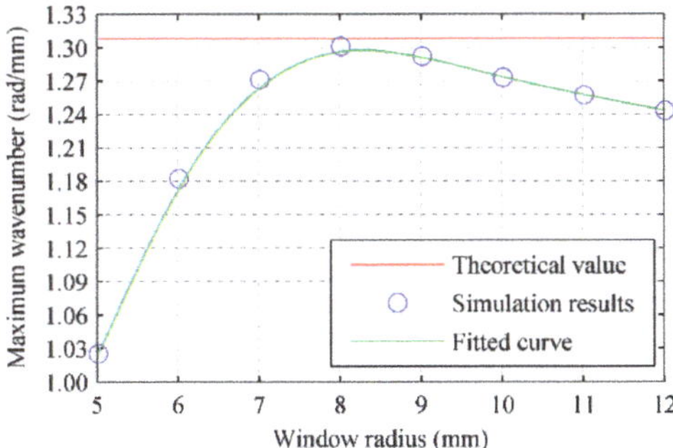

Figure 25. The maximum wavenumber obtained under various window radii. The central frequency of the excitation signal is 400 kHz.

6. Future Work

Given that the local wavenumber approach has many advantages in structural health monitoring (SHM), we will apply this approach for damage detection in adhesive-bonded concrete structures. The main work will be focused on identifying the local debonding and the weakened adhesion. A non-contact system will be composed of two low frequency air coupled transducers. One will be located at a fixed place for actuating a desired guided wave mode by setting an appropriate incidence angle. The other will be used for receiving the signals with the same inclined angle and for scanning over the target area with a spatial interval. The work will be full of challenges as separating the desired mode from multi-modes is not easy and the attenuation existing in the concrete structure may lead to a low signal-to-noise ratio. The detailed solutions will be introduced in the future work.

7. Conclusions

The Lamb wave local wavenumber approach for characterizing flat bottom defects (including two circular flat bottom holes and a rectangular groove) in an isotropic thin plate was presented in this paper. The hybrid and non-contact system is composed of an air-coupled transducer (ACT) and a laser Doppler vibrometer (LDV). On the one hand, the A0 mode is easily actuated by the ACT with a special incidence angle. On the other hand, the out-of-plane velocity over the scanning area can be recorded by an LDV.

Both the simulation results and the experimental results demonstrated that the damage can be characterized in shape, location, and size. Compared with other methods, the proposed approach can be used to reconstruct the thickness distribution over the scanning area when the wavenumber is combined with the phase velocity dispersion curve. Consequently, the depth of the flat bottom defect also can be quantified. In addition, adjacent defects are clearly characterized and the direction of the inclined rectangular groove can be obtained by calculating the image moments under grayscale. It must be noted that an appropriate window size contributes to improving the test quality.

Author Contributions: G.F. wrote this article and accomplished the FEM analysis; H.Z. (Haiyan Zhang) was the principle researcher on this project and provided many valuable suggestions. H.Z. (Hui Zhang) designed the three-dimensional finite element model and performed the simulation; W.Z. established the experimental system and X.C. analyzed the experimental data.

Acknowledgments: The work in this paper is supported by the National Natural Science Foundation of China (Grant Nos. 11674214, 11874255, 11474195, and 51478258), and the Key Technology R&D Project of Shanghai Committee of Science and Technology (Grant No. 16030501400).

Conflicts of Interest: The authors declare no conflict of interest.

References

1. Park, I.; Jun, Y.; Lee, U. Lamb wave mode decomposition for structural health monitoring. *Wave Motion* **2014**, *51*, 335–347. [CrossRef]
2. Wang, W.; Bao, Y.; Zhou, W.; Li, H. Sparse representation for Lamb-wave-based damage detection using a dictionary algorithm. *Ultrasonics* **2018**, *87*, 48–58. [CrossRef] [PubMed]
3. Lu, G.; Li, Y.; Wang, T.; Xiao, H.; Huo, L.; Song, G. A multi-delay-and-sum imaging algorithm for damage detection using piezoceramic transducers. *J. Intell. Mater. Syst. Struct.* **2017**, *28*, 1150–1159. [CrossRef]
4. Khodaei, Z.S.; Aliabadi, M.H. Assessment of delay-and-sum algorithms for damage detection in aluminum and composite plates. *Smart Mater. Struct.* **2014**, *23*, 628–634.
5. Zhang, Y.; Li, D.; Zhou, Z. Time Reversal Method for Guided Waves with Multimode and Multipath on Corrosion Defect Detection in Wire. *Appl. Sci.* **2017**, *7*, 424. [CrossRef]
6. Zhang, H.Y.; Cao, Y.P.; Sun, X.L.; Chen, X.H.; Yu, J.B. A time reversal damage imaging method for structure health monitoring using Lamb waves. *Chin. Phys. B* **2010**, *19*, 448–455. [CrossRef]
7. Zhang, H.Y.; Ruan, M.; Zhu, W.F.; Chai, X.D. Quantitative damage imaging using Lamb wave diffraction tomography. *Chin. Phys. B* **2016**, *25*, 25–31. [CrossRef]
8. Ohara, Y.; Takahashi, K.; Ino, Y.; Yamanaka, K.; Tsuji, T.; Mihara, T. High-selectivity imaging of closed cracks in a coarse-grained stainless steel by nonlinear ultrasonic phased array. *NDT E Int.* **2017**, *91*, 139–147. [CrossRef]
9. Ambrozinski, L.; Piwakowski, B.; Stepinski, T.; Uhl, T. Evaluation of dispersion characteristics of multimodal guided waves using slant stack transform. *NDT E Int.* **2014**, *68*, 88–97. [CrossRef]
10. Sohn, H.; Kim, S.B. Development of dual PZT transducers for reference-free crack detection in thin plate structures. *IEEE Trans. Ultrason. Ferroelectr. Freq. Control* **2010**, *57*, 229–240. [CrossRef] [PubMed]
11. Marks, R.; Clarke, A.; Featherston, C.; Paget, C.; Pullin, R. Lamb Wave Interaction with Adhesively Bonded Stiffeners and Disbonds Using 3D Vibrometry. *Appl. Sci.* **2016**, *6*, 12. [CrossRef]
12. Xu, K.; Ta, D.; Hu, B.; Laugier, P.; Wang, W. Wideband Dispersion Reversal of Lamb Waves. *IEEE Trans. Ultrason. Ferroelectr. Freq. Control* **2014**, *61*, 997–1005. [CrossRef] [PubMed]
13. Ayers, J.; Apetre, N.; Ruzzene, M.; Sabra, K. Measurement of Lamb wave polarization using a one-dimensional scanning laser vibrometer (L). *J. Acoust. Soc. Am.* **2011**, *129*, 585–588. [CrossRef] [PubMed]
14. Wu, T.Y.; Ume, I.C. Fundamental study of laser generation of narrowband Lamb waves using superimposed line sources technique. *NDT E Int.* **2011**, *44*, 315–323. [CrossRef]
15. Xu, K.; Ta, D.; Moilanen, P.; Wang, W. Mode separation of Lamb waves based on dispersion compensation method. *J. Acoust. Soc. Am.* **2012**, *131*, 2714–2722. [CrossRef] [PubMed]
16. Tian, Z.; Yu, L. Lamb wave frequency–wavenumber analysis and decomposition. *J. Intell. Mater. Syst. Struct.* **2014**, *25*, 1107–1123. [CrossRef]
17. Yu, L.; Tian, Z. Lamb wave Structural Health Monitoring Using a Hybrid PZT-Laser Vibrometer Approach. *Struct. Health Monit.* **2013**, *12*, 469–483. [CrossRef]
18. Yu, L.; Leckey, C.A.C.; Tian, Z. Study on crack scattering in aluminum plates with lamb wave frequency-wavenumber analysis. *Smart Mater. Struct.* **2013**, *22*, 1–12. [CrossRef]
19. Yu, L.; Tian, Z.; Leckey, C.A.C. Crack imaging and quantification in aluminum plates with guided wave wavenumber analysis methods. *Ultrasonics* **2015**, *62*, 203–212. [CrossRef] [PubMed]
20. Rogge, M.D.; Leckey, C.A.C. Characterization of impact damage in composite laminates using guided wavefield imaging and local wavenumber domain analysis. *Ultrasonics* **2013**, *53*, 1217–1226. [CrossRef] [PubMed]
21. Rao, J.; Ratassepp, M.; Fan, Z. Quantification of thickness loss in a liquid-loaded plate using ultrasonic guided wave tomography. *Smart Mater. Struct.* **2017**, *26*. [CrossRef]
22. Rao, J.; Ratassepp, M.; Fan, Z. Guided Wave Tomography Based on Full Waveform Inversion. *IEEE Trans. Ultrason. Ferroelectr. Freq. Control* **2016**, *63*, 737–745. [CrossRef] [PubMed]

23. Chronopoulosa, D.; Droz, C.; Apalowo, R.; Ichchou, M.; Yan, W.J. Accurate structural identification for layered composite structures, through a wave and finite element scheme. *Compos. Struct.* **2017**, *182*, 566–578. [CrossRef]
24. Li, H.; Zhou, Z. Air-Coupled Ultrasonic Signal Processing Method for Detection of Lamination Defects in Molded Composites. *J. Nondestruct. Eval.* **2017**, *36*, 1–13. [CrossRef]
25. Sunarsa, T.Y.; Aryan, P.; Jeon, I.; Park, B.; Liu, P.; Sohn, H. A Reference-Free and Non-Contact Method for Detecting and Imaging Damage in Adhesive-Bonded Structures Using Air-Coupled Ultrasonic Transducers. *Materials* **2017**, *10*, 1402. [CrossRef] [PubMed]
26. Jeon, J.Y.; Gang, S.; Park, G.; Flynn, E.; Kang, T.; Han, S.W. Damage detection on composite structures with standing wave excitation and wavenumber analysis. *Adv. Compos. Mater.* **2017**, *26*, 53–65. [CrossRef]
27. Harb, M.S.; Yuan, F.G. A rapid, fully non-contact, hybrid system for generating Lamb wave dispersion curves. *Ultrasonics* **2015**, *61*, 62–70. [CrossRef] [PubMed]
28. Urban, M.W.; Nenadic, I.Z.; Qiang, B.; Bernal, M.; Chen, S.; Greenleaf, J.F. Characterization of material properties of soft solid thin layers with acoustic radiation force and wave propagation. *J. Acoust. Soc. Am.* **2015**, *138*, 2499–2507. [CrossRef] [PubMed]
29. Lee, C.; Park, S. Flaw Imaging Technique for Plate-Like Structures Using Scanning Laser Source Actuation. *Shock Vib.* **2014**, *9*, 1–14. [CrossRef]
30. Tian, Z.; Yu, L.; Leckey, C.A.C. Delamination detection and quantification on laminated composite structures with Lamb waves and wavenumber analysis. *J. Intell. Mater. Syst. Struct.* **2015**, *26*, 1723–1738. [CrossRef]
31. Juarez, P.D.; Leckey, C.A.C. Multi-frequency local wavenumber analysis and ply correlation of delamination damage. *Ultrasonics* **2015**, *62*, 56–65. [CrossRef] [PubMed]
32. Nabili, M.; Geist, C.; Zderic, V. Thermal safety of ultrasound-enhanced ocular drug delivery: A modeling study. *Med. Phys.* **2015**, *42*, 5604–5615. [CrossRef] [PubMed]
33. Bhuiyan, M.Y.; Shen, Y.; Giurgiutiu, V. Interaction of Lamb waves with rivet hole cracks from multiple directions. *J. Mech. Eng. Sci.* **2016**, *231*, 2974–2987. [CrossRef]
34. Bhuiyan, M.Y.; Shen, Y.; Giurgiutiu, V. Guided Wave Based Crack Detection in the Rivet Hole Using Global Analytical with Local FEM Approach. *Materials* **2016**, *9*, 602. [CrossRef] [PubMed]
35. Testoni, N.; Marchi, L.D.; Marzani, A. Detection and characterization of delaminations in composite plates via air-coupled probes and warped-domain filtering. *Compos. Struct.* **2016**, *153*, 773–781. [CrossRef]
36. Xiao, B.; Li, L.; Li, Y.; Li, W.; Wang, G. Image analysis by fractional-order orthogonal moments. *Inf. Sci.* **2017**, *382*, 135–149. [CrossRef]
37. Martín, J.A.H.; Santos, M.; Lope, J.D. Orthogonal variant moments features in image analysis. *Inf. Sci.* **2010**, *180*, 846–860. [CrossRef]
38. Zhou, C.; Zhang, B.; Lin, K.; Xu, D.; Chen, C.; Yang, X.; Sun, C. Near-infrared imaging to quantify the feeding behavior of fish in aquaculture. *Comput. Electron. Agric.* **2017**, *135*, 233–241. [CrossRef]
39. Losada, R.A.; Pellisier, V. Designing IIR Filters with a Given 3-dB Point. *IEEE Signal Process. Mag.* **2005**, *22*, 95–98. [CrossRef]
40. Senyurek, V.Y.; Baghalian, A.; Tashakori, S.; McDaniel, D.; Tansel, I.N. Localization of multiple defects using the compact phased array (CPA) method. *J. Sound Vib.* **2018**, *413*, 383–394. [CrossRef]

applied sciences

Article

Ultrasonic Guided Wave-Based Circumferential Scanning of Plates Using a Synthetic Aperture Focusing Technique

Jianjun Wu [1], Zhifeng Tang [2,*], Keji Yang [1], Shiwei Wu [1] and Fuzai Lv [1]

[1] State Key Laboratory of Fluid Power and Mechatronic Systems, School of Mechanical Engineering, Zhejiang University, 38 Zheda Road, Hangzhou 310027, China; 11425023@zju.edu.cn (J.W.); yangkj@zju.edu.cn (K.Y.); 06jxlwsw@zju.edu.cn (S.W.); lfzlfz@zju.edu.cn (F.L.)

[2] Institute of Advanced Digital Technologies and Instrumentation, School of Biomedical Engineering and Instrument Science, Zhejiang University, 38 Zheda Road, Hangzhou 310027, China

* Correspondence: tangzhifeng@zju.edu.cn; Tel.: +86-138-5713-1010

Received: 11 July 2018; Accepted: 6 August 2018; Published: 7 August 2018

Featured Application: Ultrasonic guided wave scanning of plates with the transducer moving along a circumferential path outside the tested area.

Abstract: Tanks are essential facilities for oil and chemical storage and transportation. As indispensable parts, the tank floors have stringent nondestructive testing requirements owing to their severe operating conditions. In this article, a synthetic aperture focusing technology method is proposed for the circumferential scanning of the tank floor from the edge outside the tank using ultrasonic guided waves. The zeroth shear horizontal (SH0) mode is selected as an ideal candidate for plate inspection, and the magnetostrictive sandwich transducer (MST) is designed and manufactured for the generation and receiving of the SH0 mode. Based on the exploding reflector model (ERM), the relationships between guided wave fields at different radii of polar coordinates are derived in the frequency domain. The defect spot is focused when the sound field is calculated at the radius of the defect. Numerical and experimental validations are performed for the defect inspection in an iron plate. The angular bandwidth of the defect spot is used as an index for the angular resolution. The results of the proposed method show significant improvement compared to those obtained by the B-scan method, and it is found to be superior to the conventional method—named delay and sum (DAS)—in both angular resolution and calculation efficiency.

Keywords: SH0 mode; circumferential scanning; synthetic aperture focusing; exploding reflector model

---

1. Introduction

Tanks are essential facilities for the storage and transportation of petroleum products. Thousands of tanks may be in service at the same time in one oil depot. Because of the harsh working conditions and corrosive storage materials, various kinds of defects, which are hidden hazards, may occur in the tanks. The structural integrity failure of the tanks may cause catastrophic accidents such as environmental pollution and explosions [1]. Thus, tanks require stringent inspection for their long-term and high-level safety. Tank floors are indispensable parts of the tanks. It is found that 80% of corrosion, which makes up the majority of the defects in the tanks, occur in the tank floors, making them key areas in the overhauls [2]. The most common non-destructive testing technologies for tank floors based on flux leakage [3–5] and ultrasonics [6] are limited for the low working efficiency of testing "step by step." The tanks must stop their service and be emptied before the inspection. The cost and time for the overhaul of the tank floors can amount to hundreds of thousands of dollars and dozens of days. Recently, a technology based on sound emission has been introduced for the structural health monitoring of tanks in service [7–9]. Elastic waves caused by the crack growth and liquid leakage are collected and analyzed for the damage detection. However, useful information is difficult to extract from the signals because of the low signal-to-noise ratios, and the localization of the damage is dissatisfactory with a limited number of transducers. Therefore, it is urgent and highly desirable to develop new methods for testing tank floors in service.

Ultrasonic guided wave methods have been proven to have tremendous potential in the rapid inspections of structures with long and slender dimensions such as pipes [10,11], plates [12–14], and rails [15]. For the inspection of plate structures such as tank floors, large areas can be inspected at the same time with one transducer using guided waves. Structural health monitoring based on guided waves has been proposed for years. The transducers are permanently installed on the plate around the monitored region. The transducers work as actuators, in turn, and all the others are sensors. The received signals are compared with those acquired when the integrity of the tank floors was confirmed. The coefficients to evaluate the signal changes can be amplitude differences of the direct waves [16] and the variations of the whole signals, named the signal difference coefficient (SDC) [17–21]. Different imaging methods are introduced for guided wave monitoring, such as the filtered back-projection method (FBP) [22], the interpolated FBP (IFBP) [23], the algebraic reconstruction technique (ART) [24], and the probabilistic reconstruction algorithm (PRA) [25], which are all basically drawn from computational tomography (CT). Recently, a method called full waveform inversion (FWI) has been introduced for guided wave monitoring [26,27]. Guided waves with high frequency dispersions are used, and the model of the monitored area is constructed and revised in the frequency domain to match the signals acquired by the model and those collected in the tests. This method performs well in corrosion imaging. However, it should be mentioned that numerous transducers are needed for high sensitivities and localization accuracies in guided wave monitoring, and multiple iterations are required in these methods for high resolutions, which may be time-consuming for large amounts of data.

Guided wave scanning is an alternative method for plate inspection. As shown in Figure 1, the tank floor in service can be evaluated with one transducer moving along the edge outside the tank [28]. The spots of the defects in the image of the original signals (i.e., the B-scan image in this study) have long lateral trailing because of the expansion of the sound fields in the plates. To improve the lateral resolution, the synthetic aperture focusing technique (SAFT) is developed. The traditional method of SAFT is called delay and sum (DAS) [29,30]. The wave packets of the defect in different signals can be superposed in the image with the distances between the defect and the transducer locations in the scanning. This method is theoretically applicable for arbitrary scanning paths of the transducer. However, the method is carried out in the time domain with poor computational efficiency, and residual trailing exists in the image. Another method of SAFT is proposed based on sound field migration in the frequency domain, providing better computational efficiency as well as imaging quality compared with the DAS method [31–33]. However, the sound field migration of this method is deduced in the Cartesian coordinate system, which can only manage signals acquired by the transducer moving along a straight path. The paths of the transducer for the scanning of the tank floors are circular, making this method inapplicable. Efficient methods of SAFT for guided wave-based circumferential scanning have not been seen yet.

In this paper, a method of SAFT for circumferential scanning (CSAFT) is proposed. In Section 2, the shear horizontal mode in the plates, named SH0, is selected for inspection, and the sound field of the magnetostrictive sandwich transducer (MST) used in this study is calculated; the imaging method is developed based on sound field migration in the polar coordinate systems. In Section 3, the proposed method is verified by numerical simulations and compared with other two methods, the B-scan and DAS methods, for different affecting factors. Experimental verifications are carried out in an iron plate in Section 4 and conclusions follow in Section 5.

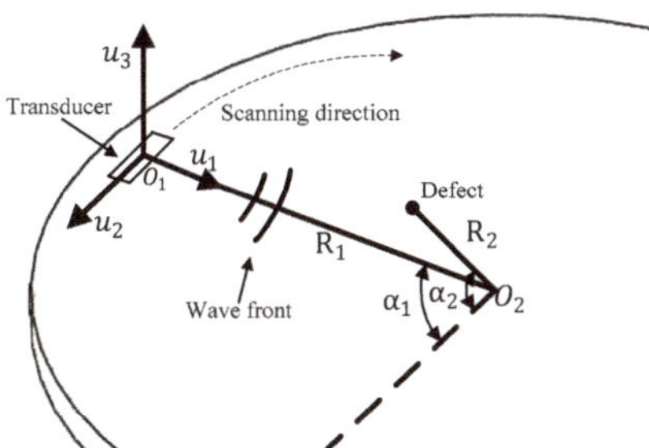

Figure 1. The diagram of the circumferential scanning using guided waves.

2. Method

2.1. Shear Horizontal Waves

As shown in Figure 1, in circumferential scanning using guided waves, the transducer to generate and receive guided waves moves around the origin point O_2 at the radius R_1. The wave front of the incident guided waves expands during the propagation. The reflection waves are generated if the defects are covered by the wave front of the incident waves. The flight time and peak values of the reflection waves are used to evaluate the location and size of the defects, respectively.

Shear horizontal waves (SH waves) and Lamb waves are two kinds of mechanical waves propagating in a solid plate. For waves propagating along the direction u_1 as in Figure 1, displacements along the direction u_3 occur in Lamb waves, which are perpendicular to the plane of the plate, but the displacements of SH waves are along the direction u_2 within the plane of the plate. Considering that the tank floors in service are in contact with liquids (storage materials, such as different kinds of oil) and that the attenuation of shear waves in liquids is much higher than that of longitudinal waves, SH waves may be better candidates for an inspection with less energy leakage. Researches of torsional waves in pipes filled with liquids and bars immersed in liquids can be found in [34,35], respectively, as references. Quantitative comparisons of the SH and Lamb waves will be carried out further under the conditions of different kinds of petroleum products with various volumes. And affecting factors such as the curvatures of the tank floors caused by the gravities of the stored materials and sludge deposition under the tank floors will be investigated as well. The SH waves are hardly affected by liquids [36]. Therefore, in this study, with the purpose of verification of the proposed imaging method, the model is simplified as an isotropic plate with the thickness d without contact of liquids, the dispersion equation of SH waves can be expressed as [37]

$$\frac{\omega^2}{c_T^2} - \frac{\omega^2}{c_p^2} = \left(\frac{n\pi}{d}\right)^2 \tag{1}$$

$$c_T = \sqrt{\frac{E}{2(1+\mu)\rho}} \tag{2}$$

where $\omega = 2\pi f$, f, c_p, and n are frequency, phase velocity, and order, respectively, and c_T is the fundamental shear wave velocity, which is determined by mechanical properties of the plate such as density ρ, Young's modulus E, and Poisson's ratio μ. Combining the definition of the wave number $k = \frac{\omega}{c_p}$, the group velocity $c_g = \frac{d\omega}{dk}$, and Equations (1) and (2), dispersion curves of phase velocity and group velocity can be calculated:

$$c_p(fd) = 2c_T \frac{fd}{\sqrt{(2fd)^2 - (nc_T)^2}} \tag{3}$$

$$c_g(fd) = c_T \sqrt{1 - \frac{(n/2)^2}{(fd/c_T)^2}} \tag{4}$$

where fd is the frequency-thickness product. Dispersion curves of phase velocity and group velocity for the iron plates with mechanical properties $\rho = 7850\ \text{kg/m}^3$, $E = 210\ \text{GPa}$, and $\mu = 0.3$ are shown in Figure 2. All the modes are named by their orders, and for the wave mode SHn, a frequency lower limit exists when n is no less than 1, which is called the cutoff frequency.

$$(fd)_n = \frac{nc_T}{2} \tag{5}$$

The thickness of the tank floor commonly used is between 6 and 19 mm [38]. In this study, the thickness d of the iron plate used for the verification in numerical simulations and experiments is selected as 4 mm for the easy operations in the laboratory. The wave structures of SH0, SH1, SH2, and SH3 modes in the direction u_2 are shown in Figure 3. It can be observed that the pure SH0 mode without dispersion can be generated when the excitation frequency is lower than the cutoff frequency of the SH1 mode. Compared with high order modes (SH1, SH2 and SH3), the wave structure of the SH0 mode is uniform along the thickness, giving the same detectability for the defects within the plates and on the surfaces. With these advantages, the SH0 mode is selected for inspection in this study.

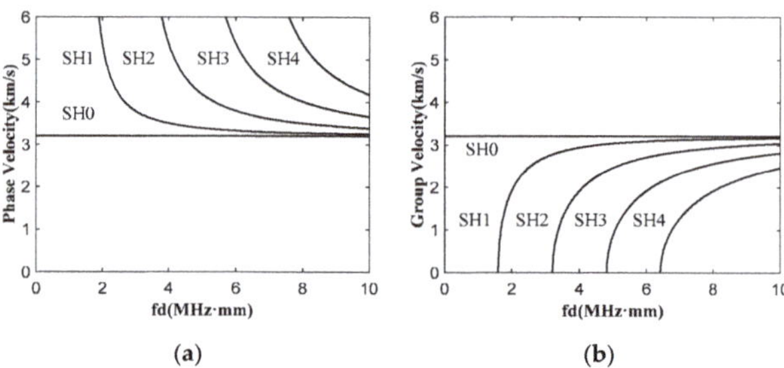

Figure 2. Dispersion curves of iron plates: (a) phase velocity curves; (b) group velocity curves.

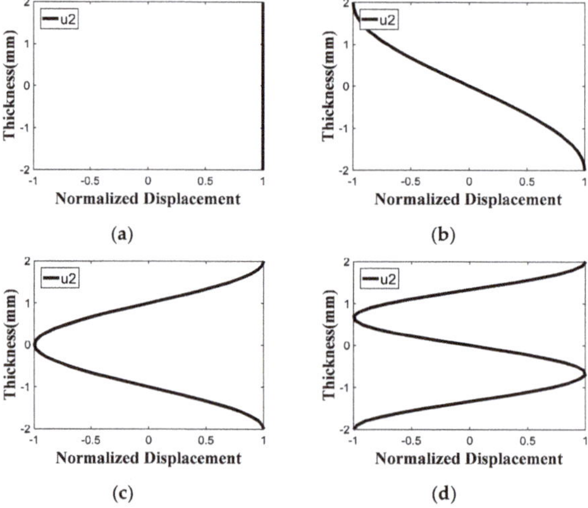

Figure 3. Wave structures of shear horizontal (SH) modes in the main vibration direction u_2: (a) SH0; (b) SH1; (c) SH2; (d) SH3.

2.2. The Sound Field of the Transducer

Piezoelectric and magnetostrictive transducers are commonly used in guided wave generation. Because of the differences in the principles of energy conversion and transducer structures, piezoelectric transducers are good at out-of-plane displacement generation, while magnetostrictive transducers have advantages in in-plane displacement generation. The magnetostrictive effect of the ferromagnetic materials is made use of in the wave generation, and the inverse magnetostrictive effect works in receiving waves. Details of the working principles of this kind of transducer can be found in [39].

The transducer used in this study is named the magnetostrictive sandwich transducer (MST), with a structure shown in Figure 4. The magnetostrictive strip is the main component for energy conversion in both wave generation and receiving, with the residual magnetism along the direction u_2 by pre-magnetization. The soft magnet is used to reduce the magnetic resistance of the dynamic magnetic fields, improving energy conversion efficiency. There are two coils in the MST to generate and sense the dynamic magnetic fields. Coil 1 is $\lambda/4$ ahead of Coil 2 in the direction u_1. λ is the wavelength of the guided wave. The burst signals in Coil 1 and Coil 2 have a phase difference of $\pi/2$. Wires in the working region of the MST are distributed with equal intervals and numbered with wn from 1 to 40. The current direction C_d and initial phase C_{ip} of all the wires are listed in Table 1.

Table 1. The current direction C_d and initial phase C_{ip} for each wire.

Name	Value							
wn	1~5	6~10	11~15	16~20	21~25	26~30	31~35	36~40
$C_d(wn)$	-1	-1	1	1	22121	-1	1	1
$C_{ip}(wn)$	$\pi/2$	0	$\pi/2$	0	$\pi/2$	0	$\pi/2$	0

Each wire in the working region can be regarded as a line sound source, and the sound field S_{wn} for the wnth wire can be derived as [40]

$$S_{wn}\left(r',\theta',t\right) \approx S_m L \sqrt{\frac{2}{\pi k r'}} \left[\frac{\sin\left(\frac{1}{2}kL\sin(\theta')\right)}{\frac{1}{2}kL\sin(\theta')} \right] e^{j\left(kr'-\omega t+\frac{\pi}{4}\right)} \tag{6}$$

where the wave number $k = \frac{2\pi}{\lambda}$, S_m is a constant coefficient, and r' and θ' denote the relative position between the wnth wire and the receiving point as in Figure 4. L is half the length of the working region. Equation (6) is valid when $r' \gg L$. The sound field of the MST can be calculated as the sound field combinations of all wires:

$$S \approx \sum_{wn=1}^{40} C_d(wn) S_{wn} e^{jC_{ip}(wn)}. \tag{7}$$

As mentioned in Section 2.1, the thickness of the iron plate used in this study is 4 mm. The group velocity c_g and phase velocity c_p of the SH0 mode are both 3208 m/s, as in Figure 2. The wave length λ of the SH0 mode is 25.1 mm as the width of the magnetostrictive strip available is 50.2 mm. Thus, the main frequency f of the designed MST is 128 kHz, which is below the cutoff frequency of the SH1 mode (401 kHz). It should be noted that, as mentioned above, the main frequency of the transducer should be determined with the consideration of both detection sensitivities and testing ranges in field tests. The normalized sound fields of the MSTs with different lengths are shown in Figure 5. The waves are controlled to propagate in one direction, and the amplitudes of the main lobes in the sound fields are much larger than those of the side lobes. It can also be seen that the widths of the main lobes decrease with the increase in the lengths of the MSTs. The divergence angle θ_d, defined as the largest angle at which the amplitudes are no less than half of the largest amplitude, is used to denote the width of the main lobes. The relationship between the divergence angles and the lengths of the MSTs from 2 mm to 150 mm are shown in Figure 6.

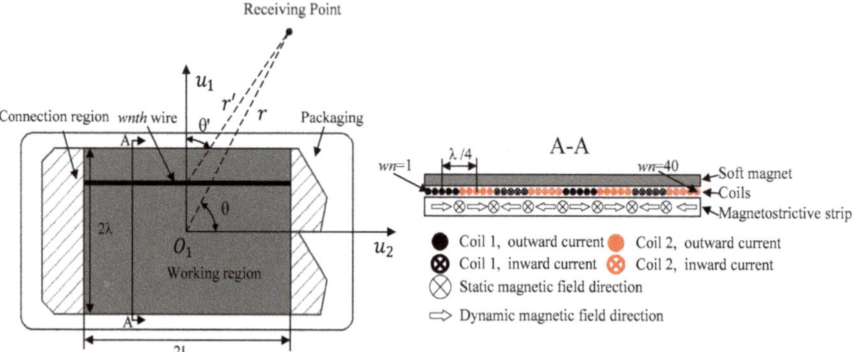

Figure 4. The structure of the magnetostrictive sandwich transducer (MST).

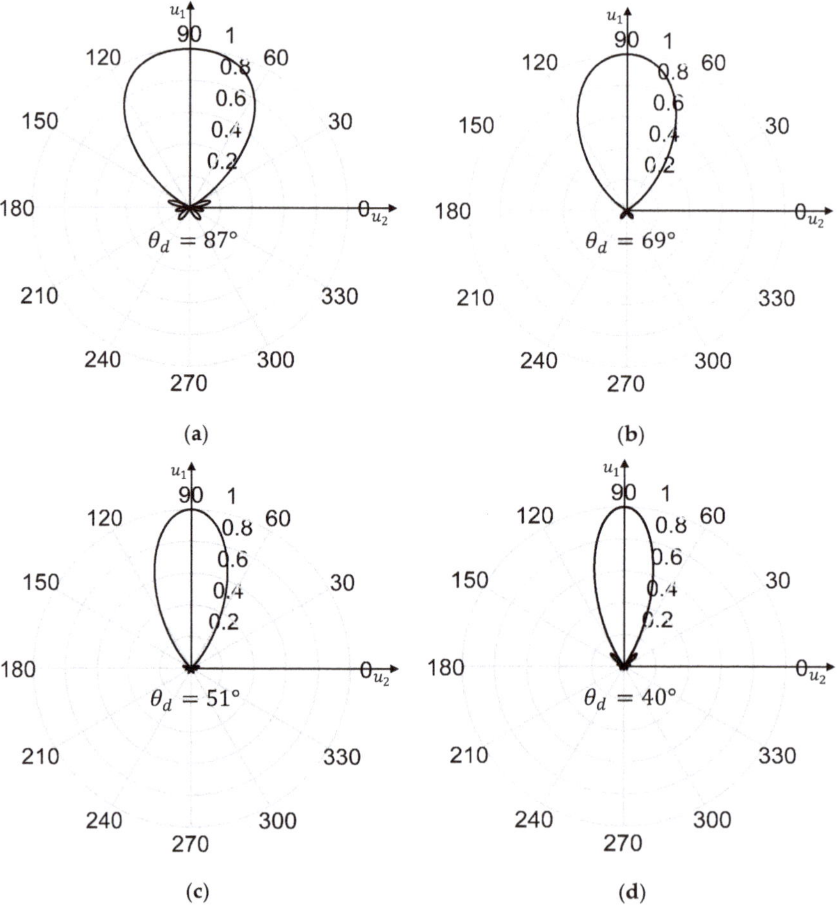

Figure 5. The sound fields and divergence angles of the MSTs with different lengths: (**a**) 2L = 30 mm; (**b**) 2L = 50 mm; (**c**) 2L = 70 mm; (**d**) 2L = 90 mm.

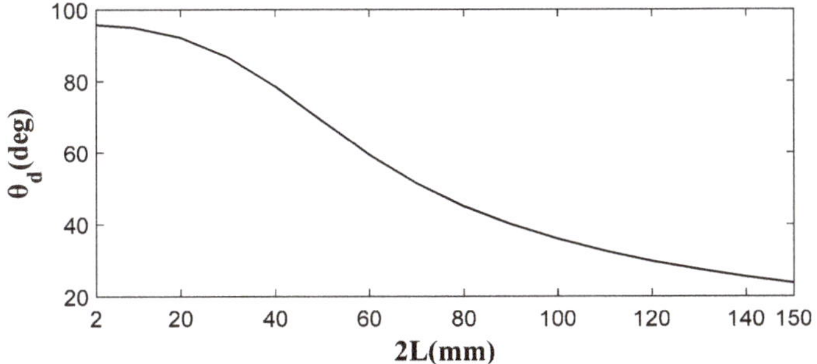

Figure 6. The relationship between the divergence angles of the main lobes and the lengths of the MSTs.

2.3. Synthetic Focusing Imaging for Circumferential Scanning

The whole procedure of the inspection with the transducer at one location can be described as follows:

Step 1: The transducer generates guided waves towards the defect;
Step 2: The defect is motivated and generates scattering waves;
Step 3: Parts of the scattering waves are received by the transducer.

When the defect is motivated by the incident pure SH0 waves, other modes may exist in the scatter waves because of the mode conversions. These modes are not dominant and they can hardly be detected by the MST in the experiments. Therefore, the mode conversion is not considered in this study. Because of the reciprocity of the sound field and the constant group velocity c_g of the SH0 mode, Steps 1 and 3 can be described in the same way. The defect can be regarded as a sound source that generates guided waves at time $t = 0$; then, the guided waves propagate to the transducer. Since the sonic path in this assumption is half of the actual path, the group velocity $\hat{c}_g$ is taken as half of the actual value c_g to maintain the flight time of the defect, i.e., $\hat{c}_g = c_g/2$. The above equivalent model is also called the exploding reflector model (ERM) [41]. As in Figure 1, the guided waves generated by the defect at the radius R_2 expand before being received by the transducer moving at the radius R_1. If R_1 is equal to R_2, there will be no expansion of the guided waves, which are simply generated from the defect at the time $t = 0$. An optimal lateral resolution can be obtained from the guided wave fields at the radius R_2 with the time term $e^{-i\omega t}\big|_{t=0} = 1$. The relationships of the guided wave fields at different radii can be deduced as follows.

In Figure 1, the polar coordinate O_2 is constructed without considering the direction along the thickness, in which the displacement of the SH0 mode is uniform. The wave equation in a polar coordinate can be written as [42]:

$$\left[\frac{\partial^2}{\partial R^2} + \frac{1}{R}\frac{\partial}{\partial R} + \frac{1}{R^2}\frac{\partial^2}{\partial \alpha^2} - \frac{1}{\hat{c}_g^2}\frac{\partial^2}{\partial t^2} \right] p(R, \alpha, t) = 0 \tag{8}$$

where $p(R, \alpha, t)$ is the wave field of the point (R, α) at time t, which can be written in the form of an inverse Fourier transform:

$$p(R, \alpha, t) = \int P(R, k_\alpha, \omega) e^{j(k_\alpha \alpha + \omega t)} dk_\alpha d\omega \tag{9}$$

where k_α and ω are the angular wave number and angular frequency, respectively. $P(R, k_\alpha, \omega)$ is the two-dimensional spectrum of $p(R, \alpha, t)$, which can be solved by substituting Equations (9) into (8):

$$P(R, k_\alpha, \omega) = A_{k_\alpha} H_{k_\alpha}^{(1)}(k_R R) + B_{k_\alpha} H_{k_\alpha}^{(2)}(k_R R) \tag{10}$$

where $A_{k_\alpha} H_{k_\alpha}^{(1)}(k_R R)$ and $B_{k_\alpha} H_{k_\alpha}^{(2)}(k_R R)$ denote the guided wave propagating away from the original point and towards the original point, respectively. $H_{k_\alpha}^{(p)}(p = 1,\ 2)$ is the pth kind of Hank function of the order k_α k_R is the wave number along the radius.

$$k_R = \begin{cases} \sqrt{\frac{\omega^2}{\hat{c}_g^2} - k_\alpha{}^2}, & \sqrt{\frac{\omega^2}{\hat{c}_g^2} - k_\alpha{}^2} \geq 0 \\ 0, & \sqrt{\frac{\omega^2}{\hat{c}_g^2} - k_\alpha{}^2} < 0 \end{cases}. \tag{11}$$

The guided waves generated by the defect propagate away from the original point to the transducer; thus, the transfer function for the guided wave fields at different radii can be derived as

$$P(R - \Delta R,\ k_\alpha,\ \omega) = P(R,\ k_\alpha,\ \omega) \frac{H_{k_\alpha}^{(1)}(k_R(R - \Delta R))}{H_{k_\alpha}^{(1)}(k_R R)} \tag{12}$$

Equation (12) can be simplified using the Rayleigh–Sommerfeld diffraction formula [43]:

$$P(R - \Delta R,\ k_\alpha,\ \omega) \approx P(R,\ k_\alpha,\ \omega) \sqrt{\frac{R - \Delta R}{R}} e^{-i\Delta R \sqrt{\frac{\omega^2}{\hat{c}_g^2} - \frac{k_\alpha{}^2}{R(R - \Delta R)}}} \tag{13}$$

This function is valid if $\cos(\alpha_1 - \alpha_2)$ is approximately equal to $1 - \frac{(\alpha_1 - \alpha_2)^2}{2}$ With a given step ΔR, the two-dimensional spectrums of the guided wave fields at the radii $(R - \Delta R,\ R - 2\Delta R, R - 3\Delta R, \ldots, R - m\Delta R)$ can be derived with those collected by the transducer at the radius R with Equation (13). The focused image can be obtained with the inverse Fourier transforms of the guided wave fields at different radii.

The main steps of the proposed algorithm are described below:

Step 1: A data matrix is constructed with scanning signals collected by the transducer at the radius R_1 as column vectors.

Step 2: The two-dimensional Fourier transform of the data matrix is performed to obtain two-dimensional spectrum $P(R_1,\ k_\alpha,\ \omega)$.

Step 3: The length ΔR and number m of the radius step are determined; The imaging range of the radius is between R_1 and $R_1 - m\Delta R$. The counter q is 1.

Step 4: Two-dimensional spectrum $P(R_1 - q\Delta R,\ k_\alpha,\ \omega)$ at the radius $R_1 - q\Delta R$ is calculated with Equation (13). Row vector $R_v(R_1 - q\Delta R)$ is constructed with the summation of each column of the two-dimensional spectrum.

Step 5: The inverse Fourier transform of Row vector $R_v(R_1 - q\Delta R)$ is performed. The result is the qth row of the image matrix. The counter q is increased by one.

Step 6: Steps 4 and 5 are executed repeatedly until q is larger than m. The image matrix is outputted at last.

The attainable angular resolution can be calculated with the similar derivation as in [41]. If the divergence angle θ_d is used to determine the angular coverage of the transducer, the thresholds of the angular wave number k_α are $-\frac{4\pi}{\lambda} R_1 \sin\left(\frac{\theta_d}{2}\right)$ and $\frac{4\pi}{\lambda} R_1 \sin\left(\frac{\theta_d}{2}\right)$, thus, the bandwidth of the angular wave number Δk_α is $\frac{2\pi}{\lambda} R_1 \sin\left(\frac{\theta_d}{2}\right)$. Attainable angular resolution θ_{ar} can be calculated as follows:

$$\theta_{ar} = \frac{2\pi}{\Delta k_\alpha} = \frac{\lambda}{R_1 \sin\left(\frac{\theta_d}{2}\right)} \tag{14}$$

For the SH0 mode at a certain frequency, the wave length is constant, and the attainable angular resolutions are affected by the divergence angles, which are determined by the structures of the transducers.

The proposed method is applicable for the circumferential scanning of isotropic plates with non-dispersive modes. For dispersive modes like the Lamb waves and higher-order SH waves, dispersion compensation can be achieved with the relationships between the group velocities $\hat{c}_g$ and the angular frequencies ω from the group velocity curves in the calculations of two-dimensional spectrums using Equation (13). For anisotropic plates, the wave equation in Equation (8) is not suitable, since the wave propagations along different directions are not the same. The proposed method may be not applicable.

The detection ability of the circumferential scanning is determined by multiple reasons such as the excitation frequency and the characteristics of the defects (shapes and sizes). With the increments of the excitation frequencies, the detection sensitivities of the waves increase while the propagation distances decrease. The shapes of the defects affect the mode conversions and propagations of the scattering waves. Details can be found in [44].

3. Numerical Simulation and Verification

3.1. Simulation Setup

For simplification, only Steps 2 and 3 in Section 2.3 are carried out in the simulation. As in Figure 7, the side length of the iron plate is 1.5 m; the thickness and material properties are the same as in Section 2.1. The original point O_2 is set in the vertex of the lower right corner; the defect is 0.825 m from O_2 with a size of 1×10 mm, and the angular position α_1 is $10°$. The defect is motivated with a tone-burst signal, which is acquired by filtering a sinusoidal signal of multiple cycles with a Hanning window, as in Figure 8. The main frequency is 128 kHz. With the angular step θ_s of $0.5°$, 91 receiving points are set at the radius 1.3 m, which are marked from RP1 to RP91. The MST is always towards the original point during the signal being received from RP1 to RP91. The sound field of the MST for receiving is the same as that for generation, and so the signals collected at these receiving points are modulated by the sound field of the MST to construct the B-scan image.

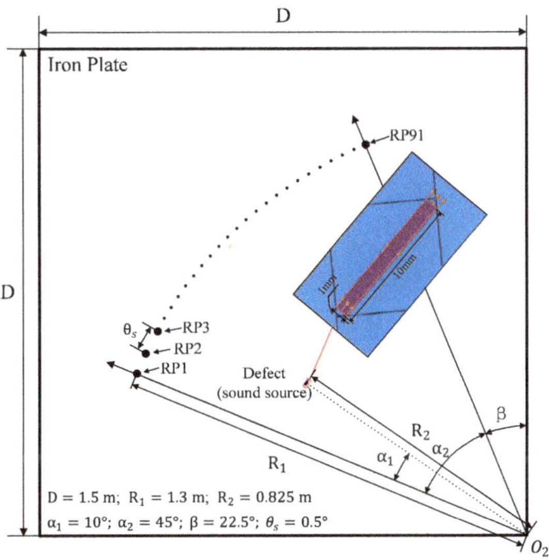

Figure 7. The diagram of the numerical model.

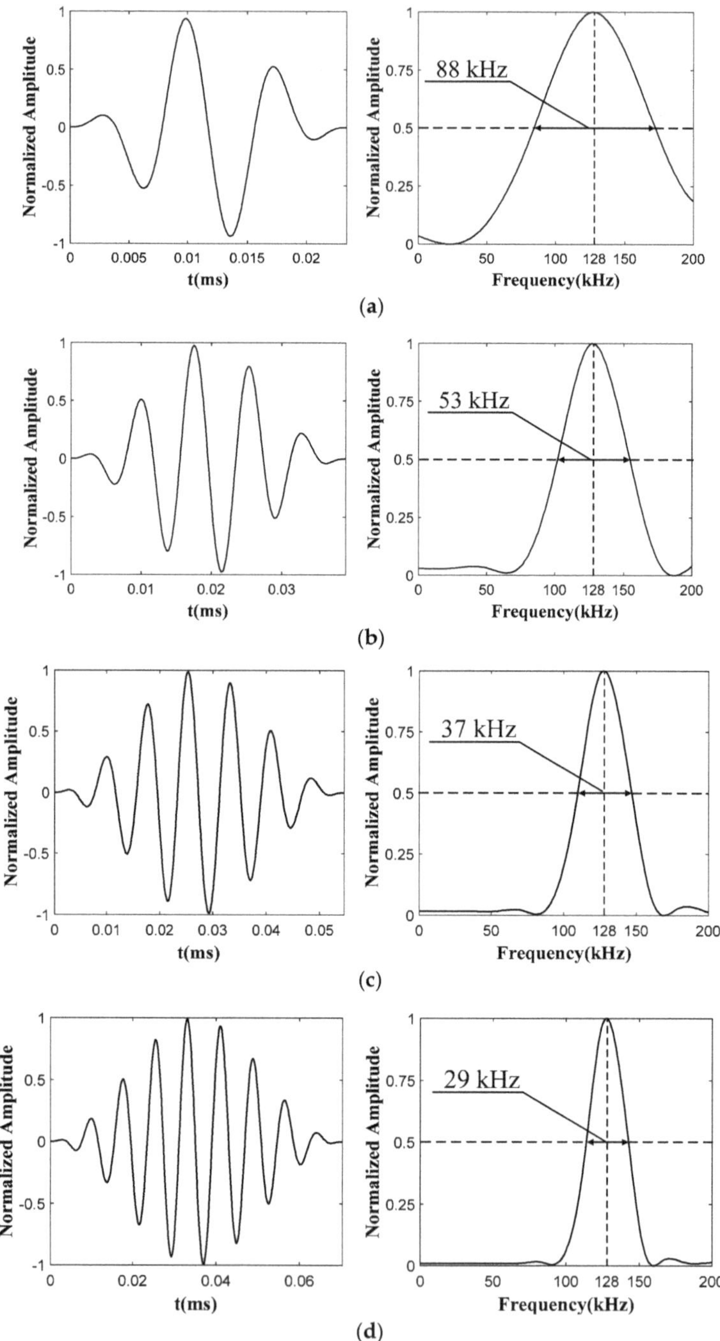

Figure 8. The excitation signals with different cycles and their amplitude spectrums: (**a**) three cycles; (**b**) five cycles; (**c**) seven cycles; (**d**) nine cycles. The half-decline frequency bandwidth of the signals is marked.

3.2. Method Comparison and Factor Analysis

The results of the proposed synthetic aperture focusing technology method for circumferential scanning (CSAFT) are compared with the B-scan image and the results by the DAS method [29]. The sound field of the MST with the length 2L = 70 mm is used to modulate the signals from the simulation. The time signals are turned into distance signals with the group velocity $\hat{c}_g$ by $R = t \times \hat{c}_g$; all the distance signals are stacked to become the B-scan image in the order of the angle positions of the receiving points. By the DAS method, the image region is first divided into pixels; then, the flight time of each pixel is calculated with the group velocity and the distance between the pixel and the transducer in the scanning. The amplitude of each pixel is the superposition of the amplitudes extracted from the time signals of the transducer at different positions according to the flight times. It should be noticed that, since only the procedure of receiving is simulated, the group velocity $\hat{c}_g$ will be the same as the group velocity $c_g = 3208$ m/s. The results in Figures 9–11 are all calculated with the excitation signal of three cycles.

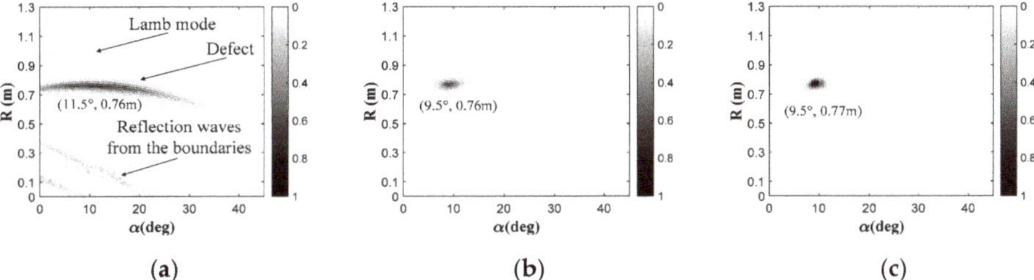

Figure 9. The results for the angular step 0.5° by different methods: (**a**) B-scan, (**b**) delay and sum (DAS), (**c**) circumferential synthetic aperture focusing technique (CSAFT). The excitation signal is a tone-burst signal with three cycles.

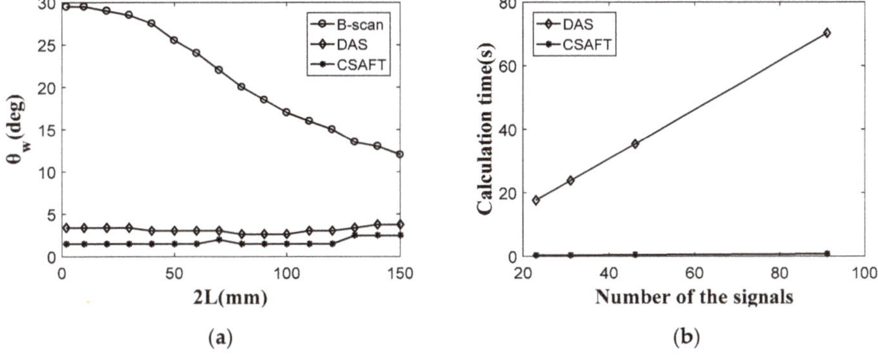

Figure 10. Comparisons of three methods with different factors: (**a**) the angular bandwidth θ_w by the transducer with different lengths; (**b**) Calculation times with different numbers of signals. The excitation signal is a tone-burst signal with three cycles.

The results for the angular step 0.5° are shown in Figure 9. In the B-scan image of Figure 9a, extra waves existing in front of the defect waves are Lamb modes generated by the defect, which travel faster than the SH0 mode. The reflection waves from the boundaries appear after the defect waves. As in Figure 6, the divergence angle of the transducer cannot be zero, and so the defect can also be detected even if it is not directly in front of the transducer. That is the reason for the tailings of the defect waves in the B-scan image, which significantly lower the angular resolution. The image is improved by the DAS method, as in Figure 9b. However, over adjustments exist as a result, leading to some residual trailing. The result of the CSAFT method is shown in Figure 9c. The best angular resolution of the defect spot with the least trailing can be obtained compared with those by other methods. The defect locations (0.76~0.77 m) in Figure 9 are calculated with the peaks of the defect spots, which results in the localization error of the distances (6.7%) for the width of the defect waves in the time signals. The denoted angular positions (9.5~11.5°) have little error compared with the actual value.

The effect of the length of the transducer is investigated in the simulation. The angular bandwidth of the defect spot θ_w, defined as the largest angle range of the region in which the amplitudes are no less than half of the peak, is used to denote the size of the defect spot in the angular direction. The divergence angle θ_d of the transducer decreases with the increment of the length of the transducer length 2L as in Figure 6. The angular bandwidth of the defect spot in the B-scan image has the same tendency of decrement as the divergence angle of the transducer, as in Figure 10a, which is 12~29.5° for a length of the transducer between 2 and 150 mm. The DAS method provides a significant improvement of the angular bandwidth (3.4~3.8°), which is basically not affected by the variations of the transducer lengths. The angular bandwidths by the CSAFT method (1.5~2.5°) are slightly superior to those by the DAS method. The CSAFT and DAS method are also compared for different angular steps, which are integral multiples of the minimum angular step 0.5°. The numbers of the receiving signals for different angular steps are listed in Table 2.

The effect of the excitation signals with different cycles is also studied and the results by the CSAFT method marked with the defect positions and angular bandwidths are shown in Figure 12. It can be observed that the widths of the defects along the radius and the angular bandwidths increase with the increment of the cycles. The defect localization accuracy decreases as well. The increment of the signal cycle is conducive to raising the energy of the incident waves for long testing ranges and narrow down the frequency bandwidth to reduce the influence of other frequencies. However, it extends the lengths of the excitation signals in the time domain, which determine the sizes of the focused defect spots. Both resolutions along the angle and radius are decreased. The cycles of the excitation signals should be selected in the consideration of both resolutions and testing ranges in field tests.

The calculation time of the CSAFT and DAS methods are tested with the same number of the signals in the same computer. As in Figure 10b, the calculation time of the CSAFT method is much less than that of the DAS method, which increases linearly with the numbers of the signals. The advantage of the proposed method in calculation efficiency can be more significant when dealing with a larger amount of data. The images by three methods for different angular steps are presented in Figures 9 and 11. Since the angular sampling rate decreases with the increment of the angular step, the defect spots are coarser in the results of the Bscan and CSAFT method. In the results by the DAS method, over-adjustments get more severe with larger trailing in the image, which may become distractions in defect recognition. The proposed CSAFT method provides the best performance for each angular step.

Table 2. The numbers of the receiving signals for different angular steps.

Name	Value			
Angular step (°)	0.5	1	1.5	2
Number of the signals	91	46	31	23

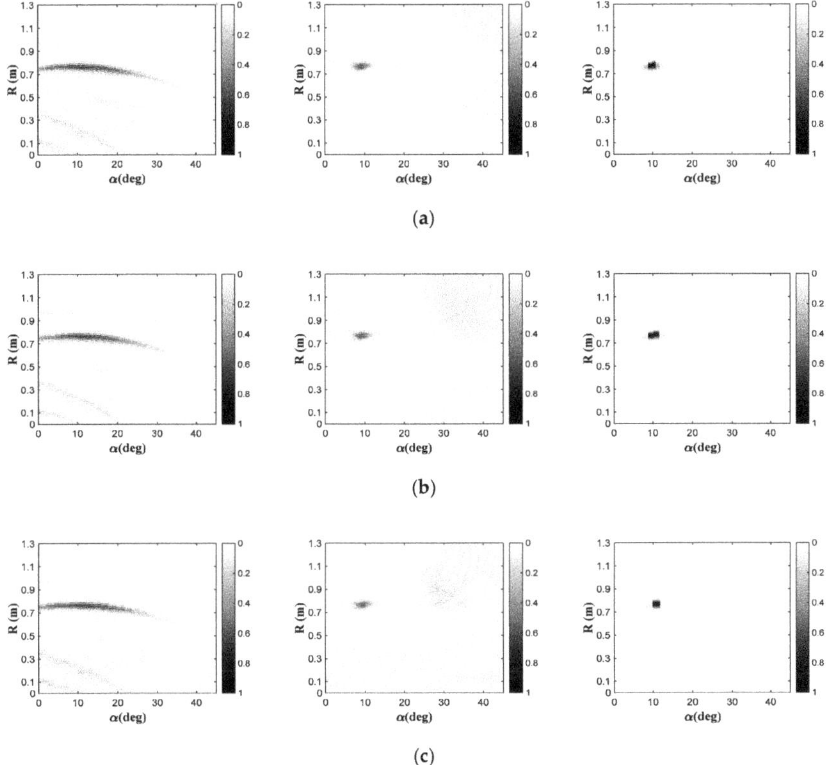

Figure 11. Comparisons of the results by the B-scan (**Left**), DAS (**Middle**), and CSAFT (**Right**) methods for different angular steps: (**a**) $\theta_s = 1°$; (**b**) $\theta_s = 1.5°$; (**c**) $\theta_s = 2.0°$. The excitation signal is a tone-burst signal with three cycles.

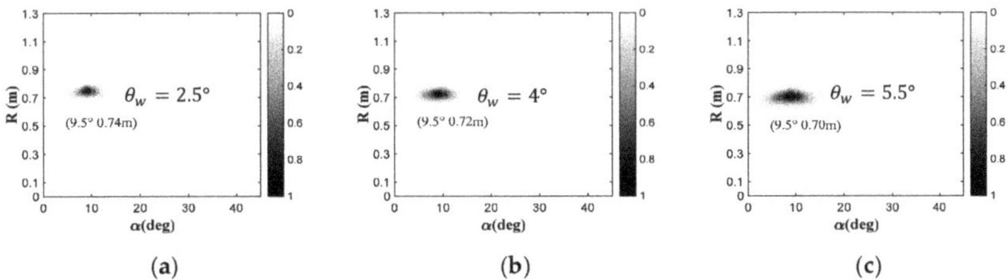

Figure 12. Comparisons of the results by the CSAFT method for excitation signals with different cycles: (**a**) five cycles; (**b**) seven cycles; (**c**) nine cycles. θ_w is the angular bandwidth of the defect spot.

4. Experimental Investigation

4.1. Experimental Setup

Experiments were carried out to verify the proposed method. As shown in Figure 13, the equipment named the magnetostrictive guided wave detector (MSGW30, Zheda Jingyi Tech, Ltd., Hangzhou, China) was used in the experiments. The equipment has the ability to generate and receive guided waves with the modules including signal generation, power amplification, pre-amplification and A/D conversion under the control of the computer. The coils in the magnetostrictive sandwich transducer (MST) were processed as a flexible printed circuit (FPC), and the material of the magnetostrictive strip is Fe-Co alloy. The circumferential scanning is shown in Figure 14. The iron plate is the same size as that in the simulation. An iron mass scatterer with the additional cross-sectional area of 100 mm^2 was coupled on the iron plate at the radius 0.825 m from the original point O_2 and 12° from the start position of the scanning as the artificial defect using the epoxy coupling agent. The MST moved around the original point O_2 at a radius of 1.3 m with the angular step 1.5° to scan the fan area with an angle of 45°. The MST was used both as an actuator and sensor. The excitation signal used in the experiment is the same as that in the simulation, which is shown in Figure 8a. The sampling ratio is 2 MHz. Thirty-one signals were collected in the scanning. It should be mentioned that the coupling agent used for the coupling between the transducer and the plate is the shear wave coupler (UGW30, Zheda Jingyi Tech, Ltd., Hangzhou, China).

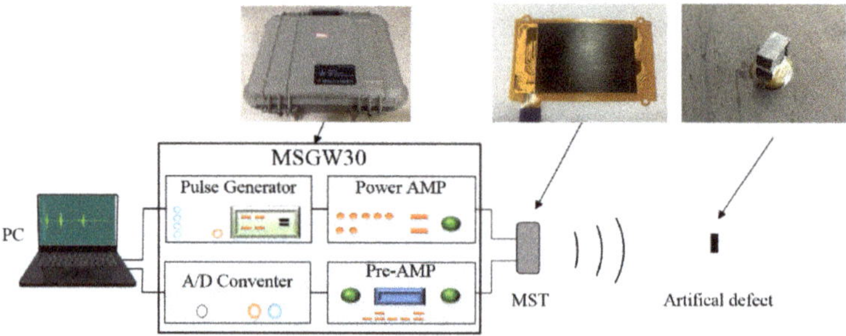

Figure 13. The magnetostrictive guided wave inspection system for circumferential scanning in the plate.

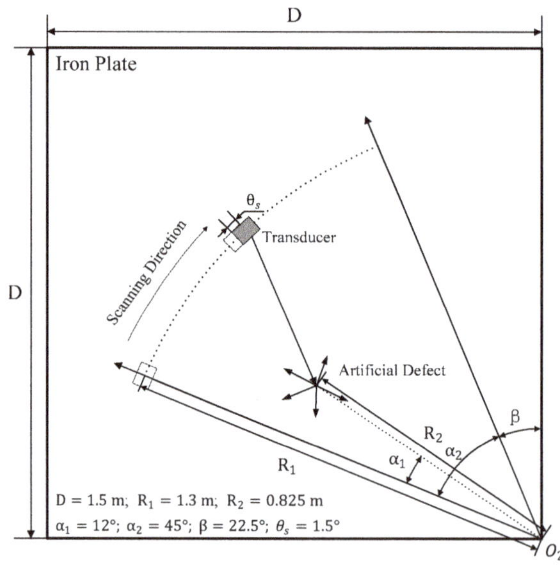

Figure 14. Circumferential scanning in the iron plate using guided waves.

4.2. Results and Discussion

The temperature variations should always be considered in guided wave inspections. The wave velocities are affected by the mechanical properties of the plates, which are changed by the temperature. Wave velocities of collected signals may be different if the scanning is carried out under large temperature variations, affecting the quality of the scanning image. Therefore, wave velocity compensation should be considered. In field tests, the scanning is usually finished within minutes and the temperature does not change rapidly in a short time. However, the situation can be different when the scanning is carried out at a different time (in the day or at night). Thus, wave velocity estimation is necessary before scanning. In this study, the guided waves were generated directly towards one boundary of the plate. The flight time of the reflection waves from the boundary and the distance between the MST and the boundary were used to calculate the group velocity, which is 2714 m/s. The reason for the error (15.4%) of the velocity may be the same as described in Section 3.2, since the peak of the reflection wave was used to determine the flight time.

The results of the three methods are shown in Figure 15. The amplitudes of the images are all normalized between 0 and 1. The estimated peak value of the environmental noise is less than 0.12. The defect spot of the B-scan method in Figure 15a has large trailing, as expected, which may lower the angular resolution. By the DAS method, the trailing around the defect spot is eliminated, as in Figure 15b; however, some trailing occurs in other regions of the image, which have similar amplitudes to the defect spot, becoming a severe distraction in defect recognition. In Figure 15c, the defect spot is improved in the angular direction despite some residual trailing. It can be observed that the defect spot by the DAS method is less obvious than that by the B-scan method, whereas the defect spot by the CSAFT method is much more recognizable. Quantitative comparisons of the defect spot by three methods are listed in Table 3. The position of the defect is denoted by the peak of the spot, and the angular bandwidth is defined as in Section 3.2. The reflection waves from the boundaries are taken as references to show the significance of the defect spots by the three methods. The peak value ratio between the defect and the boundaries by the proposed method is 580% larger than that by the B-scan method, whereas the result with the DAS method does not give any improvement. Apart from the largest significance of the defect, the results by the CSAFT method have the least error in the angular localization and smallest angular bandwidth of the defect compared with the other two methods.

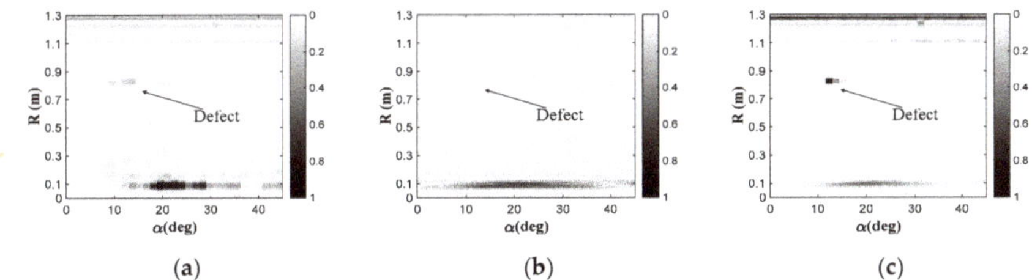

Figure 15. Experimental images by the three methods: (**a**) B-scan; (**b**) DAS; (**c**) CSAFT.

Table 3. Defect estimation by the three methods.

	Name	Actual Values	B-Scan	DAS	CSAFT
Position	Angle (°)	12	13.5	12.5	12
	Distance (m)	0.825	0.823	0.815	0.821
Angular bandwidth (°)		NAN [1]	4.5	3.8	2.0
Peak value ratio (defect/boundaries)		NAN [1]	0.25	0.14	1.71

[1] NAN means no value for this option.

5. Conclusions

A synthetic aperture focusing technique method is proposed for circumferential scanning of plates using guided waves. The SH0 mode is selected for the inspection, which can be effectively generated by the magnetostrictive sandwich transducer (MST). The proposed imaging method is based on the exploding reflector model (ERM) by solving the wave equation in a polar coordinate system and deriving the relationships between guided wave fields at different radii in the frequency domain. Numerical and experimental validation was successfully performed for the circumferential scanning of the defects in an iron plate. The main findings are summarized as follows:

1. The SH0 mode has the same detection sensitivity for the cross section of the plate with a uniform wave structure along its thickness; it is nondispersive in frequency and has little energy leakage for the inspection of plates contacted with liquids, which can be generated purely at the frequency below the cutoff frequency of the SH1 mode. The SH0 mode is ideal for the inspection of tank floors in service.
2. The MST used for SH0 wave generation can control the wave propagation direction. The main structure of the MST is designed for the SH0 mode at a certain frequency, and the divergence angle can be adjusted with the length of the transducer.
3. The angular bandwidth of the defect spot by the proposed method ($1.5 \sim 2.5°$) in the numerical simulations are found to be better than those by the traditional delay and sum (DAS) method ($3.4 \sim 3.8°$) for transducers with different lengths ($2 \sim 150$ mm), both of which have a significant improvement compared with those by the B-scan image ($12 \sim 29.5°$). The proposed method has better calculation efficiency than the DAS method, particularly for a large amount of data. No additional trailing occurs in the image of the proposed method with the increment of the angular step. In circumferential scanning, there should not be too many cycles of the excitation signal when the testing range is sufficient.
4. By the proposed method, the peak amplitude of the defect spot can be increased, providing a significant improvement for the defect spot compared with those by the B-scan and DAS methods.

The CSAFT method is expected to be effective for the non-destructive scanning of tank floors in service to minimize the risks and costs of the maintenance operations. For future study, actual oil tank floors installed with accessories will be examined. Defects of different kinds, sizes, and numbers will be inflicted on the tank floor, which will be tested when the oil tank is empty and filled with different kinds of liquids in the field.

Author Contributions: J.W. and S.W. conceived and designed the models and methods presented; Z.T., F.L., and K.Y. helped with the conception and experiments; J.W. performed the simulations, experiments, and data analysis; J.W. and Z.T. wrote this paper.

Funding: The work was supported by the National Natural Science Foundation of China (Grant Nos. 61271084, 51275454, and U1709216) and the Technique Plans of Zhejiang Province (2017C01042).

Conflicts of Interest: The authors declare no conflict of interest.

References

1. Tetteh-Wayoe, D. Shell Corrosion Allowance for Aboveground Storage Tanks. In Proceedings of the 2008 7th ASME International Pipeline Conference, Calgary, AB, Canada, 29 September–3 October 2008; American Society of Mechanical Engineers: New York, NY, USA, 2009.
2. Yang, T.; Zhang, X.; Huang, Z.S.; Yang, F.; Zhang, T.; Wu, J.W. Vertical Tank Bottom Line Integrated Acoustic Defect Detection Technology Research and Application. *Pipeline Tech. Equip.* **2016**, *4*, 21–23. (In Chinese) [CrossRef]
3. Liu, Z.; Kang, Y.; Wu, X.; Yang, S. Study on local magnetization of magnetic flux leakage testing for storage tank floors. *Insight* **2003**, *45*, 328–331. [CrossRef]
4. Ramírez, A.R.; Mason, J.S.; Pearson, N. Experimental study to differentiate between top and bottom defects for MFL tank floor inspections. *NDT E Int.* **2009**, *42*, 16–21. [CrossRef]

5. Usarek, Z.; Warnke, K. Inspection of Gas Pipelines Using Magnetic Flux Leakage Technology. *Adv. Mater. Sci.* **2017**, *17*, 37–45. [CrossRef]

6. Murayama, R.; Makiyama, S.; Kodama, M.; Taniguchi, Y. Development of an ultrasonic inspection robot using an electromagnetic acoustic transducer for a Lamb wave and an SH-plate wave. *Ultrasonics* **2004**, *42*, 825–829. [CrossRef] [PubMed]

7. Kwon, J.R.; Lyu, G.J.; Lee, T.H.; Kim, J.Y. Acoustic emission testing of repaired storage tank. *Int. J. Press. Vessel. Pip.* **2001**, *78*, 373–378. [CrossRef]

8. Riahi, M.; Shamekh, H.; Khosrowzadeh, B. Differentiation of leakage and corrosion signals in acoustic emission testing of aboveground storage tank floors with artificial neural networks. *Russ. J. Nondestruct. Test.* **2008**, *44*, 436–441. [CrossRef]

9. Paulauskiene, T.; Zabukas, V.; Vaitiekūnas, P. Investigation of volatile organic compound (VOC) emission in oil terminal storage tank parks. *J. Environ. Eng. Landsc. Manag.* **2009**, *17*, 81–88. [CrossRef]

10. Zhang, X.W.; Tang, Z.F.; Lv, F.Z.; Pan, X.H. Helical comb magnetostrictive patch transducers for inspecting spiral welded pipes using flexural guided waves. *Ultrasonics* **2017**, *74*, 1–10. [CrossRef] [PubMed]

11. Zhang, X.W.; Tang, Z.F.; Lv, F.Z.; Yang, K.J. Scattering of torsional flexural guided waves from circular holes and crack-like defects in hollow cylinders. *NDT E Int.* **2017**, *89*, 56–66. [CrossRef]

12. Sharma, S.; Mukherjee, A. Ultrasonic guided waves for monitoring corrosion in submerged plates. *Struct. Control Health Monit.* **2015**, *22*, 19–35. [CrossRef]

13. Ostachowicz, W.; Kudela, P.; Malinowski, P.; Wandowski, T. Damage localisation in plate-like structures based on PZT sensors. *Mech. Syst. Signal. Proc.* **2009**, *23*, 1805–1829. [CrossRef]

14. Wilcox, P.D. Omni-directional guided wave transducer arrays for the rapid inspection of large areas of plate structures. *IEEE Trans. Ultrason. Ferroelectr. Freq. Control* **2003**, *50*, 699–709. [CrossRef] [PubMed]

15. Evans, M.; Lucas, A.; Ingram, I. The inspection of level crossing rails using guided waves. *Constr. Build. Mater.* **2018**, *179*, 614–618. [CrossRef]

16. Mazeika, L.; Kazys, R.; Raisutis, R.; Demcenko, A.; Sliteris, R.; Cantore, C. Non-Destructive Testing of Fuel Tanks Using Long-Range Ultrasonic. In Proceedings of the 4th International Conference on Emerging Technologies in Non-Destructive Testing, Stuttgart, Germany, 2–4 April 2007; Busse, G., VanHemelrijck, D., Solodov, I., Anastasopoulos, A., Eds.; Taylor & Francis Ltd.: London, UK, 2011.

17. Hay, T.R.; Royer, R.L.; Gao, H.D.; Zhao, X.; Rose, J.L. A comparison of embedded sensor Lamb wave ultrasonic tomography approaches for material loss detection. *Smart Mater. Struct.* **2006**, *15*, 946. [CrossRef]

18. Giridhara, G.; Rathod, V.T.; Naik, S.; Mahapatra, D.R.; Gopalakrishnan, S. Rapid localization of damage using a circular sensor array and Lamb wave based triangulation. *Mech. Syst. Signal. Proc.* **2010**, *24*, 2929–2946. [CrossRef]

19. Rathod, V.T.; Mahapatra, D.R. Ultrasonic Lamb wave based monitoring of corrosion type of damage in plate using a circular array of piezoelectric transducers. *NDT E Int.* **2011**, *44*, 628–636. [CrossRef]

20. Chakraborty, N.; Rathod, V.T.; Mahapatra, D.R.; Gopalakrishnan, S. Guided wave based detection of damage in honeycomb core sandwich structures. *NDT E Int.* **2012**, *49*, 27–33. [CrossRef]

21. Ravi, N.B.; Rathod, V.T.; Chakraborty, N.; Mahapatra, D.R.; Sridaran, R.; Boller, C. Modeling Ultrasonic NDE and Guided Wave based Structural Health Monitoring. In Proceedings of the Conference on Structural Health Monitoring and Inspection of Advanced Materials, Aerospace, and Civil Infrastructure, San Diego, CA, USA, 9–12 March 2015; Shull, P.J., Ed.; SPIE—The Internaitional Society for Optical Engineering: Bellingham, WA, USA, 2015.

22. Zhao, X.; Royer, R.L.; Owens, S.E.; Rose, J.L. Ultrasonic lamb wave tomography in structural health monitoring. *Smart Mater. Struct.* **2011**, *20*, 105002. [CrossRef]

23. Zhao, X.; Rose, J.L. Ultrasonic tomography for density gradient determination and defect analysis. *J. Acoust. Soc. Am.* **2005**, *117*, 2546. [CrossRef]

24. Zhao, X.; Rose, J.L.; Gao, F.D. Determination of density distribution in ferrous powder compacts using ultrasonic tomography. *IEEE Trans. Ultrason. Ferroelectr. Freq. Control* **2006**, *53*, 360–369. [CrossRef] [PubMed]

25. Zhao, X.L.; Gao, H.D.; Zhang, G.F.; Ayhan, B.; Yan, F.; Kwan, C.; Rose, J.L. Active health monitoring of an aircraft wing with embedded piezoelectric sensor/actuator network: I. Defect detection, localization and growth monitoring. *Smart Mater. Struct.* **2007**, *16*, 1208–1217. [CrossRef]

26. Rao, J.; Ratassepp, M.; Fan, Z. Guided wave tomography based on full waveform inversion. *IEEE Trans. Ultrason. Ferroelectr. Freq. Control* **2016**, *63*, 737–745. [CrossRef] [PubMed]

27. Rao, J.; Ratassepp, M.; Fan, Z. Limited-view ultrasonic guided wave tomography using an adaptive regularization method. *J. Appl. Phys.* **2016**, *120*, 113–127. [CrossRef]

28. Puchot, A.R.; Cobb, A.C.; Duffer, C.E.; Light, G.M. Inspection Technique for above Ground Storage Tank Floors using MsS Technology. In Proceedings of the 10th International Conference on Barkhausen and Micro-Magnetics, Baltimore, ML, USA, 21–26 July 2013; Chimenti, D.E., Bond, L.J., Thompson, D.O., Eds.; American Institute Physics: New York, NY, USA, 2014.

29. Wang, C.H.; Rose, J.T.; Chang, F.K. A synthetic time-reversal imaging method for structural health monitoring. *Smart Mater. Struct.* **2004**, *13*, 415–423. [CrossRef]

30. Pulthasthan, S.; Pota, H.R. Detection, localization and characterization of damage in plates with an, in situ array of spatially distributed ultrasonic sensors. *Smart Mater. Struct.* **2008**, *17*, 035035. [CrossRef]

31. Sicard, R.; Goyette, J.; Zellouf, D. A saft algorithm for lamb wave imaging of isotropic plate-like structures. *Ultrasonics* **2002**, *39*, 487–494. [CrossRef]

32. Sicard, R.; Chahbaz, A.; Goyette, J. Guided lamb waves and l-saft processing technique for enhanced detection and imaging of corrosion defects in plates with small depth-to-wavelength ratio. *IEEE Trans. Ultrason. Ferroelectr. Freq. Control* **2004**, *51*, 1287–1297. [CrossRef] [PubMed]

33. Davies, J.; Cawley, P. The application of synthetic focusing for imaging crack-like defects in pipelines using guided waves. *IEEE Trans. Ultrason. Ferroelectr. Freq. Control* **2009**, *56*, 759–771. [CrossRef] [PubMed]

34. Aristegui, C.; Lowe, M.J.S.; Cawley, P. Guided waves in fluid-filled pipes surrounded by different fluids. *Ultrasonics* **2001**, *39*, 367–375. [CrossRef]

35. Fan, Z.; Lowe, M.J.S.; Castaings, M.; Bacon, C. Torsional waves propagation along a waveguide of arbitrary cross section immersed in a perfect fluid. *J. Acoust. Soc. Am.* **2008**, *124*, 2002–2010. [CrossRef] [PubMed]

36. Kwun, H.; Kim, S.Y.; Light, G.M. Long-range guided wave inspection of structures using the magnetostrictive sensor. *J. Korean Soc. NDT* **2001**, *21*, 383–390.

37. Rose, J.L. *Ultrasonic Guided Waves in Solid Media*, 1st ed.; Cambridge University Press: Cambridge, UK, 2014; pp. 272–275, ISBN 978-1-107-04895-9.

38. GB 50341-2014. Code for Design of Vertical Cylindrical Welded Steel Oil Tanks. China National Petroleum Corporation, 2012. Available online: www.csres.com/detail/243638.html (accessed on 29 May 2014).

39. Kim, I.K.; Kim, Y.Y. Shear horizontal wave transduction in plates by magnetostrictive gratings. *J. Mech. Sci. Technol.* **2007**, *21*, 693–698. [CrossRef]

40. Lee, J.S.; Kim, Y.Y.; Cho, S.H. Beam-focused shear-horizontal wave generation in a plate by a circular magnetostrictive patch transducer employing a planar solenoid array. *Smart Mater. Struct.* **2008**, *18*, 015009. [CrossRef]

41. Guerra, C.; Biondi, B. Fast 3D migration-velocity analysis by wavefield extrapolation using the prestack exploding-reflector model. *Geophysics* **2011**, *76*, WB151–WB167. [CrossRef]

42. Wu, S.W.; Skjelvareid, M.H.; Yang, K.J.; Chen, J. Synthetic aperture imaging for multilayer cylindrical object using an exterior rotating transducer. *Rev. Sci. Instrum.* **2015**, *86*, 083703. [CrossRef] [PubMed]

43. Haun, M.A.; Jones, D.L.; O'Brien, W.D. Efficient three-dimensional imaging from a small cylindrical aperture. *IEEE Trans. Ultrason. Ferroelectr. Freq. Control* **2002**, *49*, 1589–1592. [CrossRef]

44. Demma, A.; Cawley, P.; Lowe, M. Scattering of the fundamental shear horizontal mode from steps and notches in plates. *J. Acoust. Soc. Am.* **2003**, *113*, 1880–1891. [CrossRef] [PubMed]

Article

Signal Strength Enhancement of Magnetostrictive Patch Transducers for Guided Wave Inspection by Magnetic Circuit Optimization

Jianjun Wu [1], Zhifeng Tang [2,*], Keji Yang [1] and Fuzai Lv [1]

[1]   State Key Laboratory of Fluid Power and Mechatronic Systems, School of Mechanical Engineering, Zhejiang University, 38 Zheda Road, Hangzhou 310027, China; 11425023@zju.edu.cn (J.W.); yangkj@zju.edu.cn (K.Y.); lfzlfz@zju.edu.cn (F.L.)
[2]   Institute of Advanced Digital Technologies and Instrumentation, School of Biomedical Engineering and Instrument Science, Zhejiang University, 38 Zheda Road, Hangzhou 310027, China
*   Correspondence: tangzhifeng@zju.edu.cn; Tel.: +86-138-5713-1010

Received: 16 March 2019; Accepted: 4 April 2019; Published: 9 April 2019

Featured Application: Signal strength enhancement of the magnetostrictive patch transducer, which can be used as an element of a phase array or scanner in pipe and plate inspection.

Abstract: Magnetostrictive patch transducers (MPT) with planar coils are ideal candidates for shear mode generation and detection in pipe and plate inspection with the advantages of flexibility, lightness and good directivity. However, the low energy conversion efficiency limits the application of the MPT in long distance inspection. In this article, a method for the enhancement of the MPT was proposed by dynamic magnetic field optimization using a soft magnetic patch (SMP). The SMP can reduce the magnetic resistance of the magnetic circuit, which increases the dynamic magnetic field intensity in the magnetostrictive patch during wave generation and restricts the induced dynamic magnetic field within the area around the coils for sensing during wave detection. Numerical simulations carried out at different frequencies verified the improvement of the dynamic magnetic fields by the SMP and influence of different affecting factors. The experimental validations of the signal enhancement in wave generation and detection were performed in an aluminum plate. The amplitude magnification could reach 12.7 dB when the MPTs were covered by the SMPs. Based on the numerical and experimental results, the SMP with a large relative permeability and thickness and close fitting between the SMP and coils were recommended when other application conditions were met.

Keywords: magnetostrictive patch transducer; shear mode; soft magnetic patch; dynamic magnetic field optimization; signal strength enhancement

---

1. Introduction

Guided waves, which are ultrasonic waves restricted by the boundaries of the solid waveguide during propagation [1], have been applied in the inspection of many industrial structures, such as pipes [2,3], rods [4], plates [5–7] and rails [8–10] with the advantages of long distance, full coverage and low cost [11]. The conversions between the electrical energy and mechanical energy are mainly involved in guide wave inspection, which are accomplished by the transducers as the key parts in the inspection system. Transducers with various characteristics, including vibration directions, frequency bandwidths and sizes, are generally customized to satisfy the demands under different conditions. Based on the physical effects, the guided wave transducers can be categorized into different types, including electromagnetic acoustic transducers (EMAT), laser transducers, piezoelectric transducers (PZT) and magnetostrictive transducers. The EMATs are able to work without contacting the waveguide using the Lorentz force [11,12] although their applications are limited due to the low conversion efficiency, inapplicability for non-metal waveguides and signal instability caused by the coil vibration. The laser can be used to generate guided waves in a contactless manner based on the thermoelastic effect. However, the high power laser may burn the spot in the waveguide and the generated waves normally have very high frequencies (several MHz). With the high price, the inspection system with laser transducers still remain in laboratories without being widely used so far [13,14]. The magnetostrictive transducers and PZTs are widely used at a low cost and with satisfactory performance [15]. The magnetostrictive transducers can be made into small patches, which are flexible and light for different conditions [16]. Compared with the PZTs, they have the advantages of good durability, low requirement in coupling, low cost and superiority in shear vibration generation. For the inspection of ferromagnetic materials, the magnetostrictive transducers can also be used without contact [17].

The dimensions of the ferromagnetic materials change during the process of magnetization and their magnetic state varies when external stress is applied to them. These effects are from the properties of ferromagnetic materials, which are called the magnetostrictive effect (Joule effect) and inverse magnetostrictive effect (Villari effect), respectively. These effects can be explained with magnetic domain theory [18]. The magnetostrictive effect works in the generation of guided waves while the inverse effect accounts for the measurement [19]. Many studies have been carried out for the improvement of energy conversion, directivity and mode selection since the first application of the magnetostriction in guided wave inspection. In early studies, the magnetostriction (λ_s) of the ferromagnetic waveguide was used to generate guided waves [20]. This method still works in cable inspections [21,22]. However, the magnetostriction of these ferromagnetic waveguides, which are commonly made of iron (λ_s = 14 ppm), are very small and thus, the patches and strips made of materials with a large magnetostriction, such as Ni (λ_s = −50 ppm), Co (λ_s = −93 ppm) and Hiperco (λ_s = 60~100 ppm), are coupled with the waveguide for wave generation and detection [23–25], which can also be used for non-ferromagnetic waveguides [26]. There has been research carried out on the application of giant magnetostrictive materials, such as Galfenol (λ_s = 200~400 ppm) and Terfenol-D (λ_s = 1500~2000 ppm) in guided wave inspection [27,28]. These materials have poor machinability and need large driving magnetic fields for an ideal performance. The undesirable temperature adaptability and high cost also restricts their applications in field tests. There are two magnetic fields named the static bias magnetic field and dynamic alternating field, which should be applied into the magnetostrictive patches during the wave generation. The bias magnetic field for the elimination of the frequency doubling effect can be provided by permanent magnets [29]. For pipe inspection, the bias magnetic field may be more utilized when the magnetostrictive patches are bonded with an angle of 45° to the longitudinal axis of the pipes, which can be improved further with the patches in the Z-shape or with V-shaped yokes [30,31]. For magnetostrictive materials with certain coercive forces (e.g., Hiperco), the bias magnetic field should be more uniform, which can be achieved by pre-magnetization [32]. A solenoid coil can also provide a strong bias magnetic field, but a large direct current is needed and the magnetostrictive strip needs to be manufactured with grooves for solenoid coils [33]. The dynamic magnetic field is normally provided by the coils. Figure 1 shows the common coils used in transducers with directivity. The figure of eight coil has a large energy conversion efficiency with multiple turns as well as the coil with a core [34]. However, a high system power is needed to drive these coils with large resistances and the directivities are not satisfactory. The transducers with these coils are too large for narrow spaces and cannot be deformed to match curving surfaces. Planar coils in the shape of meanders or racetracks have good directivities. They can be manufactured as flexible printed circuits (FPC), which are small, light and flexible. As a result, they can be used as scanners in pipe and plate inspections or elements in phased arrays. The distance between the adjacent wires in planar coils depends on the generation of the wavelength from the mode, resulting in the sparse arrangement of the wires. Furthermore, these coils work in open magnetic circuits with dynamic magnetic fields dispersed in the air [19]. These characteristics result in the relative low energy conversion efficiency of the MPT with planar coils, limiting their applications in long distance inspection.

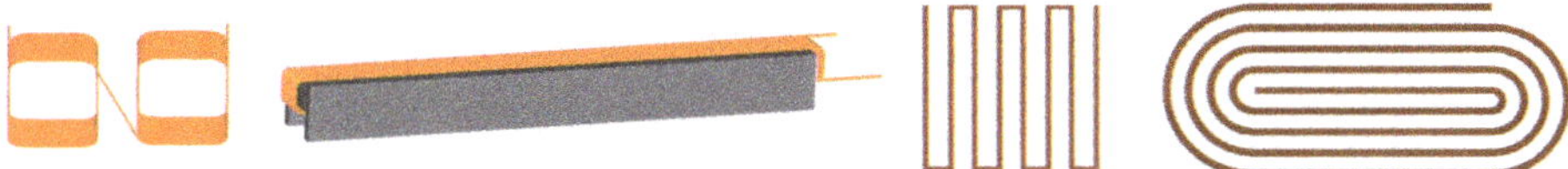

Figure 1. Coils commonly used in the magnetostrictive transducers.

In this paper, a method for the enhancement of the magnetostrictive patch transducer (MPT) with planar coils was proposed by the dynamic magnetic field optimization using a soft magnetic patch (SMP). In Section 2, the relationship between the magnetic fields and mechanical properties of the magnetostrictive materials is illustrated and the theoretical explanations of the improvement for the dynamic magnetic field are given based on the magnetic Ohm's law. As described in Section 3, a 2D model of the transducer was built to verify the dynamic magnetic field improvement and to study the effect of different affecting factors, including the thickness, magnetic conductivity, lift-off distance of the SMP with the excitation currents and induced magnetic fields at different frequencies. Experimental investigations were carried out in an aluminum plate as described in Section 4 and the conclusions follow in Section 5.

2. Theory

2.1. Magnetostrictive Equation

In ferromagnetic materials, adjacent atoms construct magnetic domains spontaneously because of the forces caused by electron spins [18]. As shown in Figure 2, without external magnetic fields, the magnetic domains distribute randomly, which results in the material having a total magnetic intensity of zero. When the external magnetic field is nonzero, the magnetic domains will move and spin to be parallel to it, changing the length of the object along the same direction. The magnetic domain theory gives a qualitative explanation of the magnetostrictive effect. The constitutive equations of magnetic and elastic properties in ferromagnetic materials are provided as follows [35]:

$$S = s\sigma + dH \tag{1}$$

$$\sigma = cS - eH \tag{2}$$

$$B = d^T\sigma + \mu H \tag{3}$$

where the superscript T denotes the transposition; S and σ are the strain and stress, respectively; H is the magnetic field intensity; B is the magnetic flux density; s and c are the elastic compliance matrix and stiffness matrix, respectively; d is the piezomagnetic matrix; e is the inverse matrix; and μ is the permeability. Equations (1) and (2) account for the magnetostrictive effect while Equation (3) contributes to the inverse magnetostrictive effect. Figure 3 shows the diagram of a MPT with planar coils. The static bias magnetic field in the magnetostrictive patch was provided by pre-magnetization, making the structure of the MPT thin and flexible without permanent magnets. The dynamic magnetic field was supplied by the coils sealed in the FPC, which sensed induced dynamic magnetic fields during the wave detection. In this paper, the model of the MPT was considered for the generation and detection of wave modes with shear vibrations in plates (shear horizontal waves) and pipes (torsional waves), which had the advantage of non-dispersion at low frequencies for inspection. The models for the longitudinal wave MPTs can be found in [36]. In shear mode generation, the dynamic magnetic field is perpendicular to the static magnetic field, of which the intensity is much smaller than that of the static magnetic field. The total strain is caused by the field superposition:

$$S_t = \mathbf{s}\sigma_t + \mathbf{d}(H_d + H_s) \tag{4}$$

where

$$\mathbf{d} = \begin{bmatrix} 0 & 0 & 0 & 0 & 0 & \frac{3\varepsilon_M}{H_s} \\ -\frac{\varphi}{2} & \varphi & -\frac{\varphi}{2} & 0 & 0 & 0 \\ 0 & 0 & 0 & \frac{3\varepsilon_M}{H_s} & 0 & 0 \end{bmatrix}^T \tag{5}$$

σ_t is the applied stress; H_d and H_s denote the intensities of the dynamic and static magnetic fields, respectively; φ is the gradient of the magnetostrictive curve of the material at H_s; and ε_M is the total magnetostrictive strain, which mostly depends on the static magnetic field since the dynamic field is much smaller. Considering that the wave amplitude is correlated with the dynamic shear strain S_d^s, Equation (4) can be simplified as:

$$S_d^s = \frac{3\varepsilon_M}{|H_s|}|H_d| \tag{6}$$

It can be concluded from Equation (6) that the dynamic shear strain is proportional to the amplitude of the dynamic magnetic field. Thus, it may be an effective way to raise the wave energy by enhancing the dynamic magnetic field.

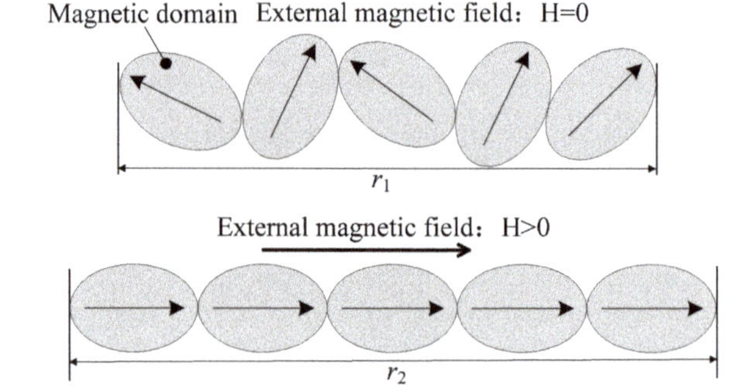

Figure 2. Magnetic domain movement in the magnetostrictive effect.

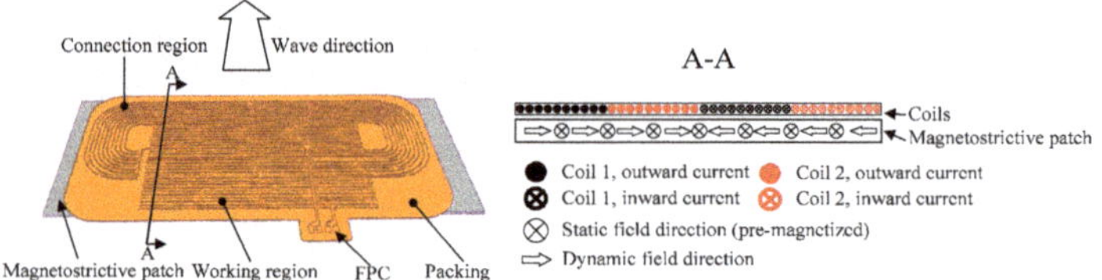

Figure 3. Magnetostrictive patch transducer for shear wave inspection.

2.2. Enhancement with a Soft Magnetic Patch

In wave generation, the dynamic magnetic field in the magnetostrictive patch is supplied by the coils loaded with sinusoidal pulses. It can be described with the magnetic Ohm's law qualitatively. As shown in Figure 4, the coils in the middle were the magnetic source, which supplied a constant magnetomotive force F when the input power of the equipment stayed unchanged. The magnetic flux circulated around the coils constructing a space named the magnetic circuit. According to the magnetic Ohm's law, the relation between the magnetomotive force F and the magnetic flux $\varnothing$ is as follows:

$$\varnothing = \frac{F}{\sum_{i=1}^{4} R_i}, \quad (i = 1, 2, 3, 4) \tag{7}$$

where

$$R_i = \frac{L_i}{\mu_i \mu_0 A_i}, \quad (i = 1, 2, 3, 4) \tag{8}$$

R_i is the magnetic resistance of part i in the magnetic circuit as marked in Figure 4; L_i and A_i is the length and sectional area, respectively; and μ_i is the relative permeability. Furthermore, $\mu_0 = 4\pi \times 10^{-7}$ N·A^{-2} is the permeability of the vacuum. Part 1 denotes the magnetostrictive patch. Parts 2, 3 and 4 were full of air. The magnetostrictive patch was not magnetized along the direction of the dynamic magnetic field and thus, the magnetic intensity H_d in part 1 could be calculated as:

$$H_d = \frac{\varnothing}{\mu_1 \mu_0 A_1} \tag{9}$$

The magnetic resistance of parts 2 and 4 were the same (i.e., $R_2 = R_4$). Combined with Equations (7) and (8), Equation (9) can be written as follows:

$$H_d = \frac{F}{L_1 + 2\frac{\mu_1 A_1}{\mu_2 A_2} L_2 + \frac{\mu_1 A_1}{\mu_3 A_3} L_3} \tag{10}$$

It could be found that without changing the material property (μ_1) and sizes (L_1 and A_1) of the magnetostrictive patch (part 1), the magnetic intensity H_d could be improved with an increase in the relative permeability of part 3 (μ_3). It could be achieved by covering a soft magnetic patch (SMP) over the coils. H_d could be improved further by increasing the thickness of the SMP (related to A_3) and decreasing the gap (L_2) between the SMP and magnetostrictive patch. It should be noted that Equation (10) qualitatively explains the improvement by the SMP in wave generation. The magnetic flux was distributed unevenly in the magnetic circuit, which also does not hold a regular geometric structure.

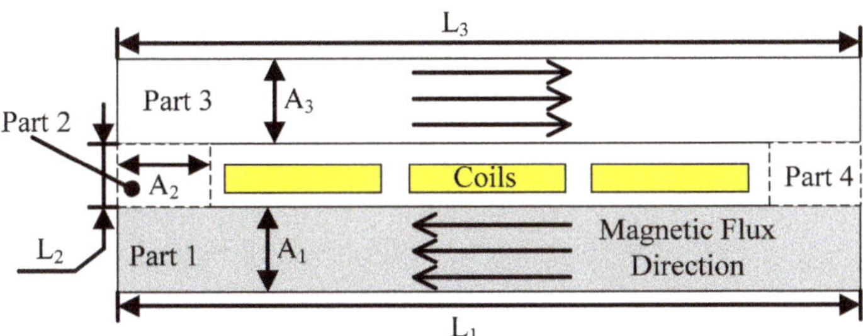

Figure 4. Diagram of dynamic magnetic fields in the MPTs (magnetostrictive patch transducers).

In wave detection, the magnetostrictive patch becomes a magnetic source because of the inverse magnetostrictive effect. The magnetic field induced by the magnetostrictive patch was proportional to the wave strength without additional magnetic fields. With the SMP above the coils, the induced magnetic field may have been restricted within the area around the coils and the induced current in the coils would be larger with an amplitude increase in the induced dynamic magnetic field. The improvement of the SMP can affect both processes when a MPT is used as an actuator and sensor at the same time. Furthermore, the electromagnetic interference, which are ubiquitous in the testing field, can be blocked by the SMP, improving the signal-to-noise ratio (SNR) of the detection signals.

3. Numerical Simulation

3.1. Simulation Setup

The dynamic magnetic field in the working region of the MPTs can be regarded as being uniform along the direction of the wires. Thus, for simplification, a two-dimensional model was built in COMSOL Multiphysics (COMSOL 5.3a, COMSOL Inc., Stockholm, Sweden) as shown in Figure 5. The coupling layer between the magnetostrictive patch and aluminum plate was neglected as it was non-conducting and non-magnetic. The coils of the two layers sealed in the FPC were made of copper. The material of the magnetostrictive patch was Fe–Co alloy. Permalloy, silicon steel, pure iron and Mn–Zn ferrite were the optional materials with a high relative permeability for the SMP. Considering the eddy-current loss, which may be caused by the dynamic magnetic field, the Mn–Zn ferrite with low conductivity was selected. In Figure 5, the real sizes of different parts are marked. The symbols represent the dimensions in the model (l: Length; t: Thickness; g: Gap distance) and the subscripts denote different parts (a: Aluminum plate; m: Magnetostrictive patch; c: Coil; s: SMS). l_{cc} was the interval length of wires in the same coil. The material properties are listed in Table 1. In this study, the generation and detection of the zeroth shear horizontal mode (SH0 mode) was considered to have the advantage of non-dispersion. To avoid the interference of high order modes, the excitation frequency should be below the cutoff frequency of the SH1 mode (312 kHz) in the aluminum plate (the thickness was 5 mm). With the consideration of the high attenuation of wave modes at high frequencies and low detection sensitivity at low frequencies in field tests, the frequency range was determined as 55–150 kHz. The frequency tests were carried out with increasing steps of 5 kHz using the FPCs named MF64 and MF128, which were designed at the main frequencies of 64 kHz and 128 kHz, respectively (MF64 for 55–90 kHz, MF128 for 95–150 kHz). Two coils were designed in one FPC for wave direction controlling as shown in Figure 3, but only one was used to avoid channel crosstalk in experiments. The simulations used the same setup and the current directions are shown in Figure 6. The red solid lines in Figure 5 are reference lines with the length of l_m to observe the magnetic intensities. For wave generation, the alternating currents with the amplitude 1A were loaded into the coils and the magnetic intensity along the *x*-axis of Line 1, which is in the middle of the magnetostrictive patch, was observed. In wave detection, the alternating magnetic fields with the amplitude 1 A/m² are applied into the magnetostrictive patch along the *x*-axis and the magnetic intensity in Line 2, which was across the middle of the coils near the magnetostrictive patch, was observed.

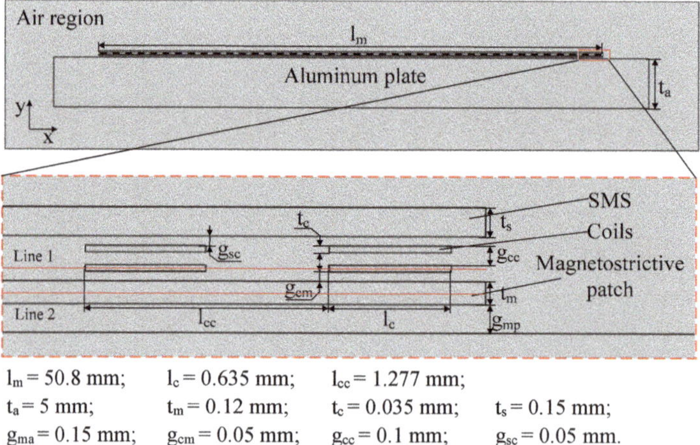

$l_m = 50.8$ mm; $l_c = 0.635$ mm; $l_{cc} = 1.277$ mm;

$t_a = 5$ mm; $t_m = 0.12$ mm; $t_c = 0.035$ mm; $t_s = 0.15$ mm;

$g_{ma} = 0.15$ mm; $g_{cm} = 0.05$ mm; $g_{cc} = 0.1$ mm; $g_{sc} = 0.05$ mm.

Figure 5. Numerical model for dynamic magnetic field simulation of magnetostrictive patch transducers.

Table 1. Material properties in the simulation.

Material Name	Air	Aluminum	Copper	Fe–Co Alloy	Mn–Zn Ferrite
Relative permeability	1	1	1	2000	2800
Conductivity (S/m)	0	3.774×10^7	5.988×10^7	2.5×10^6	0.17

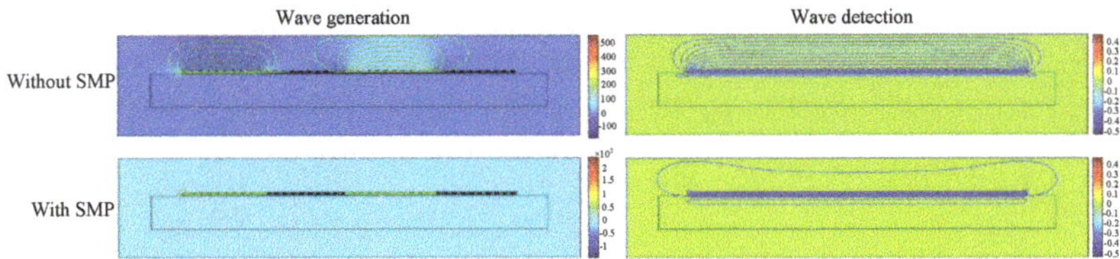

Figure 6. Current directions for different types of magnetostrictive patch transducers.

3.2. Simulation Verification and Factor Analysis

The typical dynamic magnetic field distributions of MPTs with and without the SMPs at 85 kHz in wave generation and detection are shown in Figure 7. With the SMP, the magnetic flux lines can be restricted in the narrow area around the coils instead of dispersing in the air region. Figure 8 shows the amplitudes of the magnetic field intensities along the x-axis from Line 1 and Line 2, respectively. The dynamic magnetic field existing in the magnetostrictive patch are right below the coils with currents. In wave detection, the high peaks are magnetic intensities in the coils. The dynamic magnetic field in the magnetostrictive patch was enhanced more than three times with the SMP and the induced magnetic fields in the coils were also stronger. The results shown in Figure 9 confirmed the enhancement of the SMPs at other frequencies. It should be mentioned that the sharp fall in Figure 9a may have been caused by different current loadings as shown in Figure 6.

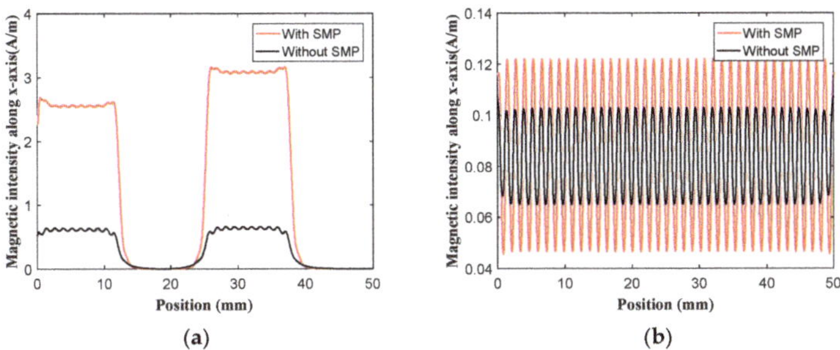

Figure 7. Magnetic field distribution in magnetostrictive patch transducers at 85 kHz. The magnetic lines are green lines in the case of wave generation and blue lines of wave detection.

Figure 8. Comparisons of magnetostrictive patch transducers with and without the SMPs at 85 kHz: (**a**) The magnetic field intensity in Line 1 in wave generation; (**b**) the magnetic field in Line 2 in wave detection.

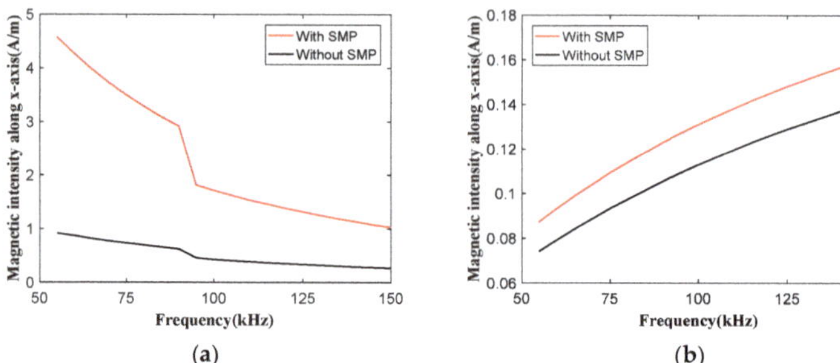

Figure 9. Comparisons of MPTs with and without the SMPs at 55~150 kHz: (**a**) The magnetic field intensity in Line 1 in wave generation; and (**b**) the magnetic field in Line 2 in wave detection.

For the proper selection of the SMP, several properties, including the relative permeability, thickness (t_s) and lift-off distance (g_{sc}), were studied with the other parameters of the model having fixed values. The typical results at 55 kHz, 85 kHz, 115 kHz and 145 kHz are shown in Figures 10–12. The magnetic fields get stronger with an increase in the relative permeability of the SMP, as shown in Figure 10a, which matches the theoretical prediction in Section 2.2. However, when the relative permeability is larger than 2000, the influence of relative permeability changes is subtle. The lower limit of the relative permeability for maximum magnetic field intensity in wave detection is 100. The dynamic magnetic field attenuates exponentially during the propagation in a good conductor so it is only distributed near the surface of the conductor (i.e., skin effect [37]). With this effect, the thickness of the SMP can be very small (0.01~0.6 mm in the simulation). The improvement of magnetic fields by an increase in the thickness was verified from the curves of 55 kHz and 85 kHz in Figure 11a, which did not increase significantly when the thickness was over 0.15 mm. For other cases at higher frequencies or in wave detection, it reaches the maximum for higher frequencies at the minimum thickness. The distance between the SMP and coils (i.e., the lift-off distance in this paper) will affect the improvement of the SMP if something necessarily has to be put in between (e.g., a second insulation layer). The influence of the SMP gets smaller naturally when the SMP gets further away from the coils. It can be explained with the increment of the magnetic resistance in the magnetic circuits over the lift-off distance as described in Section 2.2. Figure 12 shows the results for different lift-off distances. It could be found that during the lift-off of the SMP, the influence attenuation of the SMP continued until a long distance (2.5 mm) in wave generation, but the influence remained constant when the SMP was 0.5 mm away in wave detection.

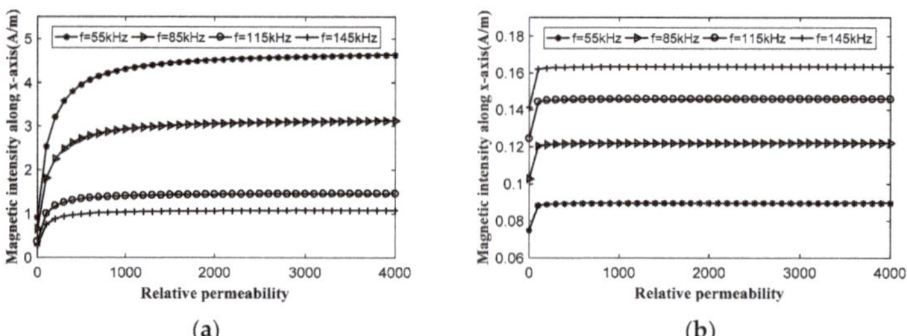

Figure 10. MPTs with the SMPs of different relative permeabilities: (**a**) The magnetic field intensity in Line 1 in wave generation; and (**b**) the magnetic field intensity in Line 2 in wave detection.

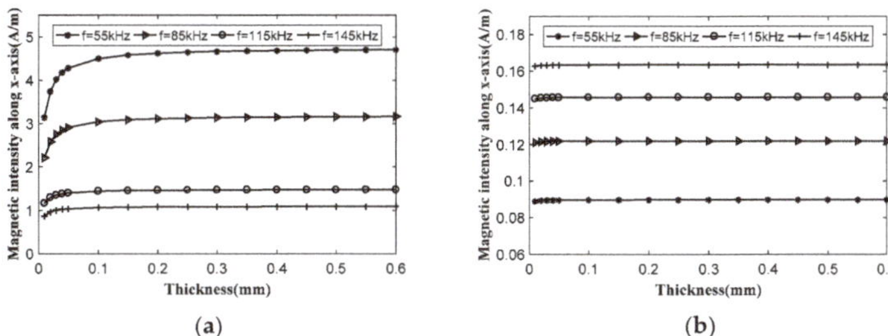

Figure 11. MPTs with the SMPs of different thicknesses: (a) The magnetic field intensity in Line 1 in wave generation; and (b) the magnetic field intensity in Line 2 in wave detection.

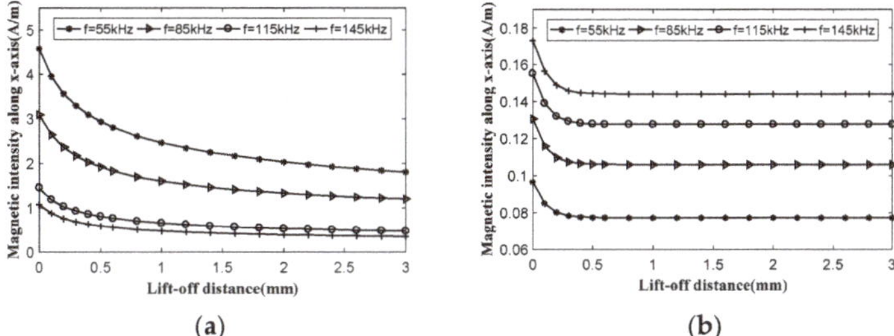

Figure 12. MPTs with the SMPs of different lift-off distances: (a) The magnetic field intensity in Line 1 in wave generation; and (b) the magnetic field intensity in Line 2 in wave detection.

4. Experimental Investigation

4.1. Experimental Setup

Experiments were carried out to verify the improvement of the SMP as shown in Figure 13. The equipment used in the experiments (MSGW30, Zheda Jingyi Tech, Ltd., Hangzhou, China) was designed for magnetostrictive guided wave inspection, which was integrated with modules, such as signal generation, power amplification, pre-amplification and A/D conversion. The equipment has two channels. Under the control of the computer, the equipment can work in the pulse-echo mode (wave generation and detection using Channel 1) and pitch-catch mode (wave generation using Channel 1 and detection using Channel 2). The inspection subject was an aluminum plate with a thickness of 5 mm. For the constant and same bonding conditions of the transducers (T1 and T2), dual adhesive tapes were used for the bonding between the magnetostrictive patches made of Fe-Co alloy and the aluminum plate. The thicknesses of the magnetostrictive patch and dual adhesive tape were 0.12 mm and 0.15 mm, respectively. The FPCs (MF64 and MF128) for the two main frequencies (64 kHz and 128 kHz) were processed with two coils and only one was used in the experiments to avoid the crosstalk. The generated guided wave propagated forward and backward without direction controlling and the wave packs from the far end of the plate in the signals were used for evaluation. The SMP was manufactured by hot-pressing with a mixture of the Mn–Zn ferrite power and rubber, which had a thickness and relative permeability of 0.15 mm and 2800, respectively. The SMP, FPC and magnetostrictive patch were all flexible and the dimensions of the SMP and magnetostrictive patch were 50.8 × 70 mm, which was the same as the working region of the FPC. The excitation signals at different frequencies were sinusoidal signals of four cycles filtered by the hanning windows and the sample frequency was 2 MHz.

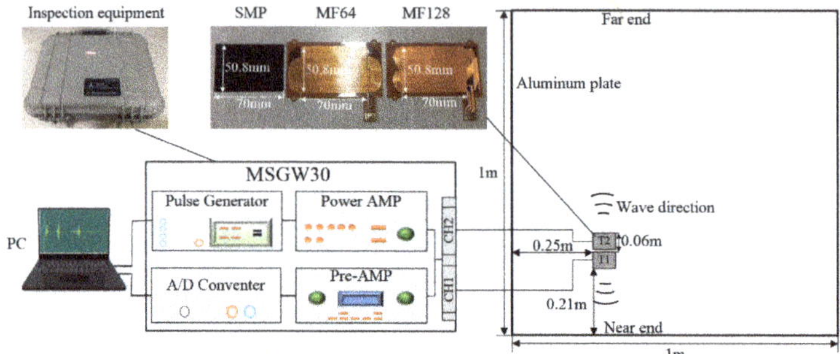

Figure 13. The diagram of the experiments for transducer improvement with the SMP.

The improvement of the SMP for wave generation and detection was first verified. In this experiment, the transducer marked as T1 in Figure 13 was used as an actuator while T2 was a sensor. The MF64 was used for frequencies of 55–90 kHz while the MF128 was used for frequencies of 95–150 kHz. The signals were set as the base data when T1 and T2 both were not covered by the SMPs. Furthermore, the signals received at the cases, such as only T1 covered by the SMP, only T2 covered by the SMP and both covered by the SMPs, were compared with the base data. The magnification was calculated using the peak values of the wave packs from the far end of the plate in the Hilbert envelopes of the signals.

The effect of thickness changes was investigated using multiple layers of SMPs. For each frequency, the signals were recorded when the MPT was covered by different layers of the SMPs (four layers at most). The last experiment was focused on the lift-off distance, which was carried out using a single layer of the SMP. The lift-off was achieved by installing multiple layers of epoxy patches between the FPC and SMP (six layers at most). The thickness of one single epoxy patch was 0.5 mm. These experiments were carried out with T1 used as an actuator and sensor in the pitch–catch mode. The signals from the cases in which the transducer was not covered by the SMP were used as the base data for amplification evaluation. It should be mentioned that in order to get rid of the void between the layers of the transducer, a nonmetal heavy weight was set above the transducer during the signal collection. The magnetostrictive patch was re-magnetized at the beginning of every experimental case to keep the uniformity of the bias magnetic field and multiple signals were collected in each case to reduce the effect of random noises.

4.2. Results and Discussions

The original signals were all processed with band-pass filters. Figure 14 shows the signals at 85 kHz when both the actuator and sensor were or were not covered by the SMPs. Comparisons were made between the ratios of the flight times in the signals and propagation distances from the transducers to the near end and far end. It can be confirmed that the wave packets at 2.25×10^{-4} s are echo waves from the near end and those at 5.48×10^{-4} s were from the far end. The far end echo waves were observed in the experiments, of which the peak values with the SMP were about two times of those without the SMP. The magnifications of signals at different frequencies were calculated by using the peak values of the far end echo waves in the envelope signals, which are shown in Figure 15. The SMP provided the enhancement for the MPT used in wave generation, which were even more than that in wave detection. The signal strength could be improved by 9.8 dB at most with both of the transducers covered by the SMPs.

Figure 16 shows the results of the MPT covered by different layers of SMPs at different frequencies. The enhancement was promoted by multiple layers of SMPs, which reached a maximum amplification of 12.7 dB at 130 kHz. However, there were no significant changes when the layer number of the SMPs was over two. The step change at the frequencies of 90 kHz and 95 kHz may have been caused by two different types of FPCs in the experiments. Considering these results, the SMP with the thickness over 0.3 mm should be recommended. The typical results for different lift-off distances of the SMP are shown in Figure 17. The decreasing trends at different frequencies match the prediction. The magnifications decrease faster at higher frequencies. A slight enhancement of 2.8 dB still exists when the SMP is moved 3 mm away from the FPC. However, for better results, the SMP should have a closer fit to the FPC under allowed conditions.

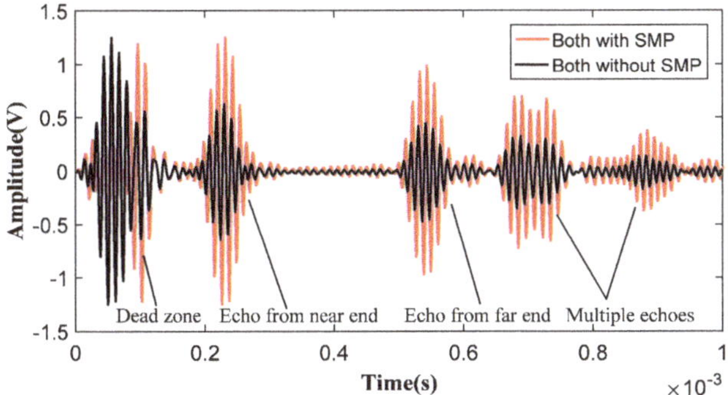

Figure 14. Original signals of the cases in which T1 and T2 both were with and without the SMPs at 85 kHz. T1 was an actuator and T2 was a sensor.

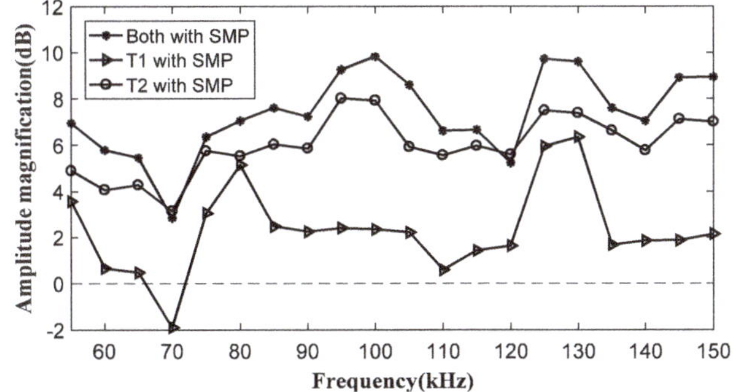

Figure 15. Amplitude magnification for the MPTs in wave generation and detection with the SMPs at different frequencies. T1 was an actuator and T2 was a sensor.

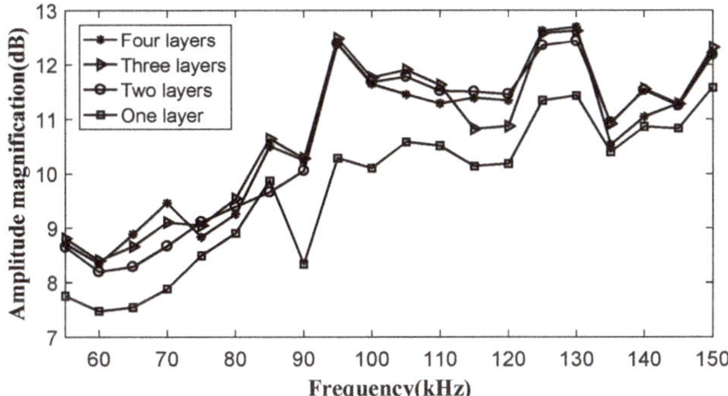

Figure 16. Amplitude magnifications for the MPT with multiple layers of SMPs at different frequencies. T1 was an actuator and sensor.

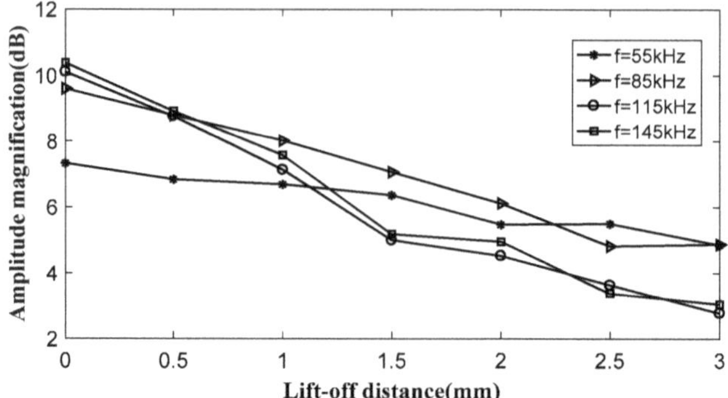

Figure 17. Amplitude magnification for the magnetostrictive patch transducer with different lift-off distances of the SMP at several typical frequencies. T1 was an actuator and sensor.

5. Conclusions

A method of signal strength enhancement was proposed for the magnetostrictive patch transducer (MPT) by dynamic magnetic field optimization with a soft magnetic patch (SMP) in guided wave inspection. The enhancement was achieved by decreasing the resistance of the magnetic circuit in wave generation and restricting the induced dynamic magnetic field in wave detection. A numerical investigation was successfully performed for the improvement of dynamic magnetic fields in the magnetostrictive patch and region around the coils. Experimental validation was carried out for the SH0 wave inspection in an aluminum plate. The main findings are summarized as follows:

1. The amplitude enhancement of the dynamic magnetic field in the magnetostrictive patch is beneficial for SH wave generation and the concentration of the induced magnetic field by the magnetostrictive patch in wave detection is good for the measurement by the coils.
2. The SMP can significantly decrease the resistance of the magnetic circuit for the dynamic magnetic field in wave generation and concentrates the induced magnetic field in wave detection. The relative permeability should be more than 2000 and the thickness should be over 0.15 mm. The increased distance between the SMP and the coils was bad for the improvement.
3. Covered by the SMPs with the relative permeability of 2800, the MPTs provided a better performance in both processes of wave generation and detection compared to those without the SMPs. The improvement of the SMP in wave detection was more than that in wave generation. The largest magnification can be 12.7 dB when the transducer is used as an actuator and sensor at the same time.
4. For ideal results, using the SMPs with the relative permeability of 2800, a thickness over 0.3 mm and close fitting between the SMP and coils is recommended.

The SMP is expected to improve the signal strength of the MPT for guided wave inspection without sacrificing the flexibility and lightness of the transducer or increasing the system power. For future studies, the SMPs made of different materials with different relative permeabilities will be tested and different types of environmental factors, including temperature and electromagnetic disturbing, will be investigated. The MPT with the SMP will also be examined in field tests of pipe and tank floor scanning.

Author Contributions: J.W. conceived and designed the models and methods presented; Z.T., F.L. and K.Y. helped with the conception and experiments; J.W. performed the simulations, experiments and data analysis; J.W. and Z.T. wrote this paper.

Funding: The work was supported by the National Natural Science Foundation of China (Grant No. 51875511) and National Key Research and Development Program of China (Grant No. 2018YFC0809000).

Conflicts of Interest: The authors declare no conflicts of interest.

References

1. Rose, J.L. *Ultrasonic Guided Waves in Solid Media*, 1st ed.; Cambridge University Press: Cambridge, UK, 2014; p. 1. ISBN 978-1-107-04895-9.
2. Heinlein, S.; Cawley, P.; Vogt, T.K. Reflection of torsional T (0, 1) guided waves from defects in pipe bends. *NDT E Int.* **2018**, *93*, 57–63. [CrossRef]
3. Leinov, E.; Lowe, M.J.S.; Cawley, P. Investigation of guided wave propagation and attenuation in pipe buried in sand. *J. Sound Vib.* **2015**, *347*, 96–114. [CrossRef]
4. Hayashi, T.; Tamayama, C.; Murase, M. Wave structure analysis of guided waves in a bar with an arbitrary cross-section. *Ultrasonics* **2006**, *44*, 17–24. [CrossRef] [PubMed]
5. Seung, H.M.; Park, C.I.; Kim, Y.Y. An omnidirectional shear-horizontal guided wave EMAT for a metallic plate. *Ultrasonics* **2016**, *69*, 58–66. [CrossRef] [PubMed]
6. Hasanian, M.; Lissenden, C.J. Second order harmonic guided wave mutual interactions in plate: Vector analysis, numerical simulation, and experimental results. *J. Appl. Phys.* **2017**, *122*, 084901. [CrossRef]
7. Yu, X.; Fan, Z.; Castaings, M.; Biateau, C. Feature guided wave inspection of bond line defects between a stiffener and a composite plate. *NDT E Int.* **2017**, *89*, 44–55. [CrossRef]
8. Wu, J.; Tang, Z.; Lv, F.; Yang, K.; Yun, C.B. Ultrasonic guided wave-based switch rail monitoring using independent component analysis. *Meas. Sci. Technol.* **2018**, *29*, 115102. [CrossRef]
9. Evans, M.; Lucas, A.; Ingram, I. The inspection of level crossing rails using guided waves. *Constr. Build. Mater* **2018**, *179*, 614–618. [CrossRef]
10. Yao, W.; Sheng, F.; Wei, X.; Zhang, L.; Ynag, Y. Propagation characteristics of ultrasonic guided waves in continuously welded rail. *Modern Phys. Lett. B* **2017**, *31*, 1740075. [CrossRef]
11. Nakamura, N.; Ogi, H.; Hirao, M. EMAT pipe inspection technique using higher mode torsional guided wave T (0, 2). *NDT E Int.* **2017**, *87*, 78–84. [CrossRef]
12. Kogia, M.; Gan, T.H.; Balachandran, W.; Livadas, M.; Kappatos, V.; Szabo, I.; Mohimi, A.; Round, A. High temperature shear horizontal electromagnetic acoustic transducer for guided wave inspection. *Sensors* **2016**, *16*, 582. [CrossRef]
13. Lee, J.R.; Chia, C.C.; Park, C.Y.; Jeong, H. Laser ultrasonic anomalous wave propagation imaging method with adjacent wave subtraction: Algorithm. *Opt. Laser Technol.* **2012**, *44*, 1507–1515. [CrossRef]
14. Lee, J.R.; Chia, C.C.; Shin, H.J.; Park, C.Y.; Yoon, D.J. Laser ultrasonic propagation imaging method in the frequency domain based on wavelet transformation. *Opt. Lasers Eng.* **2011**, *49*, 167–175. [CrossRef]
15. Sevillano, E.; Sun, R.; Perera, R. Damage detection based on power dissipation measured with PZT sensors through the combination of electro-mechanical impedances and guided waves. *Sensors* **2016**, *16*, 639. [CrossRef]
16. Vinogradov, S.; Cobb, A.; Fisher, J. New Magnetostrictive Transducer Designs for Emerging Application Areas of NDE. *Materials* **2018**, *11*, 755. [CrossRef]
17. Kwun, H.; Bartels, K.A. Magnetostrictive sensor technology and its applications. *Ultrasonics* **1998**, *36*, 171–178. [CrossRef]
18. Wang, W.B.; Cao, S.Y.; Huang, W.M. *Magnetostrive Materials and Devices*, 1st ed.; Metallurgical Industry Press: Beijing, China, 2008; pp. 8–10. ISBN 9787502445324. (In Chinese)
19. Kim, Y.Y.; Kwon, Y.E. Review of magnetostrictive patch transducers and applications in ultrasonic nondestructive testing of waveguides. *Ultrasonics* **2015**, *62*, 3–19. [CrossRef]
20. Tzannes, N.S. Joule and Wiedemann effects-The simultaneous generation of longitudinal and torsional stress pulses in magnetostrictive materials. *IEEE Trans. Sonics Ultrason.* **1966**, *13*, 33–40. [CrossRef]
21. Xu, J.; Wu, X.; Cheng, C.; Ben, A. A magnetic flux leakage and magnetostrictive guided wave hybrid transducer for detecting bridge cables. *Sensors* **2012**, *12*, 518–533. [CrossRef]
22. Xu, J.; Wu, X.; Sun, P. Detecting broken-wire flaws at multiple locations in the same wire of prestressing strands using guided waves. *Ultrasonics* **2013**, *53*, 150–156. [CrossRef]
23. Park, C.; Kim, Y.Y. Nonferromagnetic material inserted magnetostrictive patch bonding technique for torsional modal testing of a ferromagnetic cylinder. *Rev. Sci. Instrum.* **2010**, *81*, 035103. [CrossRef]
24. Kumar, K.S.; Murthy, V.; Balasubramaniam, K. Improvement in the signal strength of magnetostrictive ultrasonic guided wave transducers for pipe inspection using a soft magnetic ribbon-based flux concentrator. *Insight* **2012**, *54*, 217–220. [CrossRef]

25. Kim, H.W.; Cho, S.H.; Kim, Y.Y. Analysis of internal wave reflection within a magnetostrictive patch transducer for high-frequency guided torsional waves. *Ultrasonics* **2011**, *51*, 647–652. [CrossRef]
26. Cho, S.H.; Lee, J.S.; Kim, Y.Y. Guided wave transduction experiment using a circular magnetostrictive patch and a figure-of-eight coil in nonferromagnetic plates. *Appl. Phys. Lett.* **2006**, *88*, 224101. [CrossRef]
27. Yoo, B.; Na, S.M.; Flatau, A.B.; Pines, D.J. Ultrasonic guided wave sensing performance of a magnetostrictive transducer using Galfenol flakes-polymer composite patches. *J. Appl. Phys.* **2015**, *117*, 17A916. [CrossRef]
28. Kim, Y.; Kim, Y.Y. A novel Terfenol-D transducer for guided-wave inspection of a rotating shaft. *Sens. Actuators A Phys.* **2007**, *133*, 447–456. [CrossRef]
29. Cho, S.H.; Kim, H.W.; Kim, Y.Y. Megahertz-range guided pure torsional wave transduction and experiments using a magnetostrictive transducer. *IEEE Trans. Ultrason. Ferroelectr. Freq. Control* **2010**, *57*, 1225–1229. [CrossRef]
30. Park, C.I.; Cho, S.H.; Kim, Y.Y. Z-shaped magnetostrictive patch for efficient transduction of a torsional wave mode in a cylindrical waveguide. *Appl. Phys. Lett.* **2006**, *89*, 174103. [CrossRef]
31. Park, C.I.; Kim, W.; Cho, S.H.; Kim, Y.Y. Surface-detached V-shaped yoke of obliquely bonded magnetostrictive strips for high transduction of ultrasonic torsional waves. *Appl. Phys. Lett.* **2005**, *87*, 224105. [CrossRef]
32. Zhang, X.; Tang, Z.; Lv, F.; Pan, X. Helical comb magnetostrictive patch transducers for inspecting spiral welded pipes using flexural guided waves. *Ultrasonics* **2017**, *74*, 1–10. [CrossRef]
33. Kim, Y.G.; Moon, H.S.; Park, K.J.; Lee, J.K. Generating and detecting torsional guided waves using magnetostrictive sensors of crossed coils. *NDT E Int.* **2011**, *44*, 145–151. [CrossRef]
34. Lee, J.S.; Kim, Y.Y.; Cho, S.H. Beam-focused shear-horizontal wave generation in a plate by a circular magnetostrictive patch transducer employing a planar solenoid array. *Smart Mater. Struct.* **2008**, *18*, 015009. [CrossRef]
35. Huang, S.L.; Wang, K.; Zhao, W. *Theory and Application of Electromagnetic Ultrasonic Guided Wave*, 1st ed.; Tsinghua University Press: Beijing, China, 2013; p. 74. ISBN 978-7-302-33673-0.
36. Zhang, X.W.; Tang, Z.F.; Lv, F.Z.; Pan, X.H. Excitation of axisymmetric and non-axisymmetric guided waves in elastic hollow cylinders by magnetostrictive transducers. *J. Zhejiang Univ.-Sci. A* **2016**, *17*, 215–229. [CrossRef]
37. Yang, Q.; Li, Y.; Zhao, Z.; Zhu, L.; Luo, Y.; Zhu, J. Design of a 3-D rotational magnetic properties measurement structure for soft magnetic materials. *IEEE Trans. Appl. Supercond.* **2014**, *24*, 1–4. [CrossRef]

Article

Calculation of Guided Wave Dispersion Characteristics Using a Three-Transducer Measurement System

Borja Hernandez Crespo [1,2]**, Charles R. P. Courtney** [1,*] **and Bhavin Engineer** [3]

[1]   Department of Mechanical Engineering, University of Bath, Claverton Down, Bath BA2 7AY, UK; bhc24@bath.ac.uk
[2]   TWI Ltd., Granta Park, Great Abington, Cambridge CB21 6AL, UK
[3]   Cranfield University, College Road, Cranfield, Bedfordshire MK43 0AL, UK; bhavin.engineer@cranfield.ac.uk
*   Correspondence: c.r.p.courtney@bath.ac.uk

Received: 3 July 2018; Accepted: 26 July 2018; Published: 29 July 2018

Featured Application: Rapid generation of dispersion curves for guided wave applications when knowledge of the material properties and thickness of the structure to be inspected are unknown.

Abstract: Guided ultrasonic waves are of significant interest in the health monitoring of thin structures, and dispersion curves are important tools in the deployment of any guided wave application. Most methods of determining dispersion curves require accurate knowledge of the material properties and thickness of the structure to be inspected, or extensive experimental tests. This paper presents an experimental technique that allows rapid generation of dispersion curves for guided wave applications when knowledge of the material properties and thickness of the structure to be inspected are unknown. The technique uses a single source and measurements at two points, making it experimentally simple. A formulation is presented that allows calculation of phase and group velocities if the wavepacket propagation time and relative phase shift can be measured. The methodology for determining the wavepacket propagation time and relative phase shift from the acquired signals is described. The technique is validated using synthesized signals, finite element model-generated signals and experimental signals from a 3 mm-thick aluminium plate. Accuracies to within 1% are achieved in the experimental measurements.

Keywords: guided waves; lamb wave; dispersion curves; phase velocity; group velocity; signal processing

---

1. Introduction

Guided waves have been extensively investigated in the recent years as a nondestructive testing (NDT) technique because of their advantages compared to ultrasonic testing (UT) inspection. The key benefit of this technology is the ability to interrogate the entire thickness of thin-walled structures over large areas from a single location.

During guided wave propagation, waves experience dispersion that distorts the wave shape as the wave propagates, due to a dependence of velocity on frequency [1,2]. Dispersion curves show the relation of phase and group velocities against frequency, for a particular geometry and material. Determination of dispersion curves is important for any guided wave application; accurate dispersion curves enable wave modes of received wavepackets to be distinguished, specific wave modes to be cancelled to clean the acquired signal, the propagation of wave modes in a particular direction using phased arrays [3] and the spatial location of damage in the structure to be determined based on time-of-flight measurements [4]. Guided waves are being used in commercial products, for instance to detect wall thickness losses due to corrosion for pipeline inspection in the oil and gas industry. These commercial products generate a unique wave mode which propagates along the structure avoiding the creation of multimodal excitations, as this would increase the complexity of the analysis of the signals [5,6]. Problems with the application of guided waves occur when the material properties or thickness of the structure to be inspected are unknown due to imprecise records, commercial confidentiality or manufacturing and maintenance uncertainty. This is a particular problem when evaluating composite structures, like wind turbine blades, where elastic constants are not often available. To inspect this kind of structure in an industrial environment, it becomes impracticable as currently there are no means of generating the dispersion curves for such situations. Therefore, there is a need to be able to calculate in situ the phase and group velocities at frequencies of interest in a quick and reliable way without requiring prior knowledge of material properties or thickness.

There has been much work published on the determination of dispersion curves describing many analytical, numerical and experimental methods. For relatively simple structures like plates or pipes, dispersion curves can be predicted using the commercially available software Disperse®, which uses the global matrix method [7]. Other methods, such as the transfer matrix method [8,9] or pseudospectral collocation method [10], have been utilized to determine dispersion curves. Finite element (FE) [11] and semi-analytical finite element [12,13] methods have been also used. All these numerical and analytical methods require knowledge of material properties and thickness of the structure to determine the dispersion curves.

Experimental methods do not suffer this limitation, since guided wave data is directly acquired from the inspected structure. The most widely used experimental technique to measure dispersion curves is the 2D fast Fourier transform (FFT) [14]. This signal processing technique requires the acquisition of many signals along the wave propagation direction in order to carry out a double FFT in time and space. The result is a wavenumber–frequency matrix, which can be rearranged to give the phase velocity dispersion map. The acquisition of the signals from all required locations is manually prohibitive; therefore, a laser scanning vibrometer (LSV) is commonly used to automatically obtain the signals from preselected points. This device is highly sensitive and bulky, being restricted to controlled areas like laboratories [15] and difficult to use in situ.

Other experimental techniques have been proposed recently. Harb and Yuan [16,17] presented a noncontact technique using an air-coupled transducer (ACT) to generate the wave mode on the plate and an LSV to acquire the propagating mode. The technique requires precise control of the incidence angle of the ultrasonic pressure from the ACT upon the surface. By relating the frequency to the incidence angle at which the wave amplitude is maximized, it is possible to calculate the phase velocity using Snell's law. However, the technique has limited applicability outside of the laboratory. In works by Mažeika et al. [18,19], a zero-crossing technique is used to calculate the phase velocity based on the measurement of the position of constant phase points of the pulse corresponding to zero amplitude as a function of time. A similar approach is taken in the phase velocity method in [20], tracking the peaks of the pulse rather than the zero crossings. To produce convincing results, both techniques require the acquisition of a large number of signals at different distances, making these techniques time-consuming. For the sake of reducing time and sampling points in space, Adams et al. [21] presented a method using an array probe from which a time–space matrix can be created. A 2D filter is applied to the matrix to extract the phase velocities. The technique was demonstrated to be valid in simulation using FE analysis. However, for realistic probe sizes, the technique has limited experimental applications. In [22], sparse wavenumber analysis was used to experimentally recover the dispersion curves from an aluminum plate; however, seventeen sensors were needed to deploy the technique. An alternative approach to getting dispersion curves from experimental signals is to use time–frequency representations [23], where group velocity can be directly determined. However, the calculation of phase velocity involves the integration of group velocity requiring precise values of ω and k at the lower limit, which are not easy to obtain experimentally. For medical purposes, two techniques, SVD and LRT, were evaluated for extracting the dispersion curves of cortical bones in [24]. Both techniques require a large number of receivers and transmitters.

Experimental calculation of bulk wave velocities in dispersive materials has been studied [25]. Sachse and Pao used the method of phase spectral analysis, where the phase delay is calculated using the real and imaginary parts of the Fourier transform of the pulse, and then obtaining the group delay through the differentiation of the phase characteristics of the pulse with respect to frequency. Pialucha et al. presented the amplitude spectrum technique overcoming the requirement of the phase spectrum technique for the separation of successive pulses in the time domain for bulk waves [26]. However, the phase spectral technique provided better results when pulses can be separated.

In this paper, an experimental technique for straightforward calculations of phase and group velocities of guided waves at frequencies of interest is presented. The technique requires just the acquisition of two signals spaced a few cm using conventional transducers, enabling its application on in-service structures located in poorly controlled environments. There is no requirement for prior knowledge of the material properties or thickness of the inspected structure. In Section 2, the formulation on which the experimental technique relies is presented. Section 3 describes in detail the methodology to extract the group delay and phase shift from experimental signals. Successive sections present validation tests using synthesized signals and simulated signals from FE analysis. Finally, experimental signals from a 3 mm-thick aluminium plate are used to validate the technique as a solution to the problem presented in the introduction.

2. Theoretical Basis

Assuming that the propagating pulse is sufficiently narrow-band that dispersion does not significantly distort the wavepacket over the propagation distance, the signal can be approximated as [27],

$$\psi(x,t) = G\left(t - \frac{x}{v_g}\right)e^{i(kx-\omega t)}, \tag{1}$$

where x is propagation distance, t is time, $\omega = 2\pi f$ is the angular frequency of the harmonic wave at frequency f and $k = 2\pi/\lambda$ is the wavenumber. $G(t)$ is a function defining the temporal pulse shape. $G(t)$ is defined such that the pulse has a peak at $t = 0$. v_g is the group velocity at which the packet propagates, and so if the trajectory of the wavepacket peak is defined as the point whereby $G(t_0 - x_0/v_g) = G(0)$, then

$$v_g = \frac{x_0}{t_0}. \tag{2}$$

The wavepacket is shown schematically in Figure 1. The harmonic part of the signal has a phase of

$$\phi = kx - \omega t, \tag{3}$$

and propagates at the phase velocity [6,27].

$$v_p = \frac{\omega}{k} \tag{4}$$

The phase of the harmonic signal at the centre of the wavepacket x_0, t_0 is

$$\phi_0 = kx_0 - \omega t_0. \tag{5}$$

Rearranging and substituting for v_p and v_g gives

$$\phi_0 = kt_0\left(\frac{x_0}{t_0} - \frac{\omega}{k}\right), \tag{6}$$

$$\phi_0 = kt_0 \left(v_g - v_p\right). \tag{7}$$

For nondispersive waves, $v_p = v_g$ and Equation (7) gives the expected result that the phase at the wavepacket center is $\phi_0 = 0$ for all propagation times. Where the system is dispersive there is a finite phase difference between the harmonic part of the signal and the wavepacket, ϕ_0, which increases linearly with propagation time and is positive if $v_g > v_p$, negative if $v_g < v_p$. Assuming that this phase difference can be measured experimentally, Equations (4) and (5) further yield two useful equations for evaluating the phase velocity:

$$v_p = \frac{x_0 - \frac{\phi_0}{k}}{t_0}, \tag{8}$$

$$v_p = \frac{x_0}{t_0 + \frac{\phi_0}{\omega}}. \tag{9}$$

In Figure 2, an illustrative example of a S_0-mode wavepacket at 150 kHz is shown. The solid line is the S_0 mode after 1.2-m propagation including the dispersion. The dashed line represents the S_0 mode after the propagation without dispersion, namely, the wavepacket was calculated using the same phase velocity for all frequencies. In this example, the phase shift is negative (−1.26 radians), which means the phase velocity is higher than the group velocity. Using (9), the calculation of the phase velocity would be 1.2 m divided by the sum of 223 µs, which is the time of the wavepacket to travel 1.2 m at group velocity, plus −1.34 µs which is the time that the signal at an angular frequency, $\omega = 2\pi \cdot 150000$, takes to cover −1.26 radians. Group velocity and phase velocity are calculated using (2) and (9), respectively: $v_g = 5381$ ms^{-1} and $v_p = 5414$ ms^{-1}.

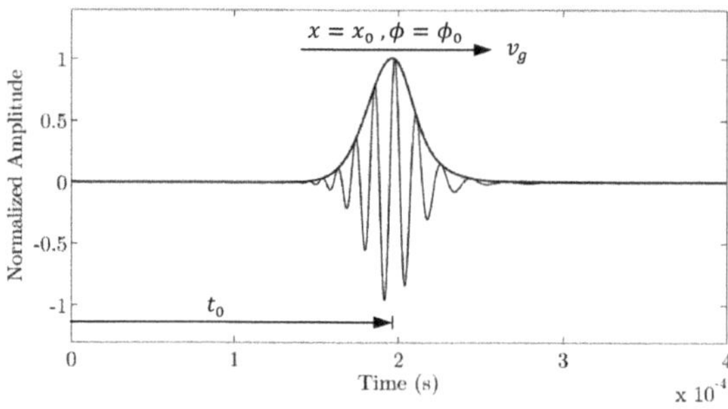

Figure 1. Wavepacket arriving at x_0 at time t_0 with phase ϕ_0 propagating at group velocity v_g. $G(t)$ is represented as the envelope of the wavepacket.

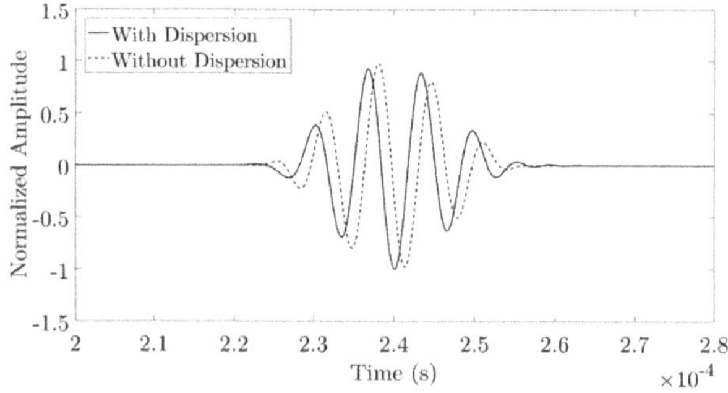

Figure 2. S_0-mode wavepacket of 5 cycles at 150 kHz after propagating 1.2 m, with dispersion and without dispersion.

3. Methodology

In this section, the methodology for determining the phase shift and hence the phase velocity is presented. This general methodology is complemented by a description of a specific experiment setup described in Section 6.1. The technique is based on the analysis of two signals acquired at different distances in the direction of wave propagation. One of them will be treated as the baseline and the other signal will be modified in time and phase in order to match with the baseline. The methodology is composed of different signal processing techniques which are presented in a block diagram in Figure 3.

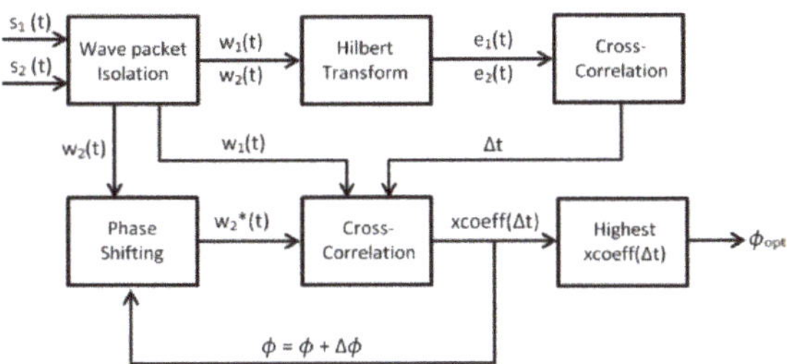

Figure 3. Methodology's block diagram to extract the optimum phase shift and time delay.

The experimental setup is composed of one transmitter and two receivers along the propagation direction of the wave to be measured, as shown in Figure 4. For isotropic systems the wave velocity is independent of direction. The technique analyses the effect of propagation between the first receiver and the second receiver. Spacing between receivers of a few centimeters is selected in order to avoid large phase shifts and waveform deformations due to dispersion.

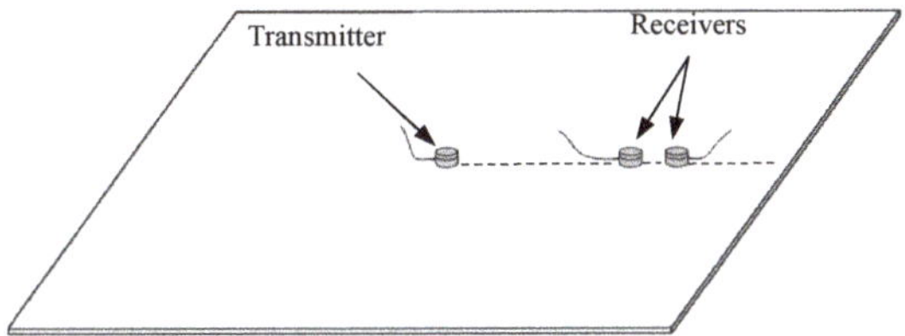

Figure 4. Schematic of the test setup.

The first step in analyzing the data is the isolation of the desired wavepacket by removing the rest of the signal. The boundaries of each wavepacket are determined by studying the slope of its envelope. The boundaries are set at the closest zero slopes before and after each peak, see example of real signals in Figure 5. Once the region containing the wavepacket is identified, the remainder of the signal is discarded. Both signals are truncated with the same boundaries in order to maintain the time difference between the wavepackets, w_1 and w_2. The boundaries used are the left limit of the signal of the closest receiver to the transmitter and the second limit of the signal of the further receiver, as can be seen in Figure 6. Afterwards, the Hilbert Transform (H(...)) [28] is applied to both wavepackets and the magnitude extracted in order to determine the envelope of each wavepacket:

$$e_1(t) = |H(w_1(t))|, \tag{10}$$

$$e_2(t) = |H(w_2(t))|. \tag{11}$$

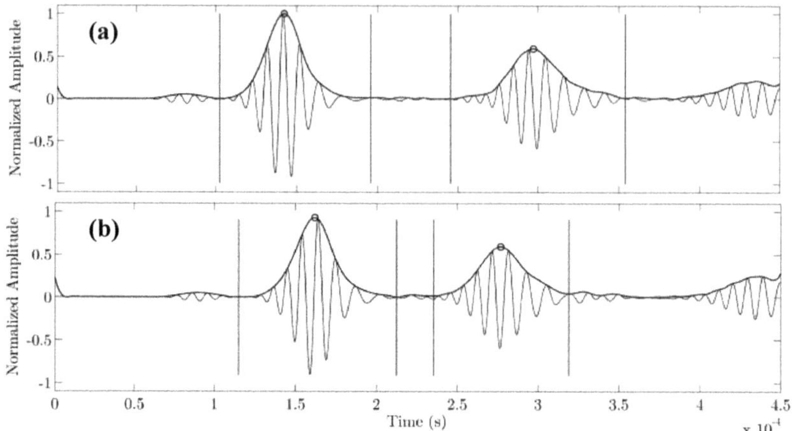

Figure 5. Wavepacket detection and boundary determination established by the algorithm for two signals acquired at: (**a**) 30 cm and (**b**) 35 cm from the transmitter using a d_{33}-type transducer.

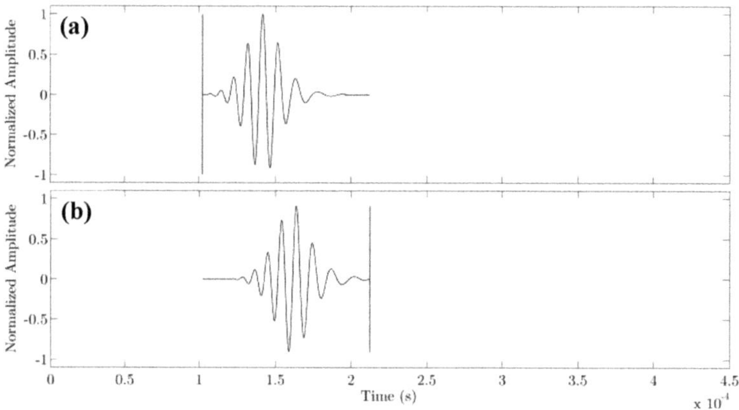

Figure 6. Shortening of the signals to reduce computational time. The same limits are used to truncate each signal in order to maintain the time difference between wavepackets. Signals acquired at: (**a**) 30 cm and (**b**) 35 cm from the transmitter.

These two envelopes are then cross-correlated to determine the time delay between them:

$$\Delta t = \arg\max_{\tau}((e_1 \star e_2)(\tau)),\tag{12}$$

where τ is the applied time delay and Δt the value that maximizes the correlation. Once this time delay is known, the phase delay to the harmonic part of $w_2(t)$ that maximizes the cross-correlation between the delayed phase-shifted wavepacket $w_2^*(t - \Delta t)$ and $w_1(t)$ is calculated. This process consists of the modification in phase of the signal from the furthest receiver by a phase angle, ϕ, and finding the phase shift that maximizes the cross correlation. The phase shift, ϕ, is applied to w_2^*; then, the cross correlation between w_1 and w_2^* is calculated. The optimum phase shift is the phase angle when the correlation coefficient with time delay, Δt, is highest.

$$\phi_{opt} = \arg\max_{\phi}((w_1 \star w_2^*)(\Delta t)(\phi))\tag{13}$$

This methodology enables the calculation of the lag time and phase shift between the two signals acquired at different distances. By knowing these three variables (distance, time and phase) and also the excitation frequency, (2) and (9) can be used to determine the group and phase velocity, respectively.

The proposed methodology in this paper is based on the acquisition of two signals spaced a few centimeters apart, since for such short distances the distortion for any wave mode is relatively small, enabling a straightforward determination of the phase shift.

4. Synthetic Signal Analysis

The first validation was carried out using synthesized signals. The signal synthesis is based on the adjustment in frequency and wavenumber of the input signal in the frequency domain. Knowing the waveform at one point, in this case the input signal at the transmitting point, and the dispersion characteristics of each wave mode, the wave modes can be reconstructed in the time domain at any distance [29]. If $u(x, t)$ is the reconstructed wave mode at a distance x from the transmitting point, t is the time, $F(\omega)$ is the Fourier transform of the input signal and $k(\omega)$ is the wavenumber as a function of the angular frequency, then

$$u(x, t) = \int_{-\infty}^{\infty} F(\omega) e^{i(k(\omega)x - \omega t)} d\omega \tag{14}$$

The relation between wavenumber and angular frequency is extracted directly from the dispersion curves provided by Disperse® and introduced in the integral.

As an initial test of the signal processing method, developed signals are synthesized that are highly pure, without noise and overlapping between modes. The higher the sampling frequency is set, the better accuracy the technique provides. The sampling frequency was set at 10 MHz, since it has been observed that it provides a good balance between computational time and accuracy.

A set of signals for symmetric (S_0), antisymmetric (A_0) and shear horizontal (SH_0) wave modes were created by evaluating equation (14) in MATLAB for two propagation distances, 30 cm and 35 cm: the analysed propagated distance (Δd) in this case is 5 cm. The analysed frequencies are from 60 to 370 kHz with steps of 10 kHz.

The results after applying the signal processing algorithm described in Section 3 to the synthesized signals are presented in Figure 7. The results (black circles) accurately reconstruct the theoretical values from Disperse® (grey lines). In Figure 8, the extracted phase shift has been represented against the frequency for the three fundamental modes. For the S_0 mode, the phase values have negative values decaying from –2 degrees at 60 kHz to –52 degrees at 370 kHz. For the A_0 mode, conversely, the phase is positive, increasing along the frequency, since the phase velocity is lower than the group velocity; from 377 degrees at 60 kHz to 673 degrees at 370 kHz. In this case, the phase values are much higher than the S_0 and SH_0 ones, because the A_0 mode is highly dispersive at these frequencies. For the SH_0 mode, the extracted phase values are 0 degrees at 60 kHz and 1 degree at 370 kHz. The SH_0 is a nondispersive mode so the phase value remains at near 0 degrees.

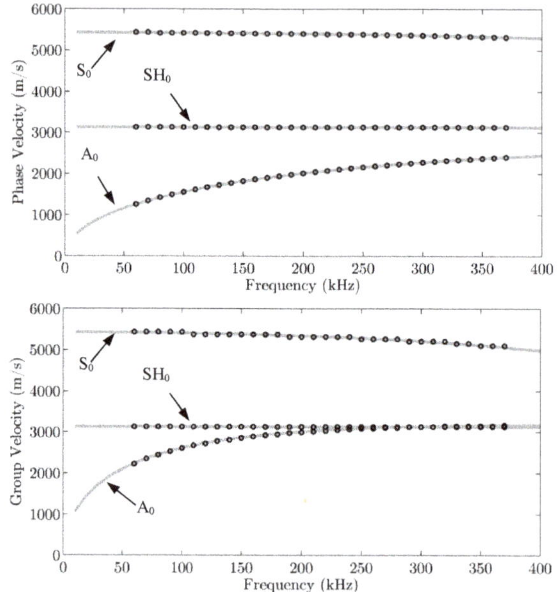

Figure 7. Comparison of the phase velocity and group velocity of S_0, A_0 and SH_0 between the results from the synthesized signals (black circles) and theoretical values (grey lines).

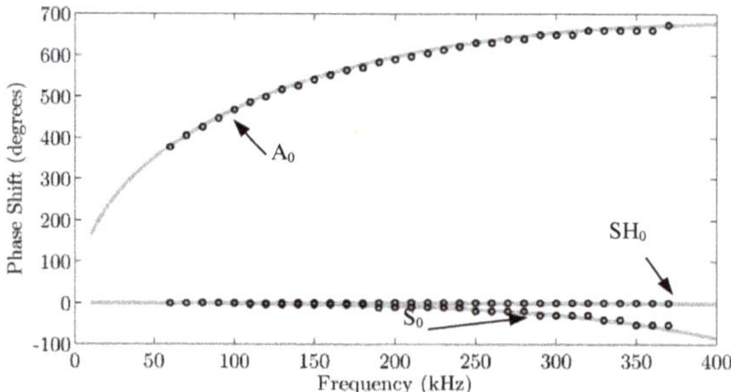

Figure 8. Comparison of the phase shift of S_0, A_0 and SH_0 between the results from the synthesized signals (black circles) and theoretical values (grey lines).

The sampling frequency is an important factor, since it determines the degree of resolution of the group velocity. As these examples are performed for short propagation distances, the time that the wave modes take to cover that distance is small. Thus, low sampling frequencies are not able to determine the group velocity accurately. Faster modes require higher sampling frequencies to get the same resolution as slower modes. In this case of 5 cm spacing, the propagation time for the S_0 mode at 200 kHz is 9.3 µs; a sampling frequency of 1 MHz will have a time resolution of 1 µs which yields a velocity resolution for this particular example of 556 m/s; with a sampling frequency of 10 MHz, the velocity resolution will be 58.4 m/s; and with a sampling frequency of 100 MHz, the velocity resolution will be 5.8 m/s. This resolution issue can be observed in Figure 7; at lower velocities the curves are smoother, and at higher velocities (S_0 mode) the curves have poorer velocity resolution.

5. Finite Element Analysis

A three-dimensional FE analysis was performed in Abaqus to simulate signals. While the previous analysis required the input of the dispersion curves to create the propagated signals, this FE analysis does not require prior knowledge of the dispersion characteristics.

FE models were created simulating guided wave propagation at five different frequencies in an aluminium plate (1000 × 1000 × 3 mm). The transmitter and receivers were modelled as ideal point transducers. The transmitting point was placed at the centre of the plate, where the coordinate system was established. A force was applied in the x-direction to simulate a shear transducer. This generated S_0 and A_0 modes along the x-axis and the SH_0 mode along the y-axis. The input signal was a 5-cycle sine with a Hanning window at central frequencies of 60, 80, 100, 120 and 140 kHz. Multiple receiver points were located on the positive x-axis and positive y-axis from the centre of the plate to the edge every 1 cm, resulting in 50 receivers at each axis. The time and spatial discretization for the finite element model was established at 10^{-2} µs and 1 mm, respectively, in order to ensure convergence and to obey the Nyquist sample-rate criterion. The mesh was formed by C3D8 elements.

The FE analysis software provides the results decoupled for each axis direction, so it is straightforward to evaluate each wave mode separately. In Figure 9, the results have been computed in cylindrical coordinates to show clearly the three wave modes of propagation at each axis. In Figure 9a, showing radial displacement, S_0 and A_0 modes are depicted, S_0 being the faster mode. In Figure 9b showing tangential displacement, the SH_0 mode propagates along the y-axis, and in Figure 9c, showing out-of-plane displacement, the A_0 mode is clearly present.

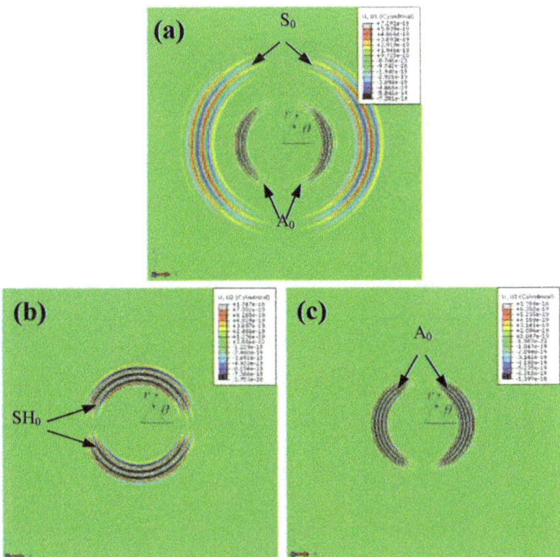

Figure 9. Images of the finite elementsimulation of the wave propagation in polar-coordinate instantaneous displacement is shown 76 µs after excitation. (**a**) Radial displacement, (**b**) tangential displacement and (**c**) out-of-plane displacement. Scale bars in meters.

Since the plate is relatively small, overlaps between wave modes and echoes from edges are produced, as can be seen in Figure 10. This is a particular problem at lower frequencies (60 and 80 kHz) due to the length of the pulses. Therefore, an additional optimisation step has been added to minimize the error introduced by the overlapping. The FE model has 50 receiver points, every 1 cm along the axis so the five least-overlapped signals are selected for analysis. The selection is performed by analyzing the amplitude at the beginning and end of the envelope of the wavepacket of interest; the signals with the lowest amplitude at those regions are selected. Then, the signal processing described in Section 3 is applied for all the pair combinations between the five selected receivers. Once the phase and group velocities are calculated for each combination, the velocities are averaged to give the final velocity estimates. Any outliers that are introduced during selection of the five signals are eliminated. As shown in Figure 11, the phase velocity matches very well with the theoretical values from Disperse®. In the group velocity graph, the results correlate well with the theoretical velocities, having a noticeable variation in the S_0 value at 60 kHz. This variation occurs at 60 kHz due to the overlap between the S_0 mode and the A_0 mode, which is more pronounced for the longer pulses at this frequency. This error does not appear for the A_0 mode because the signals used for the analysis were out-of-plane displacements, where the S_0 mode is practically imperceptible. Note that the results improve as the frequency increases, since the wavelength decreases and mode separation increases. Overlapping of pulses is an issue for this technique, as it changes the phase and waveform of the analyzed mode, highly distorting the results and impeding its application.

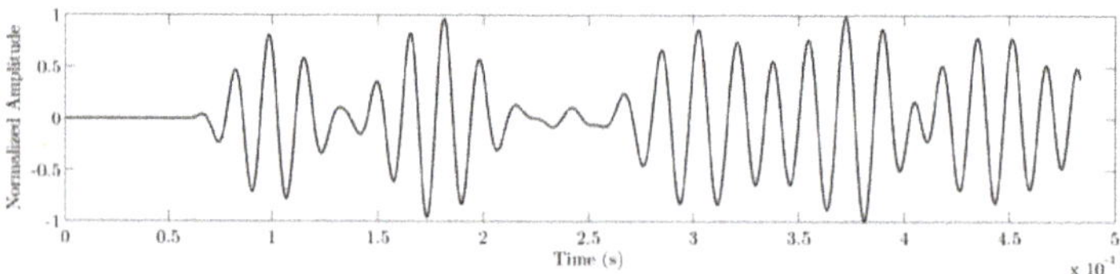

Figure 10. Signal acquired at 33 cm from excitation point at 60 kHz.

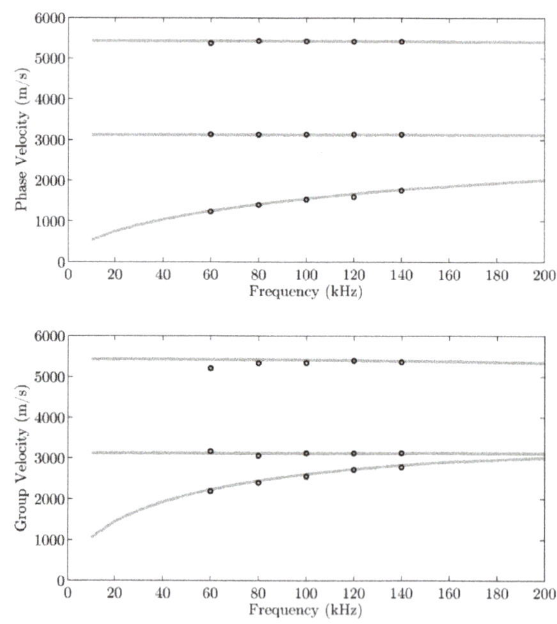

Figure 11. Comparison of the phase velocity and group velocity of S_0, A_0 and SH_0 between the results from the synthesized signals (black circles) and theoretical values (grey lines).

The results for the phase velocity are better than for the group velocity, more notably when overlapping occurs. This is because of the phase-shift term in the denominator in (9). This contribution in the phase velocity equation minimizes the erroneous calculated value of the propagation time and also provides more resolution on the calculation of the phase velocity, removing the stepped shape seen in the group velocity.

6. Experimental Demonstration

6.1. Test Setup

The specimen is a 4 m by 2 m aluminium plate of 3 mm thickness. Two piezoelectric transducers were used, one transmitter and one receiver. The transmitter was fixed at the centre of the plate using a tool that provides force over the transducer; and the receiver was manually attached at various distances from the transmitter. For the evaluation of the S_0 mode, the receiver was placed at 45, 50 and 55 cm from the transmitter. For the case of the A_0 and SH_0 modes, the receiver was set at 25, 30 and 35 cm from the transmitter. Those distances were selected to minimize the wavepacket distortion due to overlaps.

A pulser-receiver (Teletest® unit) was used to excite the transducer and acquire the signals. Two independent channels were used for the excitation and acquisition. A PC with the Teletest® software was used to control the pulser-receiver and configure the experimental parameters. Input signals of five-cycle Hanning-windowed bursts with centre frequencies from 40 kHz to 140 kHz in steps of 5 kHz were used to excite the transmitter. This frequency range was chosen since it is the operating frequency range of the transducers used in the experiment. Signals were recorded with a 1 MHz sampling frequency; 32 averages were taken, and the record length of the signals was 1 ms. The experimental setup is presented in Figure 12.

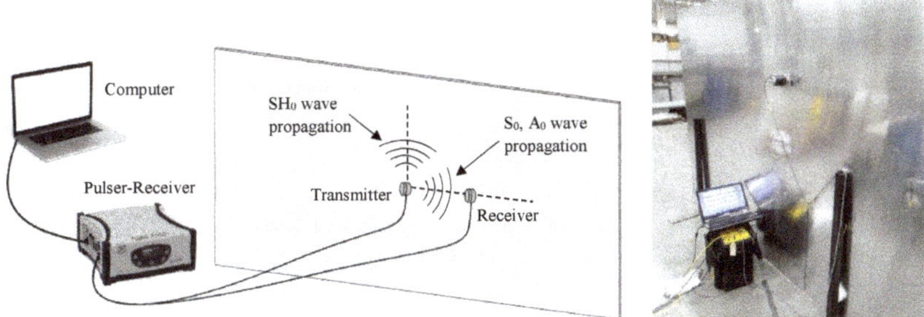

Figure 12. Experiment setup: **Left**–diagram of the experiment, **Right**–photo of the real experiment environment.

Two types of piezoelectric transducers (PI Ceramic GmbH) were used in the experiment, a shear piezoelectric element (d_{15} type) and a compressional piezoelectric element (d_{33} type), both with an active contact area of 13 mm by 3 mm. Figure 13 shows the direction of poling and corresponding shear deformation for the shear transducer. The shear transducer was used to evaluate the S_0 and SH_0 wave modes, and the A_0 wave mode was analyzed using the compressional transducer, which mainly produces an out-of-plane vibration. Using the shear transducer, S_0 waves are generated in the direction of the poling axis, and SH_0 in the perpendicular axis. A small amount of A_0 is also generated on the poling axis. In the case of the compressional transducer, S_0 and A_0 are created omnidirectionally. The averaged amplitudes of the S_0 and SH_0 modes using the shear transducer at the propagation distances specified above are 13 mV and 33 mV, respectively; and the amplitude of the A_0 mode is 63 mV using the compressional transducer.

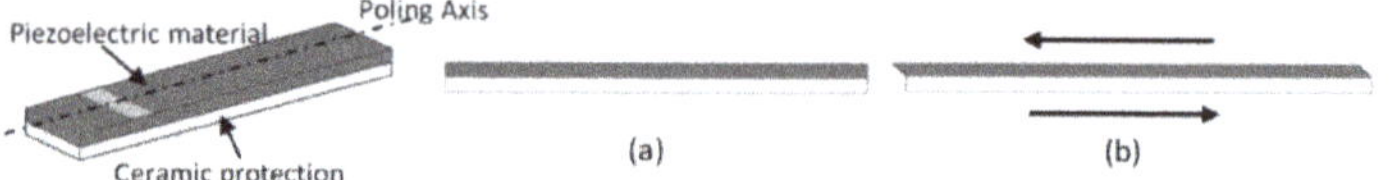

Figure 13. Schematics of the shear transducer. (**a**) Undeformed, (**b**) deformed.

Experimental signals were acquired at a sampling frequency of 1 MHz providing a poor time resolution for the wave velocities of interest. Therefore, the sampling rate of the signals was increased by a factor of 10 using cubic spline interpolation to give a time accuracy of 0.1 μm (10 MHz). The interpolation is a necessary step when the propagation distance is a few cm, as the propagation time of the faster modes, such as S_0, is of the same order of magnitude as the original sampling period. By reducing the sampling period by means of interpolation, the accuracy of the group velocity is increased. However, a small error is introduced in the signal due to the interpolation. Overall, the improvements gained for the velocity calculation are more significant than the errors introduced in the signals. The signal-to-noise ratios of the S_0, A_0 and SH_0 signals were around 30 dB, 37 dB and 35 dB, respectively.

6.2. Experimental Results

The results presented in Figures 14 and 15 have been extracted from signals at two different spacings: 5 cm and 10 cm. The velocities extracted from the experimental signals correlate quite well with the theoretical velocities. However, there are slight variations around the theoretical velocities. These variations can be caused by a number of factors, such as the aforementioned interpolation and overlapping, but also by the manipulation of the transducers, which in this case were handled manually. Small variations in the correct spacing distance between transducers lead to errors in the calculation of the velocities. The velocity–distance relationship is linear so the relative error in the calculated velocity due to misplacement is the same as the relative distance error. An error of 1 mm in an intended 5 cm spacing distance will cause a 2% error in the velocity or an error of 110 ms^{-1} in the S_0 mode ($\approx$ 5400 m/s), for example. On the other hand, with a 1 mm error transducer placement for a 10 cm spacing, the error will be of 1% of the velocity. Therefore, large errors can occur where small propagation distances are used. A_0 and SH_0 exhibit less absolute error variation than S_0 due to their slower velocities, the generated errors being proportional to the velocity.

In this paper, two propagation distances have been chosen: 5 and 10 cm. From the S_0 mode in Figure 15, it can be seen that the calculated velocities at 10 cm are more accurate, and the variance is also smaller. This is for two reasons: the first is the reduction in the error caused by an incorrect transducer placement, and the second is the implicit increase of the propagation time, which reduces the error in determining that time. Other factors that can introduce errors in the velocity calculation are the variations of the boundary conditions of the transducer, such as the applied pressure over the transducer at each receiving location, as well as the difference of transfer function of each transducer. This problem was overcome utilizing the same transducer to acquire the signals at all locations. Similarly, a change in temperature between measurements will cause velocity changes leading to erroneous values. This can be avoided by acquiring results in quick succession to minimize the impact of the temperature in the results. Using three transducers, one transmitter and two receivers would avoid the issues relating to moving the transducers.

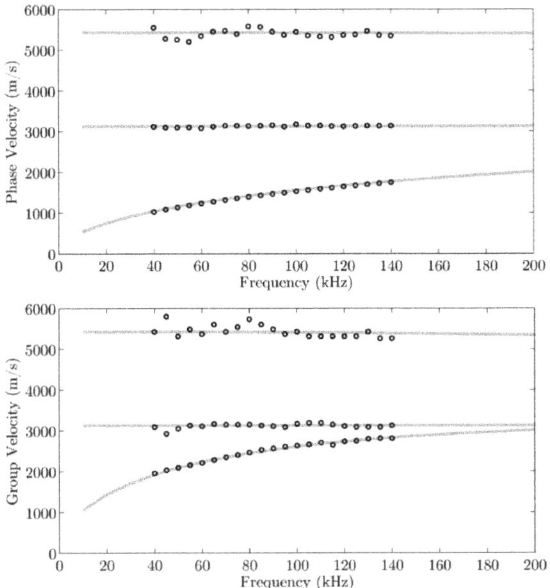

Figure 14. Phase velocity and group velocity created from experimental signals for 5 cm propagation distance (black circles) and theoretical values (grey lines).

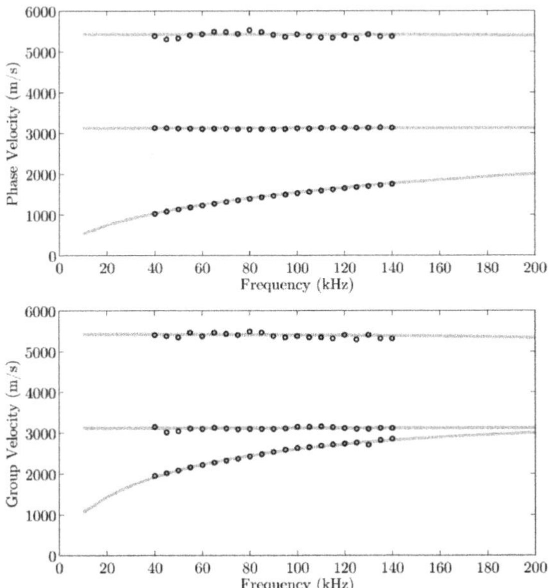

Figure 15. Phase velocity and group velocity created from experimental signals for 10 cm propagation distance (black circles) and theoretical values (grey lines).

7. Conclusions

This paper has presented a new methodology for creating dispersion curves based on measuring the phase difference and time lag between two pulses acquired at different distances. Firstly, the theoretical basis of the method has been presented where the phase velocity has been related to the phase at the pulse centre for systems of limited dispersion; then, a signal processing methodology to extract the phase shift and time lag between two signals and subsequently calculate the phase and group velocities has been outlined. Three different tests were performed to validate the formulation and methodology; using synthesized signals, signals from an FE model and experimental signals. In the three cases the performance achieved agrees well with the theoretical values.

There are some limitations to the proposed method for the extraction of the phase shift to calculate the dispersion curves. Highly dispersive modes, like high-order modes, are more challenging to evaluate; to mitigate this issue, much shorter spacing distances between transducers should be established to minimize the mode-shape distortion. Overlapping between wavepackets limits the method's applicability, since it distorts the apparent phase of the mode being analyzed. Hence, relatively large samples are required to avoid overlapping between wavepackets and edge echoes. The size of the sample can be smaller at higher frequencies. The velocity–frequency spectrum that can be evaluated using this method will be determined by the excitation mechanism used to generate the guided waves in the solid media. With appropriate experimental design, errors of less than 1% are obtained.

For future work, the proposed technique will be applied to more complex structures, such as composite plates where the elastic constants are not easy to obtain. In this case, by studying the phase and group velocities as a function of propagation direction, the full angular dependency of the dispersion curves will be obtained.

Author Contributions: B.H.C. conducted the modeling, analysis, experiments and drafted the manuscript. C.R.P.C. drafted the theory and helped edit the manuscript. B.E. helped edit the manuscript.

Funding: This research was funded by the European Commission under the Marie Curie project grant number FP7-PEOPLE-2012 ITN 309395 "MARE-WINT".

References

1. Rose, J.L. *Ultrasonic Waves in Solid Media*; Cambridge University Press: Cambridge, UK, 1999.
2. Wilcox, P.; Lowe, M.; Cawley, P. The effect of dispersion on long-range inspection using ultrasonic guided waves. *NDT E Int.* **2001**, *34*, 1–9. [CrossRef]
3. Wang, W.; Zhang, H.; Lynch, J.P.; Cesnik, C.E.; Li, H. Experimental and numerical validation of guided wave phased arrays integrated within standard data acquisition systems for structural health monitoring. *Struct. Control Health Monit.* **2018**, *25*, e2171. [CrossRef]
4. Wang, W.; Bao, Y.; Zhou, W.; Li, H. Sparse representation for Lamb-wave-based damage detection using a dictionary algorithm. *Ultrasonics* **2018**, *87*, 48–58. [CrossRef] [PubMed]
5. Cawley, P.; Lowe, M.; Alleyne, D.; Pavlakovic, B.; Wilcox, P. Practical long range guided wave inspection-applications to pipes and rail. *Mater. Eval.* **2003**, *61*, 66–74.
6. Rose, J.L. *Ultrasonic Guided Waves in Solid Media*; Cambridge University Press: Cambridge, UK, 2014.
7. Pavlakovic, B.; Lowe, M.; Alleyne, D.; Cawley, P. Disperse: A general purpose program for creating dispersion curves. In *Review of Progress in Quantitative Nondestructive Evaluation*; Springer: Boston, MA, USA, 1997; pp. 185–192.
8. Lowe, M.J. Matrix techniques for modeling ultrasonic waves in multilayered media. *IEEE Trans. Ultrason. Ferroelectr. Freq. Control* **1995**, *42*, 525–542. [CrossRef]
9. Nayfeh, A.H. The general problem of elastic wave propagation in multilayered anisotropic media. *J. Acoust. Soc. Am.* **1991**, *89*, 1521–1531. [CrossRef]

10. Quintanilla, F.H.; Fan, Z.; Lowe, M.; Craster, R. Dispersion loci of guided waves in viscoelastic composites of general anisotropy. In *AIP Conference Proceedings*; AIP Publishing: Melville, NY, USA, 2016; Volume 1706, p. 120014.

11. Yu, X.; Ratassepp, M.; Fan, Z. Damage detection in quasi-isotropic composite bends using ultrasonic feature guided waves. *Compos. Sci. Technol.* **2017**, *141*, 120–129. [CrossRef]

12. Hayashi, T.; Song, W.-J.; Rose, J.L. Guided wave dispersion curves for a bar with an arbitrary cross-section, a rod and rail example. *Ultrasonics* **2003**, *41*, 175–183. [CrossRef]

13. Cong, M.; Wu, X.; Liu, R. Dispersion analysis of guided waves in the finned tube using the semi-analytical finite element method. *J. Sound Vib.* **2017**, *401*, 114–126. [CrossRef]

14. Alleyne, D.; Cawley, P. A two-dimensional Fourier transform method for the measurement of propagating multimode signals. *J. Acoust. Soc. Am.* **1991**, *89*, 1159–1168. [CrossRef]

15. Ostachowicz, W.; Güemes, A. *New Trends in Structural Health Monitoring*; Springer Science & Business Media: London, UK, 2013; Volume 542.

16. Harb, M.; Yuan, F. A rapid, fully non-contact, hybrid system for generating Lamb wave dispersion curves. *Ultrasonics* **2015**, *61*, 62–70. [CrossRef] [PubMed]

17. Harb, M.; Yuan, F. Non-contact ultrasonic technique for Lamb wave characterization in composite plates. *Ultrasonics* **2016**, *64*, 162–169. [CrossRef] [PubMed]

18. Mažeika, L.; Draudvilienė, L. Analysis of the zero-crossing technique in relation to measurements of phase velocities of the Lamb waves. *Ultragarsas* **2010**, *65*, 7–12.

19. Draudvilienė, L.; Aider, H.A.; Tumsys, O.; Mažeika, L. The Lamb waves phase velocity dispersion evaluation using an hybrid measurement technique. *Compos. Struct.* **2018**, *184*, 1156–1164. [CrossRef]

20. Moreno, E.; Galarza, N.; Rubio, B.; Otero, J.A. Phase Velocity Method for Guided Wave Measurements in Composite Plates. *Phys. Procedia* **2015**, *63*, 54–60. [CrossRef]

21. Adams, C.; Harput, S.; Cowell, D.; Freear, S. A phase velocity filter for the measurement of Lamb wave dispersion. In Proceedings of the 2016 IEEE International Ultrasonics Symposium (IUS), Tours, France, 18–21 September 2016; pp. 1–4.

22. Harley, J.B.; Moura, J.M. Sparse recovery of the multimodal and dispersive characteristics of Lamb waves. *J. Acoust. Soc. Am.* **2013**, *133*, 2732–2745. [CrossRef] [PubMed]

23. Latif, R.; Aassif, E.; Maze, G.; Moudden, A.; Faiz, B. Determination of the group and phase velocities from time-frequency representation of Wigner-Ville. *NDT E Int.* **1999**, *32*, 415–422. [CrossRef]

24. Xu, K.; Ta, D.; Cassereau, D.; Hu, B.; Wang, W.; Laugier, P.; Minonzio, J.-G. Multichannel processing for dispersion curves extraction of ultrasonic axial-transmission signals: Comparisons and case studies. *J. Acoust. Soc. Am.* **2016**, *140*, 1758–1770. [CrossRef] [PubMed]

25. Sachse, W.; Pao, Y.-H. On the determination of phase and group velocities of dispersive waves in solids. *J. Appl. Phys.* **1978**, *49*, 4320–4327. [CrossRef]

26. Pialucha, T.; Guyott, C.; Cawley, P. Amplitude spectrum method for the measurement of phase velocity. *Ultrasonics* **1989**, *27*, 270–279. [CrossRef]

27. Graff, K.F. *Wave Motion in Elastic Solids*; Courier Corporation: Chelmsford, MA, USA, 2012; pp. 62–65.

28. Michaels, J.E.; Lee, S.J.; Croxford, A.J.; Wilcox, P.D. Chirp excitation of ultrasonic guided waves. *Ultrasonics* **2013**, *53*, 265–270. [CrossRef] [PubMed]

29. Wilcox, P.D. A rapid signal processing technique to remove the effect of dispersion from guided wave signals. *IEEE Trans. Ultrason. Ferroelectr. Freq. Control* **2003**, *50*, 419–427. [CrossRef] [PubMed]

Article

Damage Imaging in Lamb Wave-Based Inspection of Adhesive Joints

Magdalena Rucka *, Erwin Wojtczak and Jacek Lachowicz

Department of Mechanics of Materials and Structures, Faculty of Civil and Environmental Engineering, Gdansk University of Technology, Narutowicza 11/12, 80-233 Gdansk, Poland; erwin.wojtczak@pg.edu.pl (E.W.); jacek.lachowicz@pg.edu.pl (J.L.)
* Correspondence: magdalena.rucka@pg.edu.pl or mrucka@pg.edu.pl; Tel.: +48-58-347-2497

Received: 5 February 2018; Accepted: 27 March 2018; Published: 29 March 2018

Abstract: Adhesive bonding has become increasingly important in many industries. Non-destructive inspection of adhesive joints is essential for the condition assessment and maintenance of a structure containing such joints. The aim of this paper was the experimental investigation of the damage identification of a single lap adhesive joint of metal plate-like structures. Nine joints with different defects in the form of partial debonding were considered. The inspection was based on ultrasonic guided wave propagation. The Lamb waves were excited at one point of the analyzed specimen by means of a piezoelectric actuator, while the guided wave field was measured with the use of a laser vibrometer. For damage imaging, the recorded out-of-plane vibrations were processed by means of the weighted root mean square (WRMS). The influence of different WRMS parameters (i.e., the time window and weighting factor), as well as excitation frequencies, were analyzed using statistical analysis. The results showed that two-dimensional representations of WRMS values allowed for the identification of the presence of actual defects in the adhesive film and determined their geometry.

Keywords: adhesive joint; single lap joint; non-destructive testing; damage identification; Lamb waves; scanning laser vibrometry

---

1. Introduction

Adhesive joints of metal elements are a kind of non-separable connections with numerous applications, e.g., in the aviation, machine, or automotive industry [1]. The strength and durability of adhesive joints are essential for the safety of engineering structures. They are highly susceptible to any defects resulting from mistakes that were committed during the preparation and joining of elements, e.g., poor condition of the adherend surface or a lack of glue on a part of the overlay. Additionally, a gradual deterioration in the quality of the adhesive material cannot be neglected. Typical adhesive joints are not capable of being inspected based on visual techniques as the place of damage occurrence (the adhesive film) is hidden between the joined parts. Destructive testing provides very accurate results [1]. This approach allows for the determination of the influence of certain parameters (e.g., damage existence [2,3] or adhesion condition [4]) on the strength of representative samples, which were usually prepared in laboratory conditions and to predict the state of existing joints. Destructive testing is not free of disadvantages, it is, in general, a high-cost solution and cannot usually be applied to existing structures. Therefore, different non-destructive testing (NDT) methods have been adopted for the inspection of adhesive joints [5] including ultrasonics [6,7], thermography [8], radiography [7], digital image correlation [9,10], or electromechanical impedance spectroscopy [11].

In recent years, one of the most promising techniques for damage assessment is the use of guided wave propagation (e.g., [12–14]). Lamb waves are a specific type of guided wave that occur in a media that is restrained by two parallel surfaces, e.g., thin plate-like structures. The ability of Lamb waves to interact with any kind of defect makes them particularly useful in the diagnostics of various elements of engineering structures, including adhesive joints. In previous studies, guided waves have been proven to be an effective technique for the inspection of adhesive lap joints [15–18], adhesive bonds between composites [19–21], adhesive butt joins of plates [22], or the detectability of the adhesion level [23].

In the guided wave propagation technique, the wave is usually excited at one point of a structure and is collected at a number of points that are spread over the whole structure. In general, two measurement strategies can be used. The first method depends on using a small number of ultrasonic or piezoelectric transducers, followed by an analysis of registered time histories. This strategy is mainly used for structural health monitoring (SHM). Another approach is directed towards non-contact techniques, such as scanning laser Doppler vibrometry (SLDV) and the sensing of the full guided wave field (e.g., [24–26]). This strategy is more appropriate for quality assurance testing and periodic nondestructive inspection. Particular signals that are registered at numerous points distributed over a monitored area provide a two-dimensional representation of propagating waves at specific time instances. If any damage exists in an analyzed structure, a disturbance in the wavefront shape can be observed. The perturbation is more or less clearly visible depending on the type, size, and position of the defect. This visual assessment of SLDV maps allows for one to determine the presence of damage, but does not give a clear image due to the multiple reflections of waves from discontinuities and element edges [25]. The determination of the actual shape and size of the existing defect requires additional processing and damage imaging [27–29]. One of the simplest but most efficient techniques is based on vibration energy distribution and utilizes the calculation of a root mean square (RMS). This method was improved to be more applicable for damage detection by defining the weighted root mean square (WRMS) method [30]. Recently, the RMS has been successfully used in various damage detection applications [25,26,31–36]. Żak et al. [30] investigated aluminum and composite plates as well as composite elements of a helicopter, all with the damage being simulated by an additional mass. They concluded that better resolution in WRMS maps corresponded to higher values of the weighting factor (linear or square). Localization of damaged rivets in a stiffened plate structure based on RMS maps was conducted by Radzieński et al. [31] where the authors used a five-cycle burst of different frequencies (5 kHz, 35 kHz, and 100 kHz). The best results were obtained in the case of an excitation signal of 100 kHz, where the wavelength was shorter than the distance between two rivets. Lee and Park [32] considered an aluminum plate with notches of various orientation (tangential, perpendicular, and diagonal to the incident wave front). Based on RMS images, only tangential damage was successfully detected without a comparison to the intact images. Saravanan et al. [33] used radially weighted and factored RMS for damage localization in an aluminum plate. Detection of delamination in a composite T-joint using RMS images was studied by Geetha et al. [34]. Marks et al. [35] presented experimental investigations on the detection of disbonds in a stiffened panel based on the RMS plots. Wave packets with three different frequencies, namely 100 kHz, 250 kHz, and 300 kHz, were applied. The authors concluded that the excitation frequency is an important consideration for the inspection of disbonds. In the conducted works, only higher frequencies (250 kHz and 300 kHz) were able to detect damage. Lee et al. [36] applied the WRMS to localize the notches and corrosion areas in aluminum plates and chose 2 as the weighting parameter. Kudela et al. [25] studied the detection of delamination in a composite plate using WRMS with the weighting factor set as 2.7. They also proposed a selective WRMS that enabled the elimination of edge reflections. Detection of disbond inserts in composite elements was performed by Pieczonka et al. [26] using RMS and three-dimensional (3D) scanning. In the conducted experimental investigations, it was noticed that in-plane RMS maps provided a better indication of damage than out-of-plane RMS maps. Summarizing all of the works that are reported above, the main problem that is connected with damage imaging using WRMS is the fact that the weighting factor is set arbitrary with no clear explanation for the choice of a specific value.

This study presents an approach to the damage identification of adhesively bonded joints that is based on the Lamb wave propagation. The guided wave field was measured with the use of scanning laser Doppler vibrometry and then processed by means of the weighted root mean square. The object of the research was a single lap adhesive joint of metal plate-like structures. To verify the feasibility of the proposed approach, nine specimens with defects that were differing in size and position were tested. The influences of different WRMS parameters (i.e., the time window and weighting factor), as well as excitation frequencies, were analyzed. A novel element in this research was the application of the statistical analysis in determining the WRMS parameters. It has been proven that the obtained WRMS maps of damaged joints were characterized by bimodal distributions that created the opportunity to determine the optimal value of the time window, weighting factor, and excitation frequency.

2. Materials and Methods

2.1. Description of Specimens

The object of the research was a single lap adhesive joint of steel plates. Dimensions of each adherend were 270 mm × 120 mm × 3 mm. The length of the overlap was 60 mm (the half-width of the plate). The geometry of the analyzed joint is shown in Figure 1. The plates were degreased with Loctite-7063 cleaner and dried. Then, the surface of the overlap was treated with abrasive paper (type P120) and was degreased again just before applying the glue. The adhesive layer was prepared with the use of the epoxy-based glue Loctite EA 9461. The mean value of the bondline thickness for all of the specimens was equal to 0.216 mm and the standard deviation had a value of 0.043 mm, which resulted in the coefficient of variation being equal to 0.198.

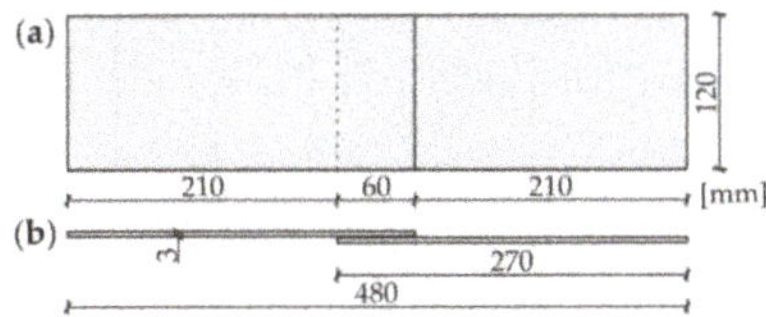

Figure 1. Geometry of the analyzed single lap joint: (**a**) plan view; and, (**b**) side view.

Investigations were conducted for nine types of specimens (Figure 2): one intact (#1) and eight (#2–#9) with different defects in the form of partial debonding. The defects were modelled as a lack of glue on a part of the overlap surface. A PTFE tape of 0.2 mm thickness was used to prevent the glue from leaking to the debonding areas of the damaged joints. The adherends of each specimen were separated after the experimental measurements to determine the exact geometry of defects in the prepared joints. Photographs of the disconnected joints are presented in Figure 3.

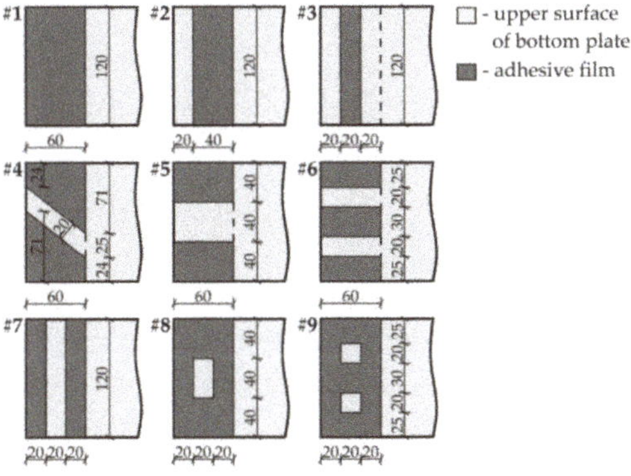

Figure 2. Investigated specimens: intact (#1) and with defects (#2–#9).

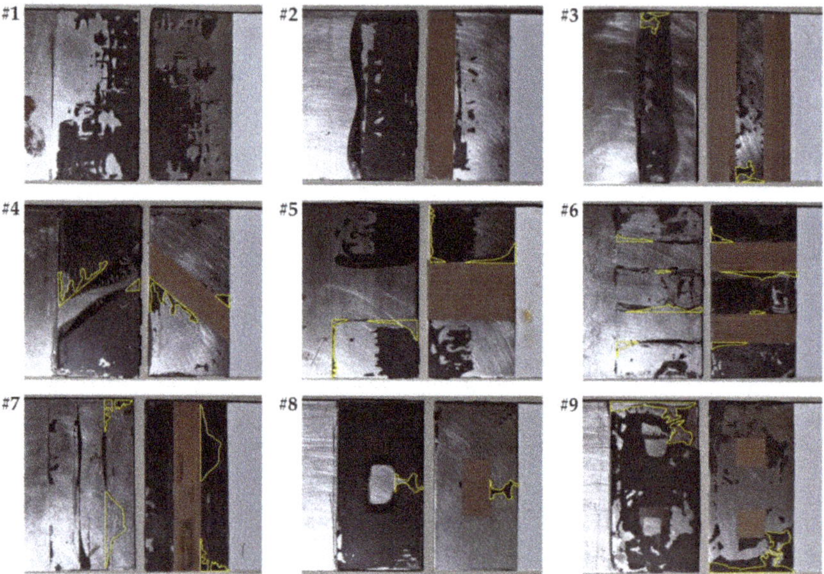

Figure 3. Photographs of the investigated specimens after separation (regions outlined with yellow line denote kissing defects).

It is visible in Figure 3 that the defects had slightly different shapes and areas than was assumed. The glue that leaked onto the surfaces covered with the PTFE tape (i.e., intended voids) did not affect the WRMS results because the tape prevented the plates from connecting even though there were some adhesive leakages. However, in addition to the intended voids, some additional areas resulting from the improper adhesion of the adhesive to the adherend (so-called kissing defects or low volume defects [8]) were identified for specimens #3 to #9. The kissing defect regions are outlined in Figure 3 by the yellow line. To quantify the real defect areas, including both the intended voids and kissing defects, the photographs seen in Figure 3 were processed using the ImageJ program [37]. The results are given in Table 1.

Table 1. Designed and determined area of defects.

No. of Sample	Design Surface of Defect Area (cm^2)	Determined Surface of Defect Area Including Kissing Defects (cm^2)	Design Relative Defect Surface (%)	Determined Relative Defect Surface (%)
#1	-	-	-	-
#2	24.00	24.00	33.33%	33.33%
#3	48.00	49.11	66.67%	68.21%
#4	15.00	17.64	20.83%	24.50%
#5	24.00	26.03	33.33%	36.16%
#6	24.00	26.33	33.33%	36.57%
#7	24.00	29.55	33.33%	41.05%
#8	8.00	9.09	11.11%	12.63%
#9	8.00	13.34	11.11%	18.53%

2.2. Experimental Setup

Experimental measurements consisted of the excitation of Lamb waves in the lower adherend and the collection of wave propagation signals in the selected area of the specimen. The experimental setup is presented in Figure 4. The excitation signal was generated by means of the arbitrary function generator AFG 3022 (Tektronix, Inc., Beaverton, OR, USA) and the high-voltage amplifier PPA 2000

(EC Electronics, Krakow, Poland). Lamb waves were excited by the plate piezoelectric actuator NAC2024 (Noliac, Kvistgaard, Denmark), with dimensions of $3 \times 3 \times 3$ mm^3 and a resonant frequency of over 500 kHz. The actuator was bonded to the bottom side of the specimen (Figure 4b) using petro wax 080A109 (PCB Piezotronics, Inc., Depew, NY, USA). The excitation signal was a five-peak wave packet obtained from a sinusoidal burst with a carrier frequency of 200 kHz modulated by the Hanning window. There were also additional excitation signals for specimens #3 and #6 that differed in the frequency value from the basic variant (100 kHz and 50 kHz, instead of 200 kHz). Acquisition of the wave propagation signals was realized by the scanning head of the laser vibrometer PSV-3D-400-M (Polytec GmbH, Berlin, Germany) with a VD-09 velocity decoder (Figure 4a). Time signals were acquired with a sampling frequency of 2.56 MHz and were averaged 40 times to improve the signal-to-noise ratio. The excitation signal was triggered as external and monitored using a reference channel of the vibrometer. To improve the backscatter of light, the scanned surface was covered with retro-reflective sheeting. Velocity values (out-of-plane components) were recorded in the time domain in 7381 points that were distributed at the top side of the specimen (on the surface of the upper plate with no PTFE tape attached, see Figure 3). The area was scanned point by point in the mesh of 61 rows and 121 columns (Figure 5), giving a scan resolution of 1.933 mm and about nine scan points per the shortest wavelength. Signals sensed at all of the points allowed for us to prepare a map of the propagating waves, i.e., a two-dimensional plane view at a specific time instance.

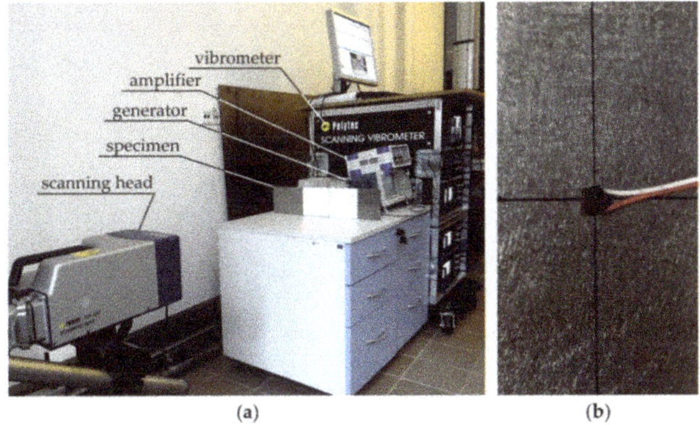

Figure 4. Experimental setup: (**a**) instrumentation for generation and acquisition of Lamb waves; and, (**b**) detail showing the piezo actuator attached to the specimen.

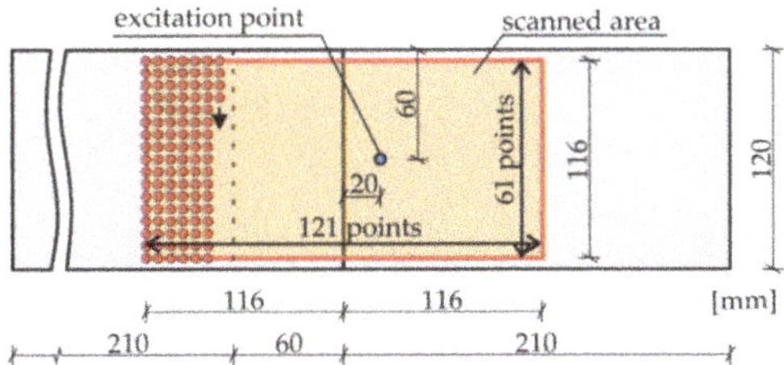

Figure 5. Investigated specimen with indicated scanning points.

2.3. Weighted Root Mean Square Damage Imaging

The guided wave field that was acquired with the use of SLDV can be further processed to more accurately identify the position and shape of defects. This processing technique depends on the calculation of the weighted root mean square for each registered time signal and the creation of a damage map. The WRMS for a time signal $s(t)$ can be defined as follows [25,26,30,33,36]

$$WRMS = \sqrt{\frac{1}{t_2 - t_1} \int_{t_1}^{t_2} \left(w(t)s(t)^2 \right) dt} \tag{1}$$

where $w(t)$ denotes a weighting factor and $(t_2 - t_1)$ is length of the time window with t_1 denoting the beginning of the time window and t_2 denoting the end of the time window. For a discrete signal $s_r = s(t_r)$ sampled with Δt interval and calculated from the interval $T = (N-1)\Delta t = t_2 - t_1$, the WRMS can be approximated by

$$WRMS = \sqrt{\frac{1}{N} \sum_{r=1}^{N} (w_r s_r^2)} \tag{2}$$

where N is the number of samples in the signal used for RMS calculation and w_r is the weighting factor defined based on the number of consecutive sample r

$$w_r = r^m, \; m \geq 0, \; r = 1, 2, \ldots, N \tag{3}$$

where m is a power of the weighting factor. The weighting factor with parameter $m > 0$ enables the decrease of importance of the samples at the moment of excitation. The interpretation of WRMS maps allows for the determination of the shape and location of the existing defects. In general, the areas with higher WRMS values can indicate the possible locations of damage that are related to the reduction of material stiffness and may be in the form of cracks, corrosion, or delamination [26].

3. Results and Discussion

3.1. Dispersion Curves

Before analyzing the defects in the lap joints, the characteristics of Lamb waves were determined for two specimens: a single steel plate with a thickness of 3 mm and dimensions of 240 mm × 300 mm, and two steel plates that were adhesively bonded with an adhesive layer with a thickness of 0.2 mm. In each plate, a five-cycle Hanning windowed sine function with center frequencies of 50 kHz, 100 kHz, and 200 kHz was used as an excitation signal. Out-of-plane vibration signals were measured in 101 points that were evenly distributed along a line with a length of 10 cm. The wavenumber-frequency relations (Figure 6) were obtained using 2D-FFT (cf. [23,38]). In the experimental results, theoretical dispersion curves that were obtained using GUIGUW software [39] based on the semi-analytical finite-element formulation [40] were superimposed.

In general, in the single plate, both S_0 and A_0 modes can propagate in the considered frequency range up to 200 kHz. However, the applied actuator utilizing the thickness-wise expansion effect enabled mainly exciting the fundamental A0 mode and the influence of the S0 mode was insignificant, as it can be seen in Figure 6a. The group velocities of A_0 mode were 2981.5 m/s, 2575.7 m/s, and 2062.7 m/s for the excitation frequencies of 200 kHz, 100 kHz, and 50 kHz, respectively, giving the wavelengths values of 14.9 mm, 25.7 mm, and 41.3 mm. Figure 6b shows the dispersion curves for two adhesively bonded plates. In this case, three modes could propagate in the frequency range up to 200 kHz (S_0, A_0 and A_1), however, in the performed experiment, the A_0 mode was dominant. It was noted that the A_0 mode dispersion curves were similar for both plates, indicating that this mode will transmit across the waveguide transitions. The group velocities of the A_0 mode propagating in two adhesively bonded plates were 2865.6 m/s, 2448.5 m/s, and 2045.8 m/s for the excitation frequencies of 200 kHz, 100 kHz, and 50 kHz, respectively, giving the wavelengths values of 14.3 mm, 24.5 mm, and 40.9 mm.

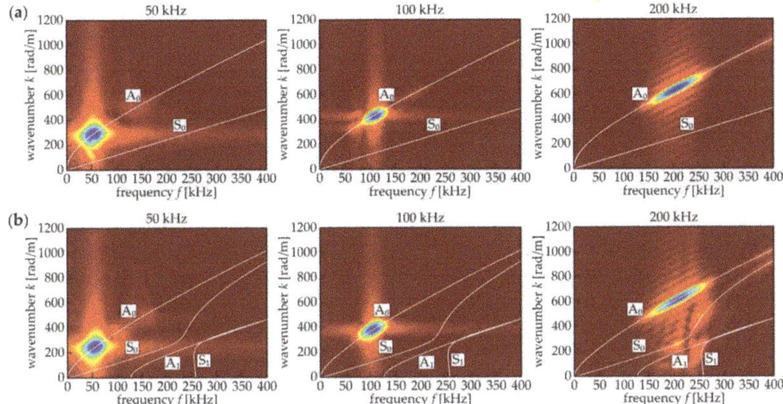

Figure 6. Experimental and theoretical dispersion curves obtained for different excitation frequencies of 50 kHz, 100 kHz, and 200 kHz: (**a**) steel 3 mm plate (E = 195.2 GPa, v = 0.3, ρ = 7741.7 kg/m^3); and (**b**) two adhesively jointed steel 3 mm plates (E = 195.2 GPa, v = 0.3, ρ = 7741.7 kg/m^3) adhesively bonded with the adhesive layer of a thickness of 0.2 mm (E = 5 GPa, v = 0.35, ρ = 1130 kg/m^3).

3.2. SLDV Maps

Figure 7 shows snapshots of the guided wave propagation signals that were obtained in the experimental scanning of the analyzed specimens. The SLDV maps are presented at the time instance of 27.7 µs. Disturbances of the wavefront shape on the disbond area were visible for each specimen from #2–#9 when compared with the intact specimen #1. The differences were more or less considerable, depending on the geometry of the defect. The most significant disturbances were observed for the specimens with stripped defects skewed and parallel to the primary direction of propagation (#4–#6). For specimens that had the disbond area in the shape of a strip perpendicular to the direction of propagation (#2, #7), the changes were not clearly visible. The exception was specimen #3, which had no adhesive near the edge of the overlap and resulted in the limited appearance of the wavefront on the overlap surface. Internal defects (#8, #9) were moderately detectable. In summary, the presence of the defects in the damaged specimens was visible in the maps, but their geometry was not determinable.

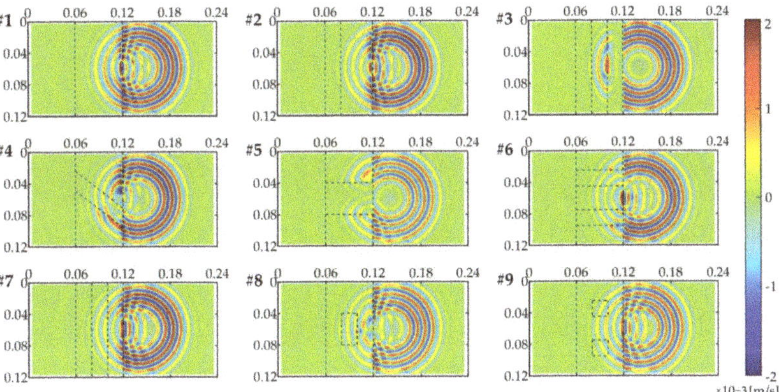

Figure 7. Scanning laser Doppler vibrometry (SLDV) maps of considered joints at the time instance 27.7 µs (dimensions of the joint are given in meters; the adhesive was 0.06–0.12 m).

3.3. WRMS Imaging

3.3.1. Influence of Time Window and Weighting Factor

Figure 8 shows the WRMS that was calculated for specimen #5 using Equation (2). Different values of the weighting factor and time window were applied. In all calculations, the beginning of the time window t_1 was always set as zero, so the length of the time window was $T = t_2$. The obtained damage maps are presented as logarithmic values of WRMS. The analysis of the results led to the conclusion that the time window is a significant factor for the readability of visualizations. According to the visual assessment, the clearness of the WRMS maps increased with the increase in the time window. For this reason, the longest time window ($T = 3.0$ ms) was tentatively chosen for the damage imaging of the other specimens (as presented further). It is also worth noting that the weighting factor was not negligible. The constant weighting factor ($w_r = 1$) was not efficient when compared with others that increased with time. The maps were not clear enough and the influence of the incident wave at the excitation point was highly visible. These effects were not observable for the weighting factors $w_r = r$ and $w_r = r^2$, as represented by the linear or square function of time, respectively. Their effectiveness was comparable and significantly higher than the constant weighting factor.

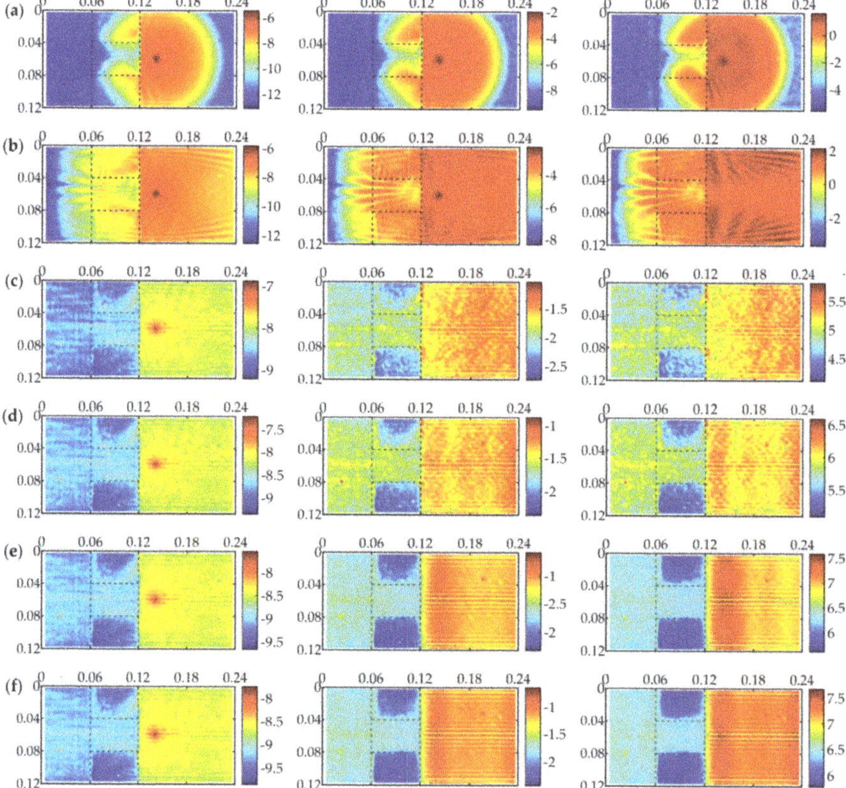

Figure 8. Weighted root mean square (WRMS) maps (values in (m/s)) for specimen #5 for different weighting factors $w_r = 1$ (first column), $w_r = r$ (second column), $w_r = r^2$ (third column) and varying time window: (**a**) $T = 0.03$ ms; (**b**) $T = 0.05$ ms; (**c**) $T = 0.50$ ms; (**d**) $T = 1.00$ ms; (**e**) $T = 2.00$ ms; and, (**f**) $T = 3.00$ ms.

To assess the clearest variant of the WRMS, damage maps for adhesive joints with internal defects (#8, #9) were prepared (Figure 9). The identification of the defect geometry for these specimens was possible, however, some differences between the WRMS maps can be observed. A comparison of the different weighting factors for both of the specimens showed the advantage of a variant with the linear function.

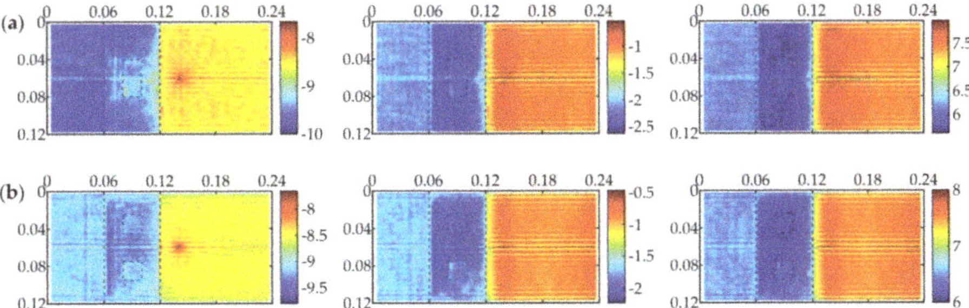

Figure 9. WRMS maps (values in (m/s)) for joints with internal defects for constant time window ($T = 3.00$ ms) and different weighting factors $w_r = 1$ (first column), $w_r = r$ (second column), $w_r = r^2$ (third column): (**a**) specimen #8; and, (**b**) specimen #9.

3.3.2. Influence of Defect Shape and Position

Figure 10 shows the WRMS maps for all of the considered specimens #1–#9. The parameters of the WRMS calculations were assumed as follows: the time window $T = 3.00$ ms and the weighting factor ($w_r = r$). The intensification of the WRMS maps could be observed at the disbond areas, where the values were greater than for the properly bonded parts. This means that the amplitudes of propagating waves vary more significantly if there is no direct connection between the bonded parts. It is worth noting that the clearness of the maps increased with the difference between values that were obtained for the disbonded and properly bonded areas.

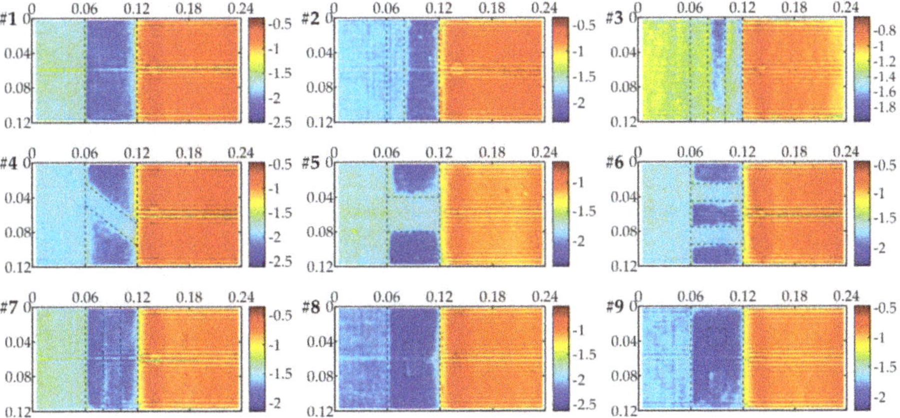

Figure 10. WRMS maps (values in (m/s)) for specimens #1–#9 ($T = 3.00$ ms, $w_r = r$).

The presence of defects was possible to determine in all cases, moreover, their shapes were also identifiable. The clearest damage image was obtained for specimens #4–#6, as well as specimens #2 and #3. Internal damages in specimens #7–#9 were also possible to identify; however, the damage maps had poorer quality. It is worth noting that the geometry of the identified defects was not in ideal agreement with the assumed one. In some cases, minor irregularities were visible, especially for samples that were intended as symmetric. This was an effect of inaccuracies in the process of the specimen preparation. The verification of the actual state of the adhesive layer was obtained after separating the joints (cf. Figure 3). It is clear that the adhesive leakages to the areas that are covered with PTFE tape were not identified in the maps, which means that the tape prevented the plates from being connected. On the other hand, the kissing defects were clearly noticeable (cf. Figure 3).

Figure 11 also shows the WRMS maps for all of the considered specimens (#1–#9), but for the square weighting factor ($w_r = r^2$). The visual assessment of the presented maps led to conclusions that were similar to the ones resulting from analysis of maps for the linear weighting factor ($w_r = r$). The value of m power affected the clearness of images: the ones for the linear weighting factor appeared to be clearer. Based only on the visual assessment of the defect images, it was not possible to judge which variant of weighting factor provided the clearest results.

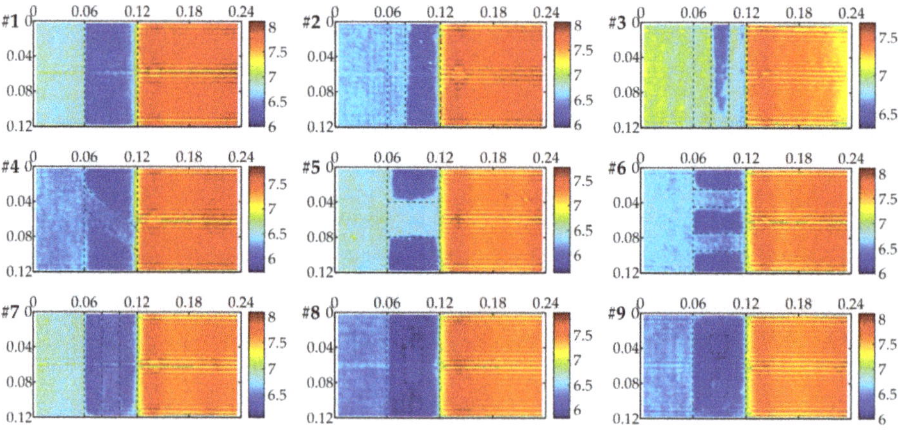

Figure 11. WRMS maps (values in (m/s)) for specimens #1–#9 (T = 3.00 ms, $w_r = r^2$).

3.3.3. Influence of Excitation Frequency

An additional aspect of the analysis was the influence of the excitation frequency on the obtained results. Figures 12 and 13 show the WRMS maps for specimens #6 and #7 for the linear and square weighting factors, respectively. The measurement of the joints was repeated three times, with different excitation frequencies, namely 200 kHz, 100 kHz, and 50 kHz. It was observed that the excitation frequency was an important factor that significantly affected the readability of the maps. The efficiency of the WRMS damage imaging increased with the increase in the excitation frequency value.

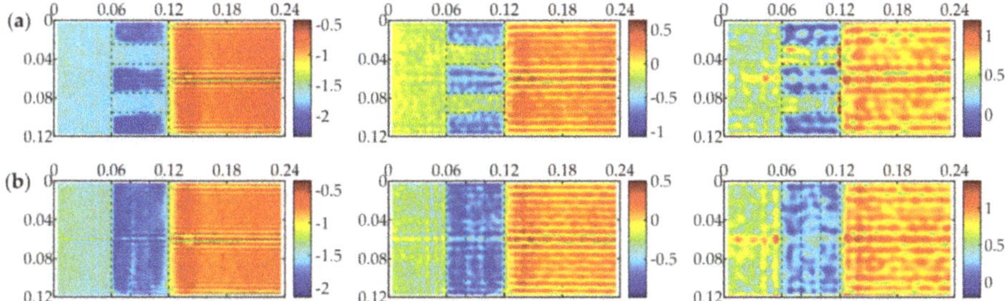

Figure 12. WRMS maps (values in (m/s)) with linear weighting factor ($w_r = r$), time window $T = 3.00$ ms and with different excitation frequencies $f = 200$ kHz (first column), $f = 100$ kHz (second column), $f = 50$ kHz (third column): (**a**) specimen #6; and (**b**) specimen #7.

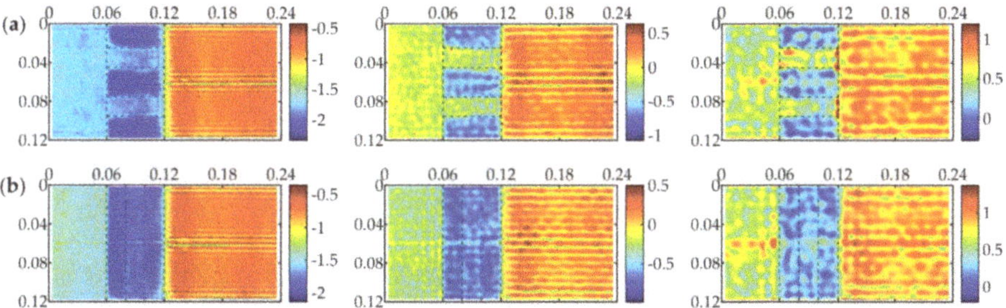

Figure 13. WRMS maps (values in (m/s)) with square weighting factor ($w_r = r^2$), time window $T = 3.00$ ms and with different excitation frequencies $f = 200$ kHz (first column), $f = 100$ kHz (second column), $f = 50$ kHz (third column): (**a**) specimen #6; and, (**b**) specimen #7.

3.3.4. WRMS Histograms

The WRMS method enables one to differentiate the areas of a properly prepared adhesive film and its defects. In addition to the visual assessment of prepared maps, a numerical analysis of its calculated values can be performed with the use of statistical analysis. The purpose is to prepare a histogram of calculated WRMS values that is expected to reveal the bimodal distribution of the sample population (concentrated on two values responding to disbonded and properly bonded areas). This presents the possibility for the quantitative determination of the optimal values of WRMS parameters. As observed from the images in Figures 10 and 11, higher WRMS values were obtained in the disbonded areas, whereas the lower ones occurred in the areas of good adhesion. To provide a clear interpretation of the WRMS histograms, the analysis was limited to the surface of the overlap (area of the adhesive joint).

In the first step, the influence of the time window on WRMS histograms was considered. Figures 14 and 15 present the WRMS histograms for specimen #5 with different values of time window and different weighting factors $w_r = r$ and $w_r = r^2$, respectively. Additionally, to emphasize the characterization of the obtained distributions, each histogram was approximated by the probability density function with the use of the kernel smoother in a MATLAB® environment (shown as a red line). It was clear that, for higher time window values (0.50 ms and more), a population of WRMS values formed a bimodal distribution that was represented by a function with two conspicuous peaks values. As concluded before, the clearness of the defect image was related to the difference between the values that were obtained for health and poor adhesion areas of the overlap. For the histograms, this difference was interpreted as the distance between the peak values that were measured along the horizontal axis. For each diagram (if possible), peak values were localized. To compare the results that were obtained for different time window values, the differences between the scales were eliminated by dividing the measured distances by the analyzed range of fitted distributions. Calculated values for both variants of weighting factors are presented in Table 2 (only if the bimodal distribution was obtained). It could be seen that the increase in time window caused the peaks to occur at longer distances. The clarity of the maps increased with the growth of the distance between peaks, which meant that the longer time window positively influenced the clearness of the maps. This fact justifies the selection of the largest value of time window (T = 3.00 ms) for the prepared WRMS maps.

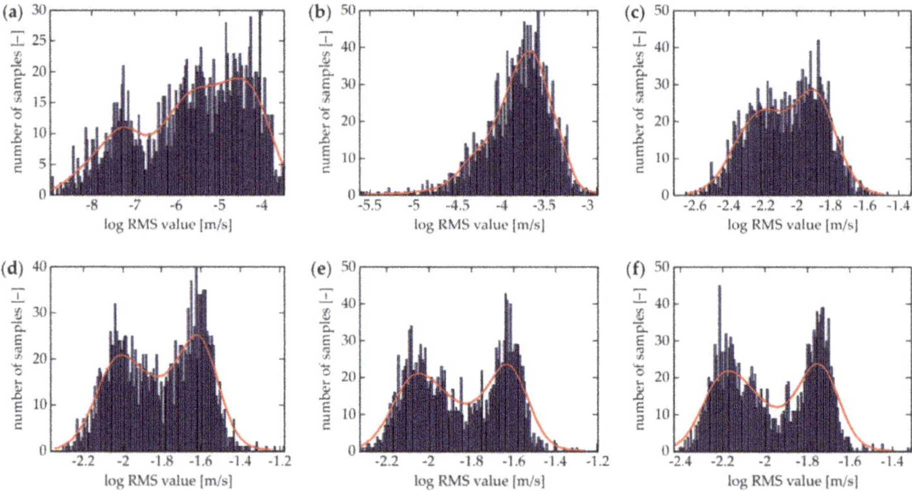

Figure 14. Histograms of WRMS values ($w_r = r$) for specimen #5 with different time windows: (**a**) T = 0.03 ms; (**b**) T = 0.05 ms; (**c**) T = 0.50 ms; (**d**) T = 1.00 ms; (**e**) T = 2.00 ms; and, (**f**) T = 3.00 ms.

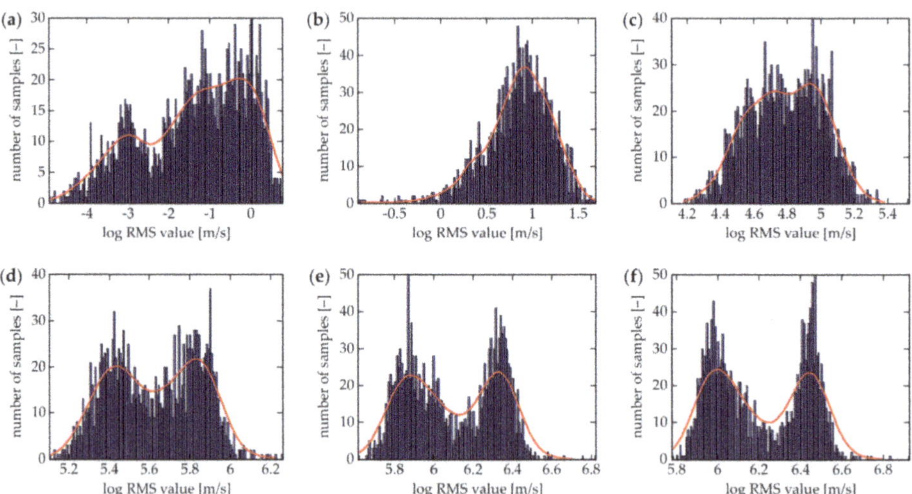

Figure 15. Histograms of WRMS values ($w_r = r^2$) for specimen #5 with different time windows: (**a**) T= 0.03 ms; (**b**) T = 0.05 ms; (**c**) T = 0.50 ms; (**d**) T = 1.00 ms; (**e**) T = 2.00 ms; and, (**f**) T = 3.00 ms.

Table 2. Calculated values of relative peaks distance of bimodal distribution for specimen #5.

Time Window (ms)	Relative Peaks Distance ($w_r = r$) (%)	Relative Peaks Distance ($w_r = r^2$) (%)
0.03	–	–
0.05	–	–
0.50	18.67	12.67
1.00	30.67	33.33
2.00	34.67	34.67
3.00	35.33	37.33

To move forward, histograms for all of the specimens were prepared with the time window $T = 3.00$ ms. Figure 16 presents the histograms with the linear weighting factor. For the intact joint (specimen #1), a common unimodal distribution was obtained. The calculated values were concentrated about a singular mode that represented the undamaged area of the adhesive joint. Typical bimodal distributions were obtained for specimens with a clear defect image (#2–#6). For specimens #7–#9, the distributions were likely to be unimodal, but when compared with specimen #1, their shapes were not that regular, and it was clear that their widths were greater than that of specimen #1. This may result in the conclusion that the defect was present, but it could not be clearly observable. Figure 17 shows the histograms with the square weighting factor. The conclusions from the analysis of each histogram were similar to ones based on Figure 16.

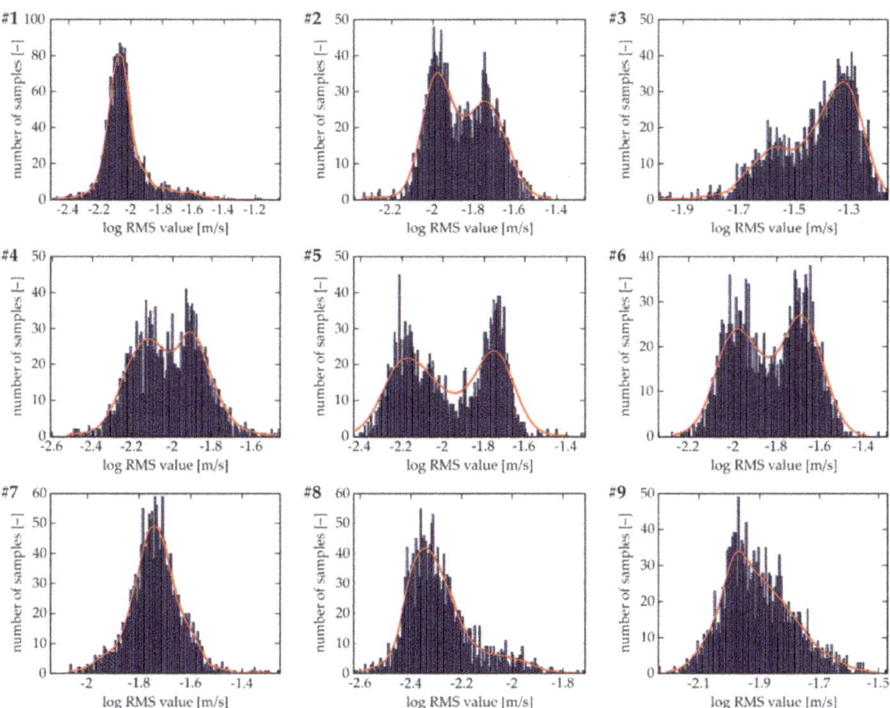

Figure 16. Histogram of WRMS values for specimens #1–#9 ($T = 3.00$ ms, $w_r = r$).

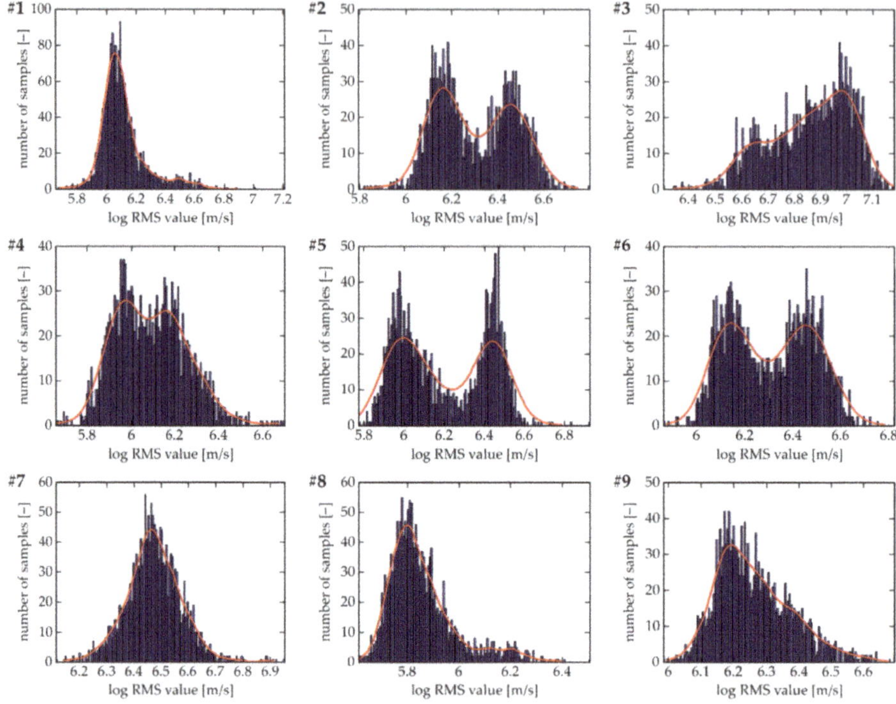

Figure 17. Histogram of WRMS values for specimens #1–#9 ($T = 3.00$ ms, $w_r = r^2$).

Figure 18 shows the relationships between the relative peak distance and the weighting factor that is represented by the power m for specimens #2–#6 that clearly presented bimodal distributions. It could be seen that for most of the specimens, higher values of power m resulted in the greater value of relative peaks distance. This means that the higher the value of m, the clearer the image of the defect on the WRMS maps. It is worth noting that the value of the relative peaks distance stabilized at a specific value of m individually for each defect, so the increase in m could not infinitely the improve the quality of maps. Furthermore, specimen #4 provided a different relationship than the others where the relative distance between the peaks at the beginning increased with power m and then decreased. Maximal values were observed for the power m between 1 and 1.5. Based on the calculations presented above, it was concluded that the proposed analysis provides the opportunity to improve the process of determining the optimal value of the weighting factor (in fact, the power m) for damage imaging using WRMS.

An attempt to quantitatively analyze the defects was performed based on the statistical analysis of the calculated WRMS values. Histograms and their approximations by probability density functions were prepared for different values of power m for specimens with a bimodal distribution (#2–#6). The limit point (here, the central point) in the peaks distance was determined as the mean value between peak values (along the horizontal axis). As previously mentioned, poor adhesion areas are related to higher values of WRMS. For that reason, the number of values that are above the central point value were counted, representing the number of points where the defect occurred. The percentage of points with higher values can be interpreted as the relative surface of the defect. Based on the relationships that are presented in Figure 19, it was clear that for most of the specimens, the relative defect surface did not vary significantly in relation to power m and it moderately decreased with its value. The exception was the defect in specimen #4 that presented the opposite relationship.

When comparing the results with measurements that were conducted on the real separated specimen (Table 1), it was clear that the good agreement of the proposed approach was obtained only for defect #3 (calculated mean value of relative surface is 68.5% and measured is 68.21%). For other specimens, the differences were much higher (calculated values ranged from 45% to 55% and measured between 33% and 37%). The reason for these discrepancies may possibly be related to the method of determining the limit point for counting the number of points that are related to the defect surface. Due to the complexity of bimodal distributions (e.g., different parameters of both modes), further analyses should be conducted.

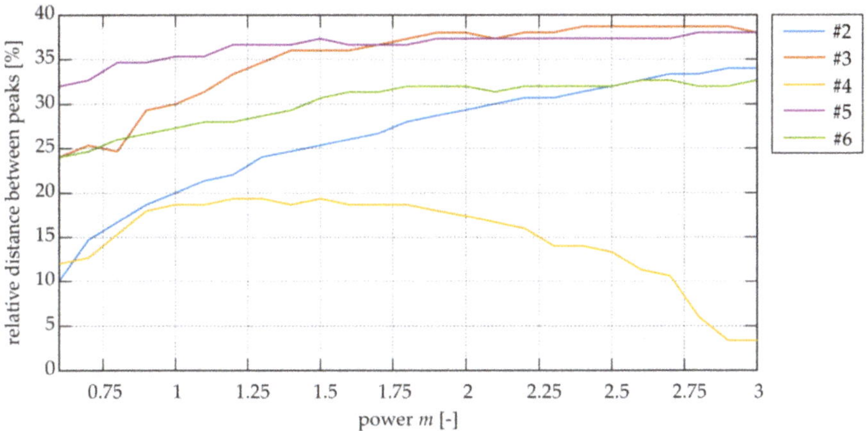

Figure 18. Relationships between the relative peak distance of WRMS histograms and weighting factor (represented by power *m*) for specimens #2–#6.

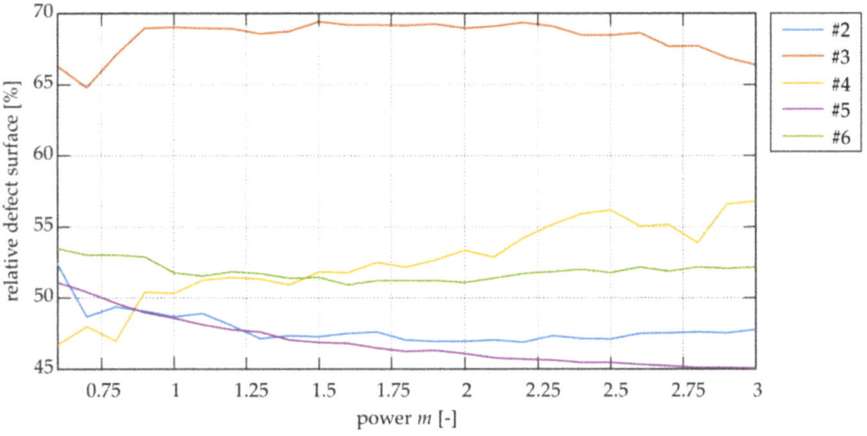

Figure 19. Relationships between the relative defect surface and weighting factor (represented by power *m*) for specimens #2–#6.

One additional aspect of this research included the statistical analysis of the influence of the excitation frequency. Figure 20 shows the WRMS histograms for specimen #6 with different excitation frequencies. For the lowest value (50 kHz), the unimodal (but asymmetric) distribution was obtained. Higher frequencies (100 kHz and 200 kHz) showed typical bimodal distributions with different characterization. To compare them, the relative peak distance was calculated. For 100 kHz, a value of 34.67% was obtained, whereas for 200 kHz, it was 27.33%. This meant that the 100 kHz frequency provided results with more differentiated WRMS values obtained in areas with good and poor quality (resulting in clearer defect image). Furthermore, the percentage of the defect surface was determined with the use of the above-presented method with the central limit point. For 100 kHz, the result was 35.67% and for 200 kHz, a value of 51.78% was achieved. When compared to the measured value (33.33%, Table 1), the nearest value was obtained for 100 kHz. However, due to the lack of results for other specimens revealing bimodal distribution, it was not possible to conclude on the supremacy of this frequency. Further research could help to solve this problem.

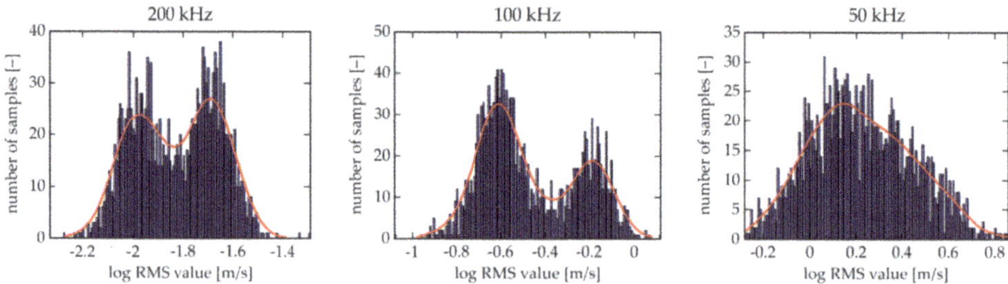

Figure 20. Histograms of WRMS values for specimen #6 with different excitation frequencies ($T = 3.00$ ms, $w_r = r$).

4. Conclusions

In this paper, an experimental analysis of the Lamb wave interaction with damage in the adhesively bonded joints was conducted. The study consisted of a visual assessment and statistical analysis of the obtained results. The presented investigations support the following conclusions:

- The analysis of maps that were obtained from the SLDV measurements provided the initial identification of defects with different efficiencies. The clearness of the SLDV maps varied depending on the shape and position of damage.
- The actual defect shape and location were determinable based on the appropriate damage imaging, where recorded out-of-plane vibrations were processed by means of the WMRS. The effectiveness of the applied method was strongly dependent on the calculation parameters (weighting factor and time window), excitation frequency, and a type of damage.
- The extension of the time window led to the increase in differences in the WRMS values between the damaged and undamaged areas of the adhesive joint. As a result, defects in the adhesive film could be identified more accurately.
- By comparing different variants of weighting factors, the constant one did not provide efficient results due to the high influence of the incident wave at the excitation point. This effect was reduced by the weighting factor, which was represented by the linear or square function that achieved comparable results. For specimens with internal defects, moderately greater effectiveness of the linear variant was observed.
- The shape and location of defects had a great impact on the effectiveness of their identification. Both the presence and the size of defects were possible to determine in all of the contested specimens based on the WRMS damage maps. The clearest results were obtained for stripe defects parallel or skewed to the primary direction of wave propagation. The disbonded areas in the shape of the strip perpendicular to the direction of wave propagation were also determinable, however, the identification was easier when such area was at the beginning or at the end of the lap. The most difficult to detect were the defects in the internal part of the adhesive film.

The differentiation of effectiveness of the method for different types of defects resulted from the occurrence of different wave phenomena at the boundaries of defects. Stripe defects, skewed, or parallel to the direction of propagation triggered a diffraction of wave fronts that provided a significant disturbance in wave propagation and resulted in a higher intensification of signals. Defects in the shape of a perpendicular strip were related to wave reflection that did not disrupt the propagation as diffraction did. Finally, internal defects were the most difficult to identify because of the local characterization of occurring wave phenomena.

- The statistical analysis of the calculated WRMS values successfully supplemented the visual assessment of the damage images. The interpretation of the histograms helped to explain the effects that were observed on the WRMS maps. The use of the relative distance between peaks of bimodal distribution was proposed as the indicator for the optimal determination of WRMS parameters. Based on this indicator, it was proven that the largest value of the time window obtained damage maps with the highest clearness. Moreover, it was possible to determine the range of the optimal value of the weighting factor and to quantify the defect areas. The statistical analysis of the influence of the excitation frequency also enabled a comparison of the effectiveness of the three different frequencies. Two higher frequencies providing bimodal distribution were proven to be more effective than the lowest excitation frequency providing unimodal distribution.

The main conclusion was that the scanning laser Doppler vibrometry and the weighted root mean square processing of guided wave signals could be successfully applied for damage imaging in adhesive films. Additionally, the statistical analysis can be useful in the interpretation of the obtained results. However, the conducted research is an initial step for further analyses that are related to condition assessment in adhesive joints of elements in real engineering structures. The proposed method to determine the optimal value of the weighting factor and to quantify the real defect areas could be further developed. The influence of different excitation frequencies could be also considered more precisely.

Acknowledgments: The research work was carried out within project No. 2015/19/B/ST8/00779, financed by the National Science Centre, Poland. This support is greatly acknowledged by the authors of the study.

Author Contributions: M.R. conceived, designed, and performed the experiments; J.L. contributed analysis tools for damage imaging; M.R., E.W. and J.L. analyzed the data; E.W. and M.R. wrote the paper.

Conflicts of Interest: The authors declare no conflict of interest.

References

1. Adams, R.D.; Wake, W.C. *Structural Adhesive Joints in Engineering*; Elsevier Applied Science Publishers: London, UK, 1986; ISBN 978-94-010-8977-7.
2. Karachalios, E.F.; Adams, R.D.; Da Silva, L.F.M. Strength of single lap joints with artificial defects. *Int. J. Adhes. Adhes.* **2013**, *45*, 69–76. [CrossRef]
3. Heidarpour, F.; Farahani, M.; Ghabezi, P. Experimental investigation of the effects of adhesive defects on the single lap joint strength. *Int. J. Adhes. Adhes.* **2018**, *80*, 128–132. [CrossRef]
4. Jeenjitkaew, C.; Guild, F.J. The analysis of kissing bonds in adhesive joints. *Int. J. Adhes. Adhes.* **2017**, *75*, 101–107. [CrossRef]
5. Adams, R.D.; Drinkwater, B.W. Nondestructive testing of adhesively-bonded joints. *NDT E Int.* **1997**, *30*, 93–98. [CrossRef]
6. Korzeniowski, M.; Piwowarczyk, T.; Maev, R.G. Application of ultrasonic method for quality evaluation of adhesive layers. *Arch. Civ. Mech. Eng.* **2014**, *14*, 661–670. [CrossRef]
7. Vijayakumar, R.L.; Bhat, M.R.; Murthy, C.R.L. Non destructive evaluation of adhesively bonded carbon fiber reinforced composite lap joints with varied bond quality. *AIP Conf. Proc.* **2012**, *1430*, 1276–1283. [CrossRef]
8. Tighe, R.C.; Dulieu-Barton, J.M.; Quinn, S. Identification of kissing defects in adhesive bonds using infrared thermography. *Int. J. Adhes. Adhes.* **2016**, *64*, 168–178. [CrossRef]
9. Vijaya Kumar, R.L.; Bhat, M.R.; Murthy, C.R.L. Evaluation of kissing bond in composite adhesive lap joints using digital image correlation: Preliminary studies. *Int. J. Adhes. Adhes.* **2013**, *42*, 60–68. [CrossRef]

10. Vijaya kumar, R.L.; Bhat, M.R.; Murthy, C.R.L. Analysis of composite single lap joints using numerical and experimental approach. *J. Adhes. Sci. Technol.* **2014**, *28*, 893–914. [CrossRef]

11. Roth, W.; Giurgiutiu, V. Structural health monitoring of an adhesive disbond through electromechanical impedance spectroscopy. *Int. J. Adhes. Adhes.* **2017**, *73*, 109–117. [CrossRef]

12. Giurgiutiu, V. *Structural Health Monitoring with Piezoelectric Wafer Active Sensors*; Elsevier: Amsterdam, The Netherlands, 2008.

13. Ostachowicz, W.; Kudela, P.; Krawczuk, M.; Zak, A. *Guided Waves in Structures for SHM*; Wiley: Hoboken, NJ, USA, 2012; ISBN 9781119965855.

14. Rose, J.L. *Ultrasonic Guided Waves in Solid Media*; Cambridge University Press: New York, NY, USA, 2014; ISBN 9781107273610.

15. Rokhlin, S. Lamb wave interaction with lap-shear adhesive joints: Theory and experiment. *J. Acoust. Soc. Am.* **1991**, *89*, 2758–2765. [CrossRef]

16. Rose, J.L.; Rajana, K.M.; Hansch, M.K.T. Ultrasonic Guided Waves for NDE of Adhesively Bonded Structures. *J. Adhes.* **1995**, *50*, 71–82. [CrossRef]

17. Lowe, M.J.S.; Challis, R.E.; Chan, C.W. The transmission of Lamb waves across adhesively bonded lap joints. *J. Acoust. Soc. Am.* **2000**, *107*, 1333–1345. [CrossRef] [PubMed]

18. Di Scalea, F.L.; Bonomo, M.; Tuzzeo, D. Ultrasonic guided wave inspection of bonded lap joints: Noncontact method and photoelastic visualization. *Res. Nondestruct. Eval.* **2001**, *13*, 153–171. [CrossRef]

19. Ren, B.; Lissenden, C.J. Ultrasonic guided wave inspection of adhesive bonds between composite laminates. *Int. J. Adhes. Adhes.* **2013**, *45*, 59–68. [CrossRef]

20. Sherafat, M.H.; Quaegebeur, N.; Lessard, L.; Masson, P.; Hubert, P. Guided Wave Propagation through Composite Bonded Joints. In Proceedings of the EWSHM 7th European Workshop on Structural Health Monitoring, Nantes, France, 10 July 2014.

21. Ren, B.; Lissenden, C.J. Modal content-based damage indicators for disbonds in adhesively bonded composite structures. *Struct. Health Monit. Int. J.* **2016**, *15*, 491–504. [CrossRef]

22. Mori, N.; Biwa, S. Transmission of Lamb waves and resonance at an adhesive butt joint of plates. *Ultrasonics* **2016**, *72*, 80–88. [CrossRef] [PubMed]

23. Gauthier, C.; El-Kettani, M.; Galy, J.; Predoi, M.; Leduc, D.; Izbicki, J.L. Lamb waves characterization of adhesion levels in aluminum/epoxy bi-layers with different cohesive and adhesive properties. *Int. J. Adhes. Adhes.* **2017**, *74*, 15–20. [CrossRef]

24. Mallet, L.; Lee, B.C.; Staszewski, W.J.; Scarpa, F. Structural health monitoring using scanning laser vibrometry: II. Lamb waves for damage detection. *Smart Mater. Struct.* **2004**, *13*, 261–269. [CrossRef]

25. Kudela, P.; Wandowski, T.; Malinowski, P.; Ostachowicz, W. Application of scanning laser Doppler vibrometry for delamination detection in composite structures. *Opt. Lasers Eng.* **2016**, *99*, 46–57. [CrossRef]

26. Pieczonka, Ł.; Ambroziński, Ł.; Staszewski, W.J.; Barnoncel, D.; Pérès, P. Damage detection in composite panels based on mode-converted Lamb waves sensed using 3D laser scanning vibrometer. *Opt. Lasers Eng.* **2017**, *99*, 80–87. [CrossRef]

27. Sunarsa, T.Y.; Aryan, P.; Jeon, I.; Park, B.; Liu, P.; Sohn, H. A reference-free and non-contact method for detecting and imaging damage in adhesive-bonded structures using air-coupled ultrasonic transducers. *Materials (Basel)* **2017**, *10*, 1402. [CrossRef]

28. Esfandabadi, Y.K.; De Marchi, L.; Testoni, N.; Marzani, A.; Masetti, G. Full wavefield analysis and Damage imaging through compressive sensing in Lamb wave inspections. *IEEE Trans. Ultrason. Ferroelectr. Freq. Control* **2017**, *65*, 269–280. [CrossRef] [PubMed]

29. Dziendzikowski, M.; Dragan, K.; Katunin, A. Localizing impact damage of composite structures with modified RAPID algorithm and non-circular PZT arrays. *Arch. Civ. Mech. Eng.* **2017**, *17*, 178–187. [CrossRef]

30. Żak, A.; Radzieński, M.; Krawczuk, M.; Ostachowicz, W. Damage detection strategies based on propagation of guided elastic waves. *Smart Mater. Struct.* **2012**, *21*, 35024. [CrossRef]

31. Radzieński, M.; Doliński, Ł.; Krawczuk, M.; Palacz, M. Damage localisation in a stiffened plate structure using a propagating wave. *Mech. Syst. Signal Process.* **2013**, *39*, 388–395. [CrossRef]

32. Lee, C.; Park, S. Flaw Imaging Technique for Plate-Like Structures Using Scanning Laser Source Actuation. *Shock Vib.* **2014**, *725030*. [CrossRef]

33. Jothi Saravanan, T.; Gopalakrishnan, N.; Prasad Rao, N. Damage detection in structural element through propagating waves using radially weighted and factored RMS. *Meas. J. Int. Meas. Confed.* **2015**, *73*, 520–538. [CrossRef]
34. Geetha, G.K.; Roy Mahapatra, D.; Gopalakrishnan, S.; Hanagud, S. Laser Doppler imaging of delamination in a composite T-joint with remotely located ultrasonic actuators. *Compos. Struct.* **2016**, *147*, 197–210. [CrossRef]
35. Marks, R.; Clarke, A.; Featherston, C.; Paget, C.; Pullin, R. Lamb Wave Interaction with Adhesively Bonded Stiffeners and Disbonds Using 3D Vibrometry. *Appl. Sci.* **2016**, *6*, 12. [CrossRef]
36. Lee, C.; Zhang, A.; Yu, B.; Park, S. Comparison study between RMS and edge detection image processing algorithms for a pulsed laser UWPI (Ultrasonic wave propagation imaging)-based NDT technique. *Sensors (Switzerland)* **2017**, *17*, 1224. [CrossRef] [PubMed]
37. Abràmoff, M.D.; Magalhães, P.J.; Ram, S.J. Image processing with imageJ. *Biophotonics Int.* **2004**, *11*, 36–41. [CrossRef]
38. Alleyne, D.N.; Cawley, P. A 2-dimensional Fourier transform method for the quantitative measurement of Lamb modes. In Proceedings of the IEEE Ultrasonics Symposium, Honolulu, HI, USA, 4–7 December 1990; Volume 2, pp. 1143–1146. [CrossRef]
39. Bocchini, P.; Asce, M.; Marzani, A.; Viola, E. Graphical User Interface for Guided Acoustic Waves. *J. Comput. Civ. Eng.* **2011**, *25*, 202–210. [CrossRef]
40. Mazzotti, M.; Bartoli, I.; Marzani, A.; Viola, E. A coupled SAFE-2.5D BEM approach for the dispersion analysis of damped leaky guided waves in embedded waveguides of arbitrary cross-section. *Ultrasonics* **2013**, *53*, 1227–1241. [CrossRef] [PubMed]

 *applied sciences*

Article

Rapid High-Resolution Wavenumber Extraction from Ultrasonic Guided Waves Using Adaptive Array Signal Processing

Shigeaki Okumura [1,*], **Vu-Hieu Nguyen** [2], **Hirofumi Taki** [3], **Guillaume Haïat** [4], **Salah Naili** [2] **and Toru Sato** [1]

[1]	Graduate School of Informatics, Kyoto University, Kyoto 606-8501, Japan; sato.toru.6e@kyoto-u.ac.jp
[2]	Université Paris-Est, Laboratoire Modélisation et Simulation Multi Echelle, MSME UMR 8208 CNRS, 94010 Créteil CEDEX, France; vu-hieu.nguyen@u-pec.fr (V.-H.N.); naili@u-pec.fr (S.N.)
[3]	Graduate School of Biomedical Engineering, Tohoku University, Sendai 980-8575, Japan; hirofumi.taki@mb6.seikyou.ne.jp
[4]	CNRS, Laboratoire Modélisation et Simulation Multi Echelle, MSME UMR 8208 CNRS, 94010 Créteil CEDEX, France; haiat@u-pec.fr
*	Correspondence: sokumura@sato-lab.0t0.jp; Tel.: +81-75-753-3394

Received: 6 March 2018; Accepted: 16 April 2018; Published: 23 April 2018

Abstract: Quantitative ultrasound techniques for assessment of bone quality have been attracting significant research attention. The axial transmission technique, which involves analysis of ultrasonic guided waves propagating along cortical bone, has been proposed for assessment of cortical bone quality. Because the frequency-dependent wavenumbers reflect the elastic parameters of the medium, high-resolution estimation of the wavenumbers is required at each frequency with low computational cost. We use an adaptive array signal processing method and propose a technique that can be used to estimate the numbers of propagation modes that exist at each frequency without the need for time-consuming calculations. An experimental study of 4-mm-thick copper and bone-mimicking plates showed that the proposed method estimated the wavenumbers accurately with estimation errors of less than 4% and a calculation time of less than 0.5 s when using a laptop computer.

Keywords: signal processing; ultrasonic guided waves; axial transmission

---

1. Introduction

The first quantitative ultrasound (QUS) technique for use in bone assessment was proposed by Langton et al. [1] in the 1980s. Since then, a number of different QUS techniques have been proposed and developed [2–4]. QUS techniques were developed for early osteoporosis detection and fracture risk evaluation screening applications. When compared with X-rays, QUS offers several advantages: it is a noninvasive process and can be performed using portable and low-cost equipment. Techniques that can evaluate the quality of cortical long bone structures have recently attracted significant research attention [2,3,5–8]. In this study, we have developed cortical long bone quality evaluation methods using the axial transmission (AT) technique [9–13].

A number of recent studies have used the AT technique, in which wide-band signals are emitted and the ultrasonic guided wave that propagates along the cortical bone is then analyzed. The cortical bone has previously been described as a transversely isotropic absorbing plate with finite thickness [14–18]. The ultrasonic guided wave that propagates in cortical bone consists of multiple propagation modes, and the frequency-dependent wavenumbers of these modes represent the elastic properties of the medium. Many techniques for frequency-wavenumber (f–k) analysis have therefore been proposed and reported.

Tran et al. proposed the Radon transform method, which estimates the phase velocity (c_p) at each frequency [12]. The phase velocity is given by $c_p = 2\pi f/k$, and the basic premises of the f–k and f–c_p analysis methods are theoretically the same. The Radon transform method uses a single transmitter with multiple receivers, estimates the phase velocity using an iterative process, and produces high-resolution estimates. However, the computational cost of this method is not low because it requires inversion of a large-scale matrix. Sasso et al. [19] proposed a singular value decomposition (SVD)-based method and used a probe composed of multiple transmitters and emitters to analyze the guided waves. Minonzio et al. extended this method and produced accurate depictions of the phase velocities of the plate [10]. Use of SVD allows the method to detect the modes with low intensity. The method was effective for estimation of the elastic properties of the medium [20]. However, the measurement resolution is determined by the aperture size, and it is necessary to estimate the number of propagation modes in the received signal. Xu et al. proposed the sparse SVD (S-SVD) method, which combines the Radon transform method proposed by Tran et al. with the SVD method [21]. The resulting method acquires super-resolution estimates using a combination of SVD and an iterative process. However, this method also requires estimation of the number of propagation modes in the received signal and multiple calculations for inversion of a large matrix.

We recently developed an adaptive beamforming technique [22]. We used the estimation of signal parameters via rotational invariance technique (ESPRIT) algorithm. This algorithm uses eigenvalue decomposition to separate the desired signal from noise. The ESPRIT algorithm also requires estimation of the number of propagation modes and the conventional method includes an iterative process with eigenvalue decomposition [22]. Because the matrix inversion and eigenvalue decomposition processes incur large computational costs, reduction of these computational costs would be useful for practical applications. Therefore, in this study, we propose a method that estimates the number of propagation modes with a low computational cost.

The number of propagation modes in the received signal, which is denoted by M in this study, is usually estimated using a thresholding process. However, the eigenvalues that correspond to the signal are not equivalent to the signal intensities. Therefore, it is not easy to determine M when using a thresholding process that only uses the intensities of the eigenvalues or singular values. Therefore, in this study, we propose a new algorithm to estimate M using information theoretic criteria. This estimation procedure would be effective for both the SVD and ESPRIT methods. While many studies on estimation of M have been reported [23–27], the resulting methods were not applied to the analysis of ultrasonic guided waves when propagating along a transversely isotropic absorbing material such as cortical bone.

In the proposed method, we do not use an iterative process with high computational cost, such as the eigenvalue decomposition or matrix inversion methods, but use a diagonal loading (DL) technique that adds the diagonal loading matrix to the covariance matrix to determine M. We compare the computational cost of the proposed method with that of a conventional method and demonstrate the effectiveness of the proposed method via simple numerical simulations and experiments using a copper plate and a bone-mimicking plate.

2. Materials and Methods

2.1. System Model

In this study, we considered the system model that is shown in Figure 1. We placed a linear array probe on top of a free isotropic or transversely isotropic plate and analyzed the guided waves that propagate along the plate. When we consider multiple propagation signals with multiple phase velocities and ignore changes in amplitude, the received signals are given by

$$S_n(\omega) = S_1(\omega) \sum_{m=1}^{M(\omega)} \exp\{-j(n-1)k_m(\omega)x\} \tag{1}$$

where $S_n(\omega)$ is the received signal at the n-th receiver in the frequency domain with an angular frequency ω and x is the receiving array pitch. k_m is the wavenumber of the m-th propagation mode, which is given by $k_m = \omega/c_m$, where c_m is the phase velocity of the m-th mode. Note that M is dependent on the frequency.

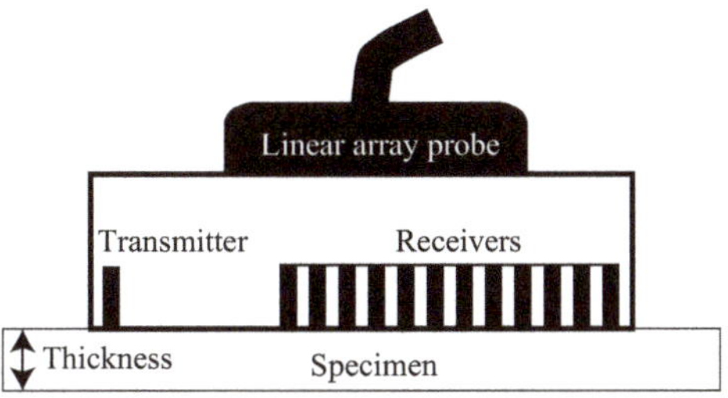

Figure 1. System model used in this study. In this study, we used a 4-mm-thick copper (homogeneous isotropic) plate and a 4-mm-thick bone mimicking (transversely isotropic) plate.

2.2. Wavenumber Estimation Using the ESPRIT Algorithm

To realize super-resolution estimates, we used the ESPRIT algorithm. The basis of this algorithm is briefly explained here. When there is a single propagating wave, i.e., $M = 1$ in Equation (1), we can then estimate the wavenumber by comparing the signals that are received at two receivers.

$$k_1(\omega) = \angle\{S_n(\omega)S_{n+1}^*(\omega)\}/x \tag{2}$$

where $\angle$ represents the angle of the complex signal.

As shown in Equation (2), the wavenumber is estimated directly without performing a peak search process. However, when $M > 1$, we cannot estimate the wavenumber by simply comparing the phases of the two received signals alone. The ESPRIT algorithm estimates the wavenumbers of multiple modes via an eigenvalue decomposition of the covariance matrix to enable the separation of multiple signals.

The covariance matrix represents a correlation between the signals at each receiver. To estimate the covariance matrix, we use the sub-array averaging technique [28]. In the sub-array averaging technique, multiple sub-arrays are situated into the full-size array. Here, we define the signal matrix as follows:

$$\mathbf{A}(\omega) = \begin{pmatrix} S_{1,1}(\omega) & \cdots & S_{N,1}(\omega) \\ \vdots & \ddots & \vdots \\ S_{1,N_{\mathrm{sub}}}(\omega) & \cdots & S_{N,N_{\mathrm{sub}}}(\omega) \end{pmatrix} \tag{3}$$

where $S_{i,j}$ is the received signal at the j-th receiver in the i-th sub-array, N is the number of sub-arrays contained within the whole array, N_{sub} is the number of receivers included within a single sub-array, i.e., the number of whole receivers is $N_{\mathrm{R}} = N + N_{\mathrm{sub}} - 1$, and $\mathbf{A}$ is the signal matrix with size $N_{\mathrm{sub}} \times N$.

The covariance matrix, $\mathbf{R}$, is then given by

$$\mathbf{R}(\omega) = \mathbf{A}(\omega)\mathbf{A}^{\mathrm{H}}(\omega) \tag{4}$$

$$= \sum_{i=1}^{N} \mathbf{R}_i(\omega) \tag{5}$$

$$= \sum_{i=1}^{N} \mathbf{S}_i(\omega)\mathbf{S}_i^{\mathrm{H}}(\omega) \tag{6}$$

where $\mathbf{S}_i(\omega) = [S_{i,1}(\omega), \cdots, S_{i,N_{\mathrm{sub}}}(\omega)]^{\mathrm{T}}$ is the signal vector at the i-th sub-array and ^{T} and ^{H} denote transpose and hermitian transpose of a matrix, respectively. As shown in Equations (5) and (6), the covariance matrix can be estimated by averaging the covariance matrices of the sub-arrays. Note that it is necessary to estimate M to enable estimation of the wavenumber.

The theoretical characteristics of the eigenvalues are expressed as follows:

$$l_1(\omega) \geq \cdots \geq l_{M(\omega)}(\omega) \geq l_{M(\omega)+1}(\omega) = \cdots = l_{N_{\mathrm{sub}}}(\omega) = \sigma_{\mathrm{n}}^2 \tag{7}$$

where $l_i(\omega)$ is i-th eigenvalue and σ_{n}^2 is the noise intensity. Note that the absolute values of these eigenvalues do not match the signal intensity directly, i.e., when two waves have the same intensity, the eigenvalues that correspond to these signals do not have the same value. Therefore, the simple thresholding process is not suitable for accurate estimation of the number of signals.

In this section, we calculate the covariance matrix and apply the eigenvalue decomposition technique. When we apply SVD to the signal matrix, $\mathbf{A}$, we theoretically obtain the same result as that obtained when using N transmitters and N_{sub} receivers with the SVD technique.

2.3. Estimation of the Number of Signals

2.3.1. Overview of the Basic Theory

In this section, we propose an estimation technique for $M(\omega)$. A method to estimate $M(\omega)$ that uses information theoretic criteria called the minimum description length (MDL) principle has previously been reported [23,24]. The evaluation index $G(m)$ that is used to estimate M is given by

$$G(m) = -\log \left(\frac{\prod_{i=m+1}^{N_{\mathrm{sub}}} l_i(\omega)^{\frac{1}{N_{\mathrm{sub}} - m}}}{\sum_{i=m+1}^{N_{\mathrm{sub}}} \frac{l_i(\omega)}{N_{\mathrm{sub}} - m}} \right)^{(N_{\mathrm{sub}} - m)N} + \frac{1}{2} m(2N_{\mathrm{sub}} - m) \log N \tag{8}$$

where $G(m)$ is the index, and the value of m that minimizes $G(m)$ represents the estimated $M(\omega)$. Here we define the estimated $M(\omega)$ as $M'(\omega)$.

2.3.2. Diagonal Loading Technique

In Equation (8), the numerator and the denominator represent the geometric and arithmetic means of the eigenvalues, respectively. Smaller eigenvalues therefore make the estimates unstable. To stabilize the estimation procedure, we thus use a diagonal loading technique that adds a diagonal matrix to the covariance matrix [25–27].

The DL process is expressed as follows:

$$\mathbf{R}'(\omega) = \mathbf{R}(\omega) + \eta(\omega)\mathbf{I} \tag{9}$$

where η is a diagonal loading factor and $\mathbf{I}$ is the identity matrix. The process that was shown in the previous equation can then be rewritten as follows:

$$l_i'(\omega) = l_i(\omega) + \eta(\omega) \tag{10}$$

where $l_i'(\omega)$ is the i-th eigenvalue that is calculated from $\mathbf{R}'(\omega)$. Therefore, even when a different DL factor is used with the same covariance matrix, we do not need to perform the eigenvalue decomposition process again. In other words, we use the DL technique to estimate M as an offset value and do not add the DL matrix to the covariance matrix directly. By replacing l_i in Equation (8) with l_i', we can then obtain the modified estimates.

2.3.3. Determination of the DL Factor

Selection of the DL factor is an important aspect of the estimation of M. In this study, we used two DL factors that were dependent on the received signal intensity. A schematic illustration of these two factors is shown in Figure 2.

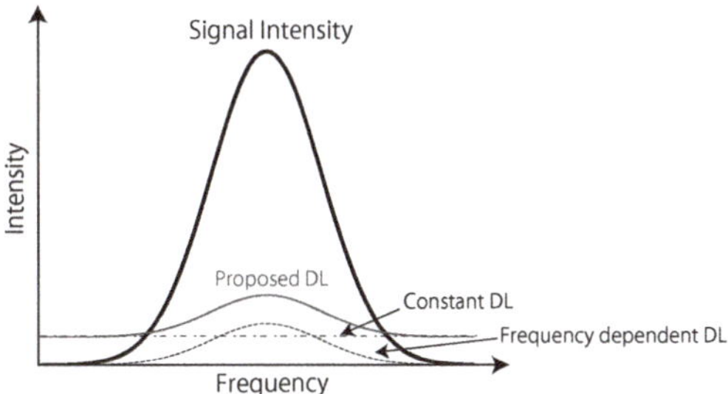

Figure 2. Proposed diagonal loading (DL) technique.

The DL factor is given by

$$\eta(\omega) = \eta_1(\omega)\delta_1(\omega) + \eta_2\delta_2 \tag{11}$$

$$\delta_1(\omega) = \text{tr}\{\mathbf{R}(\omega)\} \tag{12}$$

$$\delta_2 = \frac{1}{\omega_2 - \omega_1} \int_{\omega_1}^{\omega_2} \text{tr}\{\mathbf{R}(\omega)\}d\omega \tag{13}$$

where η_1 and η_2 are the two component DL factors. $\eta_2\delta_2$ is used for stabilization and is called the constant DL in Figure 2 because, at low signal intensities, $\eta_1(\omega)\delta_1(\omega)$, which is called the frequency-dependent DL in Figure 2, approaches zero, and the estimates would thus be unstable.

To determine $\eta_1(\omega)$ for the wavenumber estimation procedure, we assumed that the optimal $\eta_1(\omega)$ is common within a specific frequency range. This assumption was made because, as mentioned below, M' shows a step-like change with respect to η_1. The optimal η_1 that gives the optimum M value has a range and is not a critical value.

We vary $\eta_1(\omega)$ and select the minimum value of η_1 that gives $M'(\omega)$ such that it satisfies the following condition:

$$\max\{M'(\omega)\} < M_{\text{th}}, \text{ with } (\omega - \omega_{\text{w}} \leq \omega \leq \omega + \omega_{\text{w}}) \tag{14}$$

where ω_{w} is the width of the frequency window that is used for stable estimation and M_{th} is a threshold M value that is sufficiently large for wavenumber estimation.

2.4. Experimental Setup

The setup of the experimental study is the same as the system model shown at Figure 1, but we used a 128-element linear array probe with an element pitch of 0.375 mm that was manufactured by Japan Probe (Kanagawa, Japan). We selected a single transmitter and 28 receivers ($N_{\text{R}} = 28$) with an element pitch of 0.75 mm. The distance between the transmitter and the first receiver was 18.75 mm. The center frequency of the transmitted wave was 1.0 MHz. We used two test specimens: (1) a 4-mm-thick copper plate (with a shear wave velocity of 2260 m/s and a longitudinal wave velocity of 4650 m/s) and (2) a 4-mm-thick transversely isotropic bone-mimicking plate (Sawbones, Vashon, WA, USA). The elastic parameters of the bone-mimicking plate have been determined in previous studies [20,21]. The density, shear wave velocity, longitudinal wave velocity along fibers, and longitudinal wave velocity orthogonal to the fibers are 1.64 g·cm^{-3}, 1620 m/s, 3570 m/s, and 2910 m/s, respectively [20,21]. We put the probe parallel to the fibers.

In the proposed method, we set $M_{\text{th}} = N_{\text{sub}} - 1$, $\omega_{\text{w}} = 0.5$ MHz, $\eta_2 = -40$ dB, and $N_{\text{sub}} = 15$. We used $\omega_1 = 4.9$ MHz and $\omega_2 = 5.5$ MHz to select a frequency range that should include only noise.

We prepared the DL factor using values of η_1 within the -100 dB to 0 dB range at intervals of 10 dB. $M_{\text{th}} = N_{\text{sub}} - 1$ represents the maximum value of M when using a sub-array size of N_{sub}.

In the conventional SVD method, we used values of $N = 5$ and $N_{\text{sub}} = 24$ to match those used in the conventional study [21]. In the previous study, multiple transmitters were used. The use of multiple transmitters and of sub-arrays is theoretically the same.

3. Results and Discussion

3.1. Evaluation of the Number of the Propagation Modes with DL

To evaluate the M estimation procedure using the DL technique, we performed a simple simulation because it is difficult to know ground truth of M in experimental study. We assumed that three waves with $k = 1000, 2000$, and 3000 rad/m were propagating. The signal-to-noise ratio (SNR) was 40 dB, $N_{\text{R}} = 28$, and $N_{\text{sub}} = 15$.

The estimation results showed step-like changes with increasing η_1. Within the -100 dB $\leq \eta_1 \leq -90$ dB, -80 dB $\leq \eta_1 \leq -20$ dB, and -10 dB $\leq \eta_1 \leq 0$ dB ranges, the corresponding estimated M values were 14, 3, and 0, respectively. The root mean square error (RMSE) of the wavenumber estimation process with η_1 that gives the correct M is 0.73 rad/m. An important point that should be noted here is that the true M value is estimated using a wide range for η_1, e.g., -80 dB $\leq \eta_1 \leq -20$ dB.

3.2. Experimental Results

We first investigate the optimal size for the sub-array. Figure 3 shows the RMSEs when we use full-array sizes of $N_{\text{R}} = 28$ and 32 and change sub-array size. The measured specimen was bone-mimicking plate. The results show that the optimum sub-array size is $N = N_{\text{R}}/2$. Therefore, with $N_{\text{R}} = 28$, the value of $N_{\text{sub}} = 15$ is the optimal size.

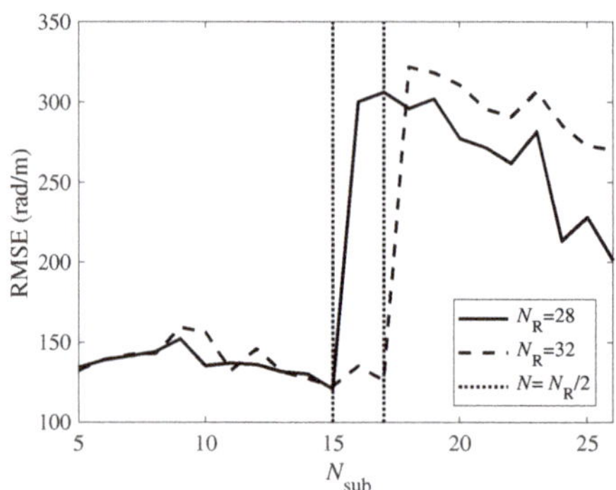

Figure 3. Dependence of the proposed method on sub-array size. The black solid line and the dotted line show the results for arrays composed of different numbers of receivers. The vertical black dotted lines show the horizontal values with $N = N_{\text{R}}/2$, i.e., $N_{\text{sub}} - 1 = N_{\text{R}}/2$.

In the previous paragraph, we described the experimental investigation of the optimal size. Here, we describe the theoretical investigation of this size [29]. To recover the rank of the covariance matrix with M modes, a minimum of M times averaging, i.e., $N = M$, is required. In addition, to measure M modes, a minimum of $N_{\text{sub}} = M + 1$ is required. Under these two conditions, when we attempt to maximize M and N_{sub}, the number of receivers N_{R} should be expressed using $N_{\text{R}} = N_{\text{sub}} + N - 1 = 2N$. Theoretically, we can confirm that the selection $N = N_{\text{R}}/2$ is reasonable.

We determined that $N_{sub} = 15$ is the optimal size. Therefore, the following results employed N_{sub} of 15. Figures 4–7 show the experimental results that were obtained for the copper plate and the bone-mimicking plate, respectively. The color maps show the results obtained using the conventional SVD method. The red dots represent estimated results obtained using the proposed method. The solid white lines show the theoretical curves. To remove any obvious false estimates, we removed all estimates with $c_p < 1200$ and $c_p > 50,000$ m/s. The proposed method thus successfully depicted the wavenumber without the need for peak search processes.

The RMSEs of the estimation results when using the proposed method with the copper plate and bone-mimicking plate were 108 rad/m and 121 rad/m, respectively. RMSEs are shown in Table 1. The wavenumbers of the shear waves that had larger effects on the theoretical curve than the longitudinal wave at the center frequency (1.0 MHz) of the copper plate and the bone-mimicking plate were 2780 rad/m and 3879 rad/m, respectively. When compared with the wavenumber of the shear wave, the RMSE was less than 4%.

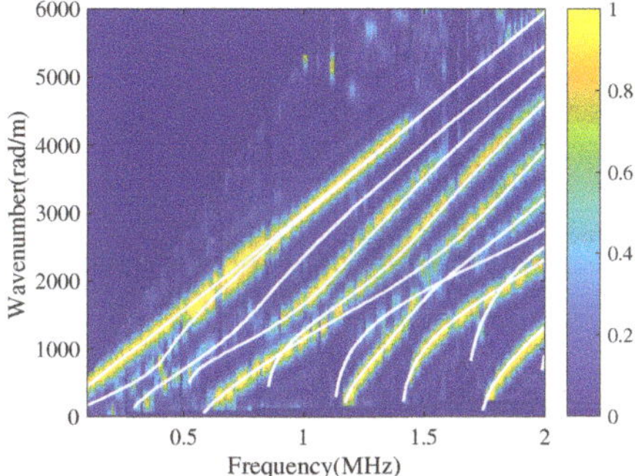

Figure 4. Wavenumbers of the 4-mm-thick copper plate estimated using the conventional singular value decomposition (SVD) method. The color intensity map shows the results obtained using the conventional SVD method. Solid white lines show the theoretical curves.

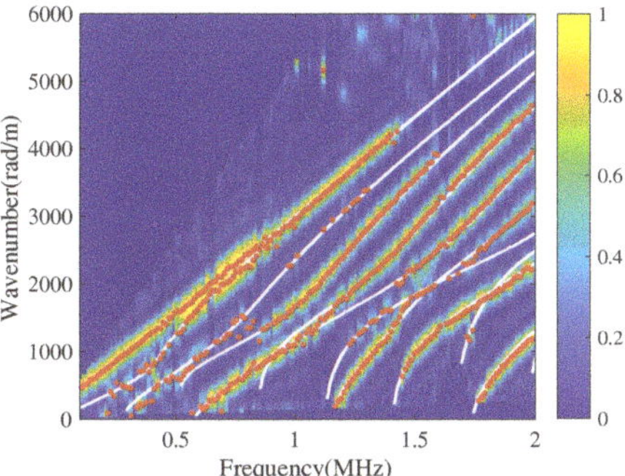

Figure 5. Wavenumbers of the 4-mm-thick copper plate estimated using the proposed method and the conventional SVD method. Red dots show the estimates obtained using the proposed method. The color intensity map shows the results obtained using the conventional SVD method. Solid white lines show the theoretical curves.

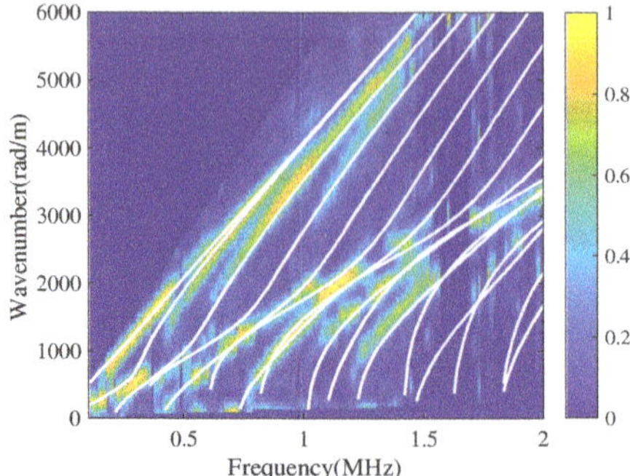

Figure 6. Wavenumbers of the 4-mm-thick bone-mimicking plate estimated using the conventional SVD method. The color intensity map shows the results obtained using the conventional SVD method. Solid white lines show the theoretical curves.

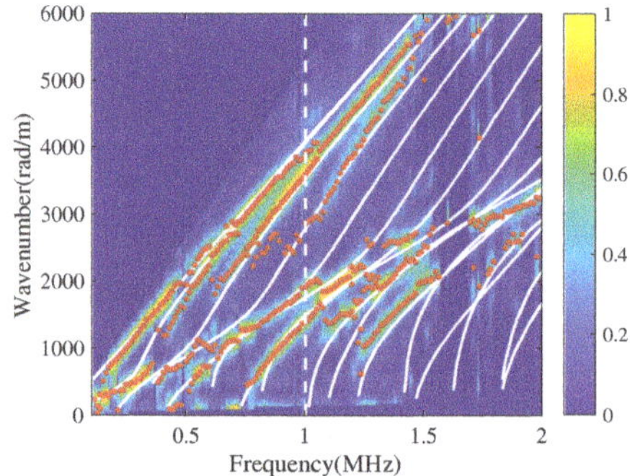

Figure 7. Wavenumbers of the 4-mm-thick bone-mimicking plate estimated using the proposed method and the conventional SVD method. Red dots show the estimates obtained using the proposed method. The color intensity map shows the results obtained using the conventional SVD method. Solid white lines show the theoretical curves.

Table 1. Root mean square errors (RMSEs) of the experimental results with different settings. The units are rad/m.

Proposed Method with Copper Plate	Proposed Method with Bone-Mimicking Plate	ESPRIT with Fixed Threshold (−40 dB)	ESPRIT with Fixed Threshold (−30 dB)	Proposed Method with $\omega_w = 0.1$ MHz	Proposed Method with $\omega_w = 0.75$ MHz
108	121	184	143	144	118

The spectrum at the frequency that is indicated by the white dotted line in Figure 7 is shown in Figure 8. As Figure 8 illustrates, the proposed method has higher resolution than the conventional method because the method can depict two modes around 4000 rad/m with small error. The wavenumbers that were estimated using the proposed method were almost the same as those at the peak position in the SVD spectrum when the resolution of the SVD method was high enough or the resolution of the ESPRIT algorithm was insufficient to separate the signals. However, it is difficult to identify some of the peaks because of their small amplitudes and low prominence when using the SVD method, such as the peaks with wavenumbers of approximately 2800 and 4300 rad/m.

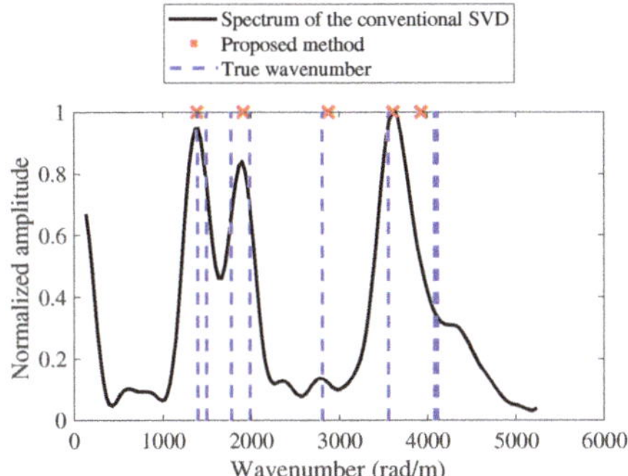

Figure 8. Depiction of the wavenumbers obtained using the proposed method and the spectrum obtained using the conventional method at 1.0 MHz. This is the cross-sectional view at the white dotted line shown in Figure 7. The vertical blue dotted lines indicate the theoretical values.

The wavenumber or phase velocity estimation steps that have high computational costs in the proposed method are eigenvalue decomposition of the covariance matrix with a size of $N_{\text{sub}} \times N_{\text{sub}}$, $\mathcal{O}(N_{\text{sub}}^3)$ and the SVD of the matrix of size $N_{\text{sub}} - 1 \times 2M$, $\mathcal{O}(\min\{(N_{\text{sub}} - 1) \times 4M^2, (N_{\text{sub}} - 1)^2 \times 2M\})$. The calculation steps are shown in detail in [30]. The corresponding steps in the conventional SVD method, the Radon transform method, and the S-SVD method are the SVD of the matrix of size $N \times N_{\text{sub}}$, $\mathcal{O}(N^2 \times N_{\text{sub}})$ [30], the inversion of the matrix of size $N_c \times N_c$, $\mathcal{O}(N_c^3)$, and the inversion of the matrix of size $N_k \times N_k$, $\mathcal{O}(N_k^3)$, respectively, where N_c and N_k are the numbers of sampling points in the phase velocity and wavenumber axes, which is normally larger than 64. Therefore, the SVD method showed the smallest computational cost with the lowest resolution. The computational cost of the proposed method was significantly lower than the corresponding costs of the Radon and S-SVD methods.

The calculation time (not including loading time) for the experimental data was less than 0.5 s when using a commercial processor (Core(TM) i7-7500U; Intel, Santa Clara, CA, USA) on a laptop computer with MATLAB (R2017a, Mathworks, Natick, MA, USA) platform. The calculation time taken for the conventional Radon transform was less than 60 s, according to [12]. While noting here that we did not use the same processor or computer, and in fact used a faster processor, the reduction in the calculation time of 99% is interesting. In addition, the computational cost of the S-SVD method is greater than that of the Radon method. The calculation time of the conventional SVD method is around 0.1 s. Note that, as mentioned above, the conventional SVD method has considerably lower computational complexity than the proposed method. Because we ran the SVD method and the proposed method on MATLAB platform, the calculation time does not directly reflect the theoretical complexity. To examine the effectiveness of the proposed method in determination of M, we compared the performance of the proposed method with that of the ESPRIT method with a fixed threshold. Figures 9 and 10 show the results obtained for a bone plate when we used eigenvalues that exceed -40 and -30 dB of the maximum eigenvalue, respectively. The red circles represent the results obtained using the proposed method. The blue cross marks show the estimates obtained using the ESPRIT algorithm with a fixed threshold. The solid gray lines show the theoretical curves. The RMSEs of the ESPRIT method with fixed thresholds of -40 and -30 dB were 184 and 143 rad/m, respectively. As shown in Figure 9, smaller thresholds tend to make the estimates unstable because the ESPRIT algorithm treated the noise as the desired signal. In contrast, larger thresholds can miss the weaker modes as shown in Figure 10. The threshold should thus be determined manually. The results are sensitive to the threshold value, so it is difficult to determine the threshold.

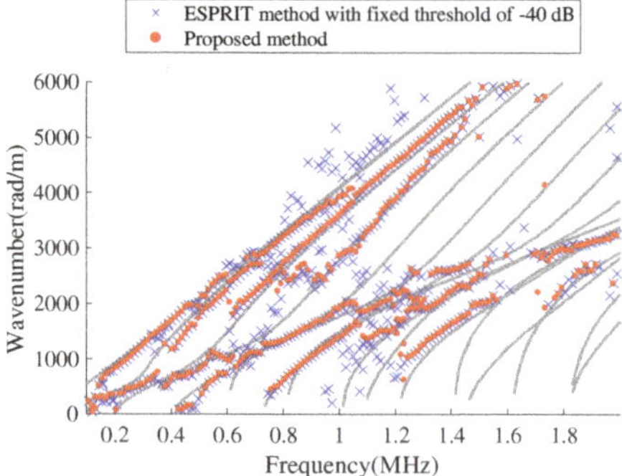

Figure 9. Wavenumbers of the 4-mm-thick bone-mimicking plate estimated using the conventional estimation of signal parameters via rotational invariance technique (ESPRIT) algorithm at a fixed threshold of -40 dB and the proposed method. Red dots show the estimates obtained using the proposed method and blue cross marks show the estimates obtained using the conventional method. Solid gray lines show the theoretical curves.

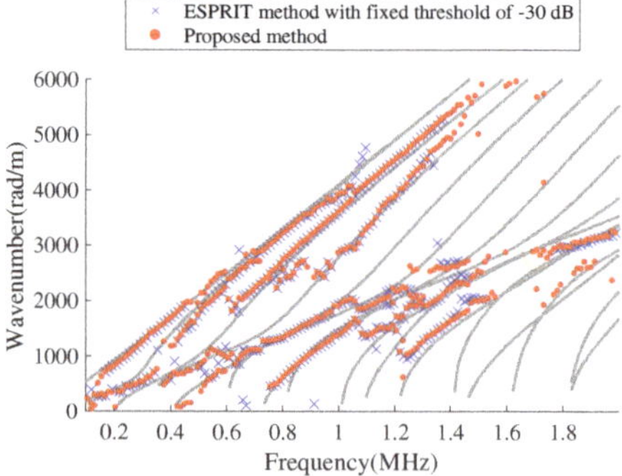

Figure 10. Wavenumbers of the 4-mm-thick bone-mimicking plate estimated using the conventional ESPRIT algorithm at a fixed threshold of -30 dB and the proposed method. Red dots show the estimates obtained using the proposed method and blue cross marks show the estimates obtained using the conventional method. Solid gray lines show the theoretical curves.

Figure 11 shows the wavenumbers for values of $N_{sub} = 5$ and 26. The red cross marks and blue circles show the results obtained for $N_{sub} = 5$ and 26, respectively. The black lines show the theoretical curves. The smaller sub-array had lower resolution and missed many of the modes, while the larger sub-array caused false estimates.

We used a frequency window, denoted by ω_w, of 0.5 MHz. When we used ω_w values of 0.1 and 0.75 MHz in the experiments with the bone-mimicking plate, the RMSEs obtained were 144 and 118 rad/m, respectively. Therefore, the dependence on ω_w was insignificant and a larger window width provided stable estimation.

We chose a DL factor of $\eta_2 = -40$ dB. Because this value was added for stabilization when we analyzed the frequency range using a significantly smaller signal intensity, a smaller value was selected. When we used a larger value, more stable estimates were obtained as a result but the weaker modes were missed by the method.

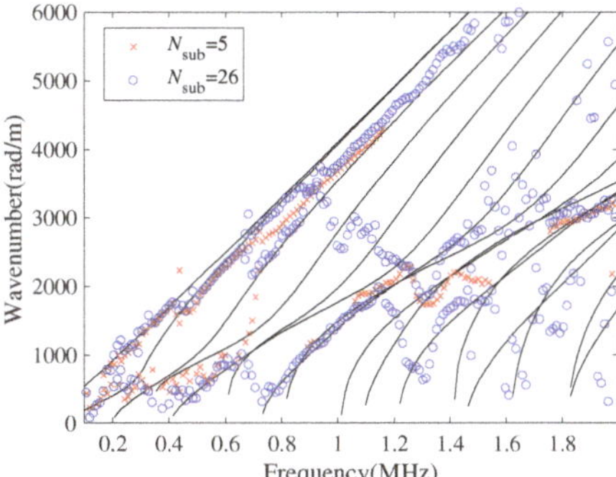

Figure 11. Wavenumbers estimated for differently sized sub-arrays. Red cross marks and blue circles show the results obtained for $N_{sub} = 5$ and 26, respectively. Black lines show the theoretical curves.

We used the ESPRIT algorithm in this study. The novel advantage of the ESPRIT algorithm when used for wavenumber estimation of a guided wave propagating along the cortical bone is that it can estimate the wavenumber without the need for a peak search process. To extract the wavenumber from the f–k spectrum image when using conventional methods such as the Radon and SVD-based methods, multiple one-dimensional peak search processes are required [6,20]. While the computational cost is not high, parameters such as the intensity threshold and prominence must be set. Thus, the performance of the conventional method such as RMSE depends on the additional parameter setting.

4. Conclusions

In this study, we proposed a high-resolution and low-computational-cost technique for an AT device. We estimated the number of propagation modes using information theoretic criteria and the DL technique. We proposed a method to estimate the optimal DL value required for guided wave estimation. The proposed method did not involve processes of high computational cost. The proposed method was evaluated experimentally using 4-mm-thick copper and bone-mimicking plates. The estimation error was less than 4% and the calculation time when using the proposed method on a laptop computer was less than 0.5 s. While the processor is different and is in fact faster than that which was used in the previous conventional study, the calculation time is less than 1% of that of the conventional method. We believe that our proposed method thus has the potential to accurately characterize the elastic properties of cortical bone.

Acknowledgments: This work was supported in part by a Grant-in-Aid for Scientific Research (A) (Grant No. 25249057), a Grant-in-Aid for JSPS Fellows (Grant No. 15J05687), and the PEPS program (15R03051 AMETCARMAT) from the Université Paris-Est. This project received funding from the European Research Council (ERC) under the European Union's Horizon 2020 Research and Innovation Program (Grant Agreement No. 682001, project ERC Consolidator Grant 2015 BoneImplant). We thank David MacDonald, MSc, from Edanz Group (www.edanzediting.com/ac) for editing a draft of this manuscript.

Author Contributions: Shigeaki Okumura analyzed the data and wrote the paper; Vu-Hieu Nguyen contributed to the basic conception and discussion and gave feedback on the manuscript; Hirofumi Taki contributed to the feedback on the signal processing algorithm and the manuscript; Guillaume Haïat contributed to the discussion and gave feedback on the manuscript from an application perspective; Salah Naili contributed to the discussion and provided feedback on the manuscript; Toru Sato contributed to the design of the signal processing algorithm and gave feedback on the manuscript.

Conflicts of Interest: The authors declare no conflict of interest. The founding sponsors had no role in the design of the study; in the collection, analyses, or interpretation of data; in the writing of the manuscript; or in the decision to publish the results.

Abbreviations

The following abbreviations are used in this manuscript:

QUS   quantitative ultrasound
AT   axial transmission
f–k frequency-wavenumber
SVD   singular value decomposition
S-SVD   sparse singular value decomposition
ESPRIT   estimation of signal parameters via rotational invariance technique
SNR   signal-to-noise ratio
MDL   minimum description length
DL   diagonal loading
RMSE   root mean square error

References

1. Langton, C.; Palmer, S.; Porter, R. The measurement of broadband ultrasonic attenuation in cancellous bone. *Eng. Med.* **1984**, *13*, 89–91. [CrossRef]
2. Mano, I.; Horii, K.; Takai, S.; Suzaki, T.; Nagaoka, H.; Otani, T. Development of novel ultrasonic bone densitometry using acoustic parameters of cancellous bone for fast and slow waves. *Jpn. J. Appl. Phys.* **2006**, *45*, 4700–4702. [CrossRef]
3. Otani, T. Quantitative estimation of bone density and bone quality using acoustic parameters of cancellous bone for fast and slow waves. *Jpn. J. Appl. Phys.* **2005**, *44*, 4578, [CrossRef]
4. Raum, K.; Grimal, Q.; Varga, P.; Barkmann, R.; Glüer, C.C.; Laugier, P. Ultrasound to assess bone quality. *Curr. Osteoporos. Rep.* **2014**, *12*, 154–162. [CrossRef]
5. Ta, D.; Liu, C. Ultrasonic backscatter theory, method, and diagnostic instrument for cancellous bone. *J. Acoust. Soc. Am.* **2016**, *140*, 3190–3190. [CrossRef]
6. Bochud, N.; Vallet, Q.; Minonzio, J.G.; Laugier, P. Predicting bone strength with ultrasonic guided waves. *Sci. Rep.* **2017**, *7*, 43628, [CrossRef]
7. Wear, K.A.; Nagaraja, S.; Dreher, M.L.; Sadoughi, S.; Zhu, S.; Keaveny, T.M. Relationships among ultrasonic and mechanical properties of cancellous bone in human calcaneus in vitro. *Bone* **2017**, *103*, 93–101. [CrossRef]
8. Okada, S.; Kawano, A.; Oue, H.; Takeda, Y.; Yokoi, M.; Koretake, K.; Tsuga, K. Preoperative evaluation of bone quality for dental implantation using an ultrasound axial transmission device in an ex vivo model. *Clin. Exp. Dent. Res.* **2017**, *3*, 81–86, [CrossRef]
9. Xu, K.; Ta, D.; Cassereau, D.; Hu, B.; Wang, W.; Laugier, P.; Minonzio, J.G. Multichannel processing for dispersion curves extraction of ultrasonic axial-transmission signals: Comparisons and case studies. *J. Acoust. Soc. Am.* **2016**, *140*, 1758–1770. [CrossRef]
10. Minonzio, J.G.; Talmant, M.; Laugier, P. Guided wave phase velocity measurement using multi-emitter and multi-receiver arrays in the axial transmission configuration. *J. Acoust. Soc. Am.* **2010**, *127*, 2913–2919. [CrossRef]
11. Minonzio, J.G.; Foiret, J.; Talmant, M.; Laugier, P. Impact of attenuation on guided mode wavenumber measurement in axial transmission on bone mimicking plates. *J. Acoust. Soc. Am.* **2011**, *130*, 3574–3582. [CrossRef]
12. Tran, T.N.; Nguyen, K.C.T.; Sacchi, M.D.; Le, L.H. Imaging ultrasonic dispersive guided wave energy in long bones using linear Radon transform. *Ultrasound Med. Biol.* **2014**, *40*, 2715–2727. [CrossRef]
13. Tran, T.N.; Le, L.H.; Sacchi, M.D.; Nguyen, V.H.; Lou, E.H. Multichannel filtering and reconstruction of ultrasonic guided wave fields using time intercept-slowness transform. *J. Acoust. Soc. Am.* **2014**, *136*, 248–259. [CrossRef]
14. Dong, X.N.; Guo, X.E. The dependence of transversely isotropic elasticity of human femoral cortical bone on porosity. *J. Biomech.* **2004**, *37*, 1281–1287. [CrossRef]
15. Naili, S.; Vu, M.B.; Grimal, Q.; Talmant, M.; Desceliers, C.; Soize, C.; Haïat, G. Influence of viscoelastic and viscous absorption on ultrasonic wave propagation in cortical bone: Application to axial transmission. *J. Acoust. Soc. Am.* **2010**, *127*, 2622–2634. [CrossRef]

16. Haïat, G.; Naili, S.; Grimal, Q.; Talmant, M.; Desceliers, C.; Soize, C. Influence of a gradient of material properties on ultrasonic wave propagation in cortical bone: Application to axial transmission. *J. Acoust. Soc. Am.* **2009**, *125*, 4043–4052. [CrossRef]

17. Nguyen, V.H.; Naili, S. Ultrasonic wave propagation in viscoelastic cortical bone plate coupled with fluids: A spectral finite element study. *Comput. Meth. Biomech. Biomed. Eng.* **2013**, *16*, 963–974. [CrossRef]

18. Nguyen, V.H.; Tran, T.N.; Sacchi, M.D.; Naili, S.; Le, L.H. Computing dispersion curves of elastic/viscoelastic transversely-isotropic bone plates coupled with soft tissue and marrow using semi-analytical finite element (SAFE) method. *Comput. Biol. Med.* **2017**, *87*, 371–381. [CrossRef]

19. Sasso, M.; Haiat, G.; Talmant, M.; Laugier, P.; Naili, S. Singular value decomposition-based wave extraction in axial transmission: Application to cortical bone ultrasonic characterization [correspondence]. *IEEE Trans. Ultrason. Ferroelectr. Freq. Control* **2008**, *55*, 1328–1332. [CrossRef]

20. Foiret, J.; Minonzio, J.G.; Chappard, C.; Talmant, M.; Laugier, P. Combined estimation of thickness and velocities using ultrasound guided waves: A pioneering study on in vitro cortical bone samples. *IEEE Trans. Ultrason. Ferroelectr. Freq. Control* **2014**, *61*, 1478–1488. [CrossRef]

21. Xu, K.; Minonzio, J.G.; Ta, D.; Hu, B.; Wang, W.; Laugier, P. Sparse SVD method for high-resolution extraction of the dispersion curves of ultrasonic guided waves. *IEEE Trans. Ultrason. Ferroelectr. Freq. Control* **2016**, *63*, 1514–1524. [CrossRef]

22. Okumura, S.; Nguyen, V.H.; Taki, H.; Haïat, G.; Naili, S.; Sato, T. Phase velocity estimation technique based on adaptive beamforming for ultrasonic guided waves propagating along cortical long bones. *Jpn. J. Appl. Phys.* **2017**, *56*, 07JF06. [CrossRef]

23. Wax, M.; Kailath, T. Detection of signals by information theoretic criteria. *IEEE Trans. Acoust. Speech Signal Process.* **1985**, *33*, 387–392. [CrossRef]

24. Wax, M. Detection and localization of multiple sources via the stochastic signals model. *IEEE Trans. Signal Process.* **1991**, *39*, 2450–2456. [CrossRef]

25. Lombardini, F.; Gini, F. Model order selection in multi-baseline interferometric radar systems. *EURASIP J. Adv. Signal Process.* **2005**, *2005*, 108784. [CrossRef]

26. Huang, Y.; Ferro-Famil, L.; Reigber, A. Under-foliage object imaging using SAR tomography and polarimetric spectral estimators. *IEEE Trans. Geosci. Remote Sens.* **2012**, *50*, 2213–2225. [CrossRef]

27. Sauer, S.; Ferro-Famil, L.; Reigber, A.; Pottier, E. Physical parameter extraction over urban areas using L-band POLSAR data and interferometric baseline diversity. In Proceedings of the 2007 IEEE International IGARSS Geoscience and Remote Sensing Symposium, Barcelona, Spain, 23–28 July 2007; pp. 5045–5048.

28. Takao, K.; Kikuma, N. An adaptive array utilizing an adaptive spatial averaging technique for multipath environments. *IEEE Trans. Antennas Propag.* **1987**, *35*, 1389–1396. [CrossRef]

29. Shan, T.J.; Wax, M.; Kailath, T. On spatial smoothing for direction-of-arrival estimation of coherent signals. *IEEE Trans. Acoust. Speech Signal Process.* **1985**, *33*, 806–811. [CrossRef]

30. Saarnisaari, H. TLS-ESPRIT in a time delay estimation. In Proceedings of the 1997 IEEE 47th Vehicular Technology Conference, Phoenix, AZ, USA, 4–7 May 1997; Volume 3, pp. 1619–1623.

Article

Forward and Inverse Studies on Scattering of Rayleigh Wave at Surface Flaws

Bin Wang, Yihui Da and Zhenghua Qian *

State Key Laboratory of Mechanics and Control of Mechanical Structures/College of Aerospace Engineering, Nanjing University of Aeronautics and Astronautics, Nanjing 210016, China; wangbin1982@nuaa.edu.cn (B.W.); dayihui@163.com (Y.D.)

* Correspondence: qianzh@nuaa.edu.cn; Tel.: +86-25-8489-5952

Received: 21 February 2018; Accepted: 9 March 2018; Published: 12 March 2018

Featured Application: Non-destructive evaluation.

Abstract: The Rayleigh wave has been frequently applied in geological seismic inspection and ultrasonic non-destructive testing, due to its low attenuation and dispersion. A thorough and effective utilization of Rayleigh wave requires better understanding of its scattering phenomenon. The paper analyzes the scattering of Rayleigh wave at the canyon-shaped flaws on the surface, both in forward and inverse aspects. Firstly, we suggest a modified boundary element method (BEM) incorporating the far-field displacement patterns into the traditional BEM equation set. Results show that the modified BEM is an efficient and accurate approach for calculating far-field reflection coefficients. Secondly, we propose an inverse reconstruction procedure for the flaw shape using reflection coefficients of Rayleigh wave. By theoretical deduction, it can be proved that the objective function of flaw depth $d(x_1)$ is approximately expressed as an inverse Fourier transform of reflection coefficients in wavenumber domain. Numerical examples are given by substituting the reflection coefficients obtained from the forward analysis into the inversion algorithm, and good agreements are shown between the reconstructed flaw images and the geometric characteristics of the actual flaws.

Keywords: surface flaw; Rayleigh wave; scattering; modified BEM; reconstruction

---

1. Introduction

The non-destructive testing and structural health monitoring are of great importance in civil, mechanical, and aerospace engineering industries. Corrosions, pittings, and cracks may occur during service years of structures, and cause threats to public life and possession safety. Ultrasonic inspection is one of the fundamental techniques for locating surface and hidden flaws. Compared with traditionally-applied bulk waves, the Rayleigh surface wave has low attenuation and high energy concentration, which makes it suitable for inspection of defects or abnormalities on or close to surface. On the other hand, Rayleigh waves are not dispersive, as compared with other guided waves such as Lamb, SH, and Love waves, thus easier for signal processing [1,2].

Due to the great potential of Rayleigh waves in NDT (Non-Destrctive Testing) applications, it is necessary to examine their scattering and diffraction behaviors, both in the forward and inverse aspects. Researchers have long been interested in forward calculation of scattering phenomena for Rayleigh waves. One of pioneering work can be attributed to Ayster and Auld's paper of calculating the scattered field of Rayleigh waves by surface crack [3], which is followed by Achenbach, who discussed the emission of surface waves from a crack by cyclic loading in detail [4]. Furthermore, Lee and Liu deduced the analytical solution of a scattered Rayleigh wave field by semi-circular canyons [5]. However, theoretical calculations are only applicable for special cases. For more general situations, numerical approaches are required. As a widely applied numerical method, finite element method (FEM) has been applied for Rayleigh wave field simulation as it passes through cracks [6], slots [7], and notches [8] on the flat surface (of a large yet finite solid block). The FEM mesh was performed within a whole elastic solid block [7,8] or a confined region of an infinite body with absorption boundaries [6].

Although FEM is a sophisticated numerical approach, it is time- and memory-consuming for scattering problems in a half-space, especially when far-field behaviors need to be examined. Thus, some researchers turned to the boundary element method (BEM), in which only the boundaries need to be meshed. Their early works include Wong [9] and Kawase's [10] applications of discrete wavenumber BEM for calculation of scattered Rayleigh wave fields in frequency domains. Sanchez-Sesma et al. [11] and Ávila-Carrera et al. [12] used indirect BEM to solve similar problems. For half-space problems, there have been two main modeling patterns: (i) analyzing and meshing the flaw surface only, with usage of half-space Green's functions, in which the interface boundary conditions are satisfied automatically; (ii) analyzing and meshing the flaw surface and the whole interface, with usage of full-space Green's functions. For the former model, the element is fewer, however, the format of Green's functions tends to be complicated, usually not in a closed form, and requires a large amount of resources for contour or numerical integration over complex wavenumber domains [10,13]. On the other hand, for the latter model, although the whole interface is incorporated, the Green's functions are simpler and easier to apply in the BEM calculation. Thus, many recent researchers used the latter modeling approach to study the scattering of Rayleigh and Lamb waves [14–17]. But researchers often encounter the problem of "spurious reflection" because of the model truncation at far ends [18,19]. Arias [18] suggested a modified BEM method in which a far-field displacement pattern had been introduced and substituted into the BEM equation set. By enforcing displacement and stress continuous relations, the researchers were able to solve the scattering wave field while suppressing spurious reflections. However, Arias' method might not be accurate for calculating far-field reflection and transmission coefficients. In this paper, we will propose an improved modified BEM method.

The inverse analysis is an important aspect and fundamental aim for studying the scattering of Rayleigh wave. Many researchers have tried to relate the amplitude and time-of-flight (TOF) of reflected wave with surface-breaking cracks [20–23] or notches [24,25], both in experimental and numerical aspects, in order to achieve flaw location and sizing effect. These researchers mainly adopted pitch-catch configurations, and were able to detect the positions and approximate sizes of defects, but lacked further information, such as depth, width, and severity.

In order to obtain more information about defects and abnormalities in the near-surface layer, geophysical scientists and engineers have tried tomographic reconstruction approaches using data from scattered Rayleigh wave fields. Xia et al. [26] firstly used Rayleigh wave phase velocities at different frequencies to reconstruct the S-wave velocity profile of multiple elastic layers on top of the half-space. Furthermore, Xia et al. [27] and Shao et al. [28] extended the tomographic approach to the detection of near-surface shallow cavities. Köhn et al. [29] used a similar inversion method to reconstruct surface layer stiffness of an ancient sandstone façade. Chen et al. [30] adopted an inversion procedure to detect residual stresses from general Rayleigh wave dispersion of a weakly anisotropic media. These methods stem from wave signal processing theories (in earlier works [27,28]) or discretized equations of motion (in later works [29,30]). Through the formulation of an "operator matrix" of inverse problem, the researchers try to find an optimum solution of the "model vector" that is feasible to the "data set". Due to strong nonlinearity of general inverse problems, one needs to perform numerous iterations to achieve an accurate solution.

In this paper, we will propose a convenient inverse method to reconstruct the location and geometric properties of surface-opening flaws. The method is subsequent research for inversion method for plate-thinning reconstruction using SH-waves [31,32] and Lamb waves [33]. We make use of the reflection coefficients of Rayleigh waves obtained from the forward analysis in different frequencies as the input data. These data can also be attained by experiments. By introduction of Born approximation and far-field expressions of half-space Green's functions, and the usage of boundary integral equation over the flaw area, it can be proved that the objective function of flaw depth $d(x_1)$ is approximately expressed as an inverse spatial Fourier transform of reflection coefficients in wavenumber domain. This inversion procedure is especially suitable for reconstruction of weak and shallow notches, and can also be used as an initial-step image for deeper and steeper canyons.

The paper will be organized as follows: In Section 2, we will discuss about the improved forward problem using modified BEM in Section 2.1, and give some numerical examples and compare with existing results in Section 2.2. In Section 3, we will elaborate on the mathematical formula for inverse reconstruction method for surface-breaking canyon-shaped flaws in Section 3.1, and parametric influences will be discussed in detail in Section 3.2. We will give conclusions in Section 4.

2. Forward Problem by Modified BEM

2.1. Formulation of Modified BEM

The problem to be solved is shown in Figure 1. An artificial canyon-shaped flaw has been cut on the surface of an isotropic elastic half-space. The incident Rayleigh wave propagates from left to the right side, and the reflected wave is observed at the far end of the left side.

For accurately computing the scattering wave field, we divide the forward analysis into two major steps, as shown in Figure 2.

(i) A time-harmonic incident Rayleigh wave field in the intact half-space is considered, and the correspondent stress σ_{ij}^{inc} is calculated along the curve Γ_1^-, which is actually the flaw surface.

(ii) Since the actual traction on Γ_1^- is zero, the stress σ_{ij}^{inc} is released by exerting a compensation traction $t_i^{sca} = -\sigma_{ij}^{inc} n_j$ along Γ_1^-. The scattering problem is transformed into the question of solving the wave field in the flawed half-space subjected to load t_i^{sca}.

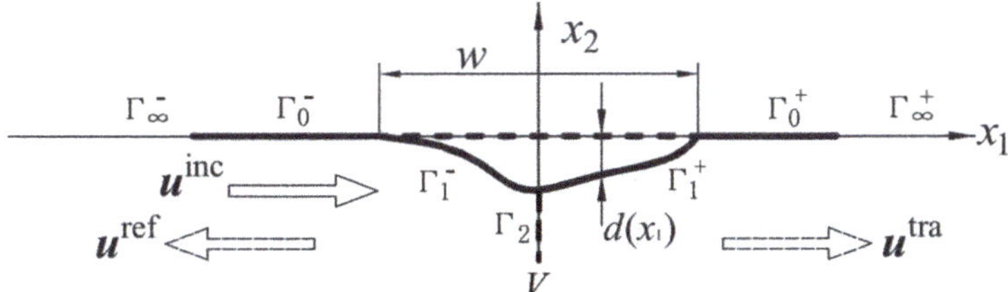

Figure 1. Problem configuration.

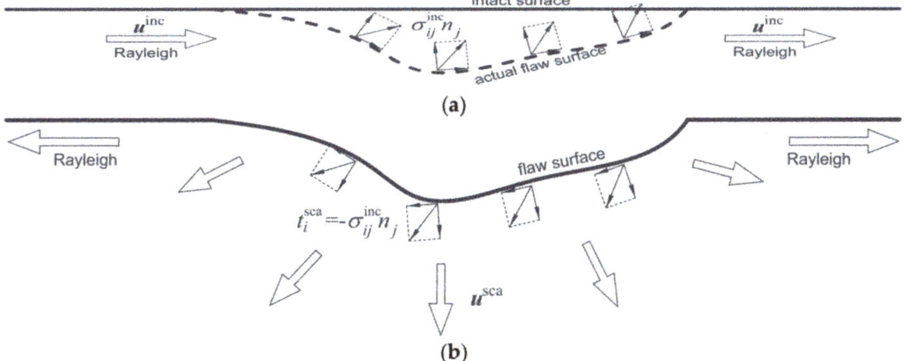

Figure 2. Two steps of calculating scattering wave field, (**a**) Step 1; (**b**) Step 2.

Then, the displacement in frequency domain of scattering wave field at an arbitrary point can be expressed in the form of boundary integral equation (BIE), as shown by Equation (1) [34].

$$c_{pi}(X)u_i^{\text{sca}}(X,\Omega) = \int_\Gamma \left[G_{ip}(x,X,\Omega)t_i^{\text{sca}}(x,\Omega) - T_{ip}(x,X,\Omega)u_i^{\text{sca}}(x,\Omega) \right]d\Gamma(x)\,(i,p=1,2) \tag{1}$$

where u_i^{sca} and t_i^{sca} are displacement and boundary traction, respectively. $G_{ip}(x,X,\Omega)$ and $T_{ip}(x,X,\Omega)$ are fundamental solutions representing the i-th component of the displacement and stress at point x due to a unit time-harmonic concentrated load of angular frequency Ω exerted in the p-th direction at point X in the elastic full-space. The coefficient $c_{pi}(X) = \delta_{pi}$ for $X \in V$, $c_{pi}(X) = \delta_{pi}/2$ for point $X \in \Gamma$ which is smooth locally, $c_{pi}(X) = 0$ for $X \notin V \cup \Gamma$. It should be noted that here the boundary Γ includes both the flaw area and the intact infinite surface, as the full-space fundamental solutions have been chosen.

Now the whole surface is divided into six parts, Γ_∞^- and Γ_∞^+ tending to infinity on the left and right side, respectively; Γ_0^- and Γ_0^+ representing the near-field boundaries; Γ_1^- and Γ_1^+ the left and right half of the flaw surface, respectively. Consequently, the scattered wave field is rewritten from Equation (1):

$$\begin{aligned}
c_{pi}(X)u_i^{\text{sca}}(X,\Omega) = &-\int_{\Gamma_\infty^- \cup \Gamma_\infty^+} T_{ip}(x,X,\Omega)u_i^{\text{sca}}(x,\Omega)d\Gamma(x) - \int_{\Gamma_0^- \cup \Gamma_0^+} T_{ip}(x,X,\Omega)u_i^{\text{sca}}(x,\Omega)d\Gamma(x) \\
&+ \int_{\Gamma_1^- \cup \Gamma_1^+} \left[G_{ip}(x,X,\Omega)t_i^{\text{sca}}(x,\Omega) - T_{ip}(x,X,\Omega)u_i^{\text{sca}}(x,\Omega) \right]d\Gamma(x), \quad (i,p=1,2)
\end{aligned} \tag{2}$$

where the traction-free boundary condition on $\Gamma_0^- \cup \Gamma_0^+ \cup \Gamma_\infty^- \cup \Gamma_\infty^+$ has been taken into account.

For traditional BEM, only the near-field boundaries $\Gamma_0^- \cup \Gamma_0^+ \cup \Gamma_1^- \cup \Gamma_1^+$ are meshed, thus, the first item on the right hand side of Equation (2) is omitted. This treatment can cause spurious reflected waves at the truncation points which do not attenuate in the 2D half-space, which are unwanted.

It is noticed that, if the truncation point is far enough from the flaw area, the displacement fields on the infinite boundaries will follow a Rayleigh wave pattern travelling out of the flaw region [18], as expressed by

$$\begin{aligned}
u_i^{\text{sca}}(x,\Omega) &\approx R^-(\Omega)\tilde{u}_i^-(x,\Omega), \quad x \in \Gamma_\infty^- \\
u_i^{\text{sca}}(x,\Omega) &\approx R^+(\Omega)\tilde{u}_i^+(x,\Omega), \quad x \in \Gamma_\infty^+
\end{aligned} \tag{3}$$

where $\tilde{u}_i^\pm$ is the displacement field of the unitary Rayleigh wave travelling in positive (+) or negative (−) directions, and $R^\pm$ the coefficient to be solved. Equation (3) makes it possible to take the far-field integrals in Equation (2) into consideration. Then, for any point on the boundary which is locally smooth, we have

$$\begin{aligned}
&\tfrac{1}{2}u_p^{\text{sca}}(X,\Omega) + R^-(\Omega)A_p^-(X,\Omega) + R^+(\Omega)A_p^+(X,\Omega) + \int_{\Gamma_0^- \cup \Gamma_0^+ \cup \Gamma_1^- \cup \Gamma_1^+} T_{ip}(x,X,\Omega)u_i^{\text{sca}}(x,\Omega)d\Gamma(x) \\
&= \int_{\Gamma_0^- \cup \Gamma_0^+} G_{ip}(x,X,\Omega)t_i^{\text{sca}}(x,\Omega)d\Gamma(x), \quad (i,p=1,2).
\end{aligned} \tag{4}$$

where the correction terms $A_p^\pm$ having the definition $A_p^-(X,\Omega) = \int_{\Gamma_\infty^-} T_{ip}(x,X,\Omega)\tilde{u}_i^-(x,\Omega)d\Gamma(x)$ and $A_p^+(X,\Omega) = \int_{\Gamma_\infty^+} T_{ip}(x,X,\Omega)\tilde{u}_i^+(x,\Omega)d\Gamma(x)$. They are constants. In paper [18], the coefficients $R^\pm$ are not explicit in the modified BEM formulation, consequently, the reflection coefficients obtained from x_1 and x_2 components of displacement might not coincide with each other. As an improvement, we will try to solve $R^\pm$ together with displacement fields.

The correction terms $A_p^\pm$ involve the integral over an infinite boundary, however, by applying the reciprocity theorem, the integral can be transferred to a bounded region [18,35].

$$\begin{aligned}
A_p^+(X,\Omega) = &\,c_{pi}(X)\tilde{u}_i^+(X,\Omega) - \int_{\Gamma_0^+} T_{ip}(x,X,\Omega)\tilde{u}_i^+(x,\Omega)d\Gamma(x) \\
&+ \int_{\Gamma_1^+ \cup \Gamma_2} \left[G_{ip}(x,X,\Omega)\tilde{t}_i^+(x,\Omega) - T_{ip}(x,X,\Omega)\tilde{u}_i^+(x,\Omega) \right]d\Gamma(x)\,(i,p=1,2)
\end{aligned} \tag{5}$$

$$A_p^-(X,\Omega) = c_{pi}(X)\tilde{u}_i^-(X,\Omega) - \int_{\Gamma_0^+} T_{ip}(x,X,\Omega)\tilde{u}_i^-(x,\Omega)d\Gamma(x)$$
$$+ \int_{\Gamma_1^+\cup\Gamma_2}\left[G_{ip}(x,X,\Omega)\tilde{t}_i^-(x,\Omega) - T_{ip}(x,X,\Omega)\tilde{u}_i^-(x,\Omega)\right]d\Gamma(x)(i,p=1,2) \tag{6}$$

where Γ_2 represents an auxiliary vertical boundary of 2–3 Rayleigh wavelengths, see Figure 2. $\tilde{t}_i^+ = -\tilde{\sigma}_{i1}^+ n_1$, and $\tilde{t}_i^- = \tilde{\sigma}_{i1}^- n_1$ are defined on Γ_2 with n_1 the unit normal vector pointing in positive x1-direction. $\tilde{\sigma}_{i1}^\pm$ is the stress field of the unitary Rayleigh wave propagating in positive or negative x1-directions, respectively.

By substitution of Equations (5) and (6) into Equation (4), we can obtain an equation set with the displacement and coefficients $R^\pm$ as the undetermined variables. Here, we adopt the constant elements (see Figure 3) in the BEM discretization, in which one element is represented by the only node in the middle, and obtain the matrix representation in the form of

$$[\mathbf{T}]\{\mathbf{u}^{\text{sca}}\} = [\mathbf{G}]\{\mathbf{t}^{\text{sca}}\} \tag{7}$$

Here, we assume that for the 1st or the N-th node, the displacements are not independent, and can be expressed as $R^-(\Omega)\tilde{u}_i^-(x_1,\Omega)$ and $R^+(\Omega)\tilde{u}_i^+(x_N,\Omega)$, respectively, where x_1 and x_N are coordinates of the 1st and the N-th node, respectively. Thus, the submatrices concerning the 1st and the N-th elements in $[\mathbf{T}]$ are combined with the correction terms $A_p^\pm$, and placed as the first and last columns. Consequently, the vector $\{\mathbf{u}^{\text{sca}}\}$ contains the displacements of nodes 2~$(N-1)$, together with coefficients $R^\pm$, as shown by

$$\{\mathbf{u}^{\text{sca}}\} = \begin{bmatrix} R^- & u_{21} & u_{22} & \cdots & u_{P1} & u_{P2} & \cdots & u_{(N-1)1} & u_{(N-1)2} & R^+ \end{bmatrix}^T \tag{8}$$

where u_{Pi} represents the displacement in ith direction at the Pth node.

The matrix $[\mathbf{T}]$ is expressed in detail as

$$[\mathbf{T}] = \begin{bmatrix}
S_{11}^- & T_{2111} & T_{2211} & \cdots & T_{p111} & T_{p211} & \cdots & T_{(N-1)111} & T_{(N-1)211} & S_{11}^+ \\
S_{12}^- & T_{2112} & T_{2212} & \cdots & T_{p112} & T_{p112} & \cdots & T_{(N-1)112} & T_{(N-1)212} & S_{12}^+ \\
\vdots & \vdots & \vdots & & \vdots & \vdots & & \vdots & \vdots & \vdots \\
S_{j1}^- & T_{21j1} & T_{22j1} & \cdots & T_{p1j1} & T_{p2j1} & \cdots & T_{(N-1)1j1} & T_{(N-1)2j1} & S_{j1}^+ \\
S_{j2}^- & T_{21j2} & T_{22j2} & \cdots & T_{p1j2} & T_{p1j2} & \cdots & T_{(N-1)1j2} & T_{(N-1)2j2} & S_{j2}^+ \\
\vdots & \vdots & \vdots & & \vdots & \vdots & & \vdots & \vdots & \vdots \\
S_{N1}^- & T_{21N1} & T_{21N1} & \cdots & T_{p1N1} & T_{p2N1} & \cdots & T_{(N-1)1N1} & T_{(N-1)2N1} & S_{N1}^+ \\
S_{N2}^- & T_{21N1} & T_{21N1} & \cdots & T_{p1N2} & T_{p1N2} & \cdots & T_{(N-1)1N2} & T_{(N-1)2N2} & S_{N2}^+
\end{bmatrix} \tag{9}$$

The elements in the first and last columns are defined as

$$S_{j1}^- = A_{j1}^- + T_{11j1}\tilde{u}_{11}^- + T_{12j1}\tilde{u}_{12}^-, S_{j2}^- = A_{j2}^- + T_{11j2}\tilde{u}_{11}^- + T_{12j2}\tilde{u}_{12}^-$$
$$S_{j1}^+ = T_{N1j1}\tilde{u}_{N1}^+ + T_{N2j1}\tilde{u}_{N2}^+ + A_{j1}^+, S_{j2}^+ = T_{N1j2}\tilde{u}_{N1}^+ + T_{N2j2}\tilde{u}_{N2}^+ + A_{j2}^+ \tag{10}$$

in which, $A_{ji}^\pm = A_i^\pm(x_J)$, obtained from Equations (5) and (6), and the elements $T_{PiJk} = (1/2)\delta_{Pj}\delta_{ik} + \int_{\Gamma_P} T_{ik}(x,x_J,\Omega)d\Gamma$, $(i,k=1,2)$. The size of matrix $[\mathbf{T}]$ is $2N \times 2(N-1)$, and $\{\mathbf{u}^{\text{sca}}\}$ $2(N-1) \times 1$.

On the other hand, $[\mathbf{G}]$ is a $2N \times 2N_0$ sized matrix, $\{\mathbf{t}^{\text{sca}}\}$ is a $2N_0 \times 1$ sized vector, where N_0 is the number of nodes on which the tractions are nonzero. The elements in $[\mathbf{G}]$ can be written as $G_{PiJk} = \int_{\Gamma_P} T_{ik}(x,x_J,\Omega)d\Gamma$, and the elements in $\{\mathbf{t}^{\text{sca}}\}$ is $t_{Pi}^{\text{sca}}(x,x_J,\Omega)$.

Since the matrix $[\mathbf{T}]$ is not square-sized, the equation is solved in the least-square sense to achieve error minimization. In the final solution, we can obtain reflection coefficient R^- at the first and transmission coefficient R^+. at the last element simultaneously, as well as the whole near-field displacements.

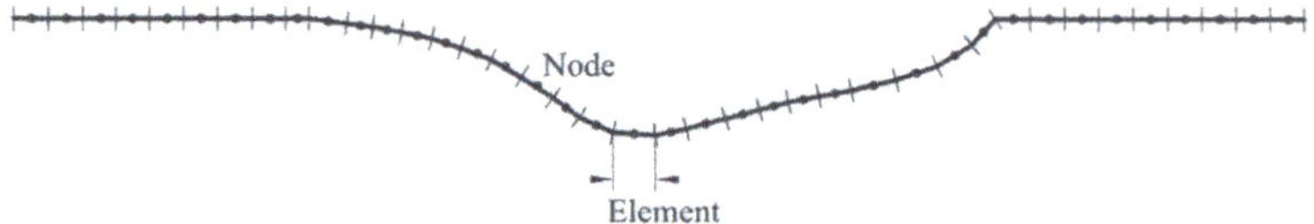

Figure 3. Boundary element method (BEM) discretization by constant elements.

2.2. Verification of Reflection Coefficients

For verification of the reflection coefficients obtained from Equation (7), we adopt the scattering equation in half-space, which uses the half-space Green's function to calculate the far-field wave pattern, as shown in [36]:

$$u_n^{\text{sca}}(X) = \int_{\Gamma_I^- \cup \Gamma_I^+} C_{ijkl} n_j(x) \frac{\partial G_{ln}^{\text{H}}(x, X)}{\partial x_k} u_n^{\text{inc}}(x) d\Gamma(x) \quad (i, j, k, l = 1, 2) \tag{11}$$

where C_{ijkl} is the stiffness tensor, n_j is the unit normal vector pointing outward the flaw area, and $G_{ln}^{\text{H}}(x, X)$ represents the Green's functions in the half-space. In the far-field, the half-space Green's function can be expressed in the Rayleigh wave form [37]:

$$G_{\alpha\beta}^{\text{H}}(x, X) = f(\zeta_0) A p_\alpha(x_2) B p_\beta(X_2) e^{+i\zeta_0(x_1 - X_1)}, (\alpha, \beta = 1, 2). \tag{12}$$

where

$$A = \begin{cases} -1 & (i = 2, x_1 < X_1) \\ 1 & \text{otherwise} \end{cases} \quad \dots B = \begin{cases} -1 & (j = 2, X_1 < x_1) \\ 1 & \text{otherwise} \end{cases}$$

$$\begin{aligned}
p_1(x_2) &= \frac{R_{\text{T0}}}{\zeta_0} e^{+R_{\text{T0}} x_2} - \frac{\zeta_0^2 + R_{\text{T0}}^2}{2 R_{\text{L0}} \zeta_0} e^{+R_{\text{L0}} x_2} \\
p_2(x_2) &= -i e^{+R_{\text{T0}} x_2} + i \frac{\zeta_0^2 + R_{\text{T0}}^2}{2 \zeta_0^2} e^{+R_{\text{L0}} x_2} \\
f(\zeta_0) &= \frac{4 i \zeta_0^4 R_{\text{L0}}}{\mu k_{\text{T}}^2 F\prime(\zeta_0)} \\
F(\zeta) &= \left(\zeta^2 + R_{\text{T0}}^2\right)^2 - 4\zeta^2 R_{\text{L0}} R_{\text{T0}}
\end{aligned} \tag{13}$$

in which $k_{\text{L}} = \Omega/c_{\text{L}}$, $k_{\text{T}} = \Omega/c_{\text{T}}$, ζ_0 are wavenumbers of S, P, and Rayleigh waves, respectively, and $R_{\text{L0}}^2 = \zeta_0^2 - k_{\text{L}}^2$, $R_{\text{T0}}^2 = \zeta_0^2 - k_{\text{T}}^2$. The Rayleigh wavenumber ζ_0 satisfies $F(\zeta_0) = 0$, and $F\prime(\zeta_0)$ is the derivative of $F(\zeta)$ at $\zeta = \zeta_0$.

The reflection coefficients derived from Equation (11) should coincide with that obtained directly from Equation (7). A numerical example will be given to show the validity of reflection coefficients. Moreover, the scattering Equation (11) is more useful in the inverse reconstruction procedure of the flaw shape.

2.3. Results of Forward Problem

Example 1. *To verify the modified BEM method, we now try to solve the classical Rayleigh wave scattering problem presented by Kawase [10], for which the solutions have already been obtained from discrete wavenumber method applying half-space Green functions.*

As shown by Figure 4a, an artificial half-circular canyon-shaped flaw is set on the surface of an isotropic elastic half-space, with Poisson's ratio of 1/3. The incident wave is a unitary Rayleigh wave travelling from left to right, whose time-domain waveform is expressed as

$$u(T) = \left[1/2\Omega^2(T - T_0)^2 - 1\right] \exp\left[-(\Omega/2)^2(T - T_0)^2\right] \tag{14}$$

in which $\Omega = \Omega r/c_{\text{T}}$ is the normalized circular frequency, where r is the radius of canyon, and c_{T} is the shear wave velocity. $T = tc_{\text{T}}/r$ represents the normalized time array. The time-domain signal is shown in Figure 5, when $\Omega = 2\pi$ and $T_0 = 2\pi$.

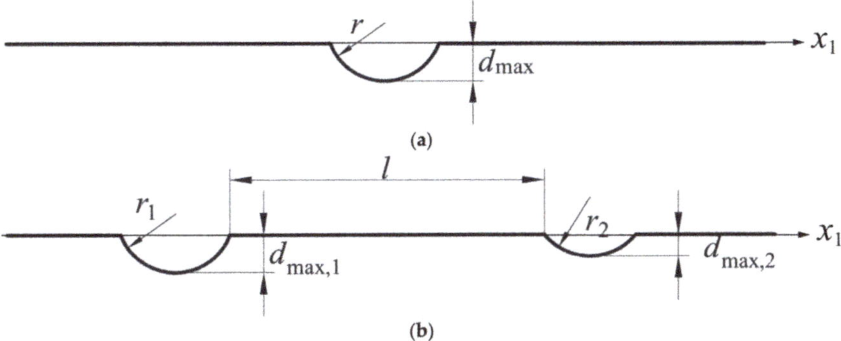

Figure 4. Numerical examples: (**a**) a half-space with an arc-shaped canyon flaw; (**b**) a half-space with double canyons.

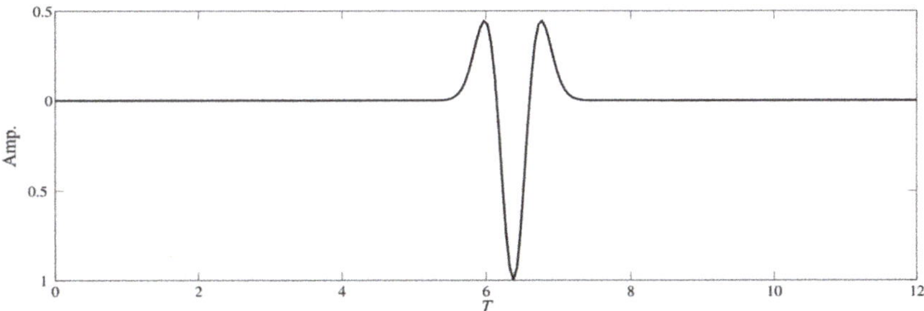

Figure 5. Time-domain signal of the incident wave.

We perform a Fourier transform to the incident wave signal in time-domain to obtain its frequency spectrum, and calculate the displacement field and reflection coefficients at each frequency point by the modified BEM. As an example, the whole displacement field along the surface at frequency $\Omega = 2\pi$ is plotted in Figure 6, and is compared with the results obtained by Kawase [10]. It is seen that the BEM displacements show good agreement with the previous results.

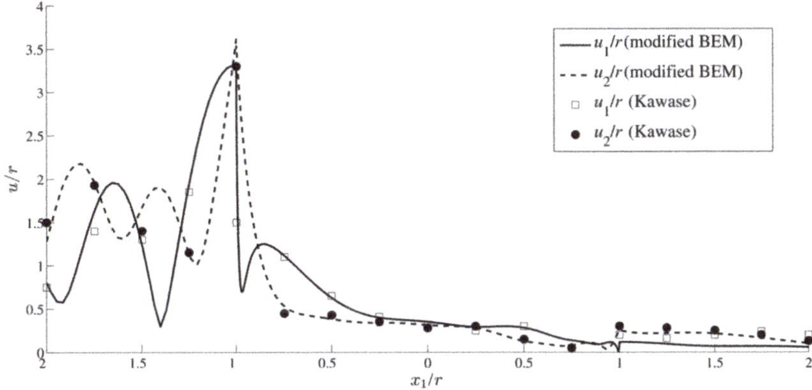

Figure 6. Comparisons of the surface displacements obtained by the modified BEM and those in Kawase [10].

Displacement fields at other frequency points have also been calculated in the same manner. An inverse Fourier transform is then applied to the frequency-domain displacements to attain time-domain wave signals for different points along the surface, as shown in Figure 7a,b, from which we can see that the spurious reflection waves are not generated.

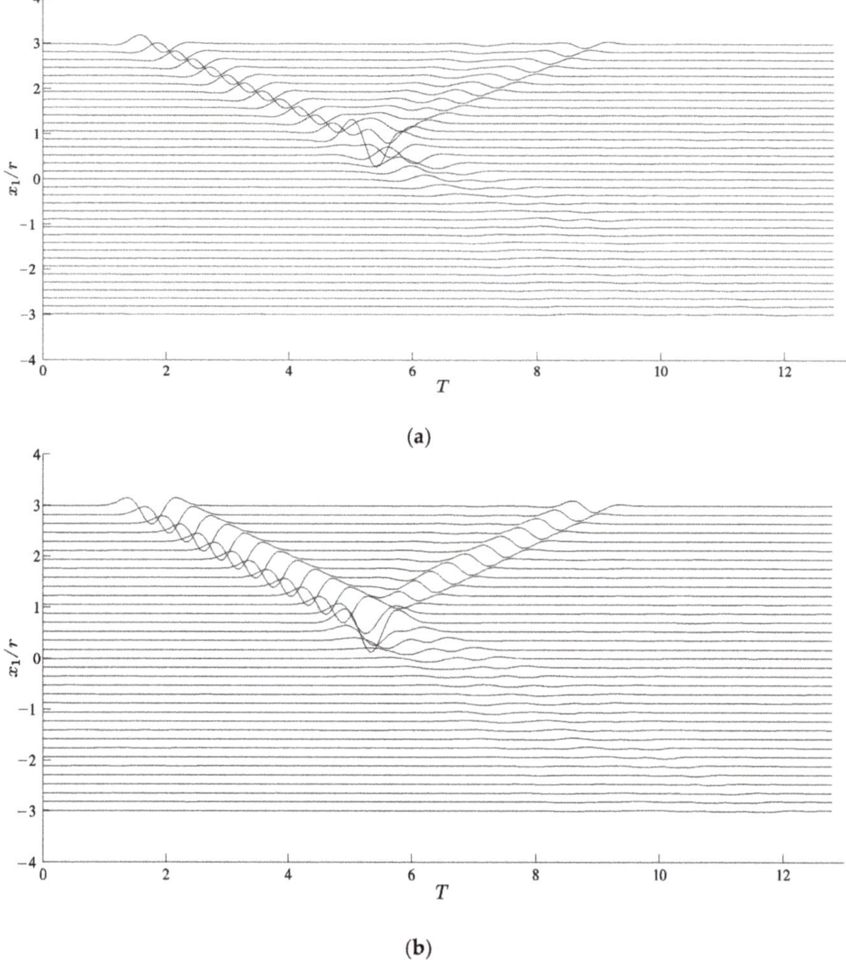

(a)

(b)

Figure 7. Displacement at different surface points in time-domain (**a**) x_1 component; (**b**) x_2 component.

Example 2. *Similarly with Example 1, an artificial arc-shaped canyon is located on the surface of an isotropic elastic half-space, whose depth–radius ratio $d_{max}/r = 0.2$. The Poisson's ratio is 1/3. We calculated the reflection coefficients in frequency domain within range $\Omega = \Omega r/c_T$ from 0.2 to 5.0 at an interval of 0.2.*

The circular dots in Figure 8 shows the reflection coefficients obtained directly from the modified BEM program, and the rectangular dots represents those derived from half-space scattering Equation (11). The good agreement illustrates the validity of the forward analysis.

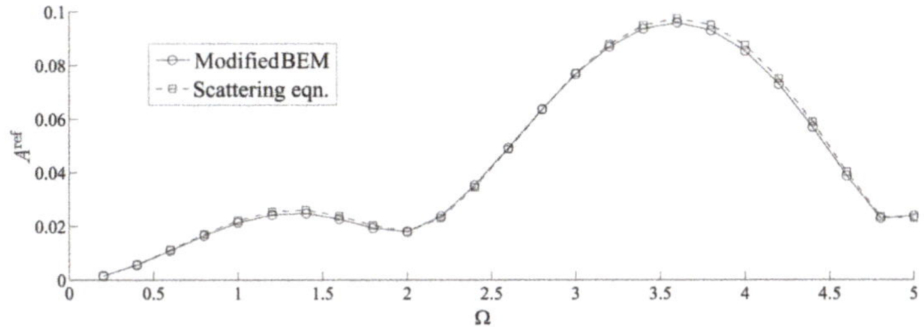

Figure 8. Reflection amplitudes obtained by BEM and half-space scattering equation respectively.

This example will be used in the following inverse reconstruction.

3. Inverse Problem by Modified BEM

3.1. Formulation of Inverse Reconstruction

We now try to reconstruct the position and geometric properties of the canyon-shaped flaw. In order to quantitatively depict the flaw's geometric characteristics, an objective function $d(x_1)$ is defined in x_1 domain, which means the flaw depth at coordinate x_1, whereas in the intact area, its value is zero.

The scattering Equation (11) in half-space is adopted for inversion. In this paper, a shallow flaw is considered, which can be thought of as a weak scatterer. In this case, the Born approximation [36] can be used for linearization, which replaces the total wave field by the incident wave field on flaw boundaries. e.g., $u_1^{\text{tot}} \approx u_1^{\text{inc}}$. Then Equation (11) becomes

$$u_n^{\text{sca}}(X) \approx \int_{\Gamma_1^- \cup \Gamma_1^+} C_{ijkl} n_j(x) \frac{\partial G_{ln}^{\text{H}}(x, X)}{\partial x_k} u_n^{\text{inc}}(x) d\Gamma(x) \quad (i, j, k, l, n = 1, 2) \tag{15}$$

It is noticed that, on the original half-space surface Γ_1' (see the dotted line in Figure 1), the traction induced by half-space Green functions is zero. Thus, the surface for integration can be extended to $\Gamma_1^- \cup \Gamma_1^+ \cup \Gamma_1'$. Furthermore, after substitution of Lamé constant λ and shear modulus μ, the scattering equation is expanded as

$$
\begin{aligned}
u_n^{\text{sca}}(X) \approx \int_{\Gamma_1^- \cup \Gamma_1^+ \cup \Gamma_1'} & \left[\left(n_1 \frac{\partial G_{1n}^{\text{H}}}{\partial x_1} u_1^{\text{inc}} + n_2 \frac{\partial G_{1n}^{\text{H}}}{\partial x_2} u_2^{\text{inc}} \right)(\lambda + 2\mu) + \left(n_2 \frac{\partial G_{1n}^{\text{H}}}{\partial x_1} u_2^{\text{inc}} + n_1 \frac{\partial G_{2n}^{\text{H}}}{\partial x_1} u_1^{\text{inc}} \right) \lambda \right. \\
& \left. + \left(n_2 \frac{\partial G_{2n}^{\text{H}}}{\partial x_1} u_1^{\text{inc}} + n_1 \frac{\partial G_{1n}^{\text{H}}}{\partial x_2} u_2^{\text{inc}} + n_2 \frac{\partial G_{2n}^{\text{H}}}{\partial x_2} u_1^{\text{inc}} + n_1 \frac{\partial G_{2n}^{\text{H}}}{\partial x_1} u_2^{\text{inc}} \right) \mu \right] d\Gamma(x) \quad (i, j, k, l, n = 1, 2)
\end{aligned}
\tag{16}
$$

Since surface $\Gamma_1^- \cup \Gamma_1^+ \cup \Gamma_1'$ is closed, we can perform Gauss's theorem to the flawed region and obtain

$$
\begin{aligned}
u_n^{\text{sca}}(X) \approx \int_V & \left[\left(\frac{\partial^2 G_{1n}^{\text{H}}}{\partial x_1^2} u_1^{\text{inc}} + \frac{\partial G_{1n}^{\text{H}}}{\partial x_1} \frac{\partial u_1^{\text{inc}}}{\partial x_1} + \frac{\partial^2 G_{2n}^{\text{H}}}{\partial x_2^2} u_2^{\text{inc}} + \frac{\partial G_{2n}^{\text{H}}}{\partial x_2} \frac{\partial u_2^{\text{inc}}}{\partial x_2} \right)(\lambda + 2\mu) \right. \\
& + \left(\frac{\partial^2 G_{1n}^{\text{H}}}{\partial x_1 \partial x_2} u_2^{\text{inc}} + \frac{\partial G_{1n}^{\text{H}}}{\partial x_1} \frac{\partial u_2^{\text{inc}}}{\partial x_2} + \frac{\partial^2 G_{2n}^{\text{H}}}{\partial x_1 \partial x_2} u_1^{\text{inc}} + \frac{\partial G_{2n}^{\text{H}}}{\partial x_2} \frac{\partial u_1^{\text{inc}}}{\partial x_1} \right) \lambda \\
& + \left(\frac{\partial^2 G_{2n}^{\text{H}}}{\partial x_1 \partial x_2} u_1^{\text{inc}} + \frac{\partial G_{2n}^{\text{H}}}{\partial x_1} \frac{\partial u_1^{\text{inc}}}{\partial x_2} + \frac{\partial^2 G_{1n}^{\text{H}}}{\partial x_1 \partial x_2} u_2^{\text{inc}} + \frac{\partial G_{1n}^{\text{H}}}{\partial x_2} \frac{\partial u_2^{\text{inc}}}{\partial x_1} \right. \\
& \left. \left. + \frac{\partial^2 G_{1n}^{\text{H}}}{\partial x_2^2} u_1^{\text{inc}} + \frac{\partial G_{1n}^{\text{H}}}{\partial x_2} \frac{\partial u_1^{\text{inc}}}{\partial x_2} + \frac{\partial^2 G_{2n}^{\text{H}}}{\partial x_1^2} u_2^{\text{inc}} + \frac{\partial G_{2n}^{\text{H}}}{\partial x_1} \frac{\partial u_2^{\text{inc}}}{\partial x_1} \right) \mu \right] d\Gamma(x) \quad (i, j, k, l, n = 1, 2)
\end{aligned}
\tag{17}
$$

It is assumed that a pure incident Rayleigh wave mode is sent from left to right, and the reflected wave is observed at the far field on the left side. Thus, the incident wave with wavenumber ζ_0 can be expressed as

$$
\begin{aligned}
u_1^{\text{inc}}(x, \zeta_0) &= A^{\text{inc}}(\zeta_0) p_1(x_2) e^{+i\zeta_0 x_1} \\
u_2^{\text{inc}}(x, \zeta_0) &= A^{\text{inc}}(\zeta_0) p_2(x_2) e^{+i\zeta_0 x_1}
\end{aligned}
\tag{18}
$$

where $p_i(x_2)$ ($i = 1, 2$) is defined in Equation (13), and A^{inc} represents the complex amplitude of the incident wave.

We use the far-field form of half-space Green's functions (12), and substitute them into Equation (17), and obtain the reflected wave of the same wave number ζ_0, as shown by Equation (19):

$$
\begin{aligned}
u_n^{\mathrm{sca}}(X, \zeta_0) \approx A^{\mathrm{inc}}(\zeta_0) \int_V &\left[\left(-2\zeta_0^2 p_1^2(x_2) + \tfrac{d^2 p_2(x_2)}{dx_2^2} p_2(x_2) + \left(\tfrac{dp_2(x_2)}{dx_2} \right)^2 \right)(\lambda + 2\mu) \right. \\
&+ \left(i\zeta_0 \tfrac{dp_1(x_2)}{dx_2} p_2(x_2) \right)(\lambda + 3\mu) + \left(i\zeta_0 \tfrac{dp_2(x_2)}{dx_2} p_1(x_2) \right)(3\lambda + \mu) \\
&\left. + \left(-2\zeta_0^2 p_2^2(x_2) + \tfrac{d^2 p_1(x_2)}{dx_2^2} p_1(x_2) + \left(\tfrac{dp_1(x_2)}{dx_2} \right)^2 \right)\mu \right] \cdot e^{+i\zeta_0 \cdot 2x_1} dV \cdot f(\zeta_0) \cdot p_n(X_2) e^{-i\zeta_0 \cdot X_1} \\
&(i, j, k, l, n = 1, 2)
\end{aligned}
$$

(19)

If one takes a thorough observation on Equation (19), he can find that $p_n(X_2) \exp(-i\zeta_0 \cdot X_1)$ represents a unitary reflected Rayleigh wave travelling from right to left, and the integral over V is taken with respect to coordinate x of field points, and can be viewed as a complex amplitude of that unitary reflected Rayleigh wave. Also, the integrand is separated into the part in the bracket concerning x_2 coordinate, and an exponential item $\exp(+i\zeta_0 \cdot 2x_1)$.

This inspires us to rewrite the expression of reflected wave into the following form

$$
\begin{aligned}
u_n^{\mathrm{sca}}(X, \zeta_0) &\approx A^{\mathrm{inc}}(\zeta_0) f(\zeta_0) \cdot \int_V \mathrm{Fun}(x_2, \zeta_0) \exp(+i\zeta_0 \cdot 2x_1) dV \cdot p_n(X_2) \exp(-i\zeta_0 \cdot X_1) \\
&\equiv A^{\mathrm{ref}}(\zeta_0) p_n(X_2) \exp(-i\zeta_0 \cdot X_1)
\end{aligned}
$$

(20)

where the $\mathrm{Fun}(x_2, \zeta_0)$ is the expression in the bracket in Equation (19), and the complex amplitude for the reflected wave is termed as A^{ref}. Now we define the reflection coefficient $C^{\mathrm{ref}}(\zeta_0)$ as the ratio between $A^{\mathrm{ref}}(\zeta_0)$ and $A^{\mathrm{inc}}(\zeta_0)$, so that

$$
C^{\mathrm{ref}}(\zeta_0) \approx f(\zeta_0) \cdot \int_V \mathrm{Fun}(x_2, \zeta_0) \exp(+i\zeta_0 \cdot 2x_1) dV
$$

(21)

Also, the volume integral is further rewritten as a dual integral:

$$
\int_V \mathrm{Fun}(x_2, \zeta_0) \exp(+i\zeta_0 \cdot 2x_1) dV = \int_{-\infty}^{\infty} \exp(+i\zeta_0 \cdot 2x_1) dx_1 \int_{-d(x_1)}^{0} \mathrm{Fun}(x_2, \zeta_0) dx_2
$$

(22)

where $d(x_1)$ is the flaw depth at coordinate x_1. The integral range has been extended from minus infinity to infinity, since $d(x_1) = 0$ in the intact areas.

After organization, the reflection coefficient is expressed as

$$
C^{\mathrm{ref}}(\zeta_0) \approx \int_{-\infty}^{\infty} e^{+i\zeta_0 \cdot 2x_1} dx_1 \int_{-d(x_1)}^{0} \left(c_{\mathrm{TT}}(\zeta_0) e^{+2R_{\mathrm{T0}} x_2} + c_{\mathrm{TL}}(\zeta_0) e^{+2(R_{\mathrm{T0}} + R_{\mathrm{L0}}) x_2} + c_{\mathrm{LL}}(\zeta_0) e^{+2R_{\mathrm{L0}} x_2} \right) dx_2
$$

(23)

where c_{TT}, c_{TL}, and c_{LL} are functions of wavenumber ζ_0, and are expressed in detail as

$$
c_{\mathrm{TT}}(\zeta_0) = 2\mu \frac{k_{\mathrm{T0}}^4}{\zeta_0^2}
$$

$$
c_{\mathrm{TL}}(\zeta_0) = \frac{\zeta_0^2 + R_{\mathrm{T0}}^2}{2\zeta_0^2 R_{\mathrm{L0}}} \left(\lambda k_{\mathrm{L0}}^2 (R_{\mathrm{T0}} - R_{\mathrm{L0}}) + \mu (7\zeta_0^2 R_{\mathrm{T0}} - 5\zeta_0^2 R_{\mathrm{L0}} + 4R_{\mathrm{L0}}^3 - R_{\mathrm{T0}}^3 - R_{\mathrm{L0}} R_{\mathrm{T0}}^2 + 6R_{\mathrm{L0}}^2 R_{\mathrm{T0}}) \right)
$$

$$
c_{\mathrm{LL}}(\zeta_0) = -(\lambda + 2\mu) \frac{k_{\mathrm{L0}}^4 (\zeta_0^2 + R_{\mathrm{T0}}^2)^2}{2\zeta_0^4 R_{\mathrm{L0}}^2}
$$

(24)

If we perform an integral over x_2 first, we can rewrite Equation (23) as

$$
C^{\mathrm{ref}}(\zeta_0) \approx \int_{-\infty}^{\infty} \left[c_{\mathrm{TT}}(\zeta_0) \frac{1 - e^{-2R_{\mathrm{T0}} d(x_1)}}{2R_{\mathrm{T0}}} + c_{\mathrm{TL}}(\zeta_0) \frac{1 - e^{-(R_{\mathrm{T0}} + R_{\mathrm{L0}}) d(x_1)}}{R_{\mathrm{T0}} + R_{\mathrm{L0}}} + c_{\mathrm{LL}}(\zeta_0) \frac{1 - e^{-2R_{\mathrm{L0}} d(x_1)}}{2R_{\mathrm{L0}}} \right] e^{+i\zeta_0 \cdot 2x_1} dx_1
$$

(25)

Since the flaw depth is considered small, it is natural to apply Taylor expansion for the exponential functions in Equation (25) at $d(x_1) = 0$, i.e.,

$$C^{\text{ref}}(\zeta_0) \approx \int_{-\infty}^{\infty} [c_{\text{TT}}(\zeta_0) + c_{\text{TL}}(\zeta_0) + c_{\text{LL}}(\zeta_0)] d(x_1) e^{+i\zeta_0 \cdot 2x_1} dx_1 \tag{26}$$

Then, it is found that the objective function $d(x_1)$ can be reconstructed using a spatial inverse Fourier transform, as illustrated by

$$d(x_1) \approx \frac{1}{2\pi} \int_{-\infty}^{\infty} \frac{C^{\text{ref}}(\zeta_0)}{c_{\text{TT}}(\zeta_0) + c_{\text{TL}}(\zeta_0) + c_{\text{LL}}(\zeta_0)} e^{-ix_1 \cdot 2\zeta_0} d(2\zeta_0) \tag{27}$$

It should be noted that the inversion formula adopts a series of assumptions for linearization. Thus, parametric analysis should be performed to show the validity range of this method.

3.2. Numerical Examples

In this sub-section, numerical examples are given to show the validity of the inversion procedure. Also, parametric studies are performed to illustrate which kind of flaw is the most suitable for this reconstruction method.

Here, we mainly focus on the arc-shaped canyons as representatives of surface flaws. It is believed that other shapes can lead to similar conclusions.

3.2.1. Choice of Parameters

The radius r, maximum depth $d_{\max}$ in Figure 4a, as well as the distance l in Figure Figure 4b for the double-flaw case, are the key geometric parameters for arc-shaped canyons; the maximum frequency $\Omega_{\max}$ is an important parameter of input data.

For generality, we perform dimensionless analysis in this paper. The normalized radius $R = r/r$ is kept constant as 1; the depth is normalized as $D = d/r$; and frequency $\Omega = \Omega r/c_{\text{T}}$, where c_{T} is the shear wave velocity in the half-space. Meanwhile, the normalized coordinates are written as $X_i = x_i/r$ $(i = 1, 2.)$

The readers can readily verify that, in dimensionless analysis, the model (i) with radius r, maximum depth $d_{\max}$, and frequency range $0 \sim \Omega_{\max}$, is equivalent to model (ii) with radius $2r$, maximum depth $2d_{\max}$, and frequency range $0 \sim 0.5\Omega_{\max}$. Thus, the absolute value of radius is of little importance. Instead, the depth–radius ratio $d_{\max}/r$ (or $D_{\max}/R$) is a more essential parameter showing the "steepness" of the canyon. Another important factor is the normalized frequency range, which reflects the incident Rayleigh wavelength in comparison with the geometric flaw scale.

3.2.2. Reconstruction Results for Models of Different Depth–Radius Ratio

We established three canyon models with depth–radius ratios $d_{\max}/r$ equaling 0.1, 0.2, and 0.5, respectively. For each model, we calculated the reflection coefficients of Rayleigh wave within the frequency range from 0 to $\Omega_{\max} = 5.0$, with a uniform interval of $d\Omega = 0.2$, as can be illustrated in Figure 8. These coefficients are substituted into the inversion Equation (27) to obtain the flaw depth function $d(x_1)$.

It should be noted, Equation (27) takes an integral from minus to positive infinity. However, the reflection coefficients are only available at positive wavenumber points. Since $d(x_1)$ is a real function, we can define $C^{\text{ref}}(-\zeta_0)$ as the complex conjugate of $C^{\text{ref}}(\zeta_0)$ (for $\zeta_0 > 0$), i.e., $C^{\text{ref}}(-\zeta_0) = C^{\text{ref}*}(\zeta_0)$. The functions $c_{\text{TT}}(\zeta_0)$, $c_{\text{TL}}(\zeta_0)$, and $c_{\text{LL}}(\zeta_0)$ are also defined similarly. The whole integrand function is defined as 0 at $\zeta_0 = 0$. Moreover, the integration range is reduced to $0 \sim \Omega_{\max}$.

After discretization, the integral of Equation (27) becomes

$$d(p\Delta x_1) = \sum_{k=0}^{N-1} \left(\frac{C^{\text{ref},k} \exp(-2i\pi pk/N)}{c_{\text{TT}}^k + c_{\text{TL}}^k + c_{\text{LL}}^k} + \frac{C^{\text{ref},k*} \exp(+2i\pi pk/N)}{c_{\text{TT}}^{k*} + c_{\text{TL}}^{k*} + c_{\text{LL}}^{k*}} \right) \cdot 2\Delta\zeta_0 \tag{28}$$

where $\Delta x_1 = \pi/N\Delta\zeta_0$ is the increment in x_1 coordinate. Integers p and k are indices in spatial and wavenumber series, respectively. The discrete inverse Fourier transform is easily performed by FFT (Fast Fourier Transform) algorithm.

The results are plotted in Figure 9. It is observed that when ratio $d_{\max}/r$ is as small as 0.1 (see Figure 9a), the geometric properties (such as location, depth, and width) and shape of the flaw can be reconstructed with good precision. When $d_{\max}/r$ value increases to 0.5 (see Figure 9c), the location of flaw can still be precisely detected, and the width and depth are slightly larger than the reality. However, the reconstructed shape is not as accurate as its shallower counterparts. The image underestimates the bottom region, and looks as if there were two local maximum depth points. The reason why the shallow and flat flaws are reconstructed better can be attributed to the application range of Born approximation, which is especially suitable for weak scatterers.

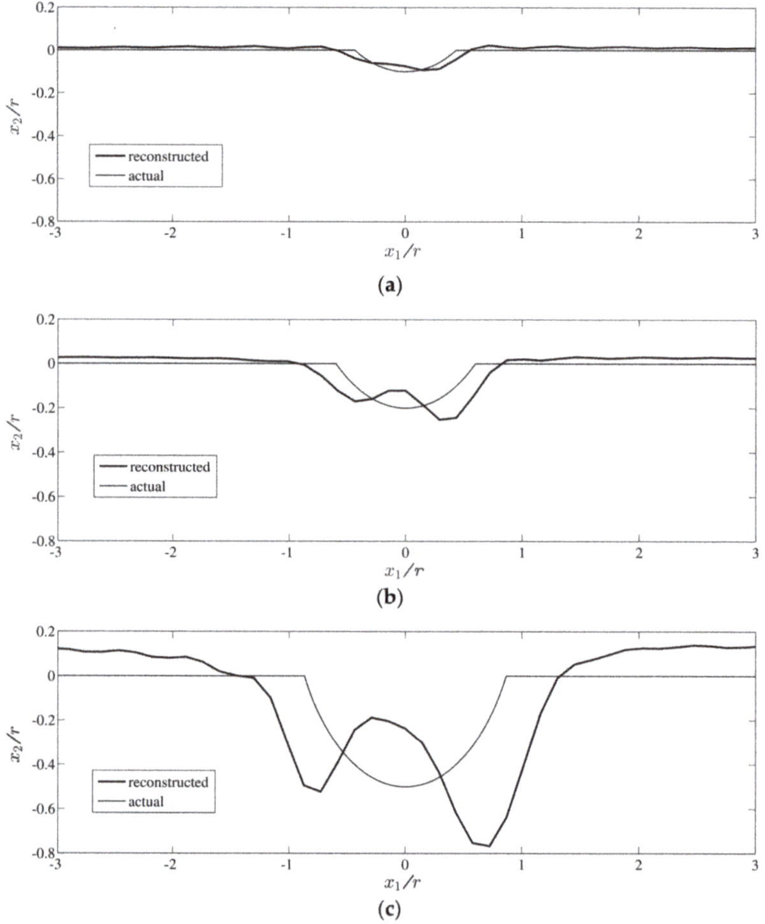

Figure 9. Inverse reconstruction results of surface canyon-shaped flaws with different depth–radius ratios (**a**) $d_{\max}/r = 0.1$; (**b**) $d_{\max}/r = 0.2$; (**c**) $d_{\max}/r = 0.5$.

Even the reconstructed image is not accurate for deeper and steeper canyons, the geometric characteristics (e.g., location, depth, and width) can be obtained with acceptable accuracy, so the image can be used as the "initial solution" in the following iterative reconstruction algorithm.

3.2.3. Reconstruction Results Using Different Frequency Range

It is necessary to discuss how the frequency range affects the inverse reconstruction results, which helps us understand the proper choice of incident wavelength in experiments. We have recalculated the inverse reconstruction for the flaws with d_{max}/r equaling 0.2 and 0.5, using frequency range $\Omega = 0 \sim 5.0$, $0 \sim 2.0$, and $0 \sim 1.0$, respectively.

The reconstruction results are plotted in Figure 10. For the case of $d_{max}/r = 0.2$ (see Figure 10a), the frequency range of $\Omega = 0 \sim 1.0$ (see the dash-dot curve) only gives a vague image, yet it still obtains the correct location. When the frequency range increases to $\Omega = 0 \sim 2.0$, the flaw shape seems to agree well with the reality, however, the image underestimates the flaw depth and overestimates the width. As the full data of reflection coefficients with $\Omega = 0 \sim 5.0$ are used in the inversion process, the flaw image becomes clearer, and the width and depth are more accurate, although the image for part of the bottom is deviated from reality.

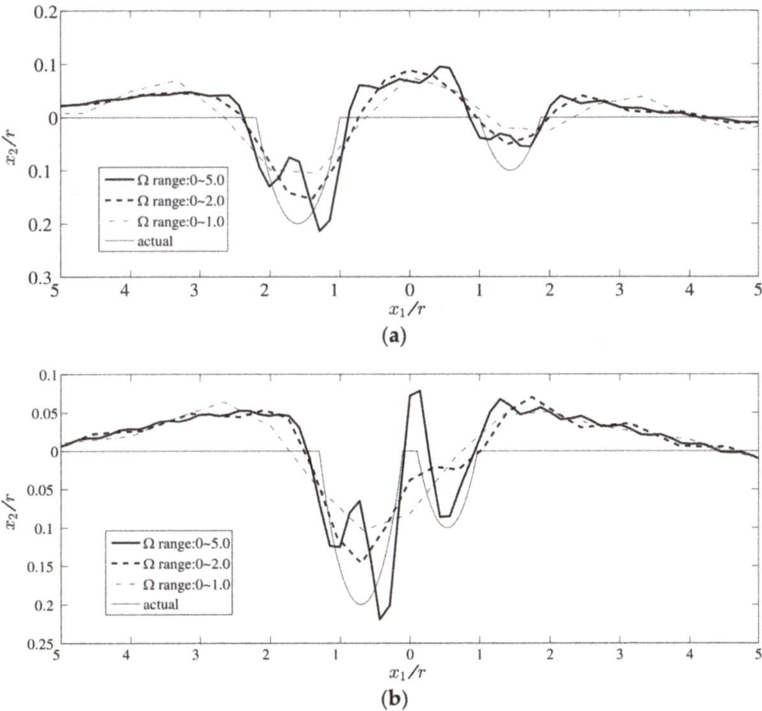

Figure 10. Inverse reconstruction results of surface canyon-shaped flaws with different frequency ranges (a) $d_{max}/r = 0.2$; (b) $d_{max}/r = 0.5$.

For the case of $d_{max}/r = 0.5$ (see Figure 10b), if only the low frequency parts of reflection data (e.g., $\Omega = 0 \sim 1.0$ and $\Omega = 0 \sim 2.0$) are taken into the inversion process, the results imply the existence of the flaw but the images are very blurred. It is only when higher frequency portions are taken into consideration that the flaw image becomes clearer and width and depth are much more accurate.

From these two examples, it can be concluded that the low-frequency or long-wavelength components of the reflection signals (e.g., $\Omega = 0 \sim 1.0$) control the canyon's basic information, such as existence and location; while the usage of higher frequency components (e.g., $\Omega = 1.0 \sim 5.0$) increases the clarity of image and gives more detailed geometric information, such as depth and width. It should be noted that the effect of high frequency components of enhancing image resolution is limited by Born approximation. As the Rayleigh wave becomes shorter in wavelength, it will penetrate less deeply into the elastic half-space, thus, the canyon becomes a large scatterer, and Born approximation loses validity. Actually, in the previous cases, if the high frequency components for $\Omega > 5.0$ are taken into consideration, the reconstructed images change little from the ones obtained from data with $\Omega = 1.0 \sim 5.0$.

3.2.4. Reconstruction Results of Double-Canyon Type Flaws

In order to examine the validity of our inversion approach in multiple flaw cases as seen in Figure 4b, we introduce here two models with double-canyon type flaws. For each model, two canyons with different geometric parameters are cut on the surface of the half-space with a distance l in between. The geometric parameters are set as (i) normalized radius: $R_1 = R_2 = 1$; (ii) normalized depth: $d_{\max,1} = 0.2$, $d_{\max,2} = 0.1$; (iii) normalized edge-to-edge distance between two flaws: ① $L = 1/r = 2.0$, ② $L = 1/r = 0.2$.

The inversion results for the two models are plotted in Figure 11, in which input data with frequency ranges $\Omega = 0 \sim 1.0$, $\Omega = 0 \sim 2.0$, and $\Omega = 0 \sim 5.0$ are shown in different curve types. It can be seen that, when the two canyons are distant (see Figure 11a), even the data with low frequency range can give correct locations of both two flaws and rough estimations of their depth and width. As the input frequency range increases to $\Omega = 0 \sim 5.0$, the geometric parameters of both two flaws are reconstructed with better accuracy.

On the other hand, if the two flaws are close with each other (see Figure 11b), the vague image reconstructed from low frequency range data cannot distinguish the two canyons. As the frequency range increases, the resolution becomes higher, and the images of two canyons are clearly separated. Thus, it is concluded that the frequency range $\Omega = 0 \sim 5.0$ is necessary for reconstruction of multiple flaws.

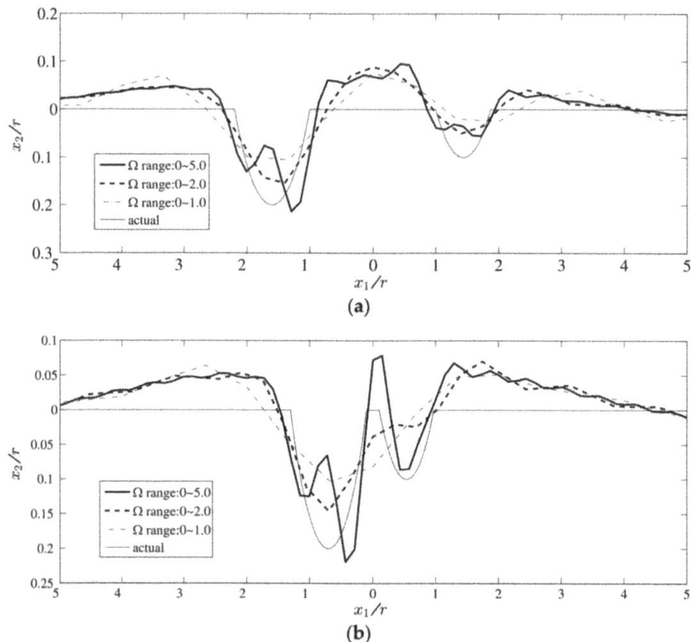

Figure 11. Inverse reconstruction results of double-canyon flaws (a) $l/r = 2.0$; (b) $l/r = 0.2$.

4. Conclusions

In this paper we present the studies of forward and inverse analysis on Rayleigh wave scattering by surface-breaking canyon-shaped flaws.

For the forward problem, we show that the modified boundary element method is an effective numerical technique to calculate the scattered wave field. By introduction of modification items concerning far-field displacement patterns in the BEM matrix, one can effectively suppress artificial reflected waves induced by model truncation. In our improved modified BEM version, the far-field reflection and transmission coefficients are solved simultaneously with the whole displacement field in the least-square sense. The improved version corrects Arias' original formulation, in which reflection coefficients obtained from x_1 and x_2 components might not coincidence with each other.

For the inverse problem, we present an inversion procedure to reconstruct the geometric properties of canyon-shaped flaw by usage of reflection coefficients of Rayleigh wave at a series of frequencies. From scattering theory for elastic waves, the scattered Rayleigh wave field is expressed as a boundary integral equation over the flawed area. By introduction of far-field form of half-space Green's function and the Born approximation, it can be proven that the objective function of flaw depth is reconstructed by inverse Fourier transform of reflection coefficients. This method gives a direct and simple inversion approach for reconstructing locations and shapes of surface flaws, which does not require baseline data and iteration.

For numerical verifications, the reflection coefficients obtained from forward analysis are used as input data in the inversion procedure. It is shown that the inversion method is effective for both single and multiple flaw cases. The method can reconstruct images especially accurately for shallow and flat notches; while for deeper flaws, the method can still represent their geometric properties (e.g., the location, depth and width), and can produce reliable "initial images" for further usage.

Acknowledgments: This work was supported by the National Natural Science Foundation of China (Nos. 51405225, 11502108), the Program for New Century Excellent Talents in Universities (No. NCET-12-0625), the Natural Science Foundation of Jiangsu Province (Nos. BK20140808, BK20140037), the Fundamental Research Funds for Central Universities (No. NE2013101), and a project Funded by the Priority Academic Program Development of Jiangsu Higher Education Institutions (PAPD).

Author Contributions: Bin Wang and Zhenghua Qian conducted theoretical analysis; Yihui Da performed numerical verifications; Bin Wang wrote the paper.

Conflicts of Interest: The authors declare no conflict of interest.

References

1. Graff, K.F. *Wave Motion in Elastic Solids*, 1st ed.; Oxford Univerisity Press: London, UK, 1975; pp. 311–393. ISBN 0-486-66745-6.
2. Achenbach, J.D. *Wave Propagation in Elasic Solids*; North-Holland: Amsterdam, The Netherlands, 1975; pp. 165–201. ISBN 0-7204-0325-1.
3. Ayter, S.; Auld, B.A. Characterization of surface wave scattering by surface breaking cracks. In Proceedings of the ARPA/AFML Review of Progress in Quantitative NDE; Thompson, D.O., Ed.; Iowa State University Press: Iowa City, IA, USA, 1979; pp. 498–504.
4. Achenbach, J.D. Acoustic emission from a surface-breaking crack under cyclic loading. *Acta Mech.* **2008**, *195*, 61–68. [CrossRef]
5. Lee, V.W.; Liu, W.-Y. Two-dimensional scattering and diffraction of p- and sv-waves around a semi-circular canyon in an elastic half-space: An analytic solution via a stress-free wave function. *Soil Dyn. Earthq. Eng.* **2014**, *63*, 110–119. [CrossRef]
6. Jian, X.; Dixon, S.; Guo, N.; Edwards, R. Rayleigh wave interaction with surface-breaking cracks. *J. Appl. Phys.* **2007**, *101*, 064906. [CrossRef]
7. Zerwer, A.; Polak, M.A.; Santamarina, J.C. Rayleigh wave propagation for the detection of near surface discontinuities: Finite element modeling. *J. Nondestruct. Eval.* **2003**, *22*, 39–52. [CrossRef]

8. Dai, Y.; Xu, B.Q.; Luo, Y.; Li, H.; Xu, G.D. Finite element modeling of the interaction of laser-generated ultrasound with a surface-breaking notch in an elastic plate. *Opt. Laser Technol.* **2010**, *42*, 693–697. [CrossRef]

9. Wong, H.L. Effect of surface topography on the diffraction of P, SV and Rayleigh waves. *Bull. Seismol. Soc. Am.* **1982**, *72*, 1167–1183.

10. Kawase, H. Time-domain response of a semi-circular canyon for incident SV, P and Rayleigh waves calculated by the discrete wavenumber boundary element method. *Bull. Seismol. Soc. Am.* **1988**, *78*, 1415–1437.

11. Sanchez-Sesma, F.J.; Ramos-Martinez, J.; Campillo, M. Seismic response of alluvial valleys for incident p, s and Rayleigh waves: A boundary integral formulation. In Proceedings of the 10th World Conference of Earthquake Engineering, Lotterdam, The Netherland, 19–24 July 1992; Ranguelov, B., Housner, G., Eds.; Balkema: Lotterdam, The Netherlands; pp. 19–24.

12. Ávila-Carrera, R.; Rodríguez-Castellanos, A.; Sánchez-Sesma, F.J.; Ortiz-Alemán, C. Rayleigh-wave scattering by shallow cracks using the indirect boundary element method. *J. Geophys. Eng.* **2009**, *6*, 221–230. [CrossRef]

13. Treyssède, F. Three-dimensional modeling of elastic guided waves excited by arbitrary sources in viscoelastic multilayered plates. *Wave Motion* **2015**, *52*, 33–53. [CrossRef]

14. Zhao, X.G.; Rose, J.L. Boundary element modeling for defect characterization potential in a wave guide. *Int. J. Solids Struct.* **2003**, *40*, 2645–2658. [CrossRef]

15. Cho, Y.; Rose, J.L. An elastodynamic hybrid boundary element study for elastic guided wave interactions with a surface breaking defect. *Int. J. Solids Struct.* **2000**, *37*, 4103–4124. [CrossRef]

16. Kamalian, M.; Gatmiri, B.; Bidar, A.S. On time-domain two-dimensional site response analysis of topographic structures by BEM. *J. Seismol. Earthq. Eng.* **2003**, *5*, 35–45.

17. Haghighat, A.E. Diffraction of Rayleigh wave by simple surface irregularity using boundary element method. *J. Geophys. Eng.* **2015**, *12*, 365–375. [CrossRef]

18. Arias, I.; Achenbach, J.D. Rayleigh wave correction for the bem analysis of two-dimensional elastodynamic problems in a half-space. *Int. J. Numer. Methods Eng.* **2004**, *60*, 2131–2146. [CrossRef]

19. Tosecký, A.; Koleková, Y.; Schmid, G.; Kalinchuk, V. Three-dimensional transient half-space dynamics using the dual reciprocity boundary element method. *Eng. Anal. Bound. Elem.* **2008**, *32*, 597–618. [CrossRef]

20. Cook, D.A.; Berthelot, Y.H. Detection of small surface-breaking fatigue cracks in steel using scattering of Rayleigh waves. *NDT E Int.* **2001**, *34*, 483–492. [CrossRef]

21. Masserey, B.; Fromme, P. Surface defect detection in stiffened plate structures using Rayleigh-like waves. *NDT E Int.* **2009**, *42*, 564–572. [CrossRef]

22. Lee, F.W.; Chai, H.K.; Lim, K.S. Assessment of reinforced concrete surface breaking crack using Rayleigh wave measurement. *Sensors* **2016**, *16*, 337. [CrossRef] [PubMed]

23. Thring, C.B.; Fan, Y.; Edwards, R.S. Focused Rayleigh wave EMAT for characterisation of surface-breaking defects. *NDT E Int.* **2016**, *81*, 20–27. [CrossRef]

24. Lee, F.W.; Chai, H.K.; Lim, K.S. Characterizing concrete surface notch using Rayleigh wave phase velocity and wavelet parametric analyses. *Constr. Build. Mater.* **2017**, *136*, 627–642. [CrossRef]

25. Wang, L.; Xu, Y.; Xia, J.; Luo, Y. Effect of near-surface topography on high-frequency Rayleigh-wave propagation. *J. Appl. Geophys.* **2015**, *116*, 93–103. [CrossRef]

26. Xia, J.; Miller, R.D.; Park, C.B. Estimation of near-surface shear-wave velocity by inversion of Rayleigh waves. *Geophysics* **1999**, *64*, 691–700. [CrossRef]

27. Xia, J.; Nyquist, J.E.; Xu, Y.; Roth, M.J.S.; Miller, R.D. Feasibility of detecting near-surface feature with Rayleigh-wave diffraction. *J. Appl. Geophys.* **2007**, *62*, 244–253. [CrossRef]

28. Shao, G.-Z.; Tsoflias, G.P.; Li, C.-J. Detection of near-surface cavities by generalized s-transform of Rayleigh waves. *J. Appl. Geophys.* **2016**, *129*, 53–65. [CrossRef]

29. Koehn, D.; Meier, T.; Fehr, M.; De Nil, D.; Auras, M. Application of 2D elastic Rayleigh waveform inversion to ultrasonic laboratory and field data. *Near Surf. Geophys.* **2016**, *14*, 461–476. [CrossRef]

30. Chen, Y.; Man, C.-S.; Tanuma, K.; Kube, C.M. Monitoring near-surface depth profile of residual stress in weakly anisotropic media by Rayleigh-wave dispersion. *Wave Motion* **2018**, *77*, 119–138. [CrossRef]

31. Wang, B.; Hirose, S. Inverse problem for shape reconstruction of plate-thinning by guided SH-waves. *Mater. Trans.* **2012**, *53*, 1782–1789. [CrossRef]

32. Wang, B.; Qian, Z.; Hirose, S. Inverse shape reconstruction of inner cavities using guided SH-waves in a plate. *Shock Vib.* **2015**, *2015*. [CrossRef]

33. Wang, B.; Hirose, S. Shape reconstruction of plate thinning using reflection coefficients of ultrasonic Lamb waves: A numerical approach. *ISIJ Int.* **2012**, *52*, 1320–1327. [CrossRef]
34. Schmerr, L.W., Jr. *Fundamentals of Ultrasonic Nondestructive Evaluation*; Plenum Publishing Corporation: New York, NY, USA, 1998.
35. Achenbach, J.D. *Reciprocity in Elastodynamics*; Cambridge University Press: Cambridge, UK, 2003; ISBN 052181734X.
36. Snieder, R. General theory of elastic wave scattering. In *Scattering: Scattering and Inverse Scattering in Pure and Applied Science*; Pike, R., Sabatier, P., Eds.; Academic Press: Cambridge, MA, USA, 2002; pp. 528–542.
37. Kobayashi, S. Elastodynamics. In *Boundary Element Methods in Mechanics, Volume 3 in Computational Methods in Mechanics*; Beskos, D.E., Ed.; North-Holland: Amsterdam, The Netherlands, 1987.

Article

The Study of Non-Detection Zones in Conventional Long-Distance Ultrasonic Guided Wave Inspection on Square Steel Bars

Lei Zhang [1,2]**, Yuan Yang** [1,]*****, Xiaoyuan Wei** [1] **and Wenqing Yao** [1]

[1] School of Automation and Information Engineering, Xi'an University of Technology, Xi'an 710048, China; leizhang830102@stu.xaut.edu.cn (L.Z.); wxy@stu.xaut.edu.cn (X.W.); yaowenqing@stu.xaut.edu.cn (W.Y.)
[2] School of Physics and Optoelectronic Technology, Baoji University of Arts and Sciences, Baoji 721016, China
* Correspondence: yangyuan@xaut.edu.cn; Tel.: +86-029-8231-2087

Received: 15 December 2017; Accepted: 15 January 2018; Published: 17 January 2018

Abstract: In a low-frequency ultrasonic guided wave dual-probe flaw inspection of a square steel bar with a finite length boundary, the flaw reflected pulse wave cannot be identified using conventional time monitoring when the flaw is located near the reflection terminal; therefore, the conventional ultrasonic echo method is not applicable and results in a non-detection zone. Using analysis and simulations of ultrasonic guided waves for the inspection of a square steel bar, the reasons for the appearance of the non-detection zone and its characteristics were analyzed and the range of the non-detection zone was estimated. Subsequently, by extending the range of the conventional detection time domain, the envelope of the specific reflected pulse signal was extracted by a combination of simulations and related envelope calculations to solve the problem of the non-detection zone in conventional inspection methods. A comparison between the simulation and the experimental results demonstrate that the solution is feasible. This study has certain practical significance for ultrasonic guided wave structural monitoring.

Keywords: ultrasonic guided wave; nondestructive testing; square steel bar; non-detection zone

---

1. Introduction

Currently, ultrasonic guided wave inspection is an effective long-distance nondestructive inspection technique [1–3]. Low-frequency ultrasonic guided waves are widely used for long-distance ultrasonic guided wave inspection [4]. The ultrasonic pulse width is wider at low frequency than at high frequency. Due to the influence of the pulse dispersion for long-distance transmissions, the reflected signal pulse is wider than in short-distance transmission [5]. The superposition state occurs easily during actual waveform measurements when the flaw is close to the terminal. In addition, the signal trailing interference aggravates the superposition state and the flaw reflected pulse is more difficult to distinguish. In the low-frequency ultrasonic guided wave flaw inspection experiments for a steel square bar as described in this study, we find that the reflected pulse wave of the flaw 'disappears' before the first terminal reflected pulse wave (in the conventional monitoring interval) when the flaw is near the reflection terminal; this results in a loss of the flaw's location information, which is referred to as the non-detection zone in this paper. Therefore, the conventional ultrasonic echo method is invalid under this condition. The literature does not provide clear details on low-frequency ultrasonic guided wave excitation pulse conditions and no studies have focused on providing a solution for the problem of the non-detection zone mentioned in this study. However, the non-detection zone cannot be ignored in studies on nondestructive testing (NDT). In this study, we designed experiments and simulations for the inspection of a square steel bar with the focus on the above-mentioned problem.

Research on ultrasonic reflected pulse waves should be concerned with solving the problem of the non-detection zone to facilitate the actual detection of flaws. To determine the ultrasonic propagation characteristics caused by different types of flaws or structural damages, the reflected pulse wave is used to analyze the reflection and transmission. In reference [5] the flaws of different opening angles were analyzed and in reference [6], the first reflected pulses in a 20–45 Hz range for four sizes and three orientations of transverse rail head flaws were investigated. Moreover, the reflected pulse wave was used to locate the flaws. For example, the flaws in a steel rail flange were detected in reference [7] and the orientation and position of small flaws in pipes were detected in reference [8]. In previous studies, researchers have provided considerable advances in long-distance non-destructive testing (NDT). However, in most studies, the flaw-reflected pulse before the first terminal reflected pulse wave was used to locate and identify the flaw, a method we refer to a "monitoring Interval I" hereafter. Different from the methods used in existing studies, the analysis time domain of the reflected pulse waves is expanded to the "monitoring Interval II" in the following sections. In order to resolve the non-detection zone problem in long-distance low-frequency ultrasonic guided wave detection, a feasible solution to locate the flaw is proposed in this study.

To describe the non-detection zone clearly, the cause of the existence of the non-detection zone is clarified by using a simulation with an equivalent model of low-frequency ultrasonic guided waves for the inspection of a square steel bar in Section 2.The experimental platform and the results are described in Section 3. The characteristics of the non-detection zone are analyzed and discussed. Based on the characteristics of the reflected pulse waveform in the non-detection zone, a solution is proposed. Finally, the experimental verification is described in Section 4.

2. Theory of the Non-Detection Zone

Based on previous research in low-frequency ultrasonic guided wave detection [9,10], we use a square steel bar as an example and analyze the objective reasons for the appearance of the non-detection zone with a finite-length boundary. Firstly, this problem associated with the ultrasonic guided wave (UGW) becomes equivalent to the reflection of waves with different modes in a plate [11] and is the 2D equivalent of an ultrasonic guided wave simulation model. By taking full advantage of the characteristics of the low-frequency pulse wave [12], an analysis is performed on the changes in the pulse signal of the flaw identification and the creation of the non-detection zone in a conventional inspection.

Prior to the analysis of the ultrasonic guided wave echo experiments and simulations, the following constraints are defined:

- A single low-frequency incident pulse is used in the experiments and simulation and the waveforms of the other frequencies are not induced by the excited incident pulse at the corresponding frequencies.
- The mode of the incident pulse wave is a single $S0$ mode and the frequency of the incident pulse is lower than the $S1$ and $A1$ cut-off frequencies so that the mode is simplified and easy to analyze [13].
- In a symmetrical plate-like structure, the reflected pulse is also a single mode $S0$; in an asymmetrical plate-like structure (such as man-made flaws), the reflected pulses in the $A0$ mode and $S0$ mode are excited [14].
- When the frequency-thickness product ($f \cdot d$) is maintained, the group velocity of the $S0$ and $A0$ modes remains unchanged under ideal conditions [15].

2.1. The Equivalent Model of Low-Frequency Ultrasonic Guided Wave in a Square Steel Bar

The square steel bar has a symmetric structure and we only focus on the propagation characteristics of the ultrasonic guided wave propagated between the front surface and the back surface, as shown in Figure 1a. We ignore the shear (SH) wave and the other two surfaces (the up surface and the down surface); therefore, this is equivalent to a two-dimensional (2D) model where the ultrasonic Lamb waves propagate along a plate [1,3].

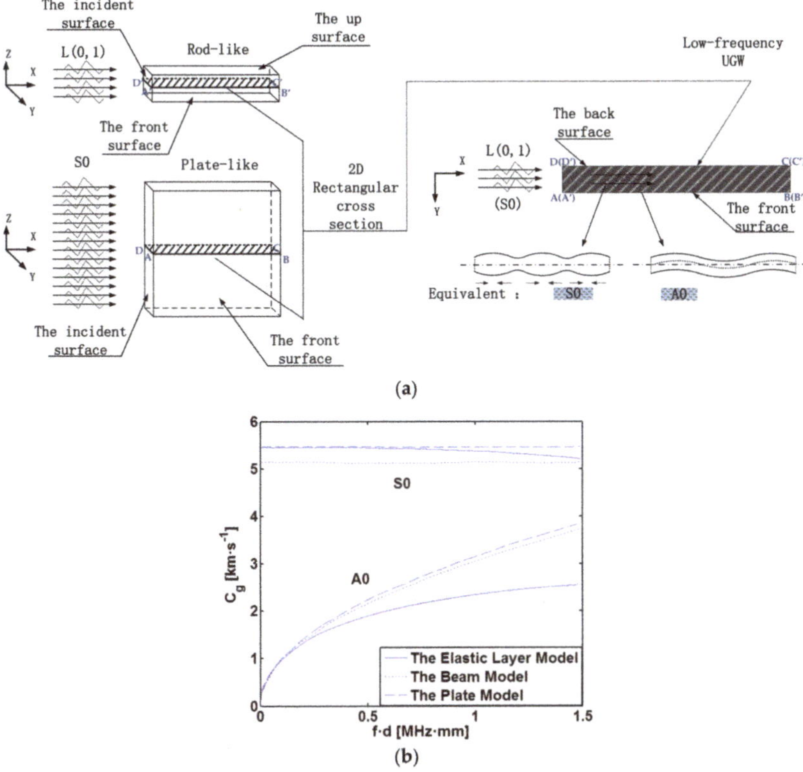

Figure 1. The equivalent model in low-frequency (**a**) The equivalent model of low-frequency ultrasonic guided wave in a square steel bar; (**b**) Low-frequency phase velocities of *S0* and *A0* modes provided by the beam, plate and elastic layer models.

2.1.1. The Mode Selection of Ultrasonic Guided Wave

Based on previous research findings, the reasons for the mode selection of ultrasonic guided wave in this study are as follows:

Firstly, the use of low-frequency ultrasound in long-distance ultrasonic guided wave inspection is preferred because the attenuation of the high-frequency ultrasound transmission over long distances is too fast, which limits the measuring distance [4]. High-frequency ultrasound waves tend to be concentrated on the surface or in localized areas of the propagation medium [1]; therefore, the meaning of detecting the ultrasonic guided wave as the body wave in the entire medium is lost. Because of the multi-mode caused by the high-frequency ultrasonic excitation, the identification of detection becomes difficult. In short, the low-frequency section and less-mode ultrasound are appropriate for the detection requirements of long-distance ultrasonic guided wave inspection. For example, a similar low-frequency ultrasonic guided wave equivalent simulation was applied to rails (a rod-like structure) by Bartoli et al. [7].

Secondly, the simulated effect of an ultrasonic equivalent model is better in the low-frequency range than in the high-frequency range. For the long wavelength in the low-frequency ultrasonic range, the equivalent model for a rod-like structure is similar to the model of a plate-like structure for the Lamb wave mode in a uniform medium. However, the mode distribution generated by the high-frequency section in a rod-like structure is not similar to the Lamb wave mode, thus there is no equivalent effect [3].

Finally, the equivalent model of low-frequency ultrasonic guided waves takes full advantage of the proven Lamb theory and can be used to theoretically control the modes of the excited pulse wave and the mode number [16,17]. The relationship between the $S0/A0$ impulsive waves and the flaw in related research can be used to analyze the mode transformation under the finite boundary conditions used in this study. Therefore, we use the ideal assumptions described in Section 2.

2.1.2. Theory on the Equivalent Model in Low-Frequency

The three-dimensional model of the steel square bar is set in the Cartesian coordinate system (x, y, z), as shown in Figure 1a. The independent of z deformation is specified by the displacement vector $\vec{u} = (u_1, u_2, 0)$ lying in the plane (x, y), where u_1, u_2, u_3 represent the displacement variable in the x, y, z directions respectively, and the model has a similar 2D plane-strain deformation in the plate-structure. The similar theoretical analysis on the equivalent model for low frequency ultrasonic wave has been proposed by Glushkov et al. [18]. We can see the similar low-frequency phase velocities of $S0$ and $A0$ modes provided by the beam and plate models for an aluminium sample in the reference [18], as shown in Figure 1b. In low frequency, this conclusion, that the axisymmetric longitudinal mode in the rod with diameter d is very similar to the symmetric mode distribution in the plate with the same thickness d, is proposed by Rose [2]. In this paper, when we focused on vibration mode $S0/A0$, without regard to other vibration modes, the ultrasonic wave propagation analysis is simplified and equivalent in low frequency.

In essence, the three-dimensional (3D) ultrasonic wave in the rod is equivalent to the two-dimensional (2D) Lamb wave in this study, which is used as the equivalent model of the low-frequency ultrasonic guided waves in a square steel bar. Both ultrasonic guided wave models are based on the same acoustic propagation control equations [1]. When low frequency ultrasonic wave excitation covers the entire incident surface, as shown in Figure 1a, the rectangular cross-section of ABCD in the plate-like structure is used to 'replace' the rectangular cross section 'A' 'B' C 'D' in the rod-like structure for the vibration mode analysis, and the vibration of ultrasonic wave between the front and back surfaces in rod-like structure can be equivalent to the vibration of ultrasonic wave between the front and back surfaces in the plate-like structure. A similar ultrasonic guided wave mode is obtained [3], the longitudinal wave L (0, 1) and the flexural wave F (1, 1) are similar to the $S0$ and $A0$ waves respectively in a 2D plate-like ultrasonic propagation. The simulations of the pulse waveform and the actual waveform exhibit good consistency. Compared with the 3D model, the estimation accuracy is slightly lower but the equivalent model reduces the simulation hardware requirements and computational costs, thereby improving the simulation efficiency.

In solid matter, as shown in Figure 1a, the vector form of the wave equation for the ultrasonic wave can be represented as follows [1]:

$$(\lambda + \mu)\nabla(\nabla \bullet \vec{u}) + \mu\nabla^2\vec{u} + (\frac{\lambda' + \mu'}{\omega})\nabla(\nabla \bullet \frac{\partial \vec{u}}{\partial t}) + (\frac{\mu'}{\omega})\nabla^2\frac{\partial \vec{u}}{\partial t} = \rho\frac{\partial \vec{u}}{\partial t} \tag{1}$$

where λ and μ are the Lame constants; λ' and μ' denote the attenuation factor in the medium.

$$\nabla^2 = \frac{\partial^2}{\partial x^2} + \frac{\partial^2}{\partial y^2} + \frac{\partial^2}{\partial z^2} \tag{2}$$

$$\nabla = \vec{i}_1\frac{\partial}{\partial x} + \vec{i}_2\frac{\partial}{\partial y} + \vec{i}_3\frac{\partial}{\partial z} \tag{3}$$

$$\vec{u} = (u_1, u_2, u_3) \tag{4}$$

The vector $\vec{u}$ is treated with the Helmholtz decomposition; u_1, u_2, u_3 represent the displacement variable in the x, y, z directions respectively; therefore, the displacement can be expressed as follows:

$$\vec{u} = \nabla\vec{\Phi} + \nabla \times \vec{H} \tag{5}$$

$$\nabla \bullet \vec{H} = 0 \tag{6}$$

where $\vec{\Phi}$ and $\vec{H}$ are the scalar and vector potentials, respectively.

Two simple wave equations can be derived as follows:

$$\left(\lambda + 2\mu + \frac{\lambda' + 2\mu'}{\omega}\frac{\partial}{\partial t}\right)\nabla^2\vec{\Phi} = \rho\frac{\partial^2\vec{\Phi}}{\partial t^2} \tag{7}$$

$$\left(\lambda + \frac{\mu'}{\omega}\right)\nabla^2\vec{H} = \rho\frac{\partial^2\vec{H}}{\partial t^2} \tag{8}$$

where ω denotes the angular frequency. If the solutions of Equations (7) and (8) are assumed to be planar harmonic waves, then

$$\vec{\Phi}(\vec{x},t) = \vec{\Phi}(\omega)e^{j\omega(\vec{N}\bullet\vec{x}/\alpha-t)} \tag{9}$$

$$\vec{H}(\vec{x},t) = \vec{H}(\omega)e^{j\omega(\vec{N}\bullet\vec{x}/\beta-t)} \tag{10}$$

where $\vec{N}$ represents the displacement vector, for the longitudinal wave, the expression of the material parameter α is as follows:

$$\alpha = \frac{\omega}{\kappa_{re1} + i|\kappa_{im1}|} \tag{11}$$

where κ_{re1}, κ_{im1} represent the real part and the imaginary part of the plural of the longitudinal wave number κ_1 respectively.

For transverse wave, the expression of the material parameter β is as follows:

$$\beta = \frac{\omega}{\kappa_{re2} + i|\kappa_{im2}|} \tag{12}$$

where κ_{re2}, κ_{im2} represent the real part and the imaginary part of the plural of the transverse wave number κ_2 respectively.

The stress in the z direction $N_3 = 0$, consequently Equations (9) and (10) can degenerate into a 2D plane strain problem, then $\frac{\partial}{\partial z} = 0$.

For the longitudinal wave, the displacement vector is

$$\vec{u}_L = \nabla\vec{\Phi} = \left\{\frac{\partial}{\partial x}, \frac{\partial}{\partial y}, 0\right\}\vec{\Phi} \tag{13}$$

The substitution of Equation (9) into Equation (13) leads to the following equations:

$$u_{L1} = \frac{\partial\vec{\Phi}(\vec{x},t)}{\partial x} = i\omega\frac{N_1}{\alpha}\vec{\Phi}(\omega)e^{i\omega(N_1x/\alpha+N_2y/\alpha-t)} \tag{14}$$

$$u_{L2} = \frac{\partial\vec{\Phi}(\vec{x},t)}{\partial y} = i\omega\frac{N_2}{\alpha}\vec{\Phi}(\omega)e^{i\omega(N_1x/\alpha+N_2y/\alpha-t)} \tag{15}$$

$$u_{L3} = 0 \tag{16}$$

where u_{L1}, u_{L2}, u_{L3} represent compressional component in x, y, z three directions respectively.

From the formula derived in reference [3], when the longitudinal wave propagates, the normal stress σ_{22} and the shear stress σ_{12} are expressed as:

$$\sigma_{22} = (\alpha - 2N_1^2\beta^2/\alpha)i\omega\beta e^{i\omega(\vec{N}\bullet\vec{x}/\alpha - t)} \tag{17}$$

$$\sigma_{12} = 2i\omega\rho N_1 N_2 \beta^2/\alpha e^{i\omega(\vec{N}\bullet\vec{x}/\alpha - t)} \tag{18}$$

where ρ represents the material density, when the transverse wave propagates, the normal stress σ_{22} and the shear stress σ_{12} are expressed as:

$$\sigma_{22} = -2i\omega\rho N_1 N_2 \beta e^{i\omega(\vec{N}\bullet\vec{x}/\beta - t)} \tag{19}$$

$$\sigma_{12} = i\omega(N_2^2 - N_1^2)\beta/\alpha e^{i\omega(\vec{N}\bullet\vec{x}/\beta - t)} \tag{20}$$

where c_p represents the phase velocity of sound, and s represents the Snell constant. The up going wave $N_1 = c_p s$, $N_2 = (1 - c_p^2 s^2)^{\frac{1}{2}}$, and the down going wave $N_1 = c_p s$, $N_2 = -(1 - c_p^2 s^2)^{\frac{1}{2}}$, when $c_p = \alpha$, longitudinal wave; when $c_p = \beta$, transverse wave.

2.2. Non-Detection Zone in Simulation

To investigate the reasons for the appearance of the non-detection zone, we focus on the simulations of the propagation of an ultrasonic guided wave in a square steel bar with a length of 6000 mm using COMSOL Multiphysics, ver.5.0.

As shown in Figure 2, during which a 10-pulse Hanning-window function pulse signal with a frequency of 30 kHz is used as the incident pulse.

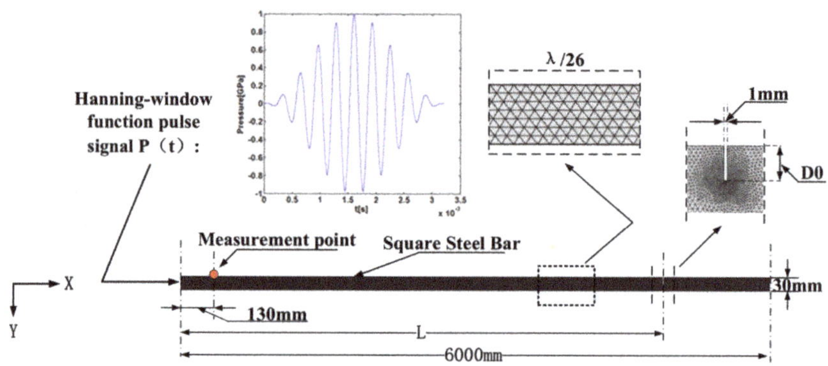

Figure 2. The simulation settings.

The Hanning function is used as the window function:

$$Hanning(t) = (1 - \cos(2\pi t/T))/2$$

The type of excitation signal:

$$p(t) = \sin(2\pi f_0 t)Hanning(t)$$

where $T = 1.6\,[\mu s]$.

The point (130, 30) is used as the measurement point and is the same as in the actual experiment in Section 3. A transverse flaw with a depth of 5 mm (*D0*) and a width of 1 mm is created at the displacement L with a variable value of $L \in [3500\ \text{mm}, 5500\ \text{mm}]$ along the x direction from the incident point and the simulation interval is set as 500 mm. ($D0$ is set as 5 mm for reducing the superposition of the flaw reflected pulses). A triangular grids ($\lambda/26$) are used in this study ($<\lambda/20$ recommended minimum mesh size [19]) and the area near the location of the flaw L is refined using a local mesh. During the simulation, the excitation energy is properly decreased so that 5 pulses are reflected. The material parameters of the steel square bar are the following:

Young's modulus: $E = 200\ [\text{GPa}]$
Density: $\rho = 7850\ [\text{kg}/\text{m}^3]$
Poisson's ratio: $\nu = 0.3$

Taking advantage of the periodicity of the terminal reflection, as shown in Figure 3a, the time domain is divided into the monitoring time Interval I, the monitoring time Interval II, the monitoring time Interval III, and so on. The pulse waves of the multiple reflections are labeled as the incident pulse wave, the first terminal reflection pulse wave, the second terminal reflection pulse wave, and so on.

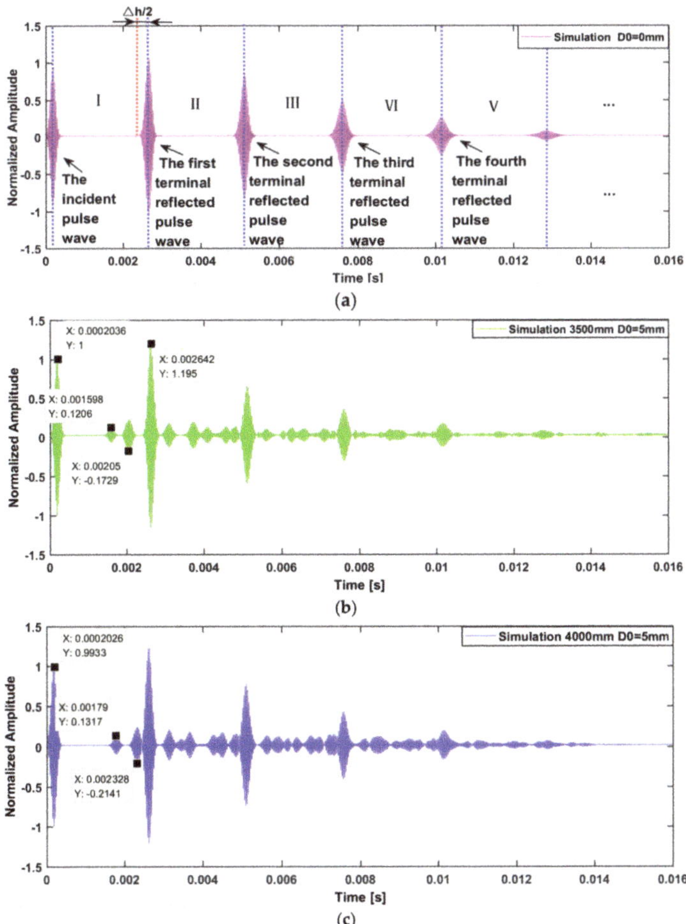

Figure 3. *Cont.*

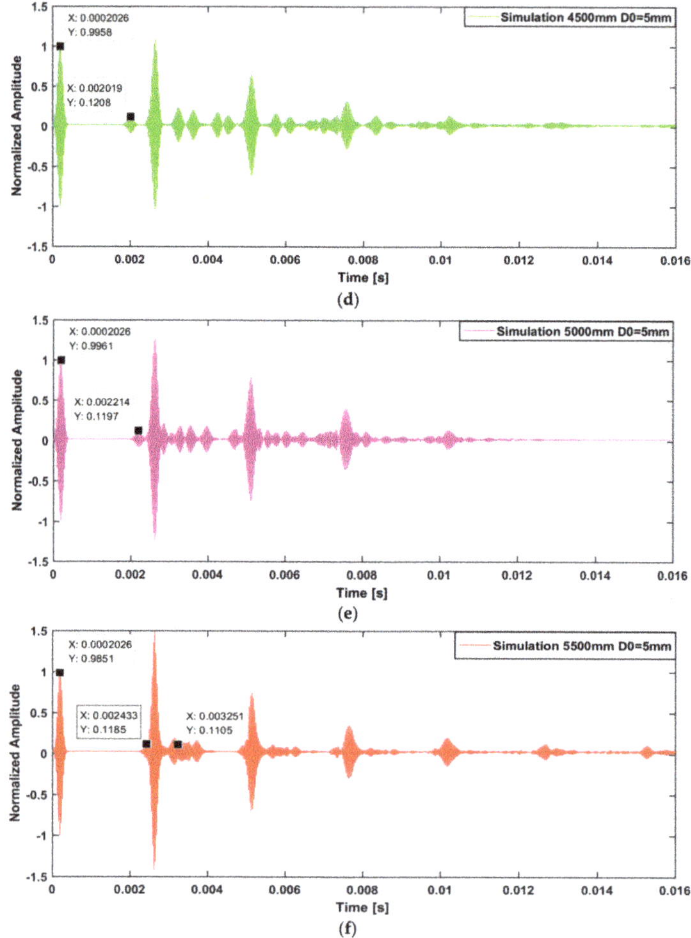

Figure 3. Waveforms of the reflected pulses in the simulation, (**a**) $D0 = 0$ mm; (**b**) $D0 = 5$ mm, $L = 3500$ mm; (**c**) $D0 = 5$ mm, $L = 4000$ mm; (**d**) $D0 = 5$ mm, $L = 4500$ mm; (**e**) $D0 = 5$ mm, $L = 5000$ mm; (**f**) $D0 = 5$ mm, $L = 5500$ mm.

As shown in Figure 2b, at $L = 3500$ mm, an $S0$ mode wave is emitted from the incident face and is then reflected by the transverse flaw [20]; then, after the reflection, two different types of waves, $S0$ and $A0$, are generated. In Interval I, the peak points of the $S0$ mode and $A0$ mode flaw-reflected pulse waves are (0.001598, 0.1206) and (0.00205, -0.1729), respectively. Based on the simulation results, the $S0$ mode flaw-reflected pulse wave is calculated to verify whether the detected location of the flaw was accurate.

It is known that the first terminal-reflected pulse wave and the initial pulse exhibit the maximum amplitude at ((0.0002036, 1) and (0.002642, 1.195)); the $S0$ wave group velocity (Cg_{S0}) was calculated as:

$$Cg_{S0} = \Delta s / \Delta t = (6 - 0.13) \times 2 / (0.002642 - 0.0002036) = 4814.6325 \, [\text{m/s}] \tag{21}$$

where Δs and Δt represent the propagation distance and the corresponding time respectively. Using the calculated Cg_{S0}, the flaw-induced $S0$ mode reflected wave is verified. The $S0$ mode pulse wave appears at 0.001598 s and L_0 denotes the distance between the transverse flaw and the initial face, the following expression can be acquired:

$$L_0 = (Cg_{S0} \times \Delta t)/2 + 0.13 = Cg_{S0} \times (0.001598 - 0.0002036)/2 + 0.13 = 3.4868\,[\text{m}] \tag{22}$$

The detected location is quite close to the actual position of the flaw (3.5 m). The relative deviation σ_0 can be calculated by:

$$\sigma_0 = (3.5 - 3.4868)/3.5 = 0.3771\% \tag{23}$$

As shown in Figure 2b, the $A0$ mode flaw-reflected pulse wave first appears at 0.00205 s. An $S0$ mode ultrasonic pulse wave was excited to the transverse flaw at 3.5 m and the mode changed from $S0$ to $A0$ when passing the asymmetric flaw. Therefore, the propagation time of the $A0$ mode ultrasonic pulse wave after mode transmission is equal to:

$$Cg_{A0} = \Delta s/\Delta t = (3.5 - 0.13)/(0.00205 - 3.5/Cg_{S0}) = 2547.2411\,[\text{m/s}] \tag{24}$$

The flaw at 4000 mm (Figure 3c) can also be calculated in the same way so that its location can be determined using the reflected $A0$ mode and $S0$ mode waves. As shown in Figure 3d,e, the flaw at 4500 mm and 5000 mm can be only located by the $S0$ mode reflected pulse in the conventional inspection because the $A0$ mode pulse wave, which is in the superposition, is not in Interval I. As shown in Figure 3f, the times of the $S0$ mode and $A0$ mode flaw-reflected pulse waves in Interval I when $D0 = 5$ mm at 5500 mm are then calculated.

The appearance time of the $S0$ mode pulse wave T_{S0} can be calculated using:

$$T_{S0} = 0.0002026 + (5.5 - 0.13) \times 2/Cg_{S0} = 0.0024333\,[\text{s}] \tag{25}$$

The appearance time of the $A0$ mode pulse wave T_{A0} can be calculated using:

$$T_{A0} = 5.5/Cg_{S0} + (5.5 - 0.13)/Cg_{A0} = 0.0032506\,[\text{s}] \tag{26}$$

This analysis shows that, when the value of L increases along the x-direction and approaches the length of the terminal, the appearance time of the $S0$ mode flaw-reflected pulse wave and the $A0$ mode flaw-reflected pulse wave gradually approach the time of the first terminal reflection pulse wave, even then the appearance time of the $A0$ mode flaw-reflected pulse appears in the monitoring time Interval II at a given value of L. For the above simulation, at $L = 5500$ mm, the $S0$ mode flaw-reflected pulse wave is superimposed by the first terminal reflection pulse wave and the $A0$ mode flaw-reflected pulse wave is superimposed by these reflection pulse waves of the monitoring time Interval II.

In particular, the calculation (see Equation (26)) indirectly suggests that, in the monitoring Interval II, the highest reflected pulse is not the delay of the $A0$ mode flaw-reflected pulse in the conventional inspections, as shown in Figure 3f. Under ideal conditions, even without the interference of the signal trailing, the actual $S0$ or $A0$ mode flaw reflection pulses cannot be easily identified and the conventional detection method based on a single mode flaw-reflected pulse wave is invalid. For actual measurements, the trailing phenomenon of the pulse wave has to be considered. This is the state of the non-detection zone referred to in this study. It is known that the information of the flaw is superimposed by the first terminal reflection pulse wave and various types of reflection pulse waves in the monitoring time Interval II described in this section. In the following section, we observe the actual waveform of the ultrasonic pulse wave under the condition of the non-detection zone in an actual experiment.

3. Experimental Section

In this section, the experimental platform and the experimental results are presented and the occurrence of the non-detection zone in square steel inspections is briefly described.

As shown in Figure 4a, a 30# (30 mm × 30 mm) square steel bar with a length of 6000 mm is used in this study; the length is sufficient to weaken the effect induced by signal trailing and facilitates the signal identification. These customized ultrasonic longitudinal-wave probes [21,22] with a central frequency of 30 kHz are installed using the dual-probe echo method. The signals are acquired and displayed using an oscilloscope (Tektronix MD04104) and the waves are recorded and the data are captured using a wave recorder (HIOKI MR8875-30), as shown in Figure 4b.

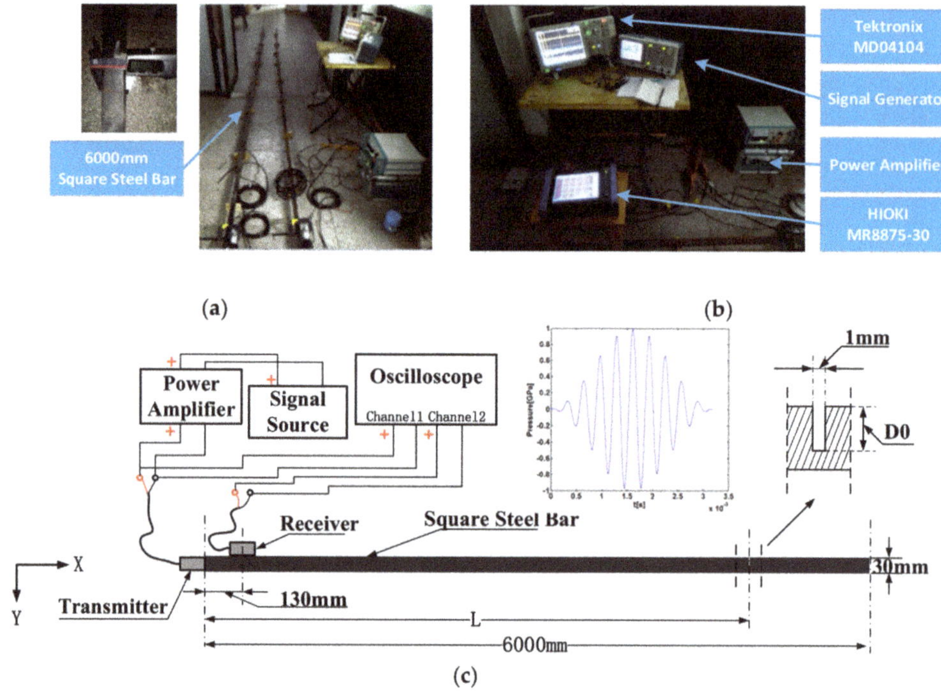

Figure 4. Establishment of the experimental platform (**a**) square steel bar with a length of 6000 mm; (**b**) picture of experimental instruments; (**c**) the diagram of the establishment of experimental platform.

The Hamming pulse signal (10 cycles at the center frequency of $f_0 = 30$ [kHz]) is set by the signal source, which is loaded to the ultrasonic probe via a power amplifier. Transverse flaws with a width of 1 mm and a depth of 0~20 mm ($D0$) are created at a displacement L along the x-direction from the incident point. Because the thickness of the steel square bar is 30 mm and the length of the actual ultrasonic probe is 10 mm, a point (130.30) is selected as the measuring point. The excitation voltage amplitude is adjusted to acquire five reflected pulses and the waveforms of the reflected pulses are monitored in the time domain. Figure 4c shows the overall layout of the experimental platform.

In a dual-probe echo long-distance inspection with the customized longitudinal-wave probe used in the experiment, the acquired reflected pulse wave of the transverse flaw occurs before reaching the first terminal; subsequently, the flaw can be located based on the modal wave velocity. However, both the experimental and the simulation results of the square steel inspection indicate that no reflected pulse is identified before the first terminal reflection pulse wave when the flaw is at a certain distance ($L \in$ (5300 mm, 6000 mm) from the excitation point. The experimental result at $L = 5500$ mm is used as an example (Figure 5). In the conventional monitoring Interval I, the reflected pulse wave with the flaw cannot be identified. According to the experimental data, even when the depth of the flaw ($D0$) reaches approximately 20 mm (i.e., the flaw represents almost 75% of the 30-mm thickness), no echo information is detected in the conventional monitoring Interval I. In actual measurements, when the ultrasonic pulse excitation ends, there are attenuation oscillations in the signal, namely the trailing phenomenon of the pulse wave. Many factors affect signal trailing, such as the manufacturing technique of the probes, the capacitance shock effect of the excitation circuit, and the sound absorption effect of the substrate. The trailing phenomenon is common in long-distance inspections with low-frequency ultrasonic guided waves [23,24]. In addition, the signal trailing compounds the difficulty of identifying the flaw-reflected pulse wave in the non-detection zone.

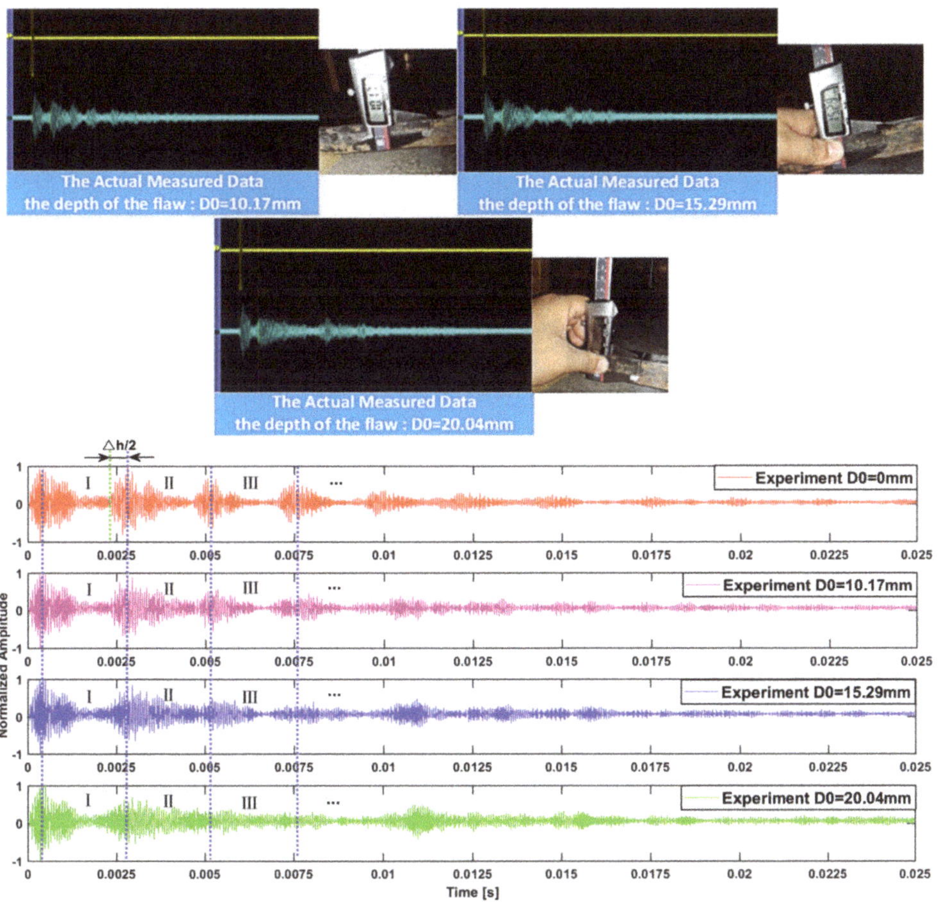

Figure 5. The reflected pulse waves with different values of $D0$ at 5500 mm.

At L = 5500 mm, the $S0$ mode and $A0$ mode flaw-reflected pulse waves are superposed by the first terminal-reflected pulse wave, various types of reflection pulse waves in the monitoring time interval II, and the pulse signal trailing. Therefore, in a conventional inspection, the non-detection zone can be easily determined in the actual experiment. A similar experimental phenomenon can also occur in the ultrasonic flaw detection of long rod-like structures and cannot be ignored for NDT.

4. Analysis and Discussion

In this section, we analyze the range of the non-detection zone and the propagation characteristics of the ultrasonic guided wave in the non-detection zone using a simulation and experiments. Subsequently, we propose a feasible solution for the practical problems.

4.1. The Range of the Non-Detection Zone

The range of the non-detection zone is a major concern in non-destructive inspections. Based on the analysis in Section 2.2, we used a dual-probe echo method for the location of the flaws and a low-frequency of the incident pulse. If the flaw is a certain distance from the terminal, the $S0$ mode flaw-reflected pulse is superposed by the first terminal-reflected pulse wave, while the $A0$ mode flaw-reflected pulse is superposed by the multiple reflected pulse wave and the pulse signal trailing in the monitoring interval II; as a result, the non-detection zone occurs. Therefore, the superposition time point of the $S0$ mode flaw-reflected pulse and the first terminal-reflected pulse wave can be considered as given the occurring conditions. Figure 6 shows the first terminal-reflected pulse wave and the $S0$ mode flaw-reflected wave.

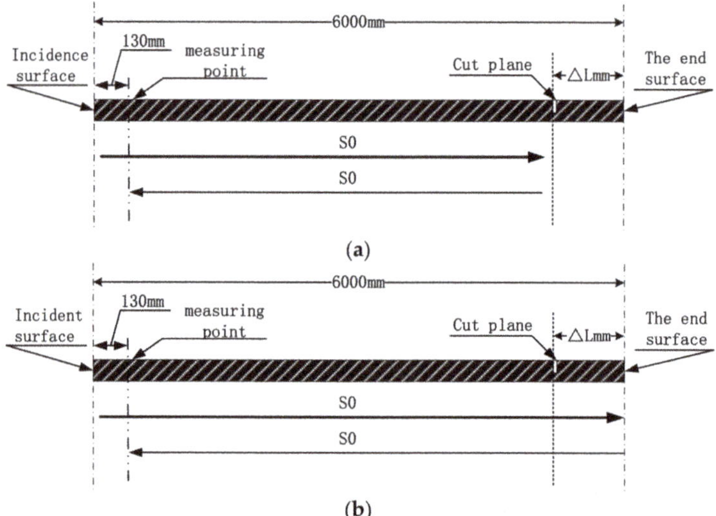

Figure 6. Schematic of the distance covered by the $S0$ mode pulse wave in (**a**) the first terminal-reflected pulse wave and (**b**) the flaw-reflected pulse wave in Interval I.

Based on these occurring conditions, the range of the non-detection zone can be described as:

$$Cg_{S0} \times (\Delta h/2)/2 \geq \Delta L > 0 \tag{27}$$

where Cg_{S0} is related to the frequency-thickness product ($f \cdot d$) of the incident pulse wave, ΔL denotes the distance between the flaw and the terminal, Δh denotes the pulse width of the first terminal-reflected pulse wave (Figures 3a and 5), which is connected with the incident pulse wave and the self-characteristic of the customized longitudinal-wave probe, and ΔL denotes the distance between the flaw and the terminal reflection boundary. It should be noted that low-frequency ultrasound is normally used as the core frequency of the excitation pulse in long-distance ultrasonic guided wave inspections [24]. This has many advantages such as reducing the excitation mode number and extending the range of the long-distance detection [25]. However, the disadvantages cannot be ignored. One disadvantage is that the ultrasonic pulse width (Δh) is wider at a low frequency than at a high frequency. In addition, in a low-frequency ultrasonic guided wave propagated over a long distance [26], the width of the first terminal reflected pulse wave (Δh) increases because of the dispersion effect, thereby expanding the range of the non-detection zone.

The first terminal reflection pulse wave is the result of the overlap of the reflected pulse waves from the end surface and the incidence surface. The width of the first terminal reflection pulse wave (Δh) varies with the transmission distance. This is related to the measuring points, the detection distance, the characteristic of the probe, and so on. In this study, the range of the non-detection zone for the inspection of the 30# square steel bar with a length of 6000 mm is used as an example; the width of the first reflected pulse (Δh) is 0.000822 s in the actual experiment (Figure 5) and 0.0005371 s in the simulation (Figure 3a).

The governing equation for the Rayleigh-Lamb waves in a plate-like structure [1] for the symmetric mode is:

$$\tan(qd)/\tan(pd) = -4k^2 pq/\left(q^2 - k^2\right)^2 \tag{28}$$

and for the anti-symmetric mode:

$$\tan(qd)/\tan(pd) = -\left(q^2 - k^2\right)^2/4k^2 pq \tag{29}$$

where $p^2 = (\omega/c_L) - k^2$, $q^2 = (\omega/c_T) - k^2$, the wave number $k = \omega/c_P$, the longitudinal wave velocity c_L, the shear velocity c_T, the phase velocity $c_P = (\omega/2\pi)\lambda$, the group velocity C_g and the thickness of the plate is d.

The formula of the group velocity theory is:

$$C_g = d\omega/dk \tag{30}$$

By using (Equations (28)–(30)) and selecting the material parameters for the iron plate and the curve of the $S0$ mode is calculated. According to the theoretical $S0$ wave group velocity (Cg_{S0}) in the dispersion curve [27] and the width of the first terminal-reflected pulse (Δh) in the experiment, the curves of the non-detection zone range are calculated and are shown in Figure 7. For the ferrous material used in this study, for which the frequency-thickness product ($f \cdot d$) is 30×30 kHz·mm, the range of the corresponding non-detection zone (ΔL) is 0.6390 m in the 6000 mm long steel square bar; This is basically coincident with the simulations. During the entire process, below a value of 2.5 MHz·mm, the range of the non-detection zone decreases with the increase in the frequency-thickness product; when the frequency-thickness product is about 2.5 MHz·mm, the minimum length of the non-detection zone is 0.2 m. In the range of 2.5 MHz·mm to 5.0 MHz·mm, the range of the non-detection zone increases with the increase in the frequency-thickness product. When the ultrasonic guided wave has a frequency-thickness product of 5.0 MHz·mm, the range of the non-detection zone varies slightly. In an actual long-distance ultrasonic guided wave propagation process, the range of the non-detection zone expands with the widening of the first terminal-reflected pulse wave, which is a new reference factor to be considered. Similarly, the simulations are conducted using different materials [1] and the other conditions remain unchanged, as the simulated non-detection zone shows in Figure 7. The materials have different properties, resulting in the increase in the minimum length of the non-detection zone. Because the minimization of the non-detection zone is desirable, alloying may present a solution in the low-frequency range.

Although the flaw's pulse signal is lost in monitoring Interval I, the reflected pulses carrying the flaw information can still be detected in other monitoring intervals, thus allowing the determination of the flaw's location. By analyzing the simulation results that satisfy the conditions for the appearance of the non-detection zone, it can be found that the overall pulse waveform exhibits apparent variations in monitoring Interval II.

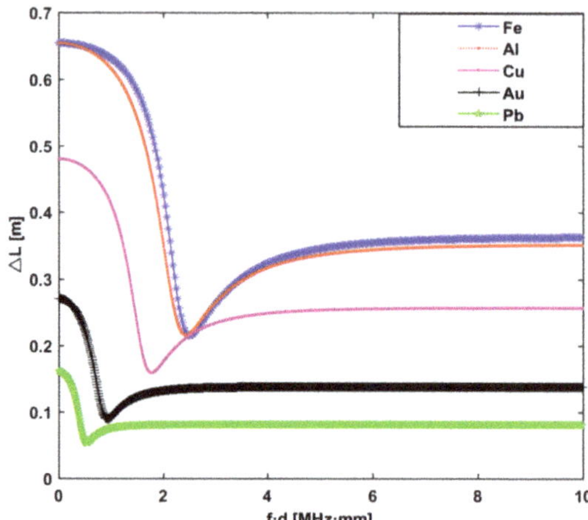

Figure 7. Curves of the non-detection zone ranges for different materials.

4.2. Characteristics of the Reflected Pulse Wave in Monitoring Interval II

This study attempts to locate the flaw in the non-detection zone using the reflected pulse wave in monitoring Interval I and monitoring Interval II based on a simulation and experimental results. The primary reasons are listed below. Firstly, the amplitude of the reflected pulse in the monitoring Interval II exceeds that encountered in the following monitoring intervals. Secondly, because the existence of the flaw, an increasing number of derivative pulse waves will occur as the propagation time increases; the condition of the pulse wave superposition is simpler in monitoring Interval II than in the following monitoring Intervals. Therefore, we used the square steel bar with a length of 6000 mm as an example and analyzed the variation of the reflected pulse waveform in monitoring Interval II.

Due to the variation of the pulse width of the terminal-reflected pulse wave and the waveform superposition in the actual experiments, the non-detection zone is observed in advance when $L = 5300$ mm. Therefore, this study focused on the simulation range of [5300 mm, 5900 mm] with $D0 = 10$ mm and the simulation interval of 100 mm. The analysis indicates that the amplitude changed with the changes in $D0$, leading to new changes in the envelope superposition of the incident pulse in Interval II and especially in the superposition of the amplitudes. We used $D0 = 10$ mm as an example of the representative and significant waveform change; other values of $D0$ can also be used for the same method.

Figure 8 shows the simulation results. It can be observed that, as the flaw location (L) changes, the amplitude of the superposed pulse wave exhibits complex changes in the time domain, especially in Interval II. Moreover, as the flaw location (L) changes, the superposition envelope of multiple reflected pulses changes significantly in monitoring Interval II, i.e., both of these factors showed a one-to-one correspondence. Within the monitoring Interval II, there exists multiple reflected wave energy; at first, at around $L = 5300{\sim}5500$ mm, the highest reflected pulse gradually moves backward; at around $L = 5600{\sim}5700$ mm, the two highest reflected pulses appear in adjacent time-domains; at around $L = 5800{\sim}5900$ mm, the flaw position is close to the terminal, the energy gradually becomes concentrated, and the amplitudes of these flaw-reflected pulse waves are significantly higher than the amplitude of the first terminal reflected pulse wave for the two conditions. Therefore, we attempted to use the above-mentioned correspondence and estimated the location of the flaw through reverse deduction based on envelope comparison.

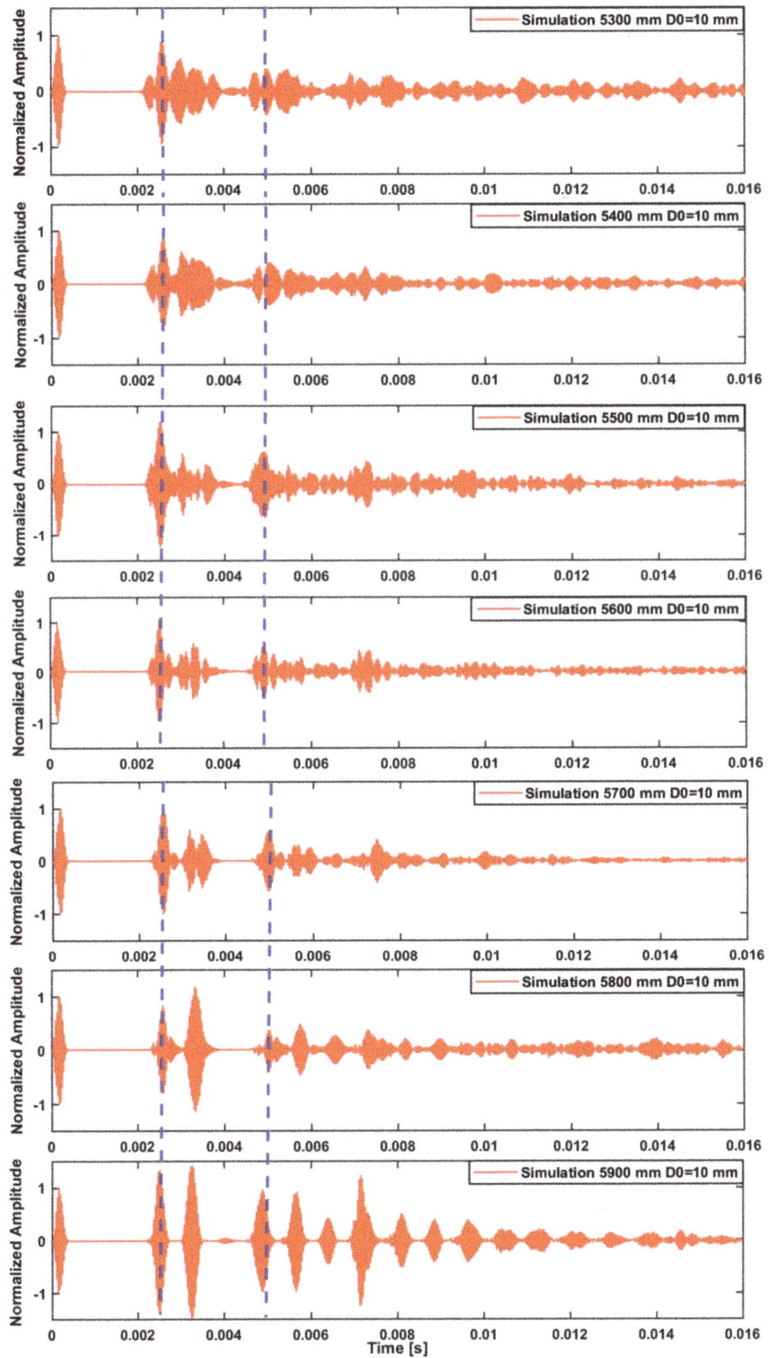

Figure 8. Simulated waveforms of the square steel bar with a length of 6000 mm at $L \in [5300$ mm, 5900 mm].

4.3. Approximation Based on Envelope Comparison

Although the reflected pulses show complex superposition patterns in Interval II, the propagation speed of the fundamental mode remains relatively stable and the superposition state varies correspondingly as the flaw location L changes along the x-direction. If the energy of the excitation pulse signal, the flaw location L, and $D0$ were fixed, the propagation energy would attenuate. Compared with the condition without any flaws, the amplitude of the pulse overall decreases but the amplitude of the pulse reflected by the flaw increases. The difference in the envelope between these two different conditions (with and without the flaw) of the pulse waves can be used to identify the reflected pulse.

Because the ultrasonic reflected pulse wave from the flaws located in monitoring Interval I is invalid, we decided to use the envelope wave of the pulse that is extended to monitoring Interval II to locate the flaws. To resolve the interference problem caused by the trailing of the actual experimental data and the unexpectedly generated waves, firstly, we use the envelope of the waveforms of two areas (for example, $D0 = 0$ mm and $D0 = 10$ mm) in the simulation and in the experiment and calculate the difference in the envelope amplitude of the two areas; these are used as basic data for comparison.

Firstly, by observing these envelopes shown in Figure 9a–f, a coarse division of the flaw areas is proposed based on the analysis in Section 4.2.

- Situation 1: in the actual experimental envelope during monitoring Interval II, the reflected pulse with only one envelope amplitude that is larger than the amplitude of the terminal reflected pulse wave is categorized into a zone ranging from $L = 5800$ mm to $L = 5900$ mm;
- Situation 2: the reflected pulses with two similar highest envelope amplitudes are categorized into a zone ranging from $L = 5600$ mm to $L = 5700$ mm;
- Situation 3: the remaining flaw detection pulses with different amplitudes are categorized into a zone from $L = 5400$ mm to $L = 5500$ mm.

Secondly, we selected the effective time domain intervals for comparison (from 7.649×10^{-4} s to 4.653×10^{-3} s intervals within the monitoring Intervals I and II) to avoid the interference caused by the trailing of the incident pulse and the second terminal-reflected pulse wave, as shown in Figure 9f.

Thirdly, the envelope theory (ET) is known as an auxiliary field method [28,29]; therefore, we could use this convenient method for the wave comparison between the simulation and the measurements. The envelope correlation algorithm is applied to compute the maximum correlation coefficient in order to choose the most similar envelope waveforms in a quantitative approach and locate the flaws.

The discretization cross-correlation function can be represented as follows:

$$R_{se}(k) = \sum_{n=0}^{N-1} s(n)e(n-k) \quad k = 0, 1, 2, \cdots, M - 1 \tag{31}$$

where $R_{se}(k)$ denotes the correlation function for the simulation data $s(n)$ and the experimental data $e(n)$. N is the cumulative average number and k is the time-lapse sequence number, M is a positive integer. The formula for the envelope correlation is expressed as follows:

$$\rho = R_{se}(0) / \sqrt{\sum_{n=0}^{N-1} s^2(n) \sum_{n=0}^{N-1} e^2(n)} \tag{32}$$

The scan comparisons are not required ($k = 0$) in Equation (31). The envelope correlation coefficient (ρ) of the experimental data and the simulation data are obtained using Equation (32), and the envelope whose simulation data is closest to the experimental data envelope is selected as the indicator for locating the corresponding flaw.

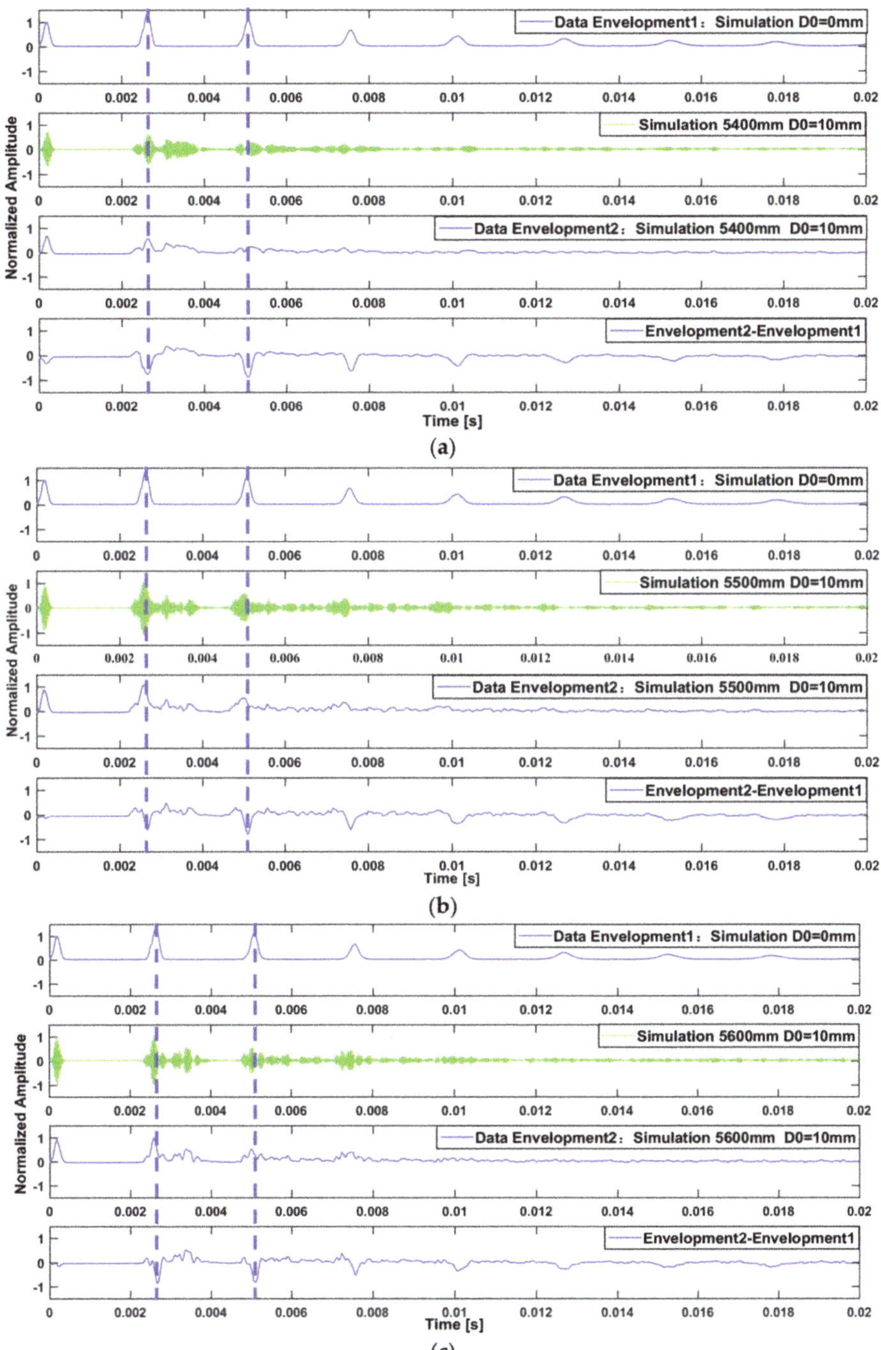

Figure 9. *Cont.*

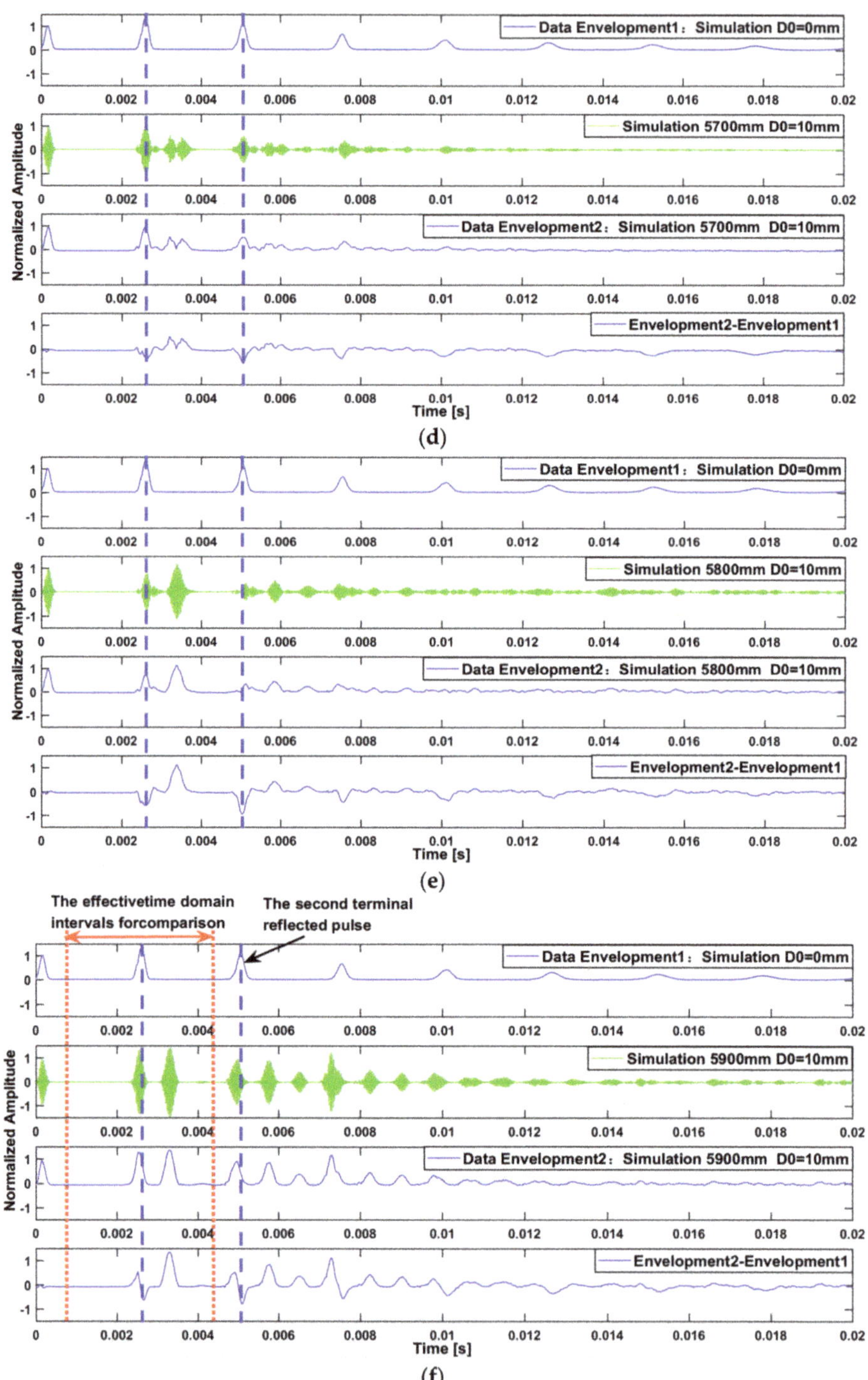

Figure 9. Envelope calculation results of the pulse wave at different locations of flaw: (**a**) $L = 5400$ mm $D0 = 10$ mm (**b**) $L = 5500$ mm $D0 = 10$ mm (**c**) $L = 5600$ mm $D0 = 10$ mm (**d**) $L = 5700$ mm $D0 = 10$ mm (**e**) $L = 5800$ mm $D0 = 10$ mm (**f**) $L = 5900$ mm $D0 = 10$ mm.

From the above analysis, the approximation is based on the inverse comparison with the simulation envelope in the non-detection zone. As shown in Figure 10, the main steps are as follows:

Data pre-processing: we prepared the envelope data in the experiment and the simulation without the flaw. The experimental data can be replaced by the experimental data of a similar square steel bar without flaws in practical measurements. For the simulation data, we can use $D0 = 0$ mm.

- Step 1: The experimental data with the flaw obtained in the non-detection zone are processed using the envelope calculation and we subtract the envelope of the experimental data without the flaw to reduce the effect of the signal trailing. Then the result of the envelope is extracted in the effective comparison interval (from 7.649×10^{-4} s to 4.653×10^{-3} s).
- Step 2: Based on the obtained experimental envelope (Envelopment 2–Envelopment 1), the simulation range ($L \in [L_1, L_2]$) is selected according to the rough classification described in Section 4.3 to reduce the computation cost.
- Step 3: According to the comparison requirements, the variable value of L is selected in the appropriate range (L_1 mm $\leq L_1 + k\Delta x \leq L_2$ mm, $\Delta x = 100$ mm, $k \in$ N), the depth of the flaw (y) in the experiment is selected as the value of $D0$ in the simulation (0 mm $\leq D0 = y < 30$ mm), and the simulation results have been processed using the envelope calculations respectively. For every simulation envelope, the envelope of the simulation without the flaw ($D0 = 0$ mm) is subtracted. Then the results of these envelopes are extracted in the comparison interval as a series of comparisons.
- Step 4: The correlation coefficients between the experimental envelope data from Step 1 and every simulation envelope data from Step 3 are determined and the result returns the maximum correlation coefficient ρ; subsequently, the estimated position of the flaw (L) is determined according to the reverse correspondence.

Final data processing: By changing the value of Δx, a similar approximation is used near the initial estimated value of L to approximate the real value of L in a further simulation.

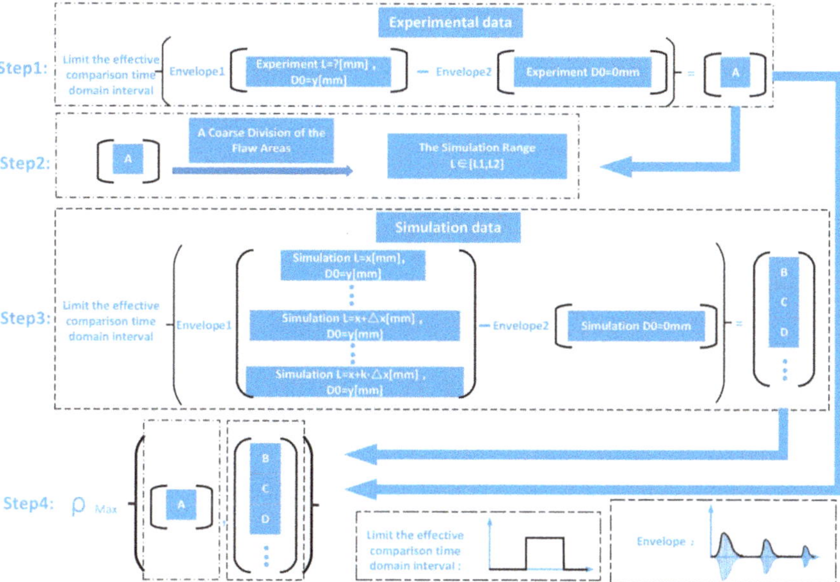

Figure 10. The schematic diagram of the envelope approximation method.

4.4. Experimental Validation

In order to localize the flaw in the non-detection zone, the different envelopes of the simulated envelope waveforms (Envelopment 2–Envelopment 1) are used as the evaluation standard to compare with those of the experimental data. This section describes the validation of the experiment. In order to depict the validation results, instead of strictly following the steps mentioned in Section 4.3, the preliminary classification in Step 2 was omitted. The variable value of L is selected from an appropriate range (L_1 = 5400 mm, L_2 = 5900 mm) in the simulation. $D0$ = 10 mm is used as an example of the experimental data to determine the location of the flaw (L).

Figure 11a shows the experimental pulse waveform of the square steel bar without the flaw ($D0$ = 0 mm) and the experimental and simulated pulse waveforms of the square steel bar with the flaw ($D0$ = 10 mm). Using the envelope calculation, it can be observed that the envelope waveform in monitoring Interval II reduces the interference of the signal trailing in the experiments. Compared with the simulation data, it is evident that the peak points of the difference envelopes in the experiment are consistent, as shown in the red dotted line. The time relative deviation σ_1 can be calculated by:

$$\sigma_1 = (0.003187 - 0.003107)/0.003187 = 2.5101\% \tag{33}$$

The flaw's reflected information that is hidden and the superposition envelope of the flaw-reflected pulse in the trailing are extracted (Envelopment 2–Envelopment 1).

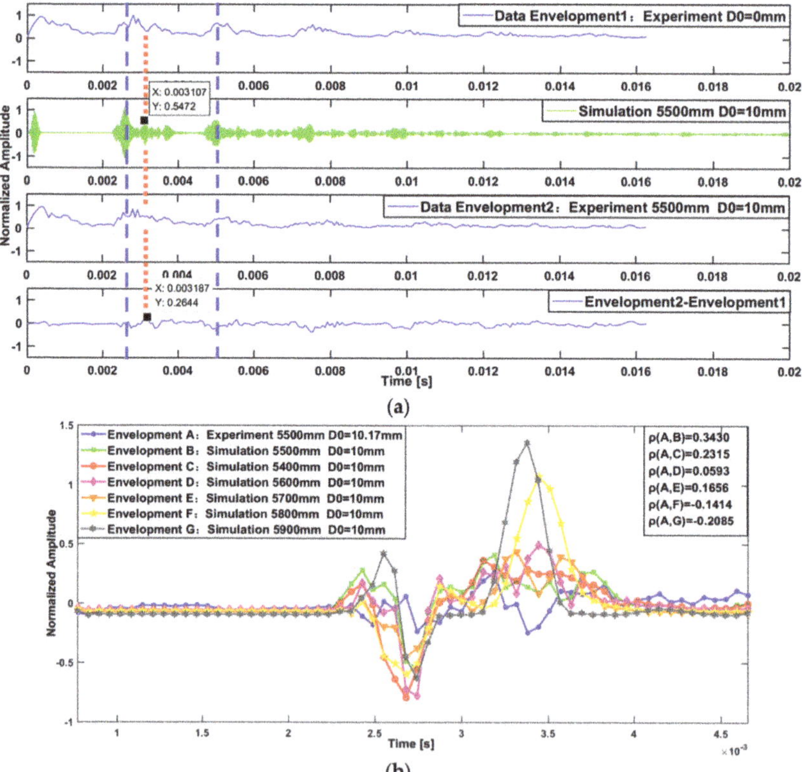

Figure 11. Location of the flaw through the envelope approximation method: (**a**) The envelope of experimental data processing (**b**) The comparison of multiple envelopes in the effective comparison interval.

Subsequently, the results of the different envelopes (Envelopment 2–Envelopment 1) in the simulation and the experiment are extracted in the comparison interval, as shown in Figure 11b. The correlation coefficients (ρ) of the experimental data and the different simulation data are determined. Because the simulated envelope waveform is used as the evaluation standard, the noise introduced by the real application cannot be eliminated by the correlation algorithm; therefore, the correlation coefficient is relatively low. However, we can still determine that the correlation coefficient, which is negative, should be discarded in the range from $L = 5800$ mm to $L = 5900$ mm; in the range from $L = 5600$ mm to $L = 5700$ mm, the envelopes of the waveform are not well matched. However, in the range from $L = 5400$ mm to $L = 5500$ mm, the approximated simulation waveforms in the flaw area ($L = 5500$ mm) have a significant correlation coefficient of 0.3430; therefore, $L = 5500$ mm is selected as the estimated value of the location of the flaw. If these calculating steps are strictly carried out according to the solution in the previous section, based on the preliminary rough classification, it is found that the actual measurement data range from $L = 5400$ mm to $L = 5500$ mm and we can obtain similar results rapidly to approximate the real value of L in further simulations.

Therefore, the flaw in the non-detection zone can be identified using the inverse correspondence. Using an equivalent mode and approximation algorithms, this methodology provides estimations of the locations of flaws relatively accurately.

Based on the analysis in Section 2.1.1, when the equivalent plate thickness (d) is constant, as long as the frequency is below the cutoff frequency of $A1$ and $S1$, the excitation frequency in this range can be used for low-frequency ultrasound mode selection. For example, the proposed analysis is applicable to the ultrasonic probe core frequency range of 20–60 kHz low-frequency ultrasonic guided wave detection. The concept of detecting flaws using simulations also applies to different types of flaws. In this study, we extended the analysis to a triangular flaw. This approach also resulted in a similar positioning effect. When $f \cdot d$ is known, the modal phase velocity and the group velocity basically remain unchanged, whereas there is a one-to-one correspondence between the type of the reflected pulse wave and the location of the flaw. Therefore, in this study, we use a triangular flaw (40 kHz, $D0 = 5$ mm, $L = 5700$ mm) as another example.

Based on the data analysis shown in Figure 12, we can obtain a similar conclusion; the flaw area ($L = 5700$ mm) has a significant correlation coefficient (ρ) of 0.3091. Therefore, $L = 5700$ mm is then selected as the estimated value of L (the location of the flaw). The solution remains valid, but the characteristics of the waveform superposition have changed, which can be used to identify the triangle type of the flaw.

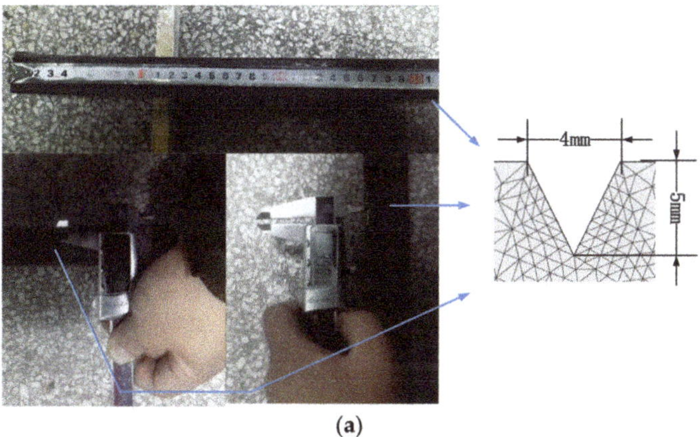

(a)

Figure 12. *Cont.*

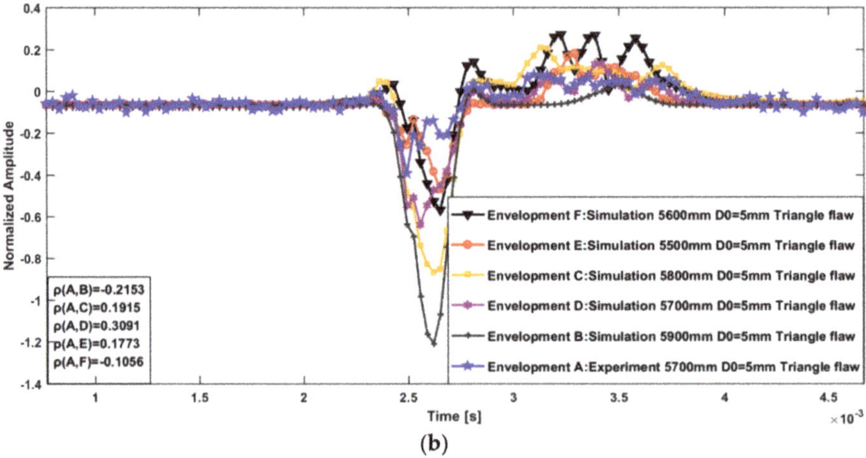

Figure 12. Location of a triangle flaw. (**a**) The triangle type of the flaw on the steel square bar; (**b**) The comparison of multiple envelopes in the effective comparison interval for the triangle flaw.

The relevant analysis and the solution presented in this paper are still valid for low-frequency ultrasonic guided wave detection in a rod-like structure medium. We can also obtain a similar conclusion for the non-detection zone problem for a rail structure. For example, a rail experiment using 60 kHz as the actual measurement frequency was conducted by Rose [9]. Although the experiment used different excitation pulse waveforms, which were affected by other derived modes wave and pulse trailing, the superimposed reflected pulse waveform obtained in the second monitoring time domain and the peak value was far higher than the peak value of the first reflected pulse; this result is very similar to the results of the characteristic analysis in this study.

5. Conclusions

Previous studies have provided no clear statements or solutions regarding the non-detection zone in low-frequency ultrasonic guided waves. However, the existence of the non-detection zone cannot be ignored in NDT. In this study, we located the non-detection zone in long-distance conventional ultrasonic guided wave inspections of a square steel bar and investigated the reasons and conditions for the appearance of the non-detection zone through simulations. Further, by analyzing the characteristics of the simulated pulse wave when the non-detection zone appears, the range of the non-detection zone and a solution are proposed. The simulation envelope is used for a relative comparison; by suppressing the interference of the ultrasonic signal trailing using the envelope calculation, the flaw is located by a comparison with the simulation results of the multi-pulse superimposed envelope in the extended time domain (the monitoring time Intervals I and II). The simulation results are verified with actual measurements. The results of this study provide a feasible solution for locating the non-detection zone and serve as a powerful supplement for conventional long-distance ultrasonic guided wave inspections. This type of non-detection zone can also occur in the ultrasonic flaw detection of long rod-like structures; therefore, this study has certain applicability and practical significance for other structures. However, due to the influence of actual field factors on the actual detection of the pulse energy loss, the amplitude deviation between the equivalent simulation and the actual measurement is large in this study and is not conducive to the prediction of the depth ($D0$) of the flaw. For this reason, it is necessary to improve the approximate model of the low-frequency ultrasonic guided wave and the estimation of various types of flaws to provide a new approach that is widely applicable.

Acknowledgments: This work was supported by the Key Research and Development Program of Shaanxi Province (Grant No. 2017ZDXM-GY-130), Xi'an City Science and Technology Project (Grant No. 2017080CG/RC043 (XALG009)).

Author Contributions: Yuan Yang and Lei Zhang performed the theoretical analysis, conceived and designed the experiments; Lei Zhang, Wenqing Yao and Xiaoyuan Wei performed the experiments and analyzed the data; Lei Zhang wrote the paper.

References

1. Rose, J.L. *Ultrasonic Guided Waves in Solid Media*, 1st ed.; Cambridge University Press: New York, NY, USA, 2014; ISBN 978-110-70489-59.
2. Rose, J.L. *Ultrasonic Waves in Solid Media*, 1st ed.; Cambridge University Press: New York, NY, USA, 1999; ISBN 0521640431.
3. Ai, C.Y.; Li, J.; Liu, Y.; Ni, T. *Ultrasonic Test Theory and Technology for Multilayered Bonding Structure*, 1st ed.; National Defence Industry Press: Beijing, China, 2014; ISBN 978-711-80958-07.
4. Verma, B.; Mishra, T.K.; Balasubramaniam, K.; Rajagopal, P. Interaction of low-frequency axisymmetric ultrasonic guided waves with bends in pipes of arbitrary bend angle and general bend radius. *Ultrasonics* **2014**, *54*, 801–808. [CrossRef] [PubMed]
5. Ryue, J.; Thompson, D.J.; White, P.R.; Thompson, D.R. Decay rates of propagating waves in railway tracks at high frequencies. *J. Sound Vib.* **2009**, *320*, 955–976. [CrossRef]
6. Gravenkamp, H.; Prager, J.; Saputra, A.A.; Song, C. The simulation of lamb waves in a cracked plate using the scaled boundary finite element method. *J. Acoust. Soc. Am.* **2012**, *132*, 1358–1367. [CrossRef] [PubMed]
7. Bartoli, I.; Scalea, F.L.D.; Fateh, M.; Viola, E. Modeling guided wave propagation with application to the long-range defect detection in railroad tracks. *NDT E Int.* **2005**, *38*, 325–334. [CrossRef]
8. Campos-Castellanos, C.; Gharaibeh, Y.; Mudge, P.; Kappatos, V. In the application of long range ultrasonic testing (LRUT) for examination of hard to access areas on railway tracks. *Railw. Cond. Monit. Non-Destr. Test.* **2012**, 1–7. [CrossRef]
9. Rose, J.L.; Avioli, M.J.; Mudge, P.; Sanderson, R. Guided wave inspection potential of defects in rail. *NDT E Int.* **2004**, *37*, 153–161. [CrossRef]
10. Wu, J.; Wang, Y.; Zhang, W.; Nie, Z.; Lin, R.; Ma, H. Defect detection of pipes using Lyapunov dimension of Duffing oscillator based on ultrasonic guided waves. *Mech. Syst. Signal Proc.* **2016**, *82*, 130–147. [CrossRef]
11. Yao, W.; Sheng, F.; Wei, X.; Zhang, L.; Yang, Y. Propagation characteristics of ultrasonic guided waves in continuously welded rail. *Mod. Phys. Lett. B* **2017**, 1740075. [CrossRef]
12. Su, Z.; Ye, L.; Lu, Y. Guided lamb waves for identification of damage in composite structures: A review. *J. Sound Vib.* **2006**, *295*, 753–780. [CrossRef]
13. Fateri, S.; Lowe, P.S.; Engineer, B.; Boulgouris, N.V. Investigation of ultrasonic guided waves interacting with piezoelectric transducers. *IEEE Sens. J.* **2015**, *15*, 4319–4328. [CrossRef]
14. Mirahmadi, S.J.; Honarvar, F. Application of signal processing techniques to ultrasonic testing of plates by lamb wave mode. *NDT E Int.* **2011**, *44*, 131–137. [CrossRef]
15. Pai, P.F.; Deng, H.; Sundaresan, M.J. Time-frequency characterization of lamb waves for material evaluation and damage inspection of plates. *Mech. Syst. Signal Proc.* **2015**, *62–63*, 183–206. [CrossRef]
16. Deng, Q.T.; Yang, Z.C. Scattering of S_0 lamb mode in plate with multiple damages. *Appl. Math. Model.* **2011**, *35*, 550–562. [CrossRef]
17. Wan, X.; Tse, P.W.; Chen, J.; Xu, G.; Zhang, Q. Second harmonic reflection and transmission from primary s0 mode lamb wave interacting with a localized microscale damage in a plate: A numerical perspective. *Ultrasonics* **2017**, *82*, 57–71. [CrossRef] [PubMed]
18. Glushkov, E.; Glushkova, N.; Eremin, A.; Giurgiutiu, V. Low-cost simulation of guided wave propagation in notched plate-like structures. *J. Sound Vib.* **2015**, *352*, 80–91. [CrossRef]
19. Rucka, M. Experimental and numerical study on damage detection in an l-joint using guided wave propagation. *J. Sound Vib.* **2010**, *329*, 1760–1779. [CrossRef]
20. Le, C.E.; Castaings, M.; Hosten, B. The interaction of the s0 lamb mode with vertical cracks in an aluminium plate. *Ultrasonics* **2002**, *40*, 187–192. [CrossRef]

21. Yang, Y.; Wei, X.; Zhang, L.; Yao, W. The effect of electrical impedance matching on the electromechanical characteristics of sandwiched piezoelectric ultrasonic transducers. *Sensors* **2017**, *17*, 2832. [CrossRef] [PubMed]
22. Wei, X.; Yang, Y.; Yao, W.; Zhang, L. Pspice modeling of a sandwich piezoelectric ceramic ultrasonic transducer in longitudinal vibration. *Sensors* **2017**, *17*, 2253. [CrossRef] [PubMed]
23. Zhao, Y.; Li, F.; Cao, P.; Liu, Y.; Zhang, J.; Fu, S.; Zhang, J.; Hu, N. Generation mechanism of nonlinear ultrasonic lamb waves in thin plates with randomly distributed micro-cracks. *Ultrasonics* **2017**, *79*, 60–67. [CrossRef] [PubMed]
24. Loveday, P.W. Simulation of piezoelectric excitation of guided waves using waveguide finite elements. *IEEE Trans. Ultrason. Ferroelectr. Freq. Control* **2008**, *55*, 2038. [CrossRef] [PubMed]
25. Ryue, J.; Thompson, D.J.; White, P.R.; Thompson, D.R. Investigations of propagating wave types in railway tracks at high frequencies. *J. Sound Vib.* **2008**, *315*, 157–175. [CrossRef]
26. Zhang, Y.; Li, D.; Zhou, Z. Time reversal method for guided waves with multimode and multipath on corrosion defect detection in wire. *Appl. Sci.* **2017**, *7*, 424. [CrossRef]
27. Zhang, H.Y.; Xu, J.; Ma, S.W.; Ta, D.A. The high-frequency scattering of the s0 lamb mode by a circular blind hole in a plate. *Phys. Procedia* **2015**, *70*, 455–458. [CrossRef]
28. Santacruz, J.; Tardón, L.; Barbancho, I.; Barbancho, A. Spectral envelope transformation in singing voice for advanced pitch shifting. *Appl. Sci.* **2016**, *6*, 368. [CrossRef]
29. Semay, C. The hellmann–feynman theorem, the comparison theorem, and the envelope theory. *Results Phys.* **2015**, *5*, 322–323. [CrossRef]

Article

Research on a Rail Defect Location Method Based on a Single Mode Extraction Algorithm

Bo Xing [1], Zujun Yu [1,2], Xining Xu [1,2,*], Liqiang Zhu [1,2] and Hongmei Shi [1,2]

[1] School of Mechanical, Electronic and Control Engineering, Beijing Jiaotong University, Beijing 100044, China; 15116333@bjtu.edu.cn (B.X.); zjyu@bjtu.edu.cn (Z.Y.); lqzhu@bjtu.edu.cn (L.Z.); hmshi@bjtu.edu.cn (H.S.)

[2] Key Laboratory of Vehicle Advanced Manufacturing, Measuring and Control Technology (Beijing Jiaotong University), Ministry of Education, Beijing 100044, China

* Correspondence: xnxu@bjtu.edu.cn

Received: 14 February 2019; Accepted: 11 March 2019; Published: 15 March 2019

Abstract: This paper proposes a rail defect location method based on a single mode extraction algorithm (SMEA) of ultrasonic guided waves. Simulation analysis and verification were conducted. The dispersion curves of a CHN60 rail were obtained using the semi-analytical finite element method, and the modal data of the guided waves were determined. According to the inverse transformation of the excitation response algorithm, modal identification under low-frequency and high-frequency excitation was realized, and the vibration displacements at other positions of a rail were successfully predicted. Furthermore, an SMEA for guided waves is proposed, through which the single extraction results of four modes were successfully obtained when the rail was excited along different excitation directions at a frequency of 200 Hz. In addition, the SMEA was applied to defect location detection, and the single reflection mode waveform of the defect was extracted. Based on the group velocity of the mode and its propagation time, the distance between the defect and the excitation point was measured, and the defect location was predicted as a result. Moreover, the SMEA was applied to locate the railhead defect. The detection mode, the frequency, and the excitation method Were selected through the dispersion curves and modal identification results, and a series of signals of the sampling nodes were obtained using the three-dimensional finite element software ANSYS. The distance between the defect and the excitation point was calculated using the SMEA result. When compared with the structure of the simulated model, the errors obtained were all less than 0.5 m, proving the efficacy of this method in precisely locating rail defects, thus providing an innovated solution for rail defect location.

Keywords: rail; ultrasonic guided wave; semi-analytical finite element; single mode extraction algorithm; defect location

---

1. Introduction

Ultrasound guided waves are widely used in the non-destructive testing of continuous welded rails because of their wide coverage and rapid propagation over large distances [1,2]. Compared with simple waveguide structures, such as plates and pipelines, rails have much more complex cross-sectional structures; thus, they require far more complicated ultrasonic wave modes for defect detection. With the increase in the frequency, the number of modes increases, making the research more difficult. However, the analytical method is no longer suitable for the analysis of the propagation characteristics of ultrasonic guided waves in rails. Therefore, numerical analysis methods have been broadly introduced, including the Boundary Element Method [3], the Finite Difference Method [4], the Finite Element Method [5], and the Semi-Analytical Finite Element Method [6]. Among them, the semi-analytical finite element method reduces the analysis dimensions and has a higher computational efficiency, thus is commonly adopted in this field of researches.

The time difference between the main wave packet of the reflected wave and the excitation wave packet in the acquisition signal must be calculated; the position of the defect can be inferred by multiplying the group velocity of the main modes. Its accuracy is related to the selected detection mode. Therefore, it is necessary to have a good understanding of the propagation characteristics of guided waves in rails to select the appropriate modal type, frequency, and excitation mode for defect detection. D. Alleyne and P. Cawley [7] used two-dimensional Fourier transform to separate the guided wave modes and calculate the corresponding phase velocities.T. Hayashi et al. [8] utilized the semi-analytical finite element (SAFE) method to calculate the group velocity and phase velocity dispersion curves of guided waves in rails at frequencies of 0–100 kHz and carried out experimental verification. He Cunfu et al. [9] used the vibration mode analysis method to analyze the guided wave modes at frequencies of 0–50 kHz; the dispersion characteristics and wave structures of five typical modes were analyzed, and the mode types and frequency ranges suitable for inspecting railhead and rail bottom defects were obtained. P.W. Loveday and C.S. Long [10–14] analyzed the guided wave mode data in rails at frequencies of 25 kHz and 35 kHz using the SAFE method and measured along the rails using a laser vibrometer. A large number of vibration data were obtained, the amplitudes of each mode were calculated, and the vibration waveforms of distant rails were successfully predicted. In 2017, P.W. Loveday et al. [15] studied the problem of the mode repulsion and crossing behavior of the approaching wave number versus frequency curves by analyzing the second derivative of the eigenvalue with respect to the wave number. These methods analyze the modal characteristics of guided waves propagating in rails, giving a modal analysis. However, it has never been possible to separate the reflected guided wave modes of defects to use them for defect location directly.

Meanwhile, a large number of research teams have applied ultrasonic guided waves for defect detection in waveguide structures. Most of the research results focus on simple cross-sectional structures such as plates and pipes, while studies on complex structures such as rails remain relatively rare. C.M. Lee et al. [2] adopted the finite element method to analyze the energy distribution characteristics of guided waves with different frequencies in rails. The simulation results showed that a high-frequency guided wave of 200 kHz is concentrated on the upper surface of the railhead, and a low-frequency guided wave of 30 kHz is distributed through the entire railhead. The guided waves at low frequencies are more sensitive to the transverse cracks of the railhead, while the guided waves at high frequencies are good at detecting defects in the shelling. These results have been experimentally verified. Lu Chao et al. [16] selected transverse and vertical vibration modes to detect oblique cracks on the rail bottom and analyzed the relationship between the angle of the oblique cracks and the scattering characteristics of the guided waves. G. Zumpano et al. [17] utilized finite element software to simulate the rail wear and applied different excitation frequencies. Location analysis shows that the location error is affected by the excitation frequency. However, as a result of the dispersion characteristics of ultrasonic guided waves and the frequency aliasing phenomenon, it is impossible to accurately locate the defect from the group velocity. For this reason, there are relatively few studies on the application of ultrasonic guided waves in rail defects location. Nevertheless, because of the results related with the prediction of defect sizes and the identification of defect types, it seems practical to study this method in terms of defect location.

To locate rail cracks accurately by using ultrasonic guided waves, it is generally necessary to obtain a single mode propagating in the rail. The research ideas can be roughly divided into two kinds: one is to directly excite a single mode in the rail, and the other is to extract a single mode from the signal propagating in the rail. In previous studies, the research group proposed a single mode excitation method, which can excite a relatively pure single mode in rails [18]. At the same time, the group also carried out research on the optimal mode selection of rail crack detection in which a selection model of crack detection mode is created. For a specific crack, the guided wave frequency and mode suitable for crack detection are quickly selected. In this paper, a single mode extraction algorithm (SMEA) is proposed to ascertain the precise location of defects. The method is as follows: Firstly, the mode and frequency of defect detection are selected and a three-dimensional model of the defect present in the railhead is established. The simulation analysis is carried out using the three-dimensional finite element analysis software ANSYS. The defect models of the railhead are stimulated with low-frequency (200 Hz) and high-frequency (60 kHz) signals. The content of each mode of guided wave is quantitatively analyzed, and the position of the railhead defect is calculated according to the single mode extraction method. The semi-analytical finite element method and accurate modal identification method are described in Section 2. In Section 3, the SMEA and verification process are discussed. The defect location method and simulation results are described in Section 4. Conclusions are given in Section 5.

2. An Accurate Modal Identification Method

2.1. Basic Characteristics of Ultrasonic Guided Waves in Rails

As a result of the complex cross-sectional structure of rails, the number of guided wave modes propagating in a rail is large. To detect the internal defects of rails based on ultrasonic guided wave technology, the most important thing is to grasp the fundamental characteristics of ultrasonic guided waves in rails, such as the frequency, wave number, phase velocity, group velocity, mode shape, and other information, so as to analyze the propagation characteristics of the guided waves. The dispersion curves of ultrasonic guided waves in rails can be obtained with the semi-analytical finite element method. Taking the rail laid on Beijing–Shanghai high-speed railway in China as the research object [19], the finite element method was used to discretize the cross-section. It was assumed that the guided waves propagate along the longitudinal direction of the rail in the form of harmonic vibrations. The wave equation was established based on the finite element method. The eigenvalues and eigenvectors were obtained by solving the eigenvalue equation. The eigenvalues contain the information of frequencies and wave numbers, and the eigenvectors contain the information of mode shapes. Thus, the dispersion curves and mode shapes of guided waves in the CHN60 rail were obtained.

First, the coordinate system of the CHN60 rail was established, as shown in Figure 1.

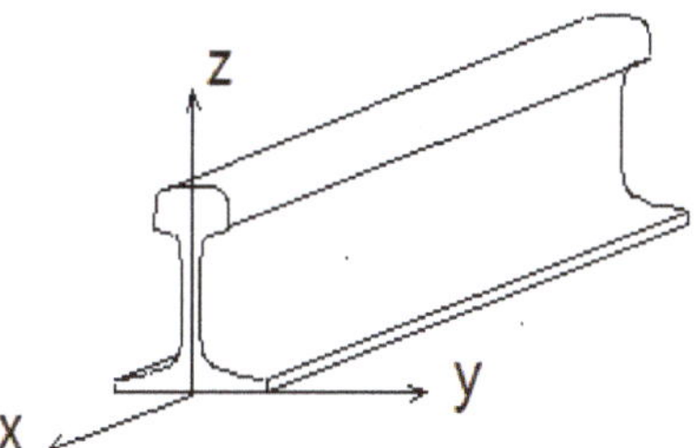

Figure 1. CHN60 rail coordinate system.

The wave number of guided waves is ζ and the frequency is ω. The displacement, stress, and strain of each node in the rail can be expressed as Equation (1).

$$u = \begin{bmatrix} u_x & u_y & u_z \end{bmatrix}^T$$

$$\sigma = \begin{bmatrix} \sigma_x & \sigma_y & \sigma_z & \sigma_{yz} & \sigma_{xz} & \sigma_{xy} \end{bmatrix}^T \tag{1}$$

$$\varepsilon = \begin{bmatrix} \varepsilon_x & \varepsilon_y & \varepsilon_z & \gamma_{yz} & \gamma_{xz} & \gamma_{xy} \end{bmatrix}^T$$

ε_x, ε_y, and ε_z are normal strains, and γ_{yz}, γ_{xz}, and γ_{xy} are shear strains. In the elastic range of materials, stress and strain conform to Hooke's Law. It can be expressed as $\sigma = C\varepsilon$, where C is the elastic constant matrix of a rail.

The relationship between the strain and displacement at any node in the rail is expressed in a matrix form as follows:

$$\varepsilon = \left[L_x \frac{\partial u}{\partial x} + L_y \frac{\partial u}{\partial y} + L_z \frac{\partial u}{\partial z} \right] \tag{2}$$

Therefore, in the form

$$L_x = \begin{bmatrix} 1 & 0 & 0 \\ 0 & 0 & 0 \\ 0 & 0 & 0 \\ 0 & 0 & 0 \\ 0 & 0 & 1 \\ 0 & 1 & 0 \end{bmatrix} L_y = \begin{bmatrix} 0 & 0 & 0 \\ 0 & 1 & 0 \\ 0 & 0 & 0 \\ 0 & 0 & 1 \\ 0 & 0 & 0 \\ 1 & 0 & 0 \end{bmatrix} L_z = \begin{bmatrix} 0 & 0 & 0 \\ 0 & 0 & 0 \\ 0 & 0 & 1 \\ 0 & 1 & 0 \\ 1 & 0 & 0 \\ 0 & 0 & 0 \end{bmatrix} \tag{3}$$

The SAFE method was used to get the dispersion curves of the ultrasonic guided waves which propagate in the form of harmonics in the longitudinal direction of the rail. Therefore, only the finite element discretization of the rail cross-section was needed. The displacement of any discrete node in the rail can be calculated as shown in Equation (4), where x is the longitudinal coordinate of the rail.

$$u(x,y,z,t) = \begin{bmatrix} u_x(x,y,z,t) \\ u_y(x,y,z,t) \\ u_z(x,y,z,t) \end{bmatrix} = \begin{bmatrix} U_x(y,z) \\ U_y(y,z) \\ U_z(y,z) \end{bmatrix} e^{-i(\xi x - \omega t)} \tag{4}$$

The triangular element was selected to discretize the cross-section of the CHN60 rail, and 177 nodes and 255 elements were obtained, as shown in Figure 2. Seven nodes are circled in the figure to illustrate the signal extraction nodes in the subsequent modal identification method.

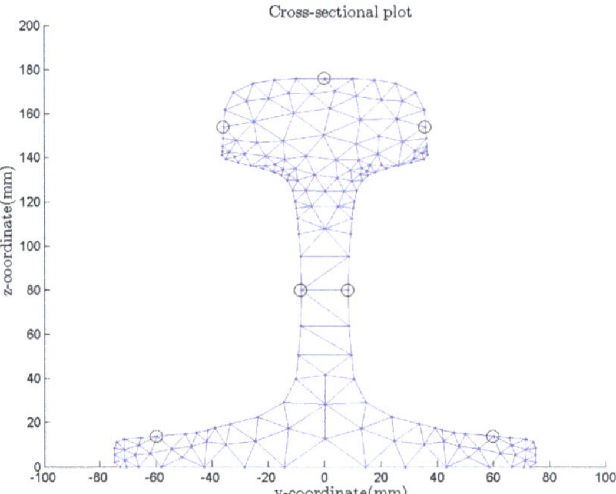

Figure 2. Discretization of the cross-section of the CHN60 rail.

The rail cross-section was discretized by triangular elements. The displacement of any particle in an element can be obtained by the displacement of the nodes and the shape function. The strain and stress vectors of the element can also be expressed by the displacement of the nodes. According to the Hamilton principle, the dynamic equation of guided waves propagating in a CHN60 rail can be obtained by calculating the strain energy and potential energy at any point simultaneously [20]:

$$\left[K_1 + i\xi K_2 + \xi^2 K_3 - \omega^2 M \right]_M U = 0 \tag{5}$$

In this form, U contains the displacements of the nodes in x, y, and z directions. M is the mass matrix of the nodes. ξ and ω are the wave number and angular frequency, respectively, and K_1, K_2, and K_3 are the stiffness matrices.

Given the frequency ω, the eigenvalue of Equation (5) is the wave number ξ of the guided waves, and the eigenvector contains the mode shapes of the rail. Normalized processing can be used to plot the corresponding mode shapes of the guided waves. The modes with pure imaginary or complex wave numbers are not considered here because these modes will exponentially decay as the distance increases and cannot propagate. The dispersion curves of the phase velocity and group velocity of ultrasonic guided waves in the CHN60 rail are plotted, as shown in Figure 3.

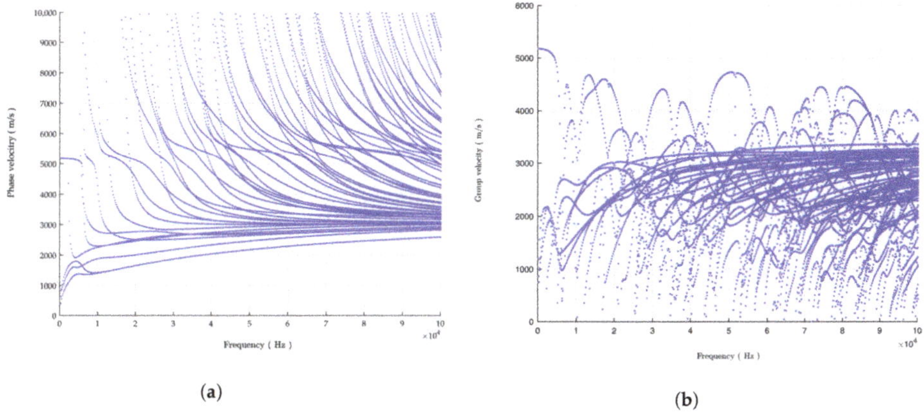

(a)   (b)

Figure 3. (**a**) Phase velocity; and (**b**) group velocity dispersion curves.

The dispersion curves of the phase velocity and group velocity are shown in the figure. At the same frequency, the rail has several propagable guided wave modes, and at the higher the frequency, it has more guided wave modes.

2.2. Excitation Response Analysis of Rails

On the basis of the excitation response analysis method, the vibration signals of a non-defective rail can be calculated at any point. Firstly, the system function model U of the rail should be obtained. Then, the frequency domain signal $\widehat{F}(f)$ of the excitation function signal $v_1(t)$ can be obtained. Thereafter, the excitation response result can be calculated $V = U \times \widehat{F}(f)$. Finally, the inverse Fourier transform of V can be determined, which is the excitation response of the time domain vibration signal.

The system function model of the rail [21] is shown in Equation (6), where m is the mode number ($m = 1, 2, \ldots, M$). Ω_m is the amplitude and phase parameters of mode m. U_m^L and U_m^R are the left eigenvector and the right eigenvector of mode m. $\widetilde{p}$ is the amplitude of the excitation signal. x_s and x are the longitudinal coordinates of the excitation point and the receiving point, and U_m^{Rup} is the mode shape of mode m.

$$U(y, z, f) = \sum_{m=1}^{M} \Omega_m U_m^{Rup} e^{-i[\xi_m(x - x_s)]} \tag{6}$$

In this form, the parameters are expressed as follow:

$$\Omega_m = -\frac{U_m^L \tilde{p}}{B_m}$$

$$B_m = U_m^L B U_m^R$$

$$B = \begin{bmatrix} K_1 - \omega^2 M & 0 \\ 0 & -K_3 \end{bmatrix}$$

The result of the frequency response analysis of the excitation response $V(f)$ can be expressed by Equation (7).

$$V(y, z, f) = \hat{F}(f) \cdot U(y, z, f)$$
$$= \hat{F}(f) \cdot \sum_{m=1}^{M} \Omega_m U_m^{Rup} e^{-i[\xi_m(x - x_s)]} \tag{7}$$

The inverse Fourier transform result is the time domain vibration signal of the node.

2.3. Modal Identification

In Figure 4, T is the transmitter of the guided waves, R is the receiver of guided waves, X is the location of a rail defect, and the arrows show the propagation paths of the guided waves. Path 1 is the direction of backward propagation after exciting the guided wave. Paths 2 and 3 are the directions of transmission and reflection after the guided wave meets the crack. According to the propagation time of the reflected wave (Path 3) and the group velocity of the main modes, the guided wave propagating distance of Path 3 can be calculated to locate the defect.

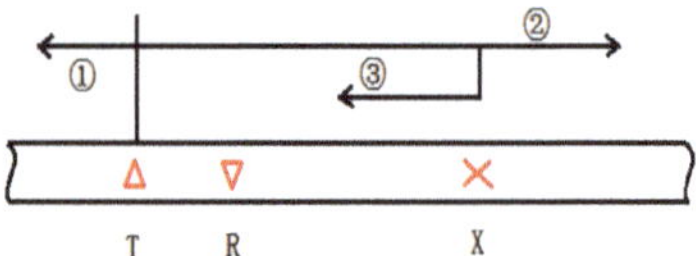

Figure 4. Schematic diagram of defect location.

To solve the problem of defect location, it is very important to analyze the mode propagation and reflection of guided waves in defective rails. The amplitude of each mode in a complete rail can be calculated using the excitation response algorithm, as described in Section 2.2, but the mechanisms involved in the interaction between modes and defects are still difficult to understand. Therefore, a method that can accurately analyze the direct mode and reflection mode of a defect is needed. In fact, the accurate modal identification method used in this paper is the inverse transformation of the excitation response analysis method.

2.3.1. Theoretical Derivation

According to the excitation response analysis method, the vibration displacement of any point of the rail is equal to the superposition of all the propagable modes' vibration displacements at that point. However, when there are both direct and reflected waves in the rail, the vibration displacement at any point is the superposition of the direct and reflected waves, that is, the superposition of all direct modes and all reflected modes at that point. Therefore, when considering the influence of the reflection wave caused by the defect, the expression of the excitation response of the node is in the form of Equation (8), in which the upper line "-" represents the relevant parameters of the reflected modes.

$$V(y, z, f) = \widehat{F}(f) \cdot U(y, z, f)$$

$$= \widehat{F}(f) \cdot \left(\sum_{m=1}^{M} \Omega_m U_m^{Rup} e^{-i[\xi_m(x-x_s)]} + \sum_{m=1}^{M} \bar{\Omega}_m \bar{U}_m^{Rup} e^{-i[\bar{\xi}_m(x-x_s)]} \right) \tag{8}$$

If we let $\Omega = \begin{bmatrix} \Omega_m & \bar{\Omega}_m \end{bmatrix}$, then with to the sampling results of a simulation or an experiment, the frequency domain signal $V(f)$ of any N points' vibration displacement on the rail can be obtained. By substituting the wave number, the mode shape, the distance between the sampling point and excitation point, and the frequency domain signals of the excitation signal at the corresponding frequency of each mode into Equation (8), the parameters Ω including the amplitude and phase of each mode can be achieved. It should be noted that the number of sampling points should be enough to ensure that the solution can be obtained relatively accurately, that is, $N > 2M$.

2.3.2. Simulation Analysis

Due to the excessive modes of high frequency guided waves, the results of modal identification are complex and difficult to study. Therefore, in the process of research, we first judged the correctness of the algorithm by studying 200 Hz low-frequency guided waves, and then increased the frequency to ultrasound band to study the modal identification results of high-frequency guided waves.

Firstly, the rail model with a railhead defect was established, and the excitation response of the rail was simulated using ANSYS. The rail was then excited by signals at a low center frequency of 200 Hz and a high center frequency of 60 kHz. The rail and defect models are shown in Figure 5.

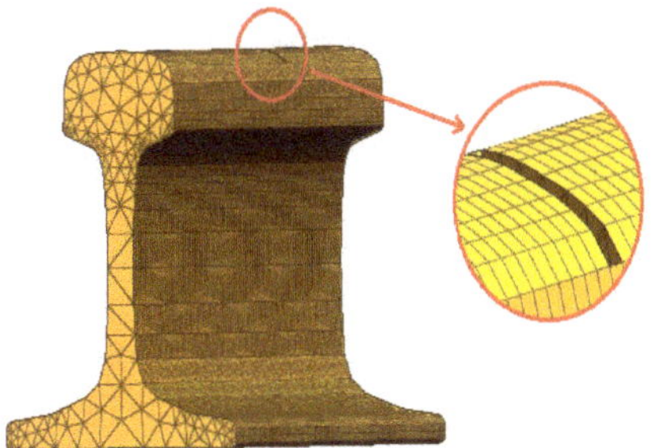

Figure 5. Rail model with head defect.

(1) Low center frequency of 200 Hz.

The length of the rail model was 200 m, and the defect was located at 120 m and had a length of 70 mm, width of 50 mm, and depth of 20 mm. The excitation signal was a sinusoidal wave modulated using Hanning window, with a center frequency of 200 Hz and five cycles. The excitation position was the center point of the side of the railhead at 90 m. A total of 32 sections were sampled from 100 m to 101.55 m. Seven nodes were selected for each section, as shown in Figure 2.

At 200 Hz, there were only four modes in the rail: the horizontal bending mode, the vertical bending mode, the torsional mode, and the extensional mode. The collected signals were substituted into the modal identification in Equation (8), which is extended into the form of Equation (9).

$$\widehat{F}(f) \begin{bmatrix} U_{1,1}^{Rup}e^{-i[\xi_1(x_1-x_s)]} & \cdots & U_{1,4}^{Rup}e^{-i[\xi_4(x_1-x_s)]} & \bar{U}_{1,1}^{Rup}e^{-i[\bar{\xi}_1(x_1-x_s)]} & \cdots & \bar{U}_{1,4}^{Rup}e^{-i[\bar{\xi}_4(x_1-x_s)]} \\ U_{1,1}^{Rup}e^{-i[\xi_1(x_2-x_s)]} & \cdots & U_{1,4}^{Rup}e^{-i[\xi_4(x_2-x_s)]} & \bar{U}_{1,1}^{Rup}e^{-i[\bar{\xi}_1(x_2-x_s)]} & \cdots & \bar{U}_{1,4}^{Rup}e^{-i[\bar{\xi}_4(x_2-x_s)]} \\ \vdots & & \vdots & \vdots & & \vdots \\ U_{1,1}^{Rup}e^{-i[\xi_1(x_{32}-x_s)]} & \cdots & U_{1,4}^{Rup}e^{-i[\xi_4(x_{32}-x_s)]} & \bar{U}_{1,1}^{Rup}e^{-i[\bar{\xi}_1(x_{32}-x_s)]} & \cdots & \bar{U}_{1,4}^{Rup}e^{-i[\bar{\xi}_4(x_{32}-x_s)]} \\ \vdots & & \vdots & \vdots & & \vdots \\ U_{7,1}^{Rup}e^{-i[\xi_1(x_1-x_s)]} & \cdots & U_{7,4}^{Rup}e^{-i[\xi_4(x_1-x_s)]} & \bar{U}_{7,1}^{Rup}e^{-i[\bar{\xi}_1(x_1-x_s)]} & \cdots & \bar{U}_{7,4}^{Rup}e^{-i[\bar{\xi}_4(x_1-x_s)]} \\ U_{7,1}^{Rup}e^{-i[\xi_1(x_2-x_s)]} & \cdots & U_{7,4}^{Rup}e^{-i[\xi_4(x_2-x_s)]} & \bar{U}_{7,1}^{Rup}e^{-i[\bar{\xi}_1(x_2-x_s)]} & \cdots & \bar{U}_{7,4}^{Rup}e^{-i[\bar{\xi}_4(x_2-x_s)]} \\ \vdots & & \vdots & \vdots & & \vdots \\ U_{7,1}^{Rup}e^{-i[\xi_1(x_{32}-x_s)]} & \cdots & U_{7,4}^{Rup}e^{-i[\xi_4(x_{32}-x_s)]} & \bar{U}_{7,1}^{Rup}e^{-i[\bar{\xi}_1(x_{32}-x_s)]} & \cdots & \bar{U}_{7,4}^{Rup}e^{-i[\bar{\xi}_4(x_{32}-x_s)]} \end{bmatrix} \begin{bmatrix} \Omega_1 \\ \Omega_2 \\ \vdots \\ \Omega_4 \\ \hat{\Omega}_1 \\ \hat{\Omega}_2 \\ \vdots \\ \hat{\Omega}_4 \end{bmatrix} = \begin{bmatrix} V_{1,1}(f) \\ V_{1,2}(f) \\ \vdots \\ V_{1,32}(f) \\ \vdots \\ V_{7,1}(f) \\ V_{7,2}(f) \\ \vdots \\ V_{7,32}(f) \end{bmatrix} \quad (9)$$

The magnitudes of each mode and the corresponding reflection mode (i.e., the modulus value of Ω) and the amplitude reflection coefficients after calculation are shown in Table 1.

Table 1. Vertical excitation mode identification results of defective rail (at a frequency of 200 Hz).

Mode	Direct Wave	Reflected Wave	Amplitude Reflection Coefficient
Horizontal bending mode	3.20×10^{-3}	3.04×10^{-4}	9.5×10^{-2}
Vertical bending mode	1.89×10^{-2}	3.52×10^{-3}	0.19
Torsional mode	1.17×10^{-2}	4.58×10^{-4}	3.9×10^{-2}
Extensional mode	4.94×10^{-4}	5.95×10^{-4}	1.2

It shows that the vertical bending mode had the highest amplitude among the direct modes, which was the main mode to be stimulated in this way. The amplitude of the extensional mode was far smaller than the amplitudes of the other modes and could be seen as zero, thus it was not considered. Meanwhile, the amplitude reflection coefficient was obtained by dividing the amplitude of the reflected mode from that of the direct mode, and the vertical bending mode was the most sensitive mode to the transverse crack of the railhead. The results of the modal identification algorithm were used to estimate the vibration of the railhead at a distance of 20 m away from the excitation point and were compared with the simulation sampling results at the same point, as shown in Figure 6a. The two methods' results almost coincide, which proves the validity of the algorithm in the modal identification of guided waves at a low frequency.

(2) High center frequency of 60 kHz

The length of the rail model was 25 m, and the defect was located at 20 m, with a length of 70 mm, a width of 50 mm, and a depth of 20 mm. A 10-cycle sinusoidal wave with a center frequency of 60 kHz was applied to excite the railhead at 12 m vertically, and 32 sections were sampled between 12.3 m and 13.23 m. The nodes were selected as above.

At 60 kHz, there were 35 modes in the rail, thus it was impossible to simply distinguish the types of modes. For this reason, they are numbered in the order of phase velocity from small to large. Using the signal with a center frequency of 60 kHz to excite the railhead, the magnitude and reflection coefficients of each mode and the corresponding reflection modes were obtained (as shown in Table 2), after the application of the accurate modal identification method.

In Table 2, it can be seen that, among the direct modes, mode No. 7, No. 8, and No. 9 had relatively high amplitudes, as they were the main modes stimulated in this way. At the same time, in the reflection mode, mode No. 7, with the largest amplitude and the highest reflection coefficient, was the most sensitive mode to the transverse crack of the railhead. Similarly, the vibration of the railhead 1.5 m away from the excitation point was estimated using the calculation results in Table 2, and compared with the simulation sampling results at that point, as shown in Figure 6b. The two main wave packets almost coincide, which shows that the algorithm is also applicable to the modal identification of guided waves at a high frequency.

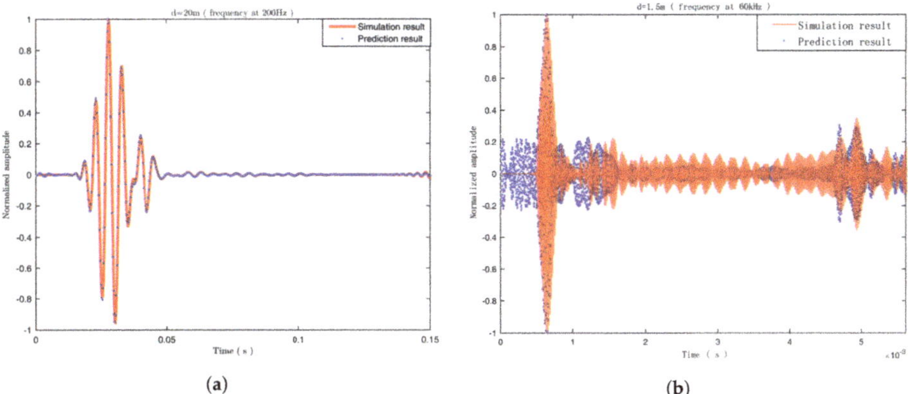

(a)

(b)

Figure 6. Comparison between the simulation results and prediction results: (**a**) 200 Hz; and (**b**) 60 kHz.

Table 2. Results of accurate modal identification (at a frequency of 60 kHz).

Mode Number	Direct Wave	Reflected Wave	Amplitude Reflection Coefficient
1	0.39	0.003	0.008
2	0.18	0.009	0.050
3	0.03	0.001	0.033
4	0.29	0.007	0.024
5	0.33	0.003	0.009
6	0.21	0.008	0.038
7	1.25	0.078	0.062
8	1.46	0.004	0.003
9	1.36	0.004	0.003
10	0.52	0.002	0.004
11	0.08	0.001	0.013
12	0.23	0.002	0.009
13	0.40	0.003	0.043
14	0.67	0.001	0.001
15	0.07	0.003	0.043
16	0.06	0.001	0.017
17	0.96	0.005	0.005
18	0.04	0	0
19	0.37	0.006	0.016
20	0.02	0.001	0.050
21	0.29	0.005	0.017
22	0.15	0.002	0.013
23	0.06	0.001	0.017
24	0.16	0.003	0.019
25	0.23	0.008	0.035
26	0.28	0.001	0.004
27	0.36	0.002	0.006
28	0.41	0.001	0.002
29	0.42	0.002	0.005
30	0.03	0.001	0.033
31	0.32	0	0
32	0.42	0	0
33	0.92	0	0
34	0.02	0	0
35	0.11	0	0

3. Single Modal Extraction Algorithm

When ultrasonic guided waves propagate in rails, dispersion occurs. That is to say, the number of the guided wave modes propagating in the rails will increase with the increase in excitation frequency; furthermore, the propagating velocity of each mode is different. With the increase in propagating distance, the wave packets of each mode will gradually stagger and overlap. Therefore, they cannot be distinguished, and so the group velocity cannot be used to locate defects. To solve this problem, an SMEA based on the accurate modal identification method is proposed.

As mentioned above, the vibration displacement at any point on the rail is the superposition of the vibration displacement of all modes at this point. Conversely, the vibration displacement of each mode at any point can be split by the total displacement of the point.

The amplitude and phase Ω_n of mode n can be obtained with the accurate modal identification algorithm, and the vibration displacement generated by mode n at the longitudinal coordinate x can be expressed by Equation (10).

$$V_n(y, z, f) = \bar{F}_f \cdot \Omega_n U_n^{Rup} e^{-i[\xi_n(x-x_s)]} \tag{10}$$

According to Equation (10), the vibration displacement of mode n at this point in a frequency domain is calculated, and the time domain waveform of the mode is obtained by inverse Fourier transform. To verify the correctness of the method, the following simulations were carried out.

A non-defective three-dimensional rail model with a length of 200 m was established. The vibrations of the railhead along the rail in a longitudinal, transverse, and vertical excitation direction were simulated using ANSYS. To show the results clearly, the excitation signal was selected as a low-frequency signal with a center frequency of 200 Hz and five cycles, which has only four modes. The modal amplitude of each mode at the 200 Hz frequency point after mode analysis is shown in Figure 7.

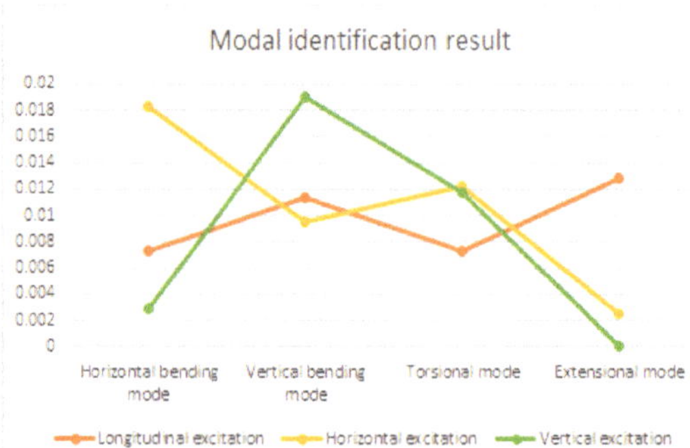

Figure 7. Modal identification results under three excitation conditions (200 Hz).

According to Figure 7, when the rail was excited longitudinally, the proportion of the extensional mode was higher. When the rail was excited horizontally, the proportions of the horizontal bending mode and torsion mode were higher. When the rail was excited vertically, the proportions of the vertical bending mode and torsion mode were higher.

Using the SMEA and choosing the frequency range 100–300 Hz, the total waveform at 30 m and four single-mode waveforms were obtained, respectively, as shown in Figure 8a–c.

It can be seen clearly in the figure that only the extensional mode existed when the railhead was excited longitudinally. The horizontal bending and torsion modes existed when the railhead was excited horizontally, and the vertical bending and torsion modes existed when the railhead was excited vertically. The results are consistent with the accurate identification results.

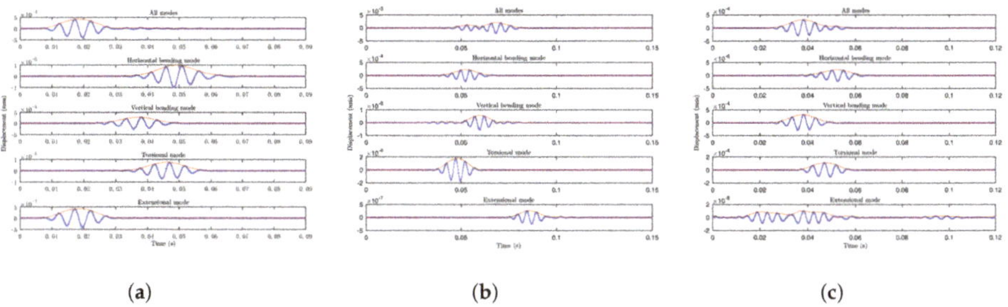

(a)   (b)   (c)

Figure 8. SMEA results: (**a**) longitudinal excitation; (**b**) horizontal excitation; and (**c**) vertical excitation.

4. Defect Location

The flow chart of the defect location method is shown in Figure 9.

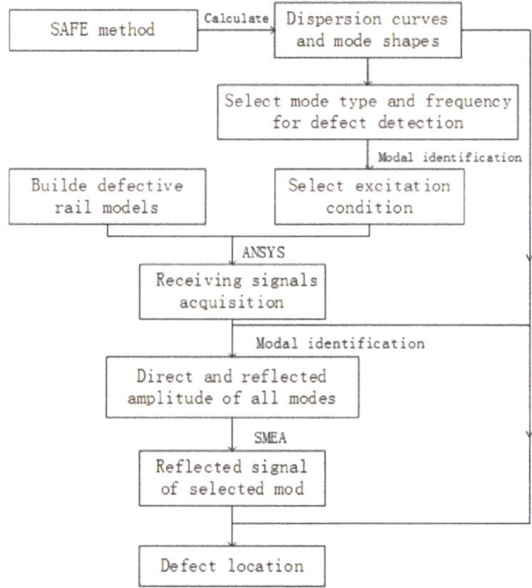

Figure 9. Defect location algorithm flow chart.

As can be seen above, the primary task of defect location is to select the mode, frequency, and excitation conditions with which to perform the defect detection. At the same time, a three-dimensional model of the rail with defects must be established. The excitation response of the rail is then simulated by ANSYS, and the vibration displacements of a series of points on the rail are obtained, so as to simulate the signals received by an experiment. According to the results of the modal identification and the SMEA, the reflected signals of the selected modes are plotted. Thereafter, the group velocity and the propagation time of the reflected mode are obtained, so that the defect location can be achieved. The following section takes a transverse crack in a railhead as an example to explain the process of defect location.

4.1. Selection of Mode, Frequency, and Excitation Conditions for Defect Detection

The following three principles are used to select the mode, frequency, and excitation conditions for railhead defect detection:

- The mode that only vibrates in the railhead with almost no movement of rail waist and rail bottom and which has a large group velocity is selected.

321

- The frequency band with better non-dispersive characteristics is selected.
- The mode with the largest amplitude is selected as the excitation condition.

The detection frequency and mode were selected in the frequency range 20–70 kHz. According to Equation (5), the mode shapes of the rail in this frequency range were calculated and drawn. The frequency of 60 kHz was taken as an example, as shown in Figure 10. Among them, the modes in which only the railhead vibrated with almost no movement in the rail waist and rail bottom were modes No. 7 and No. 14.

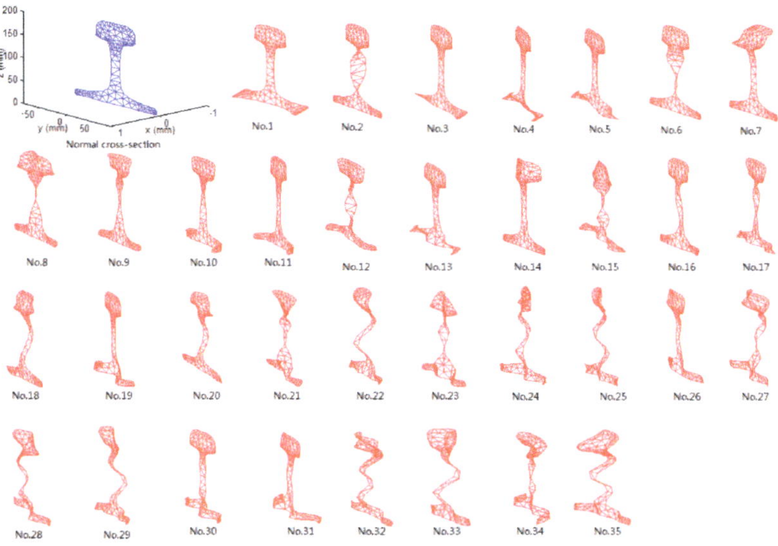

Figure 10. Rail mode shapes (60 kHz).

According to Figure 3b, the group velocity dispersion curves of modes No. 7 and No. 14 were extracted, as shown in Figure 11.

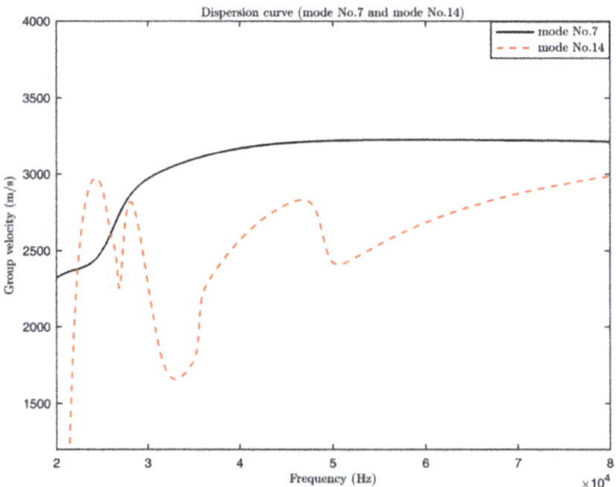

Figure 11. Group velocity dispersion curves of modes No. 7 and No. 14.

In Figure 11, we can see that mode No. 7 had almost the same group velocity and excellent non-dispersive characteristics in the proximity of 60 kHz. The group velocity of mode No. 14 was greatly influenced by the frequency and the non-dispersion characteristic was poor. Therefore, mode No. 7 at 60 kHz frequency was selected to detect the defect in the railhead.

A complete three-dimensional rail model with a length of 15 m was established. A simulation whereby the railhead was excited at 7 m along the longitudinal, transverse, and vertical directions by ANSYS was performed. The excitation signal was a sinusoidal wave modulated by Hanning window at the center frequency of 60 kHz with 10 cycles. The co-directional displacement signals between 8 m and 9 m were collected, and the amplitude of each mode under three excitation conditions were calculated using the accurate modal identification method, as shown in Figure 12.

As shown in Figure 12, the highest amplitude for the selected mode is that of vertical excitation. Therefore, vertical excitation was selected.

In summary, mode No. 7 was selected as the detection mode, 60 kHz was selected as the frequency, and vertical excitation was selected as the excitation condition.

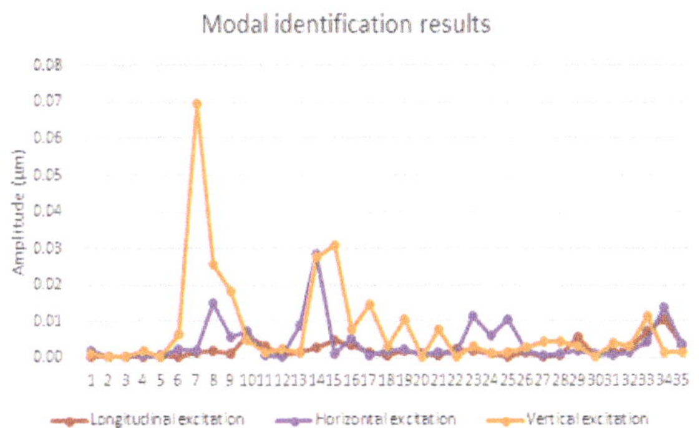

Figure 12. Modal identification results under three excitation conditions (60 kHz).

4.2. Simulation Analysis of Defect Location

According to the results in Section 4.1, the rail with a defect in the railhead was stimulated. As shown in Figure 13, the rail length was 25 m; T represents the excitation point, which was located at 12 m along the rail. X represents the defect position, and the distance between the excitation point and X was $l = 8$ m. The simulation condition used was the same as the high-frequency excitation presented in Section 2.3.2. Table 2 shows the results of the modal identification.

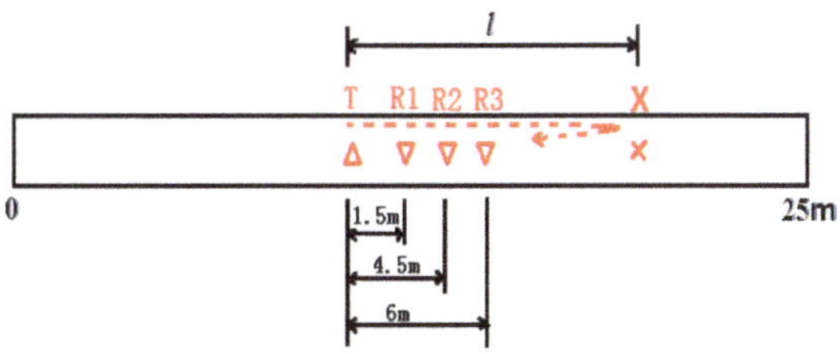

Figure 13. Schematic diagram of rail defect location.

The spacings d between the sampling points and the excitation point were 1.5 m, 4.5, m and 6 m, and the collected signal waveforms are shown in Figure 14a–c.

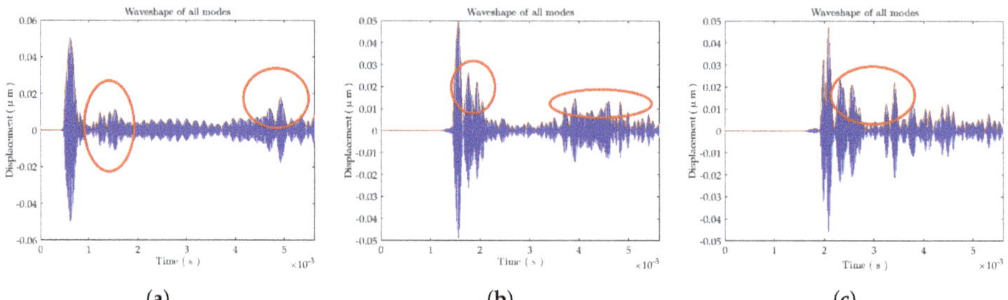

Figure 14. Acquisition waveforms for all modes with distances of: (**a**) 1.5 m; (**b**) 4.5 m; and (**c**) 6 m between excitation points and sampling points.

As a result of the complex propagation modes of guided waves in rails, only the first direct wave packet could be distinguished from the waveform. It was impossible to pick out which wave packet in the red circle was the reflected wave packet, and it was also impossible to know for certain which mode or modes each wave packet corresponded to; therefore, it was difficult to give the specific propagation time and the accurate mode group velocity. For this reason, the defect could not be precisely located.

According to the SMEA, the reflected mode waveform of mode No. 7 at 4.5 m from excitation point was calculated in the frequency range 42–78 kHz. As shown in Figure 15b, the peak time t_1 of the wave packet was 3.966×10^{-3} s. With the peak time t_0 of the excitation wave packet (8.333×10^{-5} s) and the group velocity v_0 of mode No. 7 (3148.2 m/s), it was possible to calculate the distance between the defect and the excitation point using Equation (11).

$$l = \frac{(t_1 - t_0) * v_0 + d}{2} \tag{11}$$

The actual interval was 8 m and the calculation result was 8.36 m, giving an error of only 0.36 m.

The reflection mode waveforms of mode No. 7 at 1.5 m and 6 m away from the excitation point were extracted, as shown in Figure 15a,c. The distances between the calculated defect and the excitation point were 8.33 m and 8.39 m. Therefore, the errors were 0.33 m and 0.39 m, respectively, both less than 0.5 m, and thus meeting the positioning accuracy requirements.

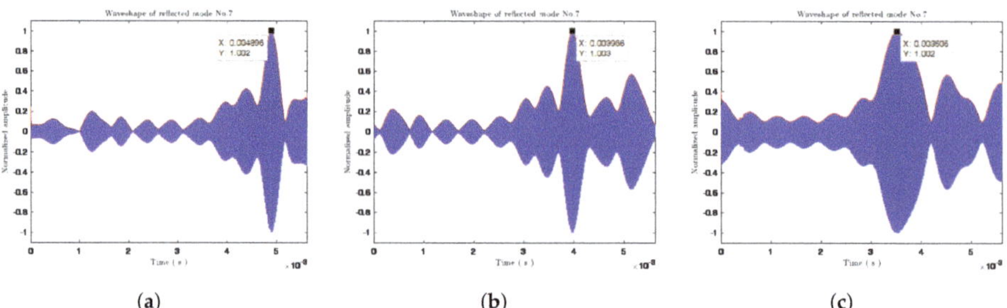

Figure 15. The reflection waveforms of mode No. 7 at a distance from the excitation point of: (**a**) 1.5 m; (**b**) 4.5 m; and (**c**) 6 m.

5. Conclusions

A rail defect location method based on an SMEA of ultrasonic guided waves was proposed. The dispersion curves of a CHN60 rail were calculated using the semi-analytical finite element method, the mode shapes of the guided waves were calculated, and the actual received signals were predicted using the simulation data provided by ANSYS. Furthermore, accurate modal identification was achieved using the inverse transformation of the excitation response method, and the amplitude of each mode of the guided waves propagating in the rail was obtained. The rail displacement curves excited by the signals at a low frequency of 200 Hz and high frequency of 60 kHz were correctly predicted.

According to the modal identification results, an SMEA was proposed. The single mode extraction results at 200 Hz were calculated and the results were consistent with those of the modal identification. Thereafter, taking a railhead defect as an example, according to certain selection principles, mode No. 7 with a signal frequency of 60 kHz and a vertical excitation condition were selected. The reflective modes of mode No. 7 at 1.5, 4.5, and 6 m from the excitation point of the railhead defect were extracted using the SMEA, and the distance between the rail defect and excitation point was obtained according to the time difference. The errors in location were 0.33, 0.36, and 0.39 m, respectively, which all fall within the 0.5 m set as the detection requirements. Hence, it can be said that the precise determination of the rail defect was accomplished.

Author Contributions: Conceptualization, B.X. and X.X.; methodology, X.X. and L.Z.; software, B.X. and H.S.; validation, Z.Y., L.Z. and H.S.; investigation, B.X.; resources, Z.Y.; writing—original draft preparation, B.X.; and writing—review and editing, B.X., X.X. and L.Z.

Funding: This research was funded by the National Key Research and Development Program of China (2016YFB1200401)

Conflicts of Interest: The authors declare no conflict of interest.

Abbreviations

The following abbreviations are used in this manuscript:

SAFE   Semi-analytical finite element
SMEA   Single mode extraction algorithm

References

1. Loveday, P.W. Guided wave inspection and monitoring of railway track. *J. Nondestruct. Eval.* **2012**, *31*, 303–309. [CrossRef]
2. Lee, C.; Rose, J.L.; Cho, Y. A guided wave approach to defect detection under shelling in rail. *NDT E Int.* **2009**, *42*, 174–180. [CrossRef]
3. Rose, J.L.; Zhu, W.; Cho, Y. Boundary element modeling for guided wave reflection and transmission factor analyses in defect classification. In Proceedings of the 1998 IEEE Ultrasonics Symposium, Sendai, Japan, 5–8 October 1998; Volume 1, pp. 885–888.
4. Harker, A.H. Numerical modelling of the scattering of elastic waves in plates. *J. Nondestruct. Eval.* **1984**, *4*, 89–106. [CrossRef]
5. Moser, F.; Jacobs, L.J.; Qu, J. Modeling elastic wave propagation in waveguides with the finite element method. *NDT E Int.* **1999**, *32*, 225–234. [CrossRef]
6. Gavrić, L. Computation of propagative waves in free rail using a finite element technique. *J. Sound Vib.* **1995**, *185*, 531–543. [CrossRef]
7. Alleyne, D.; Cawley, P. A two-dimensional Fourier transform method for the measurement of propagating multimode signals. *J. Acoust. Soc. Am.* **1991**, *89*, 1159–1168. [CrossRef]
8. Hayashi, T.; Song, W.; Rose, J.L. Guided wave dispersion curves for a bar with an arbitrary cross-section, a rod and rail example. *Ultrasonics* **2003**, *41*, 175–183. [CrossRef]
9. He, C. Propagation characteristics of ultrasonic guided wave in rails based on vibration modal analysis. *J. Vib. Shock* **2014**, *33*, 9–13. [CrossRef]
10. Loveday, P.W.; Long, C.S. Laser vibrometer measurement of guided wave modes in rail track. *Ultrasonics* **2015**, *57*, 209–217. [CrossRef] [PubMed]
11. Loveday, P.W. Measurement of modal amplitudes of guided waves in rails. *Proc. SPIE* **2008**, *6935*, 1J-1–1J-8. [CrossRef]
12. Loveday, P.W. Modeling and Measurement of Piezoelectric Ultrasonic Transducers for Transmitting Guided Waves in Rails. In Proceedings of the 2008 IEEE Ultrasonics Symposium, Beijing, China, 2–5 November 2008; pp. 410–413.

13. Loveday, P.W.; Long, C.S. Modal amplitude extraction of guided waves in rails using scanning laser vibrometer measurements. *Am. Inst. Phys. Conf. Ser.* **2012**, *1430*, 182–189. [CrossRef]
14. Loveday, P.W.; Long, C.S. Field measurement of guided wave modes in rail track. *Am. Inst. Phys. Conf. Ser.* **2013**, *1511*, 230–237. [CrossRef]
15. Loveday, P.W.; Long, C.S.; Ramatlo, D.A. Mode repulsion of ultrasonic guided waves in rails. *Ultrasonics* **2017**, *84*, 341–349. [CrossRef] [PubMed]
16. Chao, L.U.; Sheng, H.J.; Song, K.; Lin, J.M.; He, F. Ultrasonic Guided Wave Scattering Characteristics of Rail Base Oblique Cracks. *Nondestruct. Test.* **2016**, *38*, 18–25. [CrossRef]
17. Zumpano, G.; Meo, M. A new damage detection technique based on wave propagation for rails. *Int. J. Solids Struct.* **2006**, *43*, 1023–1046. [CrossRef]
18. Xu, X.; Zhuang, L.; Xing, B.; Yu, Z.; Zhu, L. An Ultrasonic Guided Wave Mode Excitation Method in Rails. *IEEE Access* **2018**, *6*, 60414–60428. [CrossRef]
19. Jian, W.; Ping, W.; Ma, D. Matching performance of CHN60N rail with various wheel profiles for Chinese high-speed railway. *IOP Conf. Ser. Mater. Sci. Eng.* **2018**, *383*, 012041. [CrossRef]
20. Zhang, Z.-T.; Yan, L.-S.; Wang, P.; Guo, L.-K.; Pan, W.; Zhang, Z.-Y. Key Techniques for Rail Strain Measurements Based on Fiber Bragg Grating Sensor. *J. China Railw. Soc.* **2012**, *34*, 65–69. [CrossRef]
21. Bartoli, I.; Coccia, S.; Phillips, R.; Srivastava, A.; Scalea, F.L.D.; Salamone, S.; Fateh, M.; Carr, G. Stress dependence of guided waves in rails. *Proc. SPIE* **2010**, *7650*, 765021. [CrossRef]

Article

Numerical and Experimental Investigation of Guided Wave Propagation in a Multi-Wire Cable

Pengfei Zhang [1], Zhifeng Tang [2,*], Fuzai Lv [1] and Keji Yang [1]

[1] State Key Laboratory of Fluid Power and Mechatronic Systems, School of Mechanical Engineering, Zhejiang University, 38 Zheda Road, Hangzhou 310027, China; zhangpengfei@zju.edu.cn (P.Z.); lfzlfz@zju.edu.cn (F.L.); yangkj@zju.edu.cn (K.Y.)

[2] Institute of Advanced Digital Technologies and Instrumentation, College of Biomedical Engineering & Instrument Science, Zhejiang University, 38 Zheda Road, Hangzhou 310027, China

* Correspondence: tangzhifeng@zju.edu.cn; Tel.: +86-0571-8767-1703

Received: 10 February 2019; Accepted: 5 March 2019; Published: 12 March 2019

Abstract: Ultrasonic guided waves (UGWs) have attracted attention in the nondestructive testing and structural health monitoring (SHM) of multi-wire cables. They offer such advantages as a single measurement, wide coverage of the acoustic field, and long-range propagation ability. However, the mechanical coupling of multi-wire structures complicates the propagation behaviors of guided waves and signal interpretation. In this paper, UGW propagation in these waveguides is investigated theoretically, numerically, and experimentally from the perspective of dispersion and wave structure, contact acoustic nonlinearity (CAN), and wave energy transfer. Although the performance of all possible propagating wave modes in a multi-wire cable at different frequencies could be obtained by dispersion analysis, it is ineffective to analyze the frequency behaviors of the wave signals of a certain mode, which could be analyzed using the CAN effect. The CAN phenomenon of two mechanically coupled wires in contact was observed, which was demonstrated by numerical guided wave simulation and experiments. Additionally, the measured guided wave energy of wires located in different layers of an aluminum conductor steel-reinforced cable accords with the theoretical prediction. The model of wave energy distribution in different layers of a cable also could be used to optimize the excitation power of transducers and determine the effective monitoring range of SHM.

Keywords: guided wave; multi-wire cable; wave structure; contact acoustic nonlinearity; energy transfer

---

1. Introduction

Multi-wire cables play an important role as the main structural components in such mechanical and civil engineering applications as load carriers in elevators, lifting machinery, and cable-stayed and suspension bridges; post-tensioners in civil structures; and as overhead transmission lines (OVTLs) in power grids. Nondestructive inspection and structural health monitoring (SHM) of these cables are crucial to ensure the proper structural performance of cable-stayed and suspension bridges and high-voltage transmission towers. Bridge cables consist of multi-layer steel wires arranged in parallel within an equilateral hexagon and a polyethylene sheath wrapped around the wire bundle. Aluminum conductor steel-reinforced (ACSR) cables are the main components of the OVTLs between the transmission towers.

Many nondestructive testing methods are available, such as visual inspection [1,2], radiography [3], computed tomography [4], acoustic emission monitoring [5], ultrasound [1,6], and magnetic flux leakage [7]. However, many of these methods have limitations when applied to multi-wire cables. For instance, conventional ultrasonic testing methods are impracticable for detecting large and long structures, such as multi-wire cables, owing to the deficiency of local detection. Ultrasonic guided wave (UGW)-based nondestructive testing (NDT) techniques are increasingly being used to inspect various multi-wire cables given their high sensitivity, long-range inspection, and full cross-sectional coverage properties. Many studies [8–18] related to guided wave dispersion properties, propagation characterization, signal processing methods, and wire breakage detection in multi-wire cables have been reported.

When UGWs encounter the end or defect of a waveguide in propagation, reflected waves propagate along the original wave path and are received by the transducer, and the transmitted waves continue to propagate forward. Defect determination methods are used to measure the size and location of the defect, and evaluate the structural health state. The cross-section of the waveguide confines UGWs to propagate in only the length direction. However, the dispersive and multimodal nature of UGW propagation has to be analyzed. UGWs with a narrow frequency bandwidth often consist of several superposed modes with different wave structures [19] and propagation velocities. Therefore, the appropriate signal processing method is needed to analyze the acquired signals of received waves [20]. Multi-wire cables are composed of a number of parallel (e.g., bridge cable) or twisted (e.g., steel wire strands, ACSR cable) wires. The factors of helical and twisted geometry, tension state, and contact and friction stresses between adjacent wires complicate the numerical models of multi-wire structures [21]. Cables are commonly composed of a number of twisted wires with different diameters and materials, resulting in different wave velocities even at the same center excitation frequency. The contact condition between adjacent wires contributes to the complexity of the dispersion and propagation characteristics of the multi-wire structures. The factors of a large number of wires, a small diameter of an individual wire, and long length present challenges to the numerical modeling of guided waves propagation in multi-wire cables, such as slow convergence and time-consuming computation. The cable sheathing and irregular wire bundle surface geometry also make it difficult to design a transducer with high coupling efficiency, which is a critical part in industrial applications in the field.

The overall dispersion characteristics of a multi-wire system with consideration of the contact stress between adjacent wires have been studied using a double-rod model, in which the guided wave mode conversion phenomenon was verified [22]. The dispersion curves could display the propagation velocities of all possible guided wave modes at different frequencies in a structure. However, as for a UGW with a given mode and center excitation frequency propagating in a multi-wire structure, dispersion curves are incapable of predicting the change in frequency components of the wave signals due to the mechanical contact of adjacent wires.

Studies of nonlinear guided waves used to detect microcracks in structures and to characterize materials have been reported [23,24]. Higher-order harmonic waves were generated in the structure with microcracks when conducting the NDT inspection using an ultrasonic wave. Similarly, the contact acoustic nonlinearity (CAN) effect is also in existence in a contact multi-wire system. A longitudinal wave is excited in one rod of the two-contact-rod system, and harmonic waves with a frequency twice the excitation frequency are generated in both of the rods due to the CAN effect.

Piezoelectric and magnetostrictive transducers have been widely used to excite and receive guided waves in a multi-wire cable. However, there is a nonmagnetostrictive effect in aluminum wires of an ACSR cable, and the magnetostriction of the innermost steel wires is too weak to generate guided waves due to the influence of the lift distance on the bias magnetic field caused by the layers of aluminum wires. Therefore, piezoelectric transducers, as practicable transducing devices, are commonly employed to carry out an inspection of an ACSR cable using UGWs. Several piezoelectric transducers, small in size, are uniformly installed on the outermost aluminum wires of the cable at the circumference [25]. The uniform coverage of the guided waves in the cable section relies on the contact between adjacent wires to transfer waves into inner layers from the outermost layer, where the transducers are installed. Models of the distribution and propagation of guided wave energy in different layers could provide guidance for determining the excitation power of transducers and estimating the monitoring range of the SHM system.

In this paper, UGW propagation in a single cylindrical waveguide and double mechanically coupled cylindrical waveguides is introduced, including the dispersion properties and CAN effect. There are many different wave modes simultaneously propagating in a structure with different velocities at different frequencies or even the same frequency, which are partially caused by the dispersion phenomenon. However, the appearance of the high-order harmonic wave in the signal of a single mode wave with a single excitation frequency is caused by the CAN effect. An efficient energy-based model of coupled cylindrical waveguides is used to predict the wave energy distribution in wires located in different layers of an ACSR cable, by which the uniformity of the cross-sectional distribution of the guided wave could be obtained. Numerical transient finite element (FE) simulation and a two-dimensional (2D) fast Fourier transform were used to verify the results of the theoretical analysis. Experimental investigations were completed to generate and receive UGWs in a two-contact-wire cable and an ACSR cable with five-layer wires. The proposed numerical simulation models were fit to the experimental data.

2. UGW Propagation in Multi-Wire Waveguides

The mechanical properties of multi-wire cables are more complicated than those of individual solid cylindrical rods. The most prominent features are the complex geometry of the cross-section and the contact and friction stresses between adjacent wires, which are largely caused by the tension loads during service. The preliminary understanding of UGW propagation in a multi-wire cable was used to study the wave propagation behavior of an individual wire using the analytical Pochhammer–Chree equations [26]. An infinite number of guided wave modes, which are classified by their particle displacement, might propagate through the structure. These modes in a cylindrical waveguide are composed of the longitudinal waves $L(m,n)$ with radial and axial vibration displacement, torsional waves $T(m,n)$ with circumferential vibration displacement, and flexural waves $F(m,n)$ with radial, axial, and circumferential vibration displacement, where $m \in \{0,1,2 \dots \}$ denotes the circumferential order and $n \in \{1,2,3 \dots \}$ stands for the nth root of the characteristic equation [27].

The performance of all possible propagating wave modes in a multi-wire cable at different frequencies could be predicted by UGW dispersion curves. However, when a guided wave was excited in one of the contact wires with a single-frequency excitation signal, it is ineffective to analyze the frequency behaviors of the wave signals for a certain mode, which could be analyzed using the CAN effect. From the perspective of UGW energy transfer, the distribution of wave energy in each layer of a cable composed of multiple layers of wires with different materials could be obtained.

2.1. Dispersion Analysis of a Single Rod

A single wire can be considered to be the same as an individual cylindrical rod, whose dispersion properties can be analytically obtained using the Pochhammer–Chree equations. UGWs tend to travel with different velocities in a structure depending on the center excitation frequency. There are two ways to define the propagation velocities of UGWs: group and phase velocity. Group velocity refers to the propagation velocity of the entire wave packet. Phase velocity is the propagation velocity at which the phase of any one frequency component of the wave travels. For a cylindrical rod, group and phase velocity dispersion curves of a 7-mm-diameter steel wire (commonly used in a bridge cable), and a steel and aluminum wire with a diameter of 2.5 mm and 3.2 mm, respectively (commonly used in ACSR cable), were acquired using an open source software package named PCDISP developed in the Matlab (MathWorks Inc., Natick, MA, USA) environment. PCDISP is described in more detail in references [28,29]. The material and geometric properties of wires used in the dispersion calculation models are summarized in Table 1. The results are shown in Figure 1.

Table 1. The material and geometric properties of wires used in the dispersion models.

Material	Diameter	Elastic Modulus	Density	Poisson Ratio
Steel	7 mm	209 GPa	7800 kg/m^3	0.3
Steel	2.5 mm	209 GPa	7800 kg/m^3	0.3
Aluminum	3.2 mm	72 GPa	2700 kg/m^3	0.35

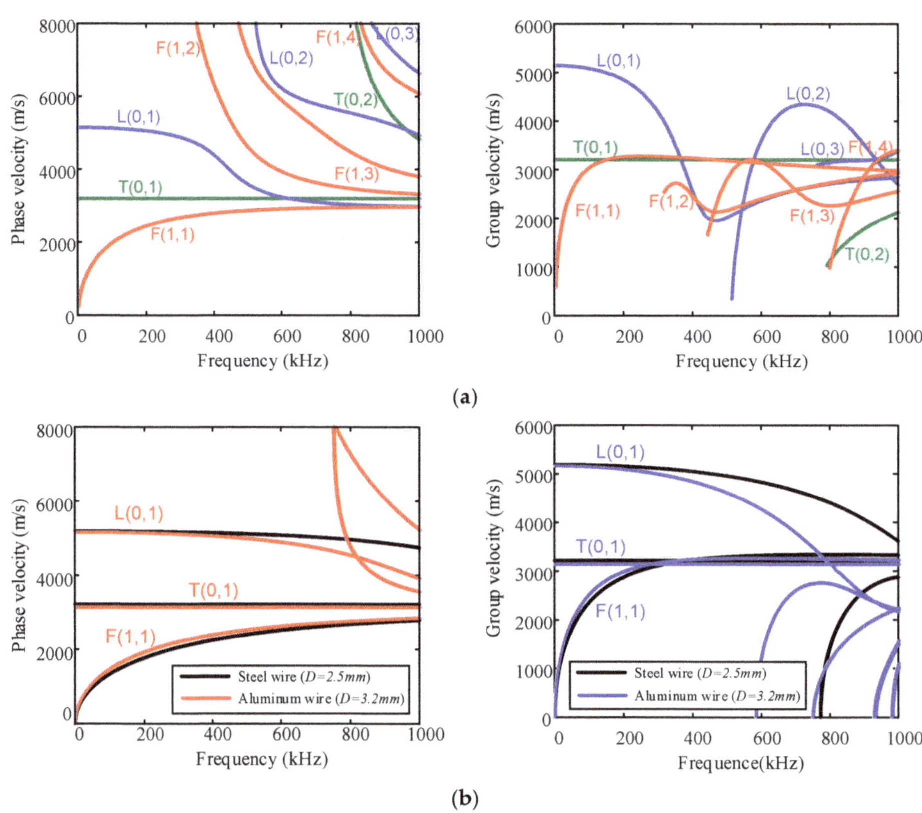

Figure 1. Phase and group velocity dispersion curves for (**a**) a single 7-mm-diameter steel wire and (**b**) a single steel and aluminum wire with a diameter of 2.5 mm and 3.2 mm, respectively.

Figure 1 shows UGW modes for three different types of wires with frequencies up to 1000 kHz. The phase and group velocities of the fundamental order flexural mode F(1,1) are slow and display severe dispersive properties in the low-frequency range (below 200 kHz). L(0,1) and T(0,1) seem to be the applicable modes for wire breakage inspection of steel and aluminum wires, though the velocity of the L(0,1) mode is faster than that of the T(0,1) mode at low frequencies, yet is slightly dispersive. Therefore, the L(0,1) mode seems ideal for wire breakage detection due to its velocity and relatively low dispersive behavior. Figure 2a–c show the vibration displacement distribution in a cross-section of a single 7-mm-diameter steel wire carrying these three types of fundamental modes: L(0,1), T(0,1), and F(1,1), at a frequency of 60 kHz. The arrows indicate the particle vibration displacements in the radial direction (in-plane). The color indicates the magnitude of the particle displacements in the z-axial direction (out-of-plane).

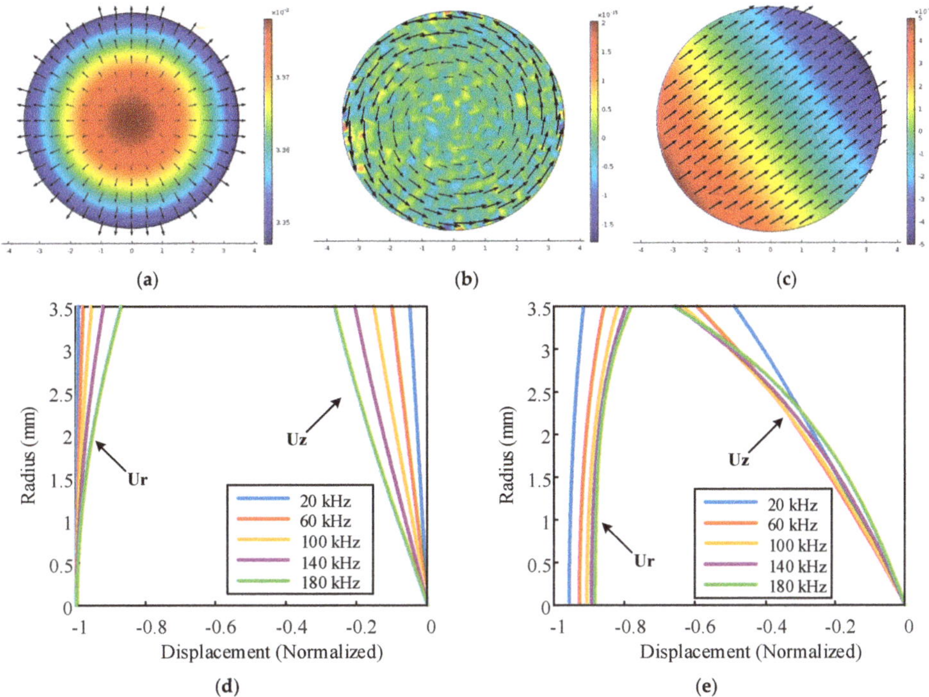

Figure 2. The vibration displacement distribution of (**a**) longitudinal, (**b**) torsional, and (**c**) flexural waves in a 7-mm-diameter steel wire. Wave structure of the (**d**) L(0,1) and (**e**) F(1,1) modes at different frequencies.

Both longitudinal and flexural guided waves create radial and axial vibration displacement. The coupling extent among the adjacent contact wires in the cable is determined by the radial vibration displacement at the wire surface (i.e., maximum radius). The larger the radial vibration displacement amplitude, the better the coupling. Therefore, the frequency of UGW wave modes with large radial vibration displacement at the wire surface become particularly critical during energy transfer. In Figure 2d,e, radial and axial vibration displacement distribution curves (wave structure) of the different radii of a 7-mm-diameter steel wire were calculated using Pochhammer–Chree equations for the L(0,1) and F(1,1) wave modes in the 20–180 kHz frequency range. As the wire radius increases, the radial displacements decrease but the axial displacements increase. This trend was significant with the increase in wave frequency below 200 kHz.

2.2. Contact Acoustic Nonlinearity of Coupled Double Rods

Higher harmonic ultrasonic waves are generated in a structure with imperfect interfaces, in which excited waves with a large amplitude are incident. This phenomenon is called the contact acoustic nonlinearity (CAN) effect [24]. This phenomenon is caused by the repeated reflection of incident waves between two surfaces. The same behavior occurs when a longitudinal UGW propagates to an imperfect interface formed by two surfaces of the contact structures. The compressional part of the wave can penetrate through the interface and propagate into another structure, but the tensile part cannot. Therefore, it is equivalent to a half-wave rectification of the transmitted wave, which is a nonlinear effect. This nonlinearity is embodied in the form of higher harmonics. In nondestructive inspection using a guided wave, when the characteristic length of the rough contact surfaces or a micro-crack is much smaller than the incident wavelength, a nonlinear phenomenon occurs.

In a pioneering work [30], the second harmonics generated by the nonlinear nature of the contact stiffness with the frequency of 2f was theoretically observed in both transmitted and reflected waves, where f is the frequency of incident waves. A pair of coupled waveguides were created with an active wire and a passive wire closely in contact with each other. An ultrasonic guided wave was excited on the left end of the active wire. The radial displacements of the active wire were partially coupled to the passive wire through contact stress in the contact interface. Additional waves were generated in the active and passive wires as a result of the local load provided by the contact stress, as shown in Figure 3.

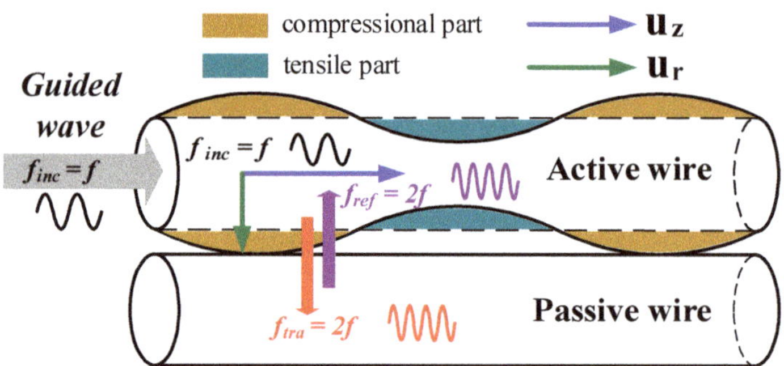

Figure 3. A schematic drawing of contact acoustic nonlinearity (CAN) at the rough contact interface of a pair of coupled cylindrical waveguides.

2.3. Energy Model of Multi-Wire Waveguides

There are many limitations of modeling accurate UGW propagation of multi-wire structures taking into account the effect of mechanical contact, such as varying contact stress among the adjacent wires, wave mode conversion, and the CAN effect. In order to simplify the propagation model of an UGW in a multi-wire cable, a model that describes the transfer of the wave between an active and a passive wire from an energy perspective was used to analyze the propagation behavior of an UGW. A multi-wire system was composed of mechanically coupled wires. They were surrounded by a spring array along the wire axis [17,18]. Thus, the wave energy of the ith wire of a cable with n wires can be mathematically described by a second-order nonlinear differential equation:

$$\frac{d^2 E_i(z)}{dz^2} = -d_m \frac{dE_i(z)}{dz} - c\sqrt{E_i(z)} \sum_{j=1}^{n} b_{ij}(\sqrt{E_i(z)} - \sqrt{E_j(z)}) \tag{1}$$

where $E_i(z)$ is the average energy transfer from all n wires to the ith wire, d_m is the damping coefficient caused by the material, c is the mechanical coupling coefficient, and a binary constant b_{ij} indicates whether the wires i and j are in contact with each other, expressed as:

$$b_{ij} = \begin{cases} 1 & \text{in contact} \\ 0 & \text{otherwise} \end{cases}. \tag{2}$$

For a two-wire model, the equation can be solved using the ode-45 built-in algorithm in Matlab. The coupling coefficient, the material damping coefficient, and the initial conditions $E_i(0) = E_{i,0}$ and $\frac{dE_i(0)}{dZ} = E'_{i,0}$ were obtained using the method of least squares fitting to process the transient finite element (FE) simulation data shown in Section 3.2. The results are shown in Figure 4. With the propagation of the guided wave, the wave energy in the active wire was transferred to the passive wire. The wave energy in the two wires gradually decayed after reaching a balance, which was caused by material damping.

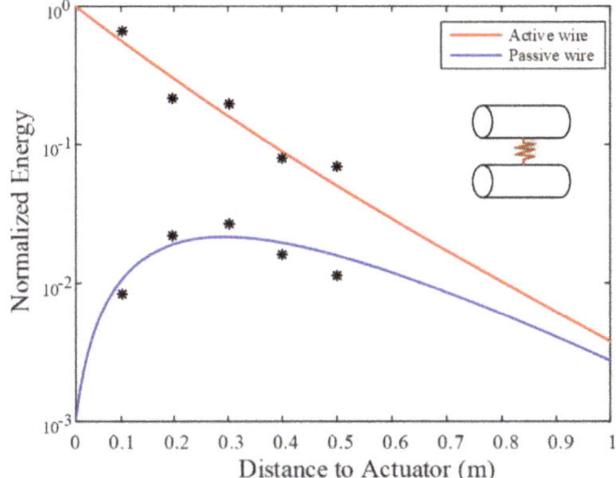

Figure 4. A model of the wave energy transfer between two adjacent wires. The results of the active wire are plotted by a red line, and the results of the passive wire are plotted by a blue line. The transient finite element (FE) simulation data of the two wires obtained in Section 3.2. are marked by asterisks.

3. Transient Finite Element Simulations

The effective performance of the conventional transient FE technique for modeling UGW propagation in various structures has previously been demonstrated. Here, the transient solver ABAQUS® (Dassault System Inc., Paris, France) was employed to visually simulate UGW propagation in two different structures: an individual wire and two contact wires. The accuracy of the results is affected by the temporal and spatial resolution of signal acquisition in the FE simulation model. The time step for dynamic analysis (i.e., temporal sampling) and finite element size (i.e., spatial sampling) were determined by the maximum frequency (i.e., minimum wavelength) of the wave in the structure, which should meet the requirements of the Nyquist–Shannon sampling theorem [31]. An optimal spatial resolution is at least eight mesh nodes per wavelength [32,33]. The temporal sampling rate should be 10 times higher than the highest frequency component of the signal. The parameters of the FE simulation model are summarized as Table 2.

Table 2. The parameters of the transient FE simulation model.

Diameter	Length	Density	Young Modulus	Poisson Ratio
7 mm	500 mm	7800 kg/m^3	209 GPa	0.3

The dispersion of the dispersive guided wave propagating in a waveguide is relevant to the bandwidth of the excitation signal. The wider the bandwidth of the excitation signal, the greater the dispersion of the wave. In this paper, two types of excitation signals were used in the transient FE simulation to investigate the dispersion performance of the guided waves in the wire. First, the time-dependent triangular pulse signal was employed to excite a broadband signal with a frequency bandwidth exceeding 500 kHz, as shown in Figure 5a,b. Second, the Hann-windowed, five-cycle, 60 kHz sinusoidal tone burst was designed to excite the narrow bandwidth signal with a specific center frequency, as shown in Figure 5c,d.

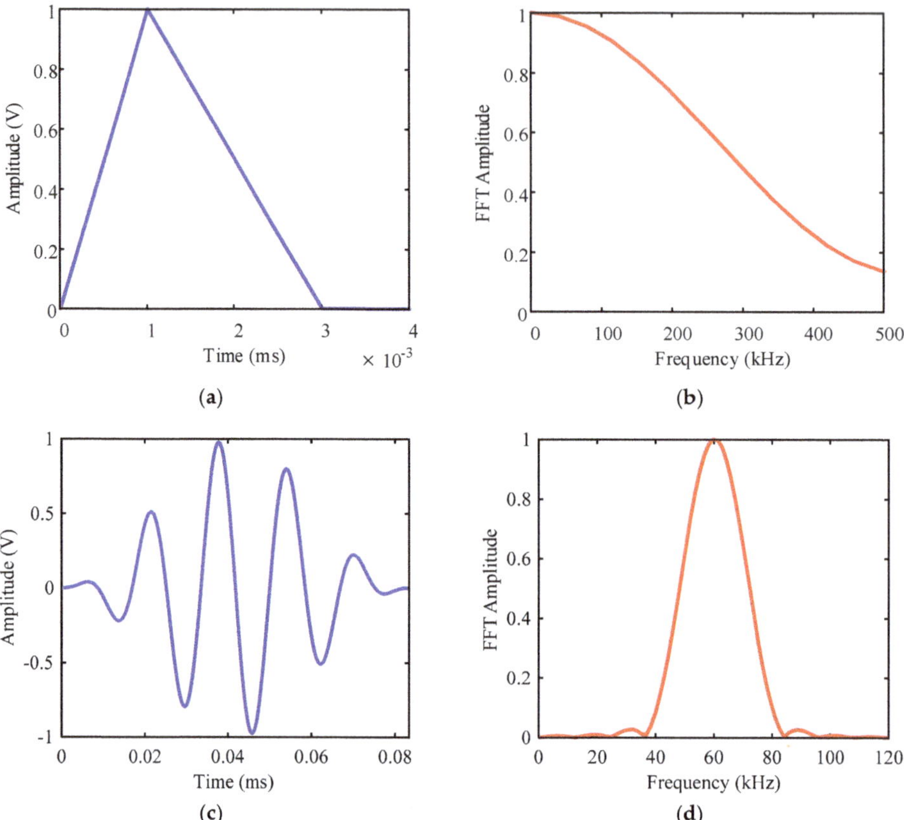

Figure 5. The excitation signal and fast Fourier transform (FFT) spectrum: the triangular pulse with a broadband of 500 kHz in the (**a**) time domain and (**b**) frequency domain; the Hann-windowed, five-cycle, 60 kHz sinusoidal tone burst in the (**c**) time domain signal and (**d**) frequency domain.

The category of the UGW mode generation in a waveguide is related to the direction of the exciting loads. In order to achieve optimal excitation performance, the direction of vibration displacement of the loading was set to be consistent with the wave structure of the wave mode to be excited. According to the analysis of the wave structures of longitudinal and flexural guided waves in Section 2.1, the oblique imposed force with uniform components in the three directions was chosen to excite longitudinal and flexural mode waves. The perpendicular imposed force with a single component along the z-direction was chosen to excite pure longitudinal mode waves, as shown in Figure 6.

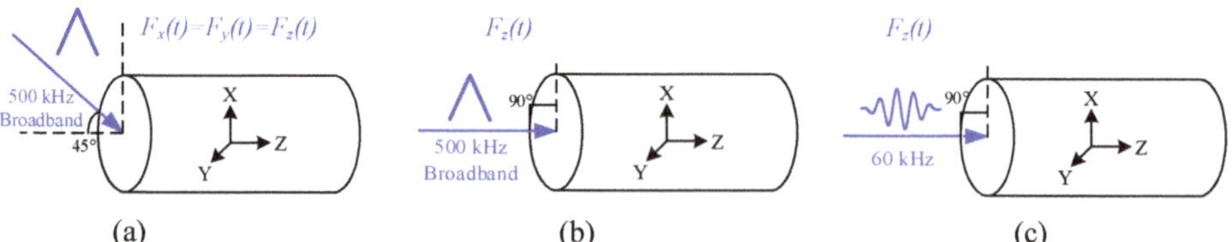

Figure 6. Guided waves in a single wire were excited using three loading methods: (**a**) an oblique triangular pulse with three equal components along the *x*, *y*, and *z* directions, (**b**) a perpendicular triangular pulse along the z-direction, and (**c**) a perpendicular sinusoidal tone burst along the z-direction.

3.1. Single Wire

In order to analyze multimodal and dispersive guided wave signals, discrete wave signals in time and space were recorded at the receiving nodes of the FE model. Signals were mapped to dispersion data in the frequency and wavenumber domains applying the two-dimensional (2D) fast Fourier transform (FFT) method [34]. The essence of the 2D FFT is to perform two fast Fourier transforms on the signal, temporally and spatially. A guided wave signal is harmonical in time and space from the perspective of wave propagation. Therefore, the time and space domain signals $f(z, t)$ can be mapped to angular frequency and wavenumber domain signals $F(k, \omega)$ as:

$$F(k, \omega) = \int_{-\infty}^{\infty} \int_{-\infty}^{\infty} f(z,t)e^{-i(kz+\omega t)} dz dt \qquad (3)$$

$$c_{ph} = \omega/k \qquad (4)$$

where $f(z, t)$ is the acceleration signal matrix recorded at the receiving nodes along the z axis of the wire, c_{ph} is the phase velocity, ω is the angular frequency, and k is the wavenumber. In this case, the accelerations of 500 series with a spatial interval of 1 mm excited by a triangular pulse were used to extract the 2D FFT spectrum, as shown in Figure 7.

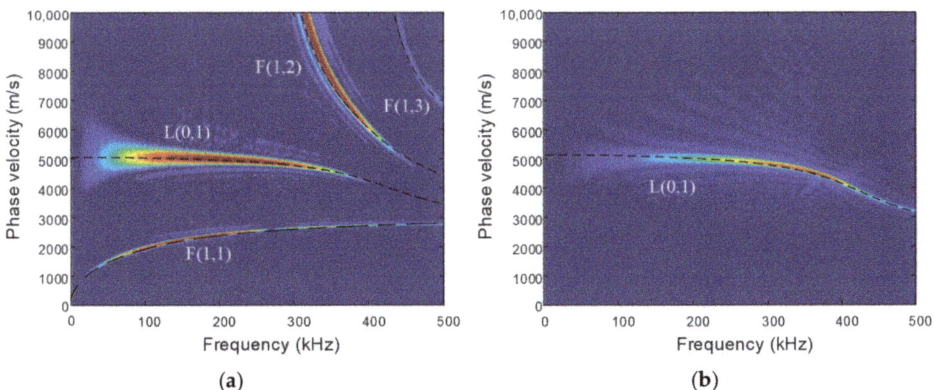

Figure 7. The phase velocity dispersion spectra obtained from nodal accelerations with different excitation methods: (**a**) an oblique triangular pulse signal with three equal components, and (**b**) a perpendicular triangular pulse signal with a z-axial component. The theoretical phase velocity dispersion curves obtained in Figure 1a are represented by the black dotted lines.

The theoretical phase velocity dispersion curves overlaid on the 2D FFT spectra in Figure 7 agreed with the transient simulation results for the longitudinal and flexural modes. As a result of the radial vibration displacement component u_r and the z-axial vibration displacement component u_z, excited by the oblique imposed forces along the x, y, and z directions, all longitudinal (i.e., L(0,1)) and flexural modes (i.e., F(1,1), F(1,2), and F(1,3)) below 500 kHz were generated in the single wire. On the other, the only the wave mode L(0,1) was excited by using the perpendicular triangular pulse along the z-direction.

In order to minimize the dispersion of guided waves during propagation in the structure, signals with a narrow bandwidth are commonly employed to excite waves for defect detection in structures. The tone burst consisting of several cycles of Hann-windowed sine waves is a typical narrow bandwidth signal, with the feature of: the larger the number of wave cycles, the narrower the bandwidth of the signal. Figure 8 shows the 2D FFT spectrum extracted from the nodal accelerations data recorded using the single wire FE model. The guided wave in the wire was excited with a Hann-windowed, five-cycle, 60 kHz sinusoidal tone burst perpendicular to the wire end face.

The dispersion nephogram shows a section of the L(0,1) mode with bandwidth of about 40 kHz, which is the same as that of the excitation signal (see Figure 5b).

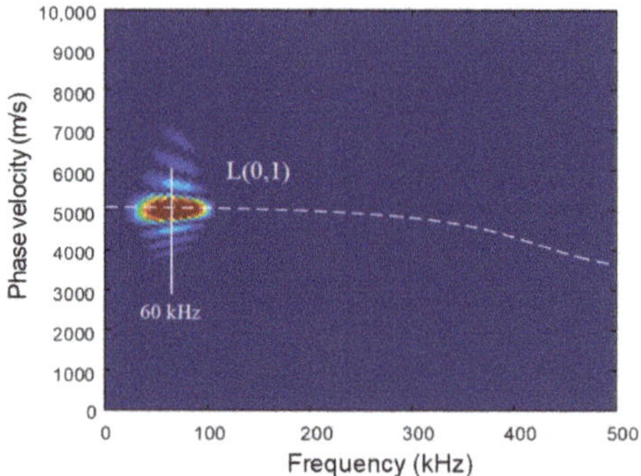

Figure 8. The phase velocity dispersion spectrum from nodal accelerations with a Hann-windowed, five-cycle, 60 kHz sinusoidal tone burst along the z-direction. The theoretical phase velocity dispersion curve of the L(0,1) mode obtained in Figure 1a is represented by a black dotted line.

3.2. Two Contact Wires

When the two ends of the multi-wire cable are subjected to an axial tension, the wire-to-wire contact stress will appear between the adjacent wires. The magnitudes of the contact stresses are proportional to the axial tension. A simplified wire-to-wire contact model was employed to study the effect of contact stresses on UGW propagation in a multi-wire cable. The FE model was made up of two 7-mm-diameter parallel straight steel wires with the same length of 500 mm. One of the wires was active and the other was passive. This pair of wires was mechanically coupled in contact using uniform stresses. The contact stresses were provided by a static radial pressure equaling 70% of the ultimate tensile strength (UTS) of the wire. These contact stresses are called Hertzian contact stresses [35]. The contact condition between the two wires was modeled using a penalty friction formulation with the friction coefficient of 0.6.

The loading conditions in the FE simulations involved two stages: static load and dynamic load. The linear ramp signals were uniformly preloaded into two wires along the wire axes with opposite radial directions of the two wires. The static wire-to-wire contact and friction stresses were generated in this load step. In the second step, guided waves were excited by the dynamic load. An investigation of the wave propagation was conducted in two cases. In Case 1, the active wire was excited with an obliquely imposed triangular pulse force on one end face with equal components ($F_x = F_y = F_z$) along three directions. In Case 2, the active wire was excited with a Hann-windowed, five-cycle, 60 kHz sinusoidal tone burst perpendicular to the end face along the z-direction.

Acceleration data were recorded at the surface nodes of the FE elements of the two wires along the z-axis. The interval of the recorded points was set as 1 mm. The 2D FFT algorithm proposed in Section 3.1 was used again to compare the two cases to observe the change in the dispersion properties of UGWs in a pair of mechanically coupled wires. The 2D FFT transform spectra for the two cases are shown in Figures 9 and 10.

In Figure 9, the dispersion nephograms in the 2D FFT spectra coincide with the theoretical phase velocity dispersion curves (see Figure 1a), which is consistent with the results obtained in the previously presented single wire case. The longitudinal mode (i.e., L(0,1)) and all flexural modes (i.e., F(1,1), F(1,2), and F(1,3)) below 500 kHz were generated in the active and passive wires. Guided waves in the passive wire are transferred by contact and friction stresses between two wires. The acoustic energy can be transferred into the adjacent wire when sufficient contact stresses occur on the interface of two wires. As a result of the leakage of energy, the guided wave modes and energy eventually trend to be the same, with the wave propagation in the two-wire system.

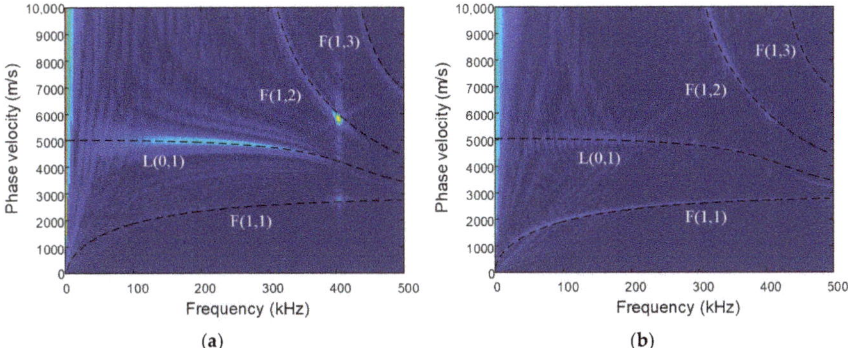

Figure 9. The phase velocity dispersion spectra of two wires from nodal accelerations with an equal components triangular pulse force along the x, y, and z directions: (**a**) the active wire and (**b**) the passive wire. The theoretical phase velocity dispersion curves obtained in Figure 1a are represented by the black dotted lines.

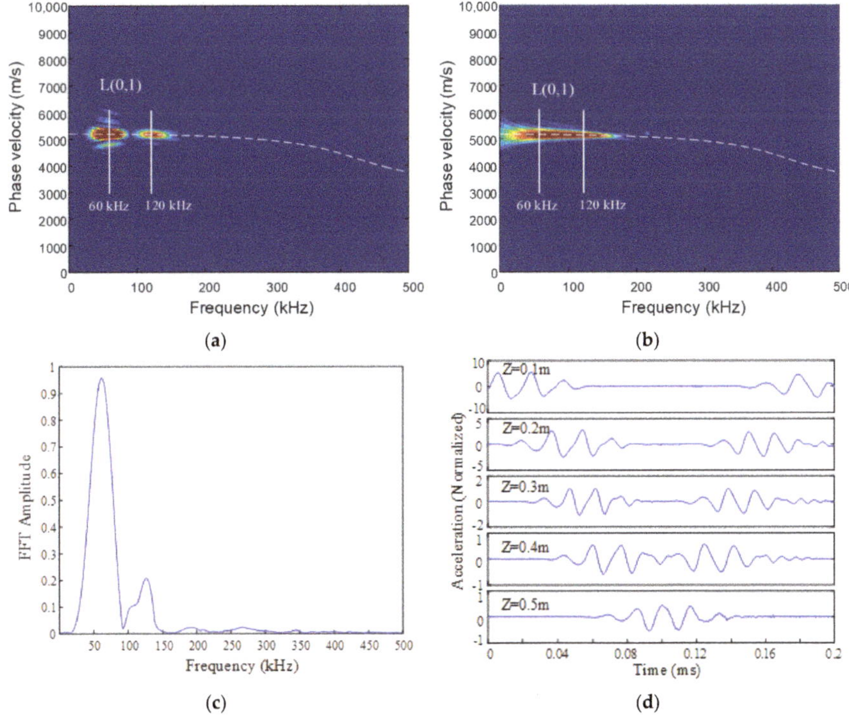

Figure 10. The phase velocity dispersion spectra of wires from nodal accelerations with a Hann-windowed, five-cycle, 60 kHz sinusoidal tone burst along the z-axial direction: (**a**) the active wire and (**b**) the passive wire. The theoretical dispersion curve of L(0,1) obtained in Figure 1a is represented by white dotted lines. (**c**) The FFT spectra of a representative received signal of the active wire and (**d**) part of the representative acceleration time-domain-received signals in the active wire.

The CAN model described in Section 2.2. and the energy transfer model described in Section 2.3. provide two different perspectives and methods to analyze the dispersion properties and propagation behaviors of the guided wave in mechanically coupled structures. Acceleration data recorded on two wire surfaces with a narrow bandwidth excitation signal loaded on the active wire end face were transformed into 2D FFT spectra, as shown in Figure 10. Dispersion nephograms in the spectra of both the active and passive wire agree with the theoretical phase velocity dispersion curve of the single L(0,1) mode below 500 kHz. The directions of the excitation loads determine the generation of guided wave mode propagation in the structure. As a result of the contact stresses between the two wires, the CAN phenomenon of the active wire was observed. The additional frequency component of 120 kHz that doubles the excitation frequency of 60 kHz appeared in the spectrum. The frequency components in the passive wire covered a bandwidth range of 60–120 kHz. This may have occurred because the dispersion of guided waves in the passive wire transferred by contact stresses was more severe than that generated by direct excitation. From the perspective of energy transfer, the guided wave energy distribution in the active and passive wires fit the results of the energy-based model described in Section 2.3, as shown in Figure 4.

4. Experimental Analysis

4.1. Experimental Setup

Experimental verification of the proposed methods was performed on three types of samples. The first sample was a single 7-mm-diameter steel wire: the main component of the multi-wire bridge cable. The second sample was a pair of mechanically coupled 7-mm-diameter steel wires in contact. The third sample was an LGJ-400/35 ACSR cable (Tianhong Electric Power Fitting Co. Ltd., Zhejiang, China) comprised of two layers of seven steel wires and three layers of 48 aluminum wires, which were arranged helically in five layers (1–6–10–16–22). The diameter of the individual steel wire was 2.5 mm and the diameter of the aluminum wire was 3.2 mm. All these samples were the same length: 1.6 m.

A piezoelectric (PZT) UGW experimental inspection system was employed, as shown in Figure 11a. The pitch-catch method was used with a broadband excitation PZT transducer pasted to one end of the cable and the same reception transducer pasted to another cable end. The coupling radius of the transducer was 10 mm and the height of the transducer was 22 mm, as shown in Figure 11b. It had a broadband characteristic (frequency range of 30–150 kHz). The directivity of the transducer was perpendicular to the contact surface. Epoxy resin glue was used to paste the transducer to the wire surface.

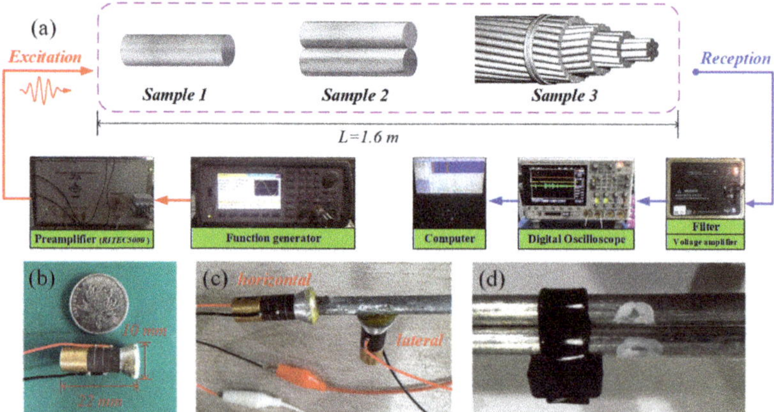

Figure 11. The experimental setup for guided wave excitation and reception in three types of samples. (**a**) The schematic of the experiment. Photos of (**b**) the piezoelectric (PZT) transducer, (**c**) the single wire experiment case, and (**d**) the two contact wires.

For excitation, the transducer was driven by a personal computer (PC)-controlled preamplifier (RAM-5000, Ritec Inc., Warwick, RI, USA) with a Hann-windowed, five-cycle, sinusoidal tone burst. The excitation signal was applied to the transducer with an amplitude of 100 V_{pp} (peak-to-peak value). The detected voltage signals were filtered by a bandpass filter with a bandwidth of 150 kHz and a center frequency of 80 kHz, and amplified by about 41 dB. Then, the signal acquisition was conducted to the filtered signals with a sampling frequency of 2000 kHz. This routine was repeated 50 times with a pulse repetition frequency (PRF) of 20 Hz to remove random noise and achieve a higher signal-to-noise ratio (SNR). The signal series were averaged and processed using Matlab.

4.2. Single Wire

In these single wire experiments, four pitch-catch configuration cases were employed for guided wave dispersion verification. The center frequency of the exciting tone burst was 60 kHz. In Case 1, the exciting transducer was installed on the end face of the left end, and the receiving transducer was similarly installed on the end face of the right end. In Case 2, the exciting transducer was installed on the end face of the left end, but the receiving transducer was installed on the lateral face of the right end. In Case 3, the exciting transducer was installed on the lateral face of the left end, but the receiving transducer was installed on the end face of the right end. In Case 4, the exciting transducer was installed on the lateral face of the left end, and the receiving transducer was similarly installed on the lateral face of the right end.

The representation of a signal in the time and frequency domain can be expressed using a spectrogram. A spectrogram is a time-frequency plot obtained using time-frequency analysis, such as a continuous wavelet transform (CWT). It is a method used to measure experimental group velocity dispersion curves. Figure 12 shows the time-frequency spectrum of the received signals in the four experimental cases using the CWT method, over which the theoretical time-frequency group velocity dispersion curves are overlaid. The theoretical propagation time was calculated using:

$$t = L/v_{group} \tag{5}$$

where $L = 1.6\ m$ is the length of the wire and v_{group} is the group velocities of all existent modes at each frequency between 0 kHz and 100 kHz. This process is equivalent to mapping the group velocity dispersion curves from the velocity-frequency domain to the time-frequency domain.

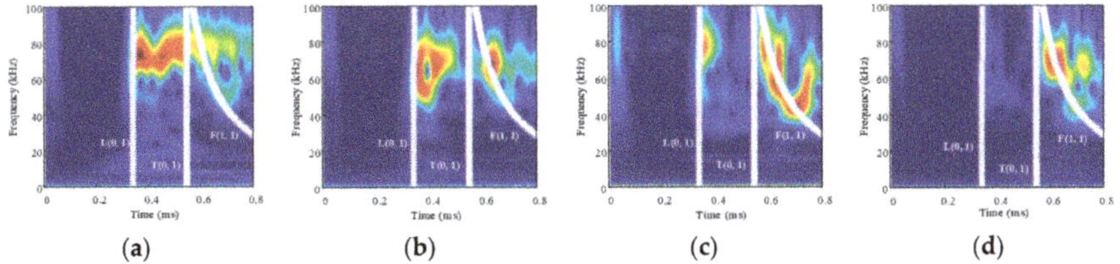

(a)   (b)   (c)   (d)

Figure 12. The time-frequency analysis spectrum using a continuous wavelet transform (CWT) for a single 7-mm-diameter steel wire at a 60-kHz frequency: (**a**) horizontal excitation and horizontal reception, (**b**) horizontal excitation but lateral reception, (**c**) lateral excitation but horizontal reception, and (**d**) lateral excitation and lateral reception. The theoretical group velocity dispersion curves obtained in Figure 1a are represented by the dotted line.

There were three guided wave modes in the 7-mm-diameter steel wire below 100 kHz, whose dispersion curves were described in Section 2.1. According to the wave structures plotted in Figure 2d,e, the radial and z-axial vibration displacements were the main components of the L(0,1) and F(1,1) modes at 60 kHz, respectively. Z-axial wave components were generated through horizontal excitation of the transducer. Radial wave components were generated through lateral excitation. Similarly, the horizontal and lateral reception methods of the transducer effortlessly received the *z*-axial and radial wave components.

Thus, the L(0,1) mode was highlighted in the measured group velocity dispersion nephograms shown in Figure 12a of Case 1. The same applied to the L(0,1) and F(1,1) modes in Cases 2 and 3 simultaneously, and F(1,1) in Case 4 individually. No torsional mode was highlighted in any of the four cases, since neither of the two excitation methods could generate a rotational displacement component.

4.3. Two Contact Wires

In order to analyze the CAN phenomena in a multi-wire cable, a wire-to-wire cable was first considered. Two identical 7-mm-diameter steel wires with a length of 1.6 m, used in previous experiments, were tied together tightly and compactly using nylon cable ties along the *z*-axial direction, as shown in Figure 11d. The wire that was excited to generate a UGW was labeled the active wire, and another one was labeled the passive wire. The installation of exciting and receiving transducers in Case 2 of Section 4.2. was employed to simultaneously receive L(0,1) and F(1,1) guided waves. The center excitation frequency of the tone burst was 60 kHz. The reception signals of the two wires were recorded and are plotted in Figure 13. The FFT spectra of the two signals were obtained. The theoretical group velocity dispersion curves calculated in Section 2.1. were used to determine the wave modes presented in the signals of two wires. There were two wave packets in the time domain signals of the active and passive wire: the direct waves of L(0,1) and F(1,1). The measured group velocities of two modes calculated from the time-of-flight (ToF) of the wave peak in two wave packets are consistent with the theoretical group velocity dispersion curves at 60 kHz.

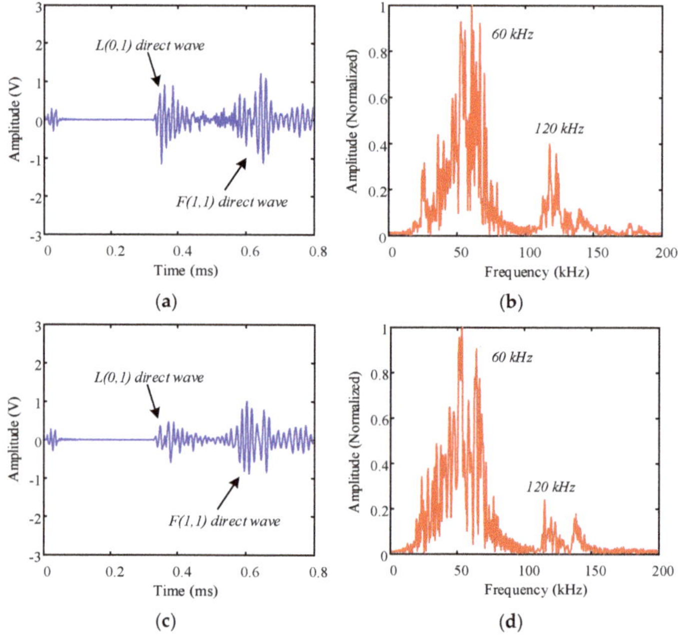

Figure 13. The wave signals and FFT spectrum for the two contact wires with an exciting center frequency of 60 kHz: (**a**) the time domain signal of the active wire, (**b**) the FFT spectrum of the active wire, (**c**) the time domain signal of the passive wire, and (**d**) the FFT spectrum of the passive wire.

The appearance of the frequency components of 120 kHz in the FFT spectrum of the active and passive wires, which was double the center frequency of the exciting signals, verifies the CAN effect of the two contact wires described in Section 2.2. and is consistent with the results of the transient FE simulation in Section 3.2.

4.4. Overhead Transmission Line

As described in Section 4.1, the third experimental sample LGJ-400/500 ACSR cable consisted of five layers (1–6–10–16–22) of wires composed of steel and aluminum. The innermost two layers were 7.5-mm-diameter seven-wire steel strands (1–6). Since the diameter of the single steel wire (D = 2.5 mm) was much smaller than that of the transducer (D = 10 mm), the entire steel strands were treated as one layer. The exciting transducer was installed on the end face of the strands' end. The receiving transducers were installed on the lateral faces of four layers (7–10–16–22) of another cable end, as shown in Figure 14a.

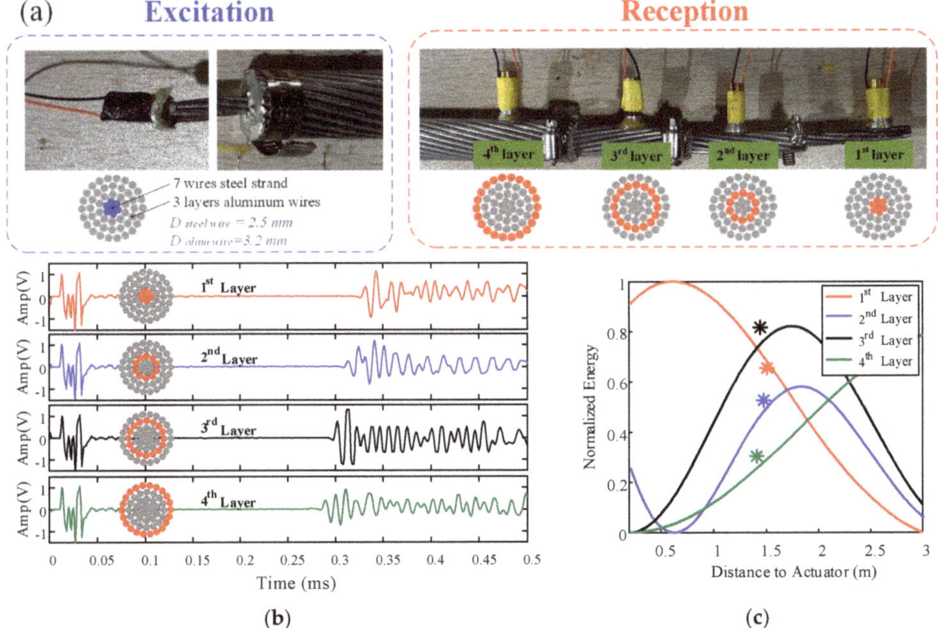

Figure 14. Guided wave propagation experiments on the aluminum conductor steel-reinforced (ACSR) cable: (**a**) a photo of the exciting and receiving transducers installation, (**b**) the received time domain signals of each layer, and (**c**) the simulated and measured time average wave energy distribution in each layer.

Precise transient FE simulations that consider contact and friction stresses are computationally expensive. Applications of the FE method are limited to the multi-wire structures with short lengths and a smaller number of wires. For the sake of simplifying the description of UGW propagation behavior in multi-wire structures, such as bridge cables and ACSR cables, the energy-based model was used to investigate the experimental data. Guided wave signals recorded by transducers installed on each layer of the cable end (Z = 1.6 m) were processed by Matlab and are plotted in Figure 14b.

In Figure 14b, the signals from the top to bottom were received by the transducers installed on the first, second, third, and fourth layers of the cable (from the inside to the outside). The wave packet marked by a circle in each signal was the direct wave excited by the exciting transducer. As a result of the contact and friction stresses between the adjacent wires in the same and adjacent layer, wave energy was transferred to each wire of the cable with the guided wave propagation. Each wire in the inner layers of the cable was in contact with four adjacent wires: left wire, right wire, wire in the upper layer, and wire in the lower layer. Thus, the wave energy of each wire can be derived from Equation (1). The initial conditions in the equation were determined using a least squares fit with recorded signals. Since the active wire (first layer) was the steel strands in the center of the cable, the normalized wave energy of the wires in the same layer was equal (second, third, and fourth layers), as shown in Figure 14c. The measured wave energy can be obtained by integrating the received signal over the time domain:

$$E(z) = \int_{t_1}^{t_2} U_r^2(z,t)\mathrm{d}t \tag{6}$$

where t_1 and t_2 correspond to the time required by the fastest and slowest modes at the frequency of 60 kHz propagating from $z = 0$ to $z = L$ (i.e., $t_1 = L/v_{group,fast}$ and $t_2 = L/v_{group,slow}$), respectively, and $U_r(z,t)$ is the radial vibration displacement at position z. The fastest and slowest mode at 60 kHz frequency in the 3.2-mm-diameter aluminum wire group velocity dispersion curves shown in Figure 1b were the L(0,1) mode and the F(1,1) mode, respectively.

5. Conclusions and Outlook

In this study, an initial theoretical analysis of UGWs in a two-contact-wire system and an ACSR cable was completed from various perspectives. A time-transient FE numerical simulation and experimental study were conducted to validate the dispersion and wave structure properties, the contact acoustic nonlinearity effect, and the energy transfer model of multilayer wires. The main findings are summarized as follows:

(1) For the wave structures of the fundamental order longitudinal and flexural modes (i.e., L(0,1) and F(1,1)) in a 7-mm-diameter steel wire, the radial vibration displacements decreased but the axial vibration displacements increased with increasing radius. This trend became apparent with the increase in guided wave frequency below 200 kHz. The severity of dispersion and number of modes in a cylindrical structure were related to the bandwidth of excitation signals and the direction of excitation loads, respectively.

(2) Guided waves with frequency f were excited to one of the two mechanically coupled wires. Received waves with frequency $2f$ were observed in both wires due to the CAN effect at the contact interface. Numerical and experimental verification were performed using a monokinetic Hann-windowed sinusoidal excitation. There are many different wave modes simultaneously propagating in a multi-wire structure with different velocities at different frequencies, which could be predicted by dispersion curves. However, it is ineffective for dispersion curves to analyze the frequency behaviors of the wave signals of a certain mode, which could be analyzed using the CAN effect.

(3) A minimally computationally expensive energy transfer model was used to predict the wave energy distribution in the wires located in different layers of the multi-wire cable. The approach was applied to analyze a pair of mechanically coupled wires using transient FE simulation. The acceleration data of the two wires were fitted to the model. The same investigation was experimentally performed on a 1.6-m-long 55-wire ACSR cable, and the wave energy of each layer was found to be consistent with the numerical analysis. Uniform distribution of the wave energy in each layer of a multi-wire cable is a basic requirement of SHM using a guided wave. The uniformity of the UGW energy depends on the contact between adjacent layers, which can be predicted by the model, when using piezoelectric transducers installed on the outermost layer of the cable. The model also could be used to optimize the excitation power of transducers and determine the effective monitoring range of SHM. In industrial applications, the results of this study could be used to improve the detection rate of the defects of wires located in the inner layers of the cable with low signal amplitudes, which are caused by the nonuniform wave energy distribution in different layers of the cable.

For future research, we plan to study the UGW propagation in multi-wire cables using more precise contact and friction models to provide guidance for the guided wave inspection of cables, and the relationship between cable force and the CAN effect.

Author Contributions: P.Z. and Z.T. conceived and designed the models and methods presented; P.Z. developed the experimental software tools; F.L. and K.Y. helped in conception and experiments; Z.T. helped in the simulations; P.Z. performed the experiments and analyzed the data; and P.Z. wrote this paper.

Funding: This research was funded by the National Natural Science Foundation of China (Grant No. 61271084 and No. U1709216), the Science and Technique Plans of Zhejiang Province (Grant No. 2017C01042), the National Key Research and Development Program of China (Grant No. 2018YFC0809000), and the Project of Inner Mongolia Power Group (Grant No. 20184733). And The APC was funded by the National Natural Science Foundation of China.

Conflicts of Interest: The authors declare no conflict of interest.

References

1. Shull, P.J. *Nondestructive Evaluation: Theory, Techniques, and Applications*; CRC Press: Boca Raton, FL, USA, 2016.
2. Ye, X.W.; Dong, C.Z.; Liu, T. Force Monitoring of Steel Cables Using Vision-Based Sensing Technology: Methodology and Experimental Verification. *Smart Struct. Syst.* **2016**, *18*, 585–599. [CrossRef]
3. Zhang, J.; Guo, Z.; Jiao, T.; Wang, M. Defect Detection of Aluminum Alloy Wheels in Radiography Images Using Adaptive Threshold and Morphological Reconstruction. *Appl. Sci.* **2018**, *8*, 2365. [CrossRef]
4. Omidi, P.; Zafar, M.; Mozaffarzadeh, M.; Hariri, A.; Haung, X.; Orooji, M.; Nasiriavanaki, M. A Novel Dictionary-Based Image Reconstruction for Photoacoustic Computed Tomography. *Appl. Sci.* **2018**, *8*, 1570. [CrossRef]
5. Qin, L.; Ren, H.; Dong, B.; Xing, F. Development of Technique Capable of Identifying Different Corrosion Stages in Reinforced Concrete. *Appl. Acoust.* **2015**, *94*, 53–56. [CrossRef]
6. Shin, E.; Kang, B.; Chang, J. Real-Time Hifu Treatment Monitoring Using Pulse Inversion Ultrasonic Imaging. *Appl. Sci.* **2018**, *8*, 2219. [CrossRef]
7. Zhang, J.; Tan, X.; Zheng, P. Non-Destructive Detection of Wire Rope Discontinuities From Residual Magnetic Field Images Using the Hilbert-Huang Transform and Compressed Sensing. *Sensors* **2017**, *17*, 608. [CrossRef] [PubMed]
8. Mijarez, R.; Baltazar, A.; Rodríguez-Rodríguez, J.; Ramírez-Niño, J. Damage Detection in ACSR Cables Based on Ultrasonic Guided Waves. *DYNA* **2014**, *81*, 226–233. [CrossRef]
9. Laguerre, L.; Treyssede, F. Non Destructive Evaluation of Seven-Wire Strands Using Ultrasonic Guided Waves. *Eur. J. Environ. Civ. Eng.* **2011**, *15*, 487–500. [CrossRef]
10. Qian, J.; Chen, X.; Sun, L.; Yao, G.; Wang, X. Numerical and Experimental Identification of Seven Wire Strand Tensions Using Scale Energy Entropy Spectra of Ultrasonic Guided Waves. *Shock Vib.* **2018**, *2018*, 6905073. [CrossRef]
11. Schaal, C.; Bischoff, S.; Gaul, L. Damage Detection in Multi-Wire Cables Using Guided Ultrasonic Waves. *Struct. Health Monit. Int. J.* **2016**, *15*, 279–288. [CrossRef]
12. Zhang, P.; Tang, Z.; Duan, Y.; Yun, C.B.; Lv, F. Ultrasonic Guided Wave Approach Incorporating Safe for Detecting Wire Breakage in Bridge Cable. *Smart Struct. Syst.* **2018**, *22*, 481–493. [CrossRef]
13. Rizzo, P. Ultrasonic Wave Propagation in Progressively Loaded Multi-Wire Strands. *Exp. Mech.* **2006**, *46*, 297–306. [CrossRef]
14. Hernandez Crespo, B.; Courtney, C.; Engineer, B. Calculation of Guided Wave Dispersion Characteristics Using a Three-Transducer Measurement System. *Appl. Sci.* **2018**, *8*, 1253. [CrossRef]
15. Raišutis, R.; Kažys, R.; Mažeika, L.; Žukauskas, E.; Samaitis, V.; Jankauskas, A. Ultrasonic Guided Wave-Based Testing Technique for Inspection of Multi-Wire Rope Structures. *NDT E Int.* **2014**, *62*, 40–49. [CrossRef]
16. Raišutis, R.; Kažys, R.; Mažeika, L.; Samaitis, V.; Žukauskas, E. Propagation of Ultrasonic Guided Waves in Composite Multi-Wire Ropes. *Materials* **2016**, *9*, 451. [CrossRef] [PubMed]
17. Schaal, C.; Bischoff, S.; Gaul, L. Energy-Based Models for Guided Ultrasonic Wave Propagation in Multi-Wire Cables. *Int. J. Solids Struct.* **2015**, *64–65*, 22–29. [CrossRef]

Appl. Sci. **2019**, *9*, 1028

18. Haag, T.; Beadle, B.M.; Sprenger, H.; Gaul, L. Wave-Based Defect Detection and Interwire Friction Modeling for Overhead Transmission Lines. *Arch. Appl. Mech.* **2009**, *79*, 517–528. [CrossRef]
19. Hayashi, T.; Tamayama, C.; Murase, M. Wave Structure Analysis of Guided Waves in a Bar with an Arbitrary Cross-Section. *Ultrasonics* **2006**, *44*, 17–24. [CrossRef] [PubMed]
20. Ibáñez, F.; Baltazar, A.; Mijarez, R. Detection of Damage in Multiwire Cables Based On Wavelet Entropy Evolution. *Smart Mater. Struct.* **2015**, *24*, 085036. [CrossRef]
21. Treyssède, F. Elastic Waves in Helical Waveguides. *Wave Motion* **2008**, *45*, 457–470. [CrossRef]
22. Li, C.; Han, Q.; Liu, Y.; Liu, X.; Wu, B. Investigation of Wave Propagation in Double Cylindrical Rods Considering the Effect of Prestress. *J. Sound Vib.* **2015**, *353*, 164–180. [CrossRef]
23. Chillara, V.K.; Lissenden, C.J. Analysis of Second Harmonic Guided Waves in Pipes Using a Large Radius Asymptotic Approximation for Axis-Symmetric Longitudinal Modes. *Ultrasonics* **2013**, *53*, 862–869. [CrossRef] [PubMed]
24. Jhang, K. Nonlinear Ultrasonic Techniques for Nondestructive Assessment of Micro Damage in Material: A Review. *Int. J. Precis. Eng. Manuf.* **2009**, *10*, 123–135. [CrossRef]
25. Legg, M.; Yücel, M.K.; Kappatos, V.; Selcuk, C.; Gan, T. Increased Range of Ultrasonic Guided Wave Testing of Overhead Transmission Line Cables Using Dispersion Compensation. *Ultrasonics* **2015**, *62*, 35–45. [CrossRef] [PubMed]
26. Gazis, D.C. Three-Dimensional Investigation of the Propagation of Waves in Hollow Circular Cylinders. I. Analytical Foundation. *J. Acoust. Soc. Am.* **1959**, *31*, 568–573. [CrossRef]
27. Ditri, J.J.; Rose, J.L. Excitation of Guided Elastic Wave Modes in Hollow Cylinders by Applied Surface Tractions. *J. Appl. Phys.* **1992**, *72*, 2589–2597. [CrossRef]
28. Seco, F.; Jiménez, A.R. Modelling the Generation and Propagation of Ultrasonic Signals in Cylindrical Waveguides. *Ultrason. Waves* **2012**. [CrossRef]
29. Seco, F.; Martín, J.M.; Jiménez, A.; Pons, J.L.; Calderón, L.; Ceres, R. Pcdisp: A Tool for the Simulation of Wave Propagation in Cylindrical Waveguides. In Proceedings of the 9th International Congress on Sound and Vibration, Orlando, FL, USA, 8–11 July 2002.
30. Biwa, S.; Nakajima, S.; Ohno, N. On the Acoustic Nonlinearity of Solid-Solid Contact with Pressure-Dependent Interface Stiffness. *J. Appl. Mech.* **2004**, *71*, 508–515. [CrossRef]
31. Abdul, J. The Shannon Sampling Theorem—Its Various Extensions and Applications: A tutorial review. *Proc. IEEE* **1977**, *65*, 1565–1596.
32. Datta, D.; Kishore, N.N. Features of Ultrasonic Wave Propagation to Identify Defects in Composite Materials Modelled by Finite Element Method. *NDT E Int.* **1996**, *29*, 213–223. [CrossRef]
33. Kirby, R. Modeling Sound Propagation in Acoustic Waveguides Using a Hybrid Numerical Method. *J. Acoust. Soc. Am.* **2008**, *124*, 1930–1940. [CrossRef]
34. Alleyne, D.; Cawley, P. A Two-Dimensional Fourier Transform Method for the Measurement of Propagating Multimode Signals. *J. Acoust. Soc. Am.* **1991**, *89*, 1159–1168. [CrossRef]
35. Johnson, K.L. *Contact Mechanics*; Cambridge University Press: Cambridge, UK, 1987.

Article

Compensation for Group Velocity of Polychromatic Wave Measurement in Dispersive Medium

Seung Jin Chang and Seung-Il Moon *

Department of Electrical and Computer Engineering, Seoul National University, Seoul 151-744, Korea;
jpromo8@gmail.com
* Correspondence: moonsi@plaza.snu.ac.kr; Tel.: +82-2-880-7257

Received: 21 November 2017; Accepted: 15 December 2017; Published: 18 December 2017

Abstract: The estimation of instantaneous frequency (IF) method is introduced to compensate for the group velocity of electromagnetic wave in dispersive medium. The location of the reflected signal can be obtained by using the time-frequency cross correlation (TFCC), following which it is used to extract the transmitted signal from the total signal acquired. The signal propagated in the dispersive medium is attenuated and distorted by the attenuation characteristics, which depend on the frequency of the medium. By using the IF curve calculated for the transmitted signal, the changed center frequency and time terms can be obtained. The obtained terms are used to compensate for the group velocity error induced by signal distortion and attenuation. Through experiments and simulation, the accuracy of the proposed method is 2% higher than that of the conventional method when the signal propagates over a long distance.

Keywords: electromagnetic wave; group velocity; time-frequency domain reflectometry; dispersive medium

---

1. Introduction

It is important to accurately measure the group velocity of electromagnetic wave in a dispersive medium to detect defects and measure distances. Since the group velocity is dependent on the frequency, and dispersion occurs when a polychromatic wave composed of multiple frequencies propagates in a dispersive medium, it is difficult to measure the exact group velocity [1]. group velocity correction is necessary when a signal propagates in any medium other than air. Detection methods based on the electromagnetic wave, called reflectometry methods, are useful for fault localization and monitoring of the health of a cable with insulator. Reflectometry methods can be categorized into time domain reflectometry (TDR), frequency domain reflectometry (FDR), and time-frequency domain reflectometry (TFDR) [2–7] depending on the incident signal type. TDR and FDR use a step pulse and sinusoidal pulse defined in the time domain and frequency domain, respectively. They have been widely used in cable diagnostics owing to their ease of implementation. However, since the incident signals of TDR and FDR defined one domain are used, it is difficult to distinguish the acquired signal from the noise. In contrast to TDR and FDR, TFDR has a higher signal-to-noise (SNR) because the incident signal of TFDR is used as a chirp signal that has characteristics in both the time and frequency domains. In TFDR, the impedance discontinuity distance is measured by obtaining the time delay of the reflected signal generated from the impedance discontinuity point for a given group velocity. Since the group velocity is determined by the permittivity and permeability of the propagation medium, the propagation can be obtained from the information about the propagation medium. The time-frequency cross correlation function (TFCC) which is a measure of the similarity of two signals, and calculates the time delay, is based on the Wigner-Ville distribution, which is an energy distribution in time-frequency analysis. However, since the group velocity depends on the frequency, if the bandwidth is wide or distortion is increased by long-distance propagation, the error of the group velocity also becomes large. Especially, in case of submarine cable lengths of several hundred km, speed compensation is essential. Our research group has been investigated the compensation for group velocity [8]. The compensation method proposed in the previous method [8] does not consider the sweep rate of the reflected signal to be changed. That is, since the frequency components included in the chirp signal have different velocities, compensation is performed without taking into account dispersion, which causes an measurement error.

In this paper, we propose a method for compensating the group velocity of poly-chromatic wave in a dispersive medium based on instantaneous frequency (IF) estimation. Using TFCC, the location of transmitted signal is obtained and the signal is extracted from the total acquired signal. IF curves of incident and transmitted signal are derived by using the phase unwrapping process. The shifted center frequency and shortened time duration of transmitted signal are obtained based on the Fast Fourier Transform (FFT) and estimation of the IF curve. The group velocity compensation is carried out through the derived terms.

2. Group Velocity Measurement Based on Electromagnetic Wave Theory

The proposed group velocity measurement method consists of two main parts. The first part comprises the transmitted signal detection method using TFCC. After the first step, the time offset and shifted center frequency of the transmitted signal are used to compensate for the group velocity. In electromagnetic theory, the group velocity of an electromagnetic plane wave can be derived as follows [1]:

$$v_p(f) = \frac{c}{\sqrt{(\epsilon(f) \cdot \mu(f))}} = \frac{c}{(\epsilon_\infty + \frac{\epsilon_s - \epsilon_\infty}{1 + j\frac{f}{f_r}})^{1/2} \cdot (1 + \frac{\mu_s - 1}{(1 + j\frac{f}{f_m})^2})^{1/2}},$$ (1)

where f, f_r, f_m are the operating frequency, relaxation frequency at which the imaginary part of the complex permittivity reaches a maximum, and resonant frequency, respectively. $\epsilon_\infty, \epsilon_s$ are the permittivity at infinity frequency and static permittivity, respectively. μ_s is the static permeability. According to (1), the group velocity depends on the operating frequency. When the polychromatic wave comprising several frequencies propagates in the dispersive media, the wave is distorted and a group velocity error occurs because of the frequency dependency of group velocity. Especially when the signal is transmitted over a long distance, the more high frequency components are attenuated, and velocity error is induced. In other words, as the signal propagates, the group velocity changes continuously with the propagation distance. In this paper, the incident signal is the signal that we injected into the cable, the reflected signal is generated at the cable termination, and transmitted signals are signals that are acquired by the oscilloscope through the inductive couplers as the incident signal flows along the cable, and the signals acquired after being reflected at the cable termination are classified as reflected signals, and the reflected signal is included in the transmitted signal. We use the Gaussian linear chirp signal as the incident polychromatic wave, and the incident signal is represented as follows:

$$s(t) = \begin{cases} Ae^{-\alpha t^2/2 + j(0.5\xi_1 t^2 + \omega t)}, & t = 0 \le t \le T_1 \\ 0, & \text{otherwise.} \end{cases}$$ (2)

where T_1 is the duration of the incident signal, A is the amplitude, ξ_1 is the normalized angular frequency sweep rate, and ω is the normalized angular frequency. The transmitted signal is expressed as follows:

$$r(t - d) = \begin{cases} \eta \cdot Ae^{-\alpha t^2/2 + j(0.5\xi_2(t-d)^2 + \omega(t-d) + \phi)}, & d \le t \le d + T_2 \\ 0, & \text{otherwise.} \end{cases}$$ (3)

where η is the magnitude of the attenuation coefficient at the travelling distance, and d is the time delay of the transmitted signal. Also, because the transmitted signal has high frequency attenuation, we define the changed parameters ζ_2 and T_2 as frequency sweep rate and time duration of transmitted signal. The normalized cross-correlation between the incident signal and the transmitted signal is used to detect the transmitted signal from the cable termination. This can be expressed as follows [3]:

$$C_{sr}(t) = \sum_{k=0}^{N-1} \frac{s_k \otimes r_{k-t}}{\sqrt{E_s E_r}} \tag{4}$$

where E_s is the energy of the incident signal in Wigner-Ville distribution, E_r is the energy of the transmitted signal, and $\otimes$ is the correlation operator. Through the normalized cross-correlation process, the transmitted signal can be extracted from the total acquired signal.

3. Compensation of Group Velocity in Dispersive Medium

For the compensation of group velocity, we measure the IF curves of incident and transmitted signals. To acquire IF curves, the incident and transmitted signals are transformed through Hilbert transform for as following analytic representation:

$$z(t) = r(t) + jH_t\{r\} = Me^{j\Phi} \tag{5}$$

where, $H_t\{x\}$ is a output at a time t of Hilbert transform filter applied the signal r, M and Φ are the magnitude and the instantaneous phase of the transmitted signal. And then, the instantaneous phase is derived as following:

$$\zeta_2(t - d)^2 + \omega(t - d) + \phi = \tan^{-1} \frac{abs(j * Hilbert(r(t)))}{abs(Hilbert(r(t)))} \tag{6}$$

where, $abs(\cdot)$ means the absolute function. Figure 1 shows the illustration of the compensation of IF curves. In Figure 1, $f_{s,c}$ and $f_{r,c}$ are the center frequency of incident signal and transmitted signal. To derive the time shifted terms, $a(f), b(f)$, of transmitted signal according to operation frequency, we obtained the shifted term using the delayed incident signal by the time delay, d, and the reflection signal as follows.

$$a(f) = \frac{f_{r,c} - f_0}{\zeta_1}, \quad b(f) = \frac{f_{r,c} - f_0}{\zeta_2}, \quad c(f) = \frac{f_{s,c} - f_{r,c}}{\zeta_1} \tag{7}$$

The concept of group velocity of polychromatic signal is associated with narrow-band pulses concentrated in the neighborhood of center frequency, ω_0, with an effective frequency band $|\omega - \omega_0| \leq \Delta\omega$, where $\Delta\omega \ll \omega_0$. Furthermore, the compensated time delay between incident signal and transmitted signal is derived as following:

$$t_{com} = d + (b(f) - a(f)) - c(f) \tag{8}$$

And then, compensated group velocity is derived as follows: $v_{com} = l/t_{com}$. where, l is the known length of target cable. In future work, we derive the frequency and group velocity, $v(f)$, curve based on (1). And then, we assume that the center frequency of transmitted signal is linearly decrease with time. According to this assumption, the center frequency of transmitted signal can be expressed as: $f_{r,c}(t) = a \cdot t + b$ where, $a < 0$ and b are the constant. The travelling distance, D can be derived as: $D = \int_{t_0}^{t_0 + t_{com}} v(f_{r,c}(t))dt$. This article compensates the speed with time delay compensation, and next it will develop this to estimate the distance through accurate propagation speed and time delay without travelling distance information.

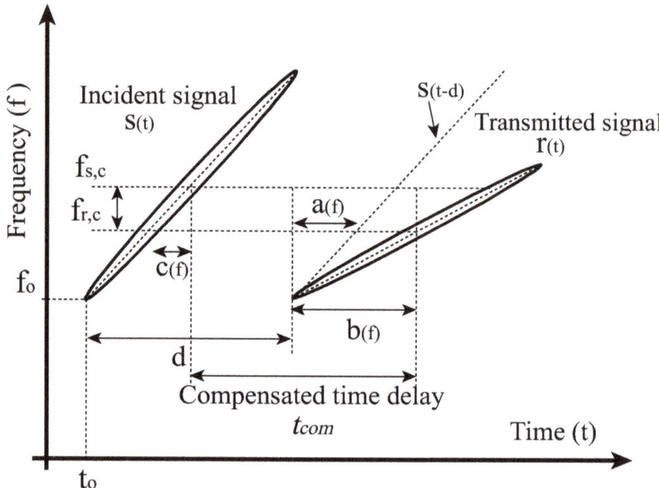

Figure 1. Illustration of compensation method based on IF curves.

4. Experimental Results

Figure 2 shows the illustration of the proposed system. The system is composed of (1) digital phosphor oscilloscope (DPO); (2) arbitrary waveform generator (AWG); and (3) inductive couplers. AWG generates the reference signal and apply the signal to target cable though the inductive coupler. The inductive coupler uses the electromagnetic induction phenomenon to apply a signal to the cable without connection between the cable core and signal line. In this paper, we use three inductive couplers and coupler 1 is used to apply a signal to cable, couplers 2 and 3 acquire the signal. An incident signal propagates along the cable and was acquired in the oscilloscope through the couplers 2 and 3. For ease of understanding, the signals were numbered according to the order in which they were acquired. The transmitted signals acquired through the couplers 2 and 3, before the signal being reflected, were numbered 1 and 2 and located at 0 m and 40 m, respectively. The distance propagated based on the first acquired signal though coupler 2 becomes the position of the signal. The signals reflected from the cable termination were in turn acquired via couplers 3 and 2, which were located at 80 m and 120 m. In order to verify the variation of group velocity due to the dispersion, we converted the experimental system into an equivalent circuit model using the simulation tool. Through the tuning of loss factor in the simulation tool, we conducted the simulation to compare the signal passing through the lossy medium with lossless medium. The comparison results are shown in Figure 3. The results consist of 3 types simulations: (a) lossless cable: 40 m; (b) lossy cable: 40 m; (c) lossy cable: 80 m. To verify the effect of loss factor, we compared the transmitted signal in lossy medium with that in lossless medium (simulations: (a) and (b)). Also, simulations were conducted with different cable lengths in order to analyze whether the group velocity of the signal varies due to the reflection (simulations: (b) and (c)). As shown in Figure 3a, the incident signals of each simulation are identical, but the reflected signals of each did not match. Figure 3b shows an enlarged view of the reflected signals. The highest peak in the time domain of the signal can be thought of as the highest energy point, and the group velocity of the signal can be determined by the time delay between these peak points, and this time delay is called the time of arrival of group velocity. Comparing the waveforms of (a) and (b), the time delay between the peak points of the reflected signal generated at 40 m and 80 m are $0.102 \cdot 10^{-6}$ s and $0.207 \cdot 10^{-6}$ s, respectively. If there was no change in group velocity depending on the travel distance, the time of arrival of group velocity of the reflected signal generated at 80 m had to be 0.204×10^{-6} m. More time delays mean slower group velocity. These results show that the group velocity is decreasing as the signal propagates. Since the incident signal is a positive chirp signal, the rear part of the signal in time domain contains a high frequency component. As shown in the reflected signals of red and black line of Figure 3b, in the front part, the zero-crossing points of each signal are matched, but the zero-crossing points in the rear part do not coincide with each other. These results indicate that the higher frequency components of the signal are attenuated as the signal propagates in the lossy medium. Comparing the reflected signals generated at 80 m, we verified that the reflection only affect the magnitude of signal, not group velocity. Figure 4a shows the total acquired signal from the inductive couplers after signal restoration process [9,10]. As seen in the fourth signal in Figure 4a, because the signal is difficult to distinguish from the noise, TFCC is used to roughly find the position of the transmitted signal. To evaluate the accuracy of proposed method, we solved the true value of group velocity using the time of arrival of group velocity of the signals in the time domain. However, the reflected signal at 120 m is too small to find out due to the attenuation. Because of this, it is very difficult to obtain the group velocity through the time delay between the highest peak points of the signal in the time domain, and the group velocity in the farther than 120 m can not be calculated. The TFCC graph is shown in the Figure 4b. As the signal propagates, attenuation of the signal occurs, which slows the group velocity and increases the time delay. Figure 4b depicts a TFCC graph based on the constant group velocity measured on 40 m. Therefore, the distance errors of 80 m and 120 m is getting larger. Based on unwrapping algorithm and Hilbert transform, the instantaneous phase of the transmitted signal can be derived and shown in Figure 5a. The signal having the positive slope of the instantaneous phase was extracted, and the frequency band of the extracted signal was obtained by FFT algorithm. In Figure 5a, the time duration of the signal is obtained by extracting the signal where the slope of the instantaneous phase is positive. The Figure 5b shows the frequency band of second signal of total signal. The changed values, time duration and frequency region, were obtained and substituted into Equation (7) to compensate the group velocity. As seen in Table 1, the group velocity in TFCC method seems to be equal regardless of the propagation distance. On the contrary, in the proposed method, the shifted terms , $a(f), b(f), c(f)$, can be obtained from the center frequency, f_{rc}, of the received signal from Figures 1 and 5. The compensation time delay term is calculated according to Equation (8). The travelling distance, D, is the known value of the cable length and is the integral of the velocity determined by the center frequency of the transmitted signal. Based on the derived compensation time delay, t_{com}, and the distance value, D, the average velocity during propagation of the transmitted signal can be obtained and are shown in Table 1. The measurement values in Table 1 were calculated by the time of arrival of group velocity of signal in Figure 5a. As the distance between each signal is set as 40 m, the group velocity, measurement value, can be obtained by dividing the distance by the measured group delay. The accuracy values in Table 1 were calculated by dividing the group velocity of proposed method by measurement value. As seen in Table 1, when the signal propagates to a short distance, the existing method is highly accurate, but the existing method does not reflect the change in group velocity when the signal propagates over a long distance. On the other hand, the proposed method compensates the group velocity change due to the dispersion, so that the accuracy of group velocity is good regardless of the distance.

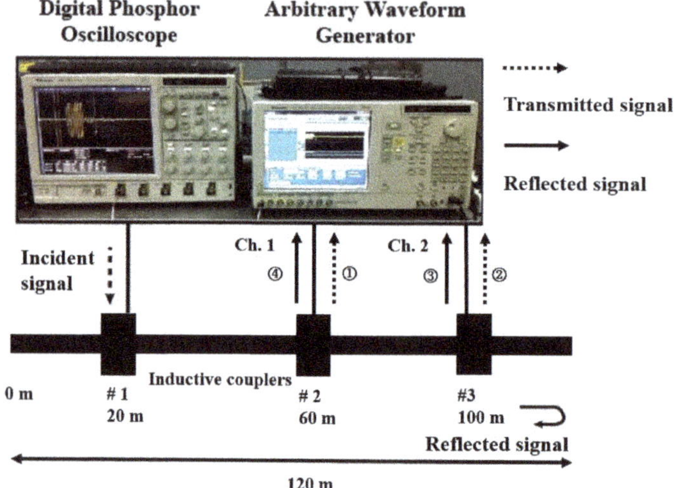

Figure 2. Experimental setup.

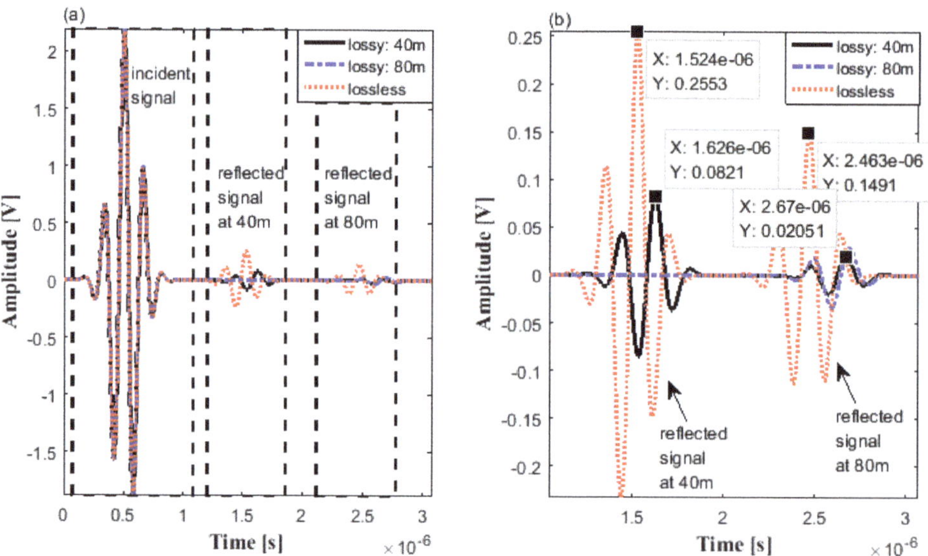

Figure 3. The results of acquired signals: 40 m lossy cable, 80 m lossy cable, lossless cable (**a**) total signal; (**b**) reflected signal.

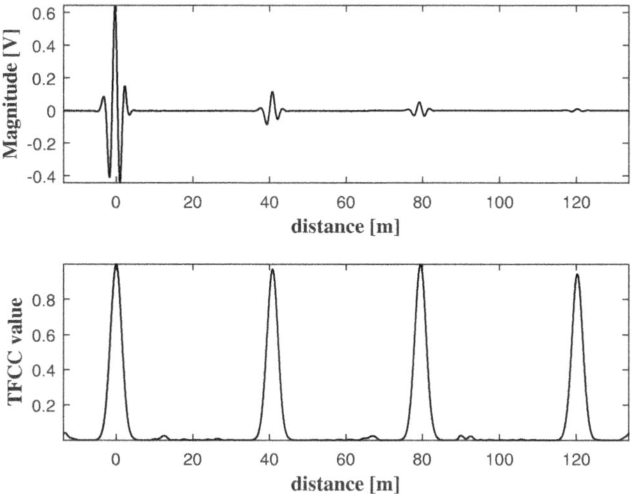

Figure 4. The results of (**a**) acquired signals, (**b**) TFCC graph.

Table 1. Estimation results of group velocity.

	Group Velocity		
	Cable Length: 40 m	**Cable Length: 80 m**	**Cable Length: 120 m**
TFCC accuracy	$1.82 \cdot 10^8$ [m/s] 100 [%]	$1.81 \cdot 10^8$ [m/s] 99 [%]	$1.81 \cdot 10^8$ [m/s] 97 [%]
proposed method accuracy	$1.83 \cdot 10^8$ [m/s] 99 [%]	$1.81 \cdot 10^8$ [m/s] 99 [%]	$1.76 \cdot 10^8$ [m/s] 99 [%]
measurement value	$1.82 \cdot 10^8$ [m/s]	$1.80 \cdot 10^8$ [m/s]	$1.77 \cdot 10^8$ [m/s]

5. Conclusions

In this paper, we proposed the group velocity compensation method using the derived time and center frequency offset terms based on the estimation of instantaneous frequency (IF). The proposed method can be divided two part. The one was the transmitted signal detector algorithm based on TFCC and multiple inductive couplers system. The second is the compensation algorithm, and the compensation terms was obtained based on the IF curve which was derived from Hilbert transform and phase unwrap algorithm of the transmitted signal. The variation of group velocity of chirp signal by dispersion in lossy medium was verified using a simulation tool. Through the comparison experiments with compensation without consideration of sweep rate change and existing methods using TFCC, superiority of the proposed method was proved. Although the group velocity of a signal propagating a short distance has similar accuracy in both the conventional method and the proposed method, when the signal propagates over a long distance (120 m), the proposed method has 2% better accuracy than the conventional method. This paper proposes a new method to compensate the group velocity error due to the dispersion of the chirp signal in the lossy medium, and this method can be applied to the detection of defect, it is possible to localize the fault in long-distance lines, such as submarine HVDC cable, without error.

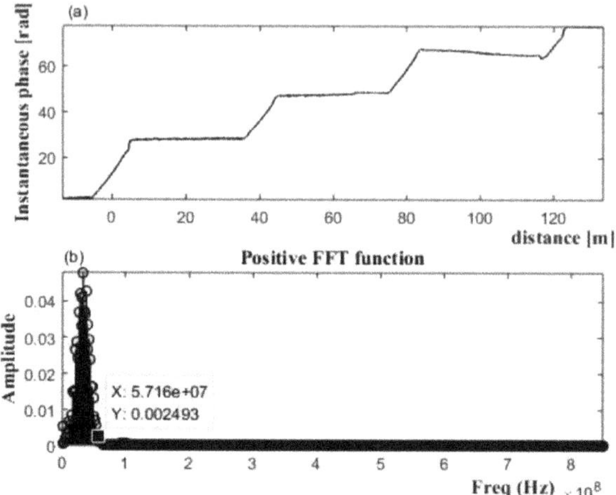

Figure 5. (**a**) Estimation of instantaneous phase, (**b**) frequency band of transmitted signal.

Author Contributions: Seung Jin Chang made a theory, designed the experiments and wrote the paper; Seung-Il Moon analyzed the data.

Conflicts of Interest: The authors declare no conflict of interest.

References

1. Cho, J.; Kim, H.; Jung, K.-Y. Simple transmission line model suitable for the electromagnetic pulse coupling analysis of twisted-wire pairs above ground. *IEICE Electron. Express* **2016**, *13*, 20160149.
2. Chowdary, K.M.; Majetich, S.A. Frequency-dependent magnetic permeability of Fe10Co90 nanocomposites. *J. Phys. D Appl. Phys.* **2014**, *47*, 175001, doi:10.1088/0022-3727/47/17/175001.
3. Chang, S.J.; Lee, C.K.; Lee, C.-K.; Han, Y.J.; Jung, M.K.; Park, J.B.; Shin, Y.-J. Condition monitoring of instrumentation cable splices using Kalman filtering. *IEEE Trans. Instrum. Meas.* **2015**, *64*, 3490–3499, doi:10.1109/TIM.2015.2444260.
4. Lee, C.K.; Park, J.B.; Shin, Y.J.; Yoon, T.S. High resolution LFMCW radar system using model-based beat frequency estimation in cable fault localization. *IEICE Electron. Express* **2014**, *11*, 20130768.
5. Lee, S.H.; Lee, C.K.; Park, J.B.; Choi, T.H. Diagnostic method for insulated power cables based on wavelet energy. *IEICE Electron. Express* **2013**, *10*, 20130335.
6. Kwak, K.S.; Doo, S.; Lee, C.K.; Yoon, T.S. Reduction of the blind spot in the time-frequency domain reflectometry. *IEICE Trans. Electron.* **2008**, *5*, 265–270.
7. Lee, C.K.; Kwak, K.S.; Yoon, T.S.; Park, J.B. Cable Fault Localization Using Instantaneous Frequency Estimation in Gaussian-Enveloped Linear Chirp Reflectometry. *IEEE Trans. Instrum. Meas.* **2013**, *62*, 129–139, doi:10.1109/TIM.2015.2444260.
8. Shin, Y.J.; Song, E.S.; Kim, J.W.; Park, J.B.; Yook, J.G.; Powers, E.J. Time-frequency domain reflectometry for smart wiring systems. *Proc. SPIE* **2002**, *4791*, doi:10.1117/12.451709.
9. Chang, S.J.; Lee, C.K.; Shin, Y.-J.; Park, J.B. Multiple Resolution Chirp Reflectometry for Fault Localization and Diagnosis High Voltage Cable in Automotive Electronics. *Meas. Sci. Technol.* **2016**, *27*, 125006, doi:10.1088/0957-0233/27/12/125006.
10. Selesnick, I.W.; Figueiredo, M.A.T. Signal restoration with overcomplete wavelet transforms: comparison of analysis and systhesis priors. *Proc. SPIE* **2009**, *7446*, doi:10.1117/12.826663.

Article

Analysis of Guided Wave Propagation in a Multi-Layered Structure in View of Structural Health Monitoring †

Yevgeniya Lugovtsova [1,*], Jannis Bulling [1], Christian Boller [2] and Jens Prager [1]

[1] Bundesanstalt für Materialforschung und –prüfung (BAM), 12205 Berlin, Germany; jannis.bulling@bam.de (J.B.); jens.prager@bam.de (J.P.)
[2] Chair of NDT and Quality Assurance (LZfPQ), Saarland University, 66125 Saarbruecken, Germany; c.boller@mx.uni-saarland.de
*   Correspondence: yevgeniya.lugovtsova@bam.de
†   This paper is an extended version of our paper published in 9th European Workshop on Structural Health Monitoring (EWSHM 2018), 10–13 July 2018 in Manchester, UK.

Received: 25 September 2019; Accepted: 25 October 2019; Published: 29 October 2019

Abstract: Guided waves (GW) are of great interest for non-destructive testing (NDT) and structural health monitoring (SHM) of engineering structures such as for oil and gas pipelines, rails, aircraft components, adhesive bonds and possibly much more. Development of a technique based on GWs requires careful understanding obtained through modelling and analysis of wave propagation and mode-damage interaction due to the dispersion and multimodal character of GWs. The Scaled Boundary Finite Element Method (SBFEM) is a suitable numerical approach for this purpose allowing calculation of dispersion curves, mode shapes and GW propagation analysis. In this article, the SBFEM is used to analyse wave propagation in a plate consisting of an isotropic aluminium layer bonded as a hybrid to an anisotropic carbon fibre reinforced plastics layer. This hybrid composite corresponds to one of those considered in a Type III composite pressure vessel used for storing gases, e.g., hydrogen in automotive and aerospace applications. The results show that most of the wave energy can be concentrated in a certain layer depending on the mode used, and by that damage present in this layer can be detected. The results obtained help to understand the wave propagation in multi-layered structures and are important for further development of NDT and SHM for engineering structures consisting of multiple layers.

Keywords: lamb waves; composite; ultrasonic testing; numerical modelling; pressure vessels

---

1. Introduction

Guided waves (GW) are of great interest for non-destructive testing (NDT) and structural health monitoring (SHM) of engineering structures such as for oil and gas pipelines, rails, aircraft components, adhesive bonds and possibly much more [1–5]. These waves can propagate over relatively long distances as long as the structure's cross-section stays constant and the difference of acoustic impedance to the surrounding environment is large. It may further allow investigation of inaccessible areas of the structure under given circumstances. To minimise complexity in GW based analysis mainly the fundamental (i.e., first symmetric and antisymmetric) modes are used in the lower frequency range, since those are well understood and can be excited, measured and consequently analysed without difficulty [6,7]. Techniques based on GWs are therefore still under investigation to explore the more complex applications such as wave propagation in hybrid structures. For instance, some structures are composed of a fibre-reinforced composite and a metal of constant cross-section [8–13]. This is a popular design of composite pressure vessels (COPV) used for gas storage. To date, a standard test procedure for this type of vessels does not exist and needs to be developed. Various approaches have been proposed to assess the current degradation state of COPVs. Destructive tests of some of the vessels running in parallel with vessels being in service were suggested in [14]. Another approach is acoustic emission monitoring for prediction of the residual lifetime of COPVs [11,15]. Many researchers have been exploring an option of monitoring pressure vessels using integrated optical fibres with Bragg grating sensors [16–20]. Another promising approach is guided wave-based monitoring [10,12,13,21–23], which has several advantages when compared to alternative monitoring techniques. One of the advantages is that the approach is non-destructive and another is the approach's holistic character with respect to detection and localisation of damage.

To examine different features of guided waves with respect to the limits of detectability of damage in a structure to be monitored, specifically also with regard to location when considering the hybrid material mentioned above, higher-order modes provide an interesting option to be explored. These modes, when considered over the whole thickness of the laminate, do not necessarily have smaller wavelengths but rather possess more complex mode shapes. This, in turn, may allow damage located in different depths to be found and differentiated. Due to dispersion and the multimodal character of GWs, development of a damage detection technique based on GWs requires careful numerical modelling such that a realistic analysis of wave propagation and mode-damage interaction can be performed.

At first, a tool for calculation of dispersion curves and mode shapes is required. For this purpose, efficient approaches have been developed and applied to layered structures, e.g., the transfer matrix method [24], the discrete layer method [25], the semi-analytical finite element method [26,27] and the scaled boundary finite element method (SBFEM) [28]. Based on these curves, appropriate modes can be identified and considered for further analysis. Next, a numerical tool for modelling the propagation of the chosen GW modes is required. Numerous methods have been applied to wave propagation problems such as finite differences [29], finite volumes [30–32], and finite element methods [11,33], to just name a few. An extensive review of simulation methods for guided wave propagation analysis can be found in [34]. In general, each method has its advantages and disadvantages depending on the given problem, and there is no universal method suitable for solving every problem efficiently.

In this article, the SBFEM is used to analyse GW propagation in a plate consisting of an isotropic aluminium layer bonded to an anisotropic laminate made of carbon fibre reinforced plastics (CFRP). Compared to other methods, the SBFEM does not require full discretisation, thus reducing the computational effort required significantly. It is suitable for calculation of dispersion curves, mode shapes and modelling of the GW propagation [28,35,36]. The dispersion curves are computed with the SBFEM by discretising the cross-section (e.g., the plate thickness) of an infinite domain using high-order elements, as shown as an example in Figure 1a. Multi-layered structures can be easily modelled using one element per layer having either isotropic or anisotropic material properties. This method is also highly efficient for modelling of the wave propagation in structures which have a constant cross-section. Because only the cross-section is discretised, the number of finite elements used in the SBFEM reduces drastically when compared to common finite element methods, as can be seen as a comparison between the SBFEM discretisation in Figure 1b and the FEM discretisation in Figure 1c, respectively. Another feature of the SBFEM is that an infinite domain, a so-called unbounded domain, as shown in Figure 1a, can be coupled with the domain of interest, thus avoiding unwanted reflections and by that simplifying the analysis of the wave propagation.

To be able to distinguish a damage in the aluminium from a damage in the CFRP in the hybrid composite addressed above, fundamental and higher-order GW modes have been determined and analysed. Additionally, to some research performed and reported in [13], a delamination case has been studied and is reported here proposing a new damage detection methodology, which allows damage localisation within the different layers. The results show that most of the wave energy (displacement) concentrates in a certain layer of the hybrid composite depending on the mode being considered and with this a damage present in the respective layer may be detected. The results obtained help to understand the wave propagation in such a multi-layered structure which may further help GW based techniques to be enhanced in the sense of NDT and SHM systems to be developed for hybrid composite structures.

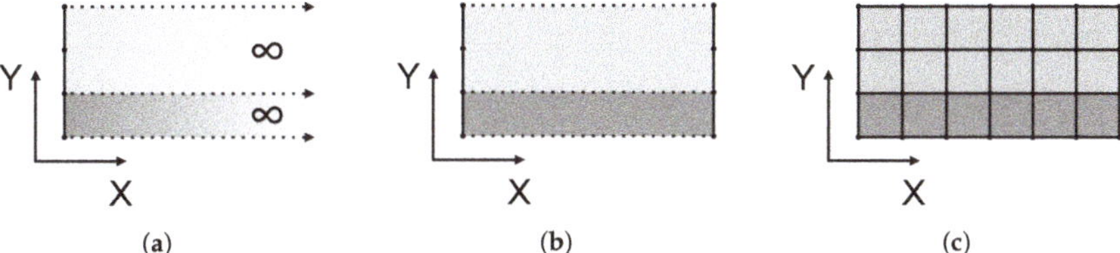

Figure 1. Schematic discretisation of: (**a**) an unbounded (infinite) domain in the scaled boundaty finite element method (SBFEM) (**b**); a domain in the SBFEM; and (**c**) in classical FEM.

2. Methodology

Tubular composite pressure vessels are designed in a way that failure mainly happens in their cylindrical part. There, axial (longitudinal) cracks in the metal liner, as well as fibre fracture and epoxy matrix cracks in the CFRP overwrap, can occur [14,37,38]. These types of damage, together with an impact damage [39,40], which may happen too, are the most critical and crucial issues with regard to a safe use of the COPV. In the work presented here and for reasons of simplification, the end caps of the vessel are neglected, and wave propagation is modelled in the circumferential (hoop) direction only. This will not only allow to detect axial cracks but also to concentrate on the circumferential CFRP plies, which are holding most of the pressure in the vessel. Furthermore, it allows damage detectability in such a hybrid material to be studied still under less complex loading conditions before moving to the more complex ones. In terms of the GW propagation, the cylindrical structure is therefore approximated to a plate, being allowed if the ratio of radius to thickness is larger than 10:1 [41]. The scope of this paper is merely to investigate the GW propagation in a multi-layered plate first while cylindrical structures need to be discussed in some future paper later. Figure 2 shows the configuration used.

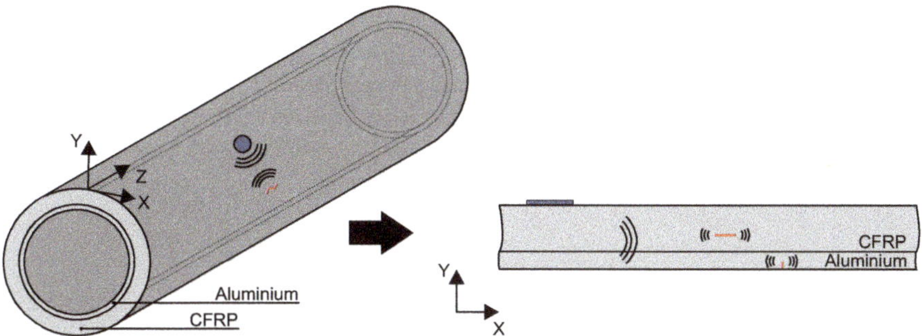

Figure 2. Sketch of a simplified structure of a COPV with corresponding sensor placement, adapted from [13].

The numerical model considered consists of a plate made of a 4 mm CFRP laminate with a [90/0/90/0] layup (from top to bottom), followed by a 2 mm aluminium layer. Modelling was performed under the two-dimensional plain strain assumption. The CFRP plies were modelled separately as a stack in accordance to their lamination direction being transversally isotropic with a 1 mm thickness of each. Material properties from the ANSYS database were used in the model and are listed in Tables 1 and 2, where ρ is density, E is Young's modulus, G is shear modulus, and ν is Poisson's ratio. Due to the production process of COPVs, a firm connection between the aluminium liner and CFRP overwrap is achieved, which allows a firm bonding between the two parts of the structure to be considered. To calculate the dispersion curves and mode shapes, the plate thickness has been approximated using finite elements of the eighth order, whereas the domain in the wave propagation direction is described analytically, as shown in Figure 1a. For the analysis of wave propagation, an excitation domain, an evaluation domain, a damage domain and two unbounded domains at both ends of the plate were used. The schematic of the numerical model is shown in Figure 3.

Table 1. Material properties of the epoxy carbon ply used (E_1 is in the fibre direction).

ρ (kg/m^3)	E_1 (GPa)	E_2, E_3 (GPa)	G_{12}, G_{13} (GPa)	ν_{12}, ν_{13}	ν_{23}
1490	121	8.6	4.7	0.27	0.4

Table 2. Material properties of aluminium used.

ρ (kg/m^3)	E (GPa)	ν
2770	70	0.33

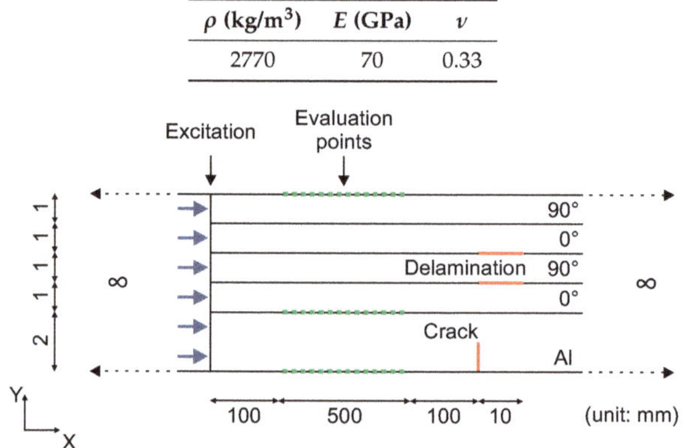

Figure 3. Schematic of the numerical model used in the SBFEM.

When actuating/deflecting the laminate at virtually any location over its cross-section, different mode shapes will be generated in accordance to the frequency being applied. To minimise the effect of dispersion by minimising a change in group velocities, some centre operation frequencies can be chosen, which have been marked with circles in Figure 4. For those, the mode shapes have been determined for which examples are shown in Figure 6. These mode shapes have further been used as the excitation to simulate a respective mode within the laminate and to understand the resulting consequences. The excitation was performed in both horizontal and vertical directions where the load distribution over the plate has been in accordance to the shape of the mode having been considered. These were pre-calculated using the unbounded domain in the SBFEM, for which an example is shown in Figure 1a. Even though such an excitation can be hardly realised in real application it has been used here for reasons of simplification and validation only and mainly to study the behaviour of a single respective mode. The excitation in time was a Hanning-windowed tone burst with the centre frequency of the desired mode. Data were evaluated using 4096 points over a 500 mm distance resulting in a 0.12 mm step. Modelling was performed in the frequency domain allowing the use of infinite domains.

For GW to be accepted for COPVs' health monitoring, damage must be detected in any part of the structure, be it the aluminium liner or the CFRP overwrap. To allow for damage detection in both parts, modes showing characteristic interaction with damage, therefore, need to be identified. For this, damage in the aluminium part was modelled by disconnecting two SBFEM domains, representing a crack of 1 mm in depth and zero width, respectively. Damage in the CFRP part was modelled by disconnecting two SBFEM boundaries, representing a delamination of 10 mm in length and zero width between two CFRP-plies. The delamination was modelled at two different positions in two separate simulation runs. The positions were between the second ply and the third ply (Position (1)), and the third ply and the fourth ply (Position (2)) (counted from top to bottom). Figure 3 schematically shows the different locations of the damages. It further shows that the excitation has been chosen to act along the thickness while the recordings have been made on the top and bottom of the hybrid composite as well as between the CFRP and aluminium part respectively being denoted as "evaluation points".

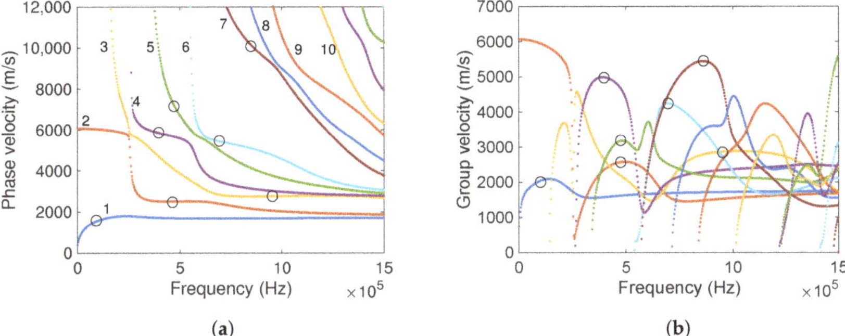

Figure 4. Dispersion curves of: (**a**) phase and (**b**) group velocities, calculated for the infinite aluminium-CFRP hybrid plate using SBFEM. The first ten guided wave modes are numbered in numerical order. The circles mark modes, as chosen for numerical modelling. Recreated from original data from [13].

3. Results

Dispersion curves for the infinite aluminium-CFRP hybrid plate are presented in Figure 4. The circles mark modes at their centre operation frequencies, as chosen for the numerical modelling. These points were selected based on the minimal change of the group velocity, which in turn allows dispersion effects of the mode to be minimised. All values from dispersion curves and the number of cycles in the tone burst used for modelling are summarised in Table 3. The number of cycles was chosen based on the frequency. The higher the frequency was, the more cycles were used in the pulse. This allows the frequency bandwidth to be kept narrow, which also minimises the dispersion effects, allows the excitation of undesired modes to be avoided beyond their cut-off frequencies and reduces the computational time. The longest computational times on the standard desktop PC used for the determination of the dispersion curves and wave propagation were 5 s and 278 s, respectively.

Table 3. Modes used for numerical modelling.

Mode	Frequency kHz	Group Velocity m/s	Phase Velocity m/s	Wavelength mm	Number of Cycles
1	100	2033	1562	15.6	10
2	475	2565	2507	5.3	20
3	950	2864	2761	2.9	30
4	400	4981	5839	14.6	20
5	475	3175	7132	15	20
6	700	4238	5431	7.8	20
7	860	5440	10,005	11.6	30

3.1. Damage in the Aluminium Liner

The signals obtained from the evaluation points, shown in Figure 3, were analysed by means of a 2D Fast Fourier Transform (FFT). The resulting dispersion map is in the frequency–wavenumber domain. It reveals different modes propagating in the plate and reflecting from the crack in the aluminium liner. The frequency–wavenumber spectra were superimposed with the dispersion curves to identify different modes, and the results are shown in Figure 5. The negative wavenumbers represent incident waves, whereas positive wavenumbers represent reflected waves. Two cases are compared in this figure, for the excitation of Mode 2 at 475 kHz (see Figure 5a,b) and of Mode 5 at 475 kHz (see Figure 5c,d), where in the cases in-plane and out-of-plane components have been determined, respectively. These examples are shown for the evaluation points positioned at the interface between aluminium and CFRP. Even though the excitation was performed by applying the corresponding mode shape at the central frequency of the desired mode to be excited, the excitation of other modes could not be avoided. However, these modes have much smaller displacements when compared to the mode having been explicitly excited. In the case of Mode 2 excitation, Mode 4 was excited too, whereas excitation of Mode 5 led to the excitation of Modes 1, 2 and 4, as shown in Figure 5. Because of the same excitation frequency and the number of cycles in the pulse, but different mode shapes, different modes were excited. The results show that, regardless of how many modes were excited in the plate, only Modes 2 and 4 reflect from the crack in the frequency range used.

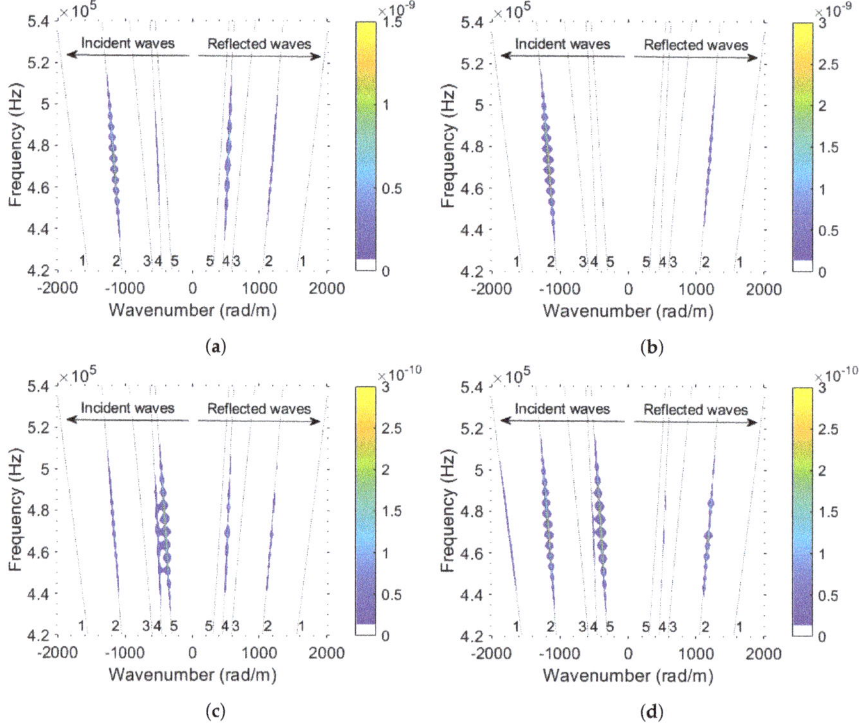

Figure 5. Frequency–wavenumber spectra of wave modes propagating in the aluminium-CFRP plate and reflecting from the 1 mm crack in the aluminium: (**a**) for in-plane and (**b**) out-of-plane components while exciting Mode 2 at 475 kHz, and (**c**) for in-plane and (**d**) out-of-plane components while exciting Mode 5 at 475 kHz.

Considering the shapes of Modes 2 and 5, shown in Figure 6a,b, respectively, it is observed that Mode 5 has a very small displacement in the aluminium at this frequency, compared to the CFRP part and Mode 2. This indicates that the energy (displacement) of Mode 5 at the frequency of 475 kHz is concentrated mainly in the CFRP layers, and thus the damage in the aluminium layer cannot be found. The frequencies for all modes chosen for modelling, including the detectability of all damages defined, are summarised in Table 4. As regards the crack in the aluminium liner only Modes 5 and 7 do not interact. Considering the mode shapes, Mode 5 has a very small displacement in the aluminium layer compared to the CFRP plies (see Figure 6b). The same holds for Mode 7, not shown here for brevity. In contrast, other modes that were investigated have comparable amplitudes in both aluminium and CFRP parts (see Figure 6a,c,d). In Figure 7a,b, the dispersion curves of the coupled

6 mm aluminium-CFRP plate are compared to the dispersion curves of a single 2 mm aluminium plate and a single 4 mm CFRP plate, respectively. The dashed lines represent four guided wave modes in the 2 mm aluminium plate for the frequency range chosen. The modes propagating in the 4 mm CFRP plate are marked with dashed-dotted lines in Figure 7b. The combination of these two structural parts being the hybrid composite considered results in the set of guided wave modes shown with solid lines. Filled and hollow circles mark modes that did and did not interact with a 1 mm crack in the aluminium layer, respectively. It is observed in Figure 7a that the modes of the hybrid composite are positioned near the modes of the pure aluminium portion (marked with red dashed lines) and showed characteristic interaction with the crack (see filled circles). Modes marked with hollow circles lay on the modes of the pure CFRP portion and did not reflect from the crack (see red dashed-dotted lines in Figure 7b). These modes have very small displacements in the aluminium, as shown in Figure 6b as an example. Thus, modes of the hybrid composite show different behaviour, with either the CFRP or the aluminium portion dominating, or both. Such an effect is very much known from the behaviour of coupled vibrations in multi-body systems. The displacement is concentrated either in the CFRP plies or in the aluminium layer, or within both parts depending on the mode and frequency chosen. Such behaviour, when carefully analysed and understood, can be advantageous, allowing for damage detection in different constituent parts of a component.

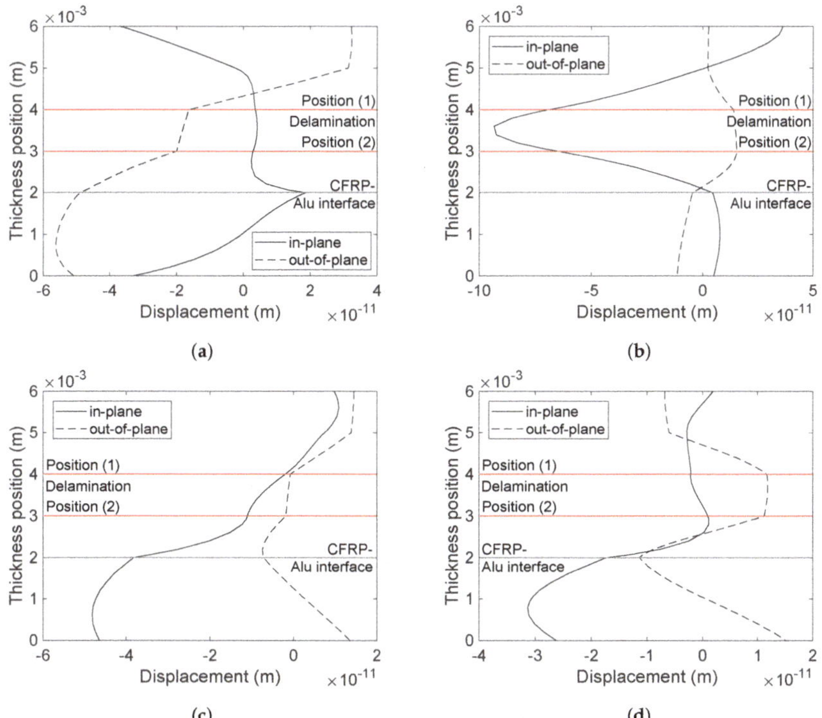

Figure 6. Mode shapes of: (**a**) Mode 2 at 475 kHz; (**b**) Mode 5 at 475 kHz; (**c**) Mode 4 at 400 kHz; and (**d**) Mode 6 at 700 kHz. Red lines mark delamination placed between the second and the third ply (Position (1)), and the third and the fourth ply (Position (2)) (counted from top to bottom).

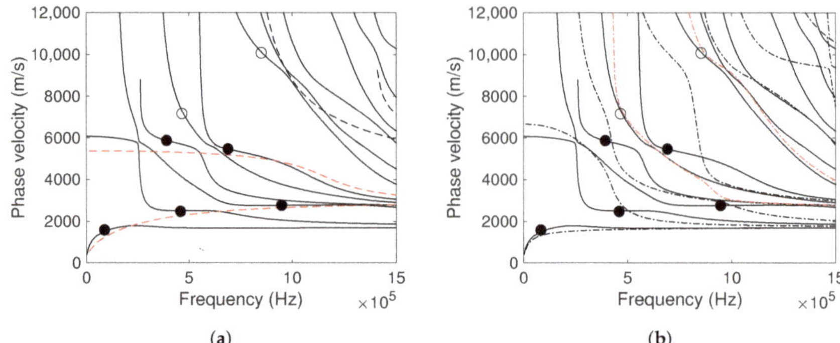

Figure 7. Combined phase velocity dispersion curves of: (**a**) the 2 mm aluminium plate (dashed lines) and the 6 mm aluminium-CFRP plate (solid lines); and (**b**) the 4 mm CFRP plate with a [90/0/90/90] layup (dash-dotted lines) and the 6 mm aluminium-CFRP plate (solid lines). The filled and hollow circles mark modes that did and did not interact with a 1 mm crack in aluminium, respectively. Red dashed lines mark two aluminium modes in the aluminium, red dash-dotted lines mark two CFRP modes. Recreated from original data from [13].

3.2. Damage in the CFRP Overwrap

Damage in the CFRP was modelled as a delamination having a 10 mm length and zero width. The delamination was placed at two positions between different CFRP plies, as shown in Figure 3. Position (1) corresponds to the delamination placed between the second and the third ply, whereas Position (2) corresponds to the delamination between the third and the fourth ply (counted from top to bottom). Modelling was performed in two separate simulation runs, and the same modes, shown in Table 3, were used. Figure 8 shows the resulting frequency–wavenumber spectra of the propagating modes and modes reflected from the delamination at Position (2). Excitation was performed for Modes 2 and 5 at 475 kHz, and signals were evaluated at points positioned at the interface between aluminium and CFRP. In contrast to the crack in the aluminium, from which only Modes 2 and 4 reflect, more modes interact with the delamination in the CFRP. These are Modes 1, 2, 3 and 5, whereas Mode 4 did not interact with the delamination at Position (2). The results for all modes being modelled are summarised in Table 4. Modes 4, 6 and 7 did not interact with the delamination at Position (2), whereas only Mode 6 did not interact with the delamination at Position (1). As in the case of the crack in the aluminium, this can be attributed to the mode shape. Every mode modelled has comparable or bigger displacement amplitudes between the aluminium and CFRP parts (see the example shown for some modes in Figure 6). For instance, Modes 2 and 5 reflect from the delamination even though the in-plane displacement of Mode 2 is almost zero at both delamination positions (see Figure 6a), whereas, for Mode 5, it is the out-of-plane displacement which is close to zero (see Figure 6b). Mode 4 interacts with the delamination placed at Position (1), but not at Position (2). The out-of-plane displacement is almost equal at these positions, and the in-plane displacement is even bigger in the latter case. A contribution to this behaviour may further be the anisotropic properties of the CFRP and its layup. Numerical modelling was also performed under 2D conditions and it could be that the guided wave modes reflect, and scatter but, in another direction, which has not been modelled so far. In general, it is difficult to draw fundamental conclusions at this stage as long as such multi-layered systems are not systematically researched and described answering the question why some modes do interact with the delamination, whereas the others do not. However, the importance of numerical modelling of guided wave propagation and analysis, including their interaction with damage, is again highly emphasised with the test case presented here.

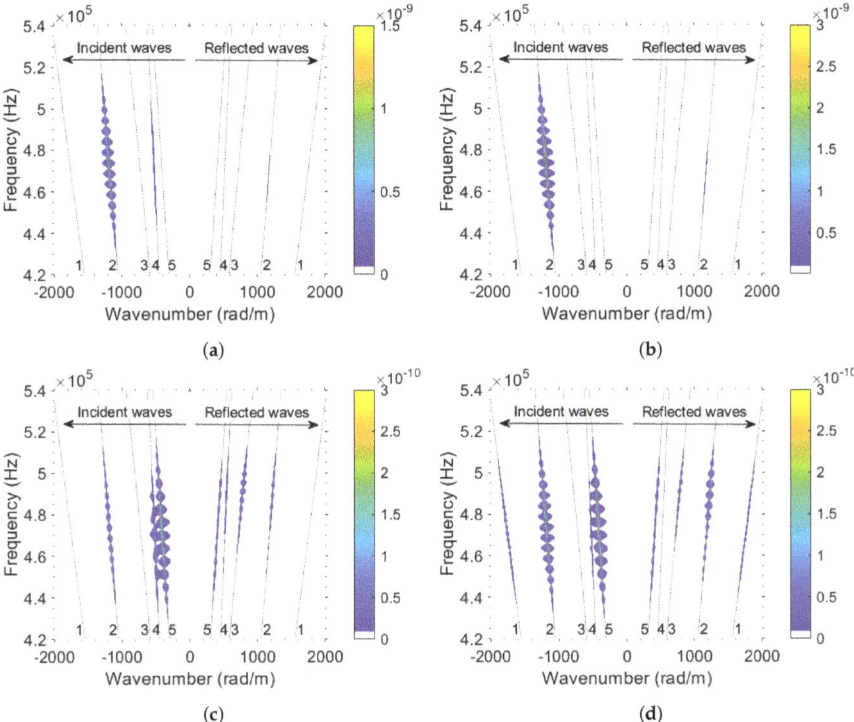

Figure 8. Frequency–wavenumber spectra of wave modes propagating in the aluminium-CFRP plate and reflecting from the 10 mm delamination between the third and the fourth ply [Position (2)]: (**a**) for in-plane and (**b**) out-of-plane components while exciting Mode 2 at 475 kHz, and (**c**) for in-plane and (**d**) out-of-plane components while exciting Mode 5 at 475 kHz.

Table 4. Comparison of the mode-damage interaction for the damage positioned in different parts of the structure. ✓, modes which reflect from the damage; ✗, modes which did not reflect from the damage.

Mode	Frequency kHz	Wavelength mm	Crack in Aluminium	Delamination at Position (1)	Delamination at Position (2)
1	100	15.6	✓	✓	✓
2	475	5.3	✓	✓	✓
3	950	2.9	✓	✓	✓
4	400	14.6	✓	✓	✗
5	475	15	✗	✓	✓
6	700	7.8	✓	✗	✗
7	860	11.6	✗	✓	✗

4. Discussion

The example described above demonstrates that multi-layered structures are rather complex when it comes to the description of guided waves travelling through them. To understand this, numerical analysis tools are of an essential need. In the case considered here, the SBFEM was used to calculate dispersion curves and mode shapes, as well as to analyse the propagation of guided waves in a plate consisting of an isotropic metal bonded to anisotropic carbon fibre reinforced layered material. The method allows appropriate modes to be identified and their interaction with different damage types to be analysed.

Results show that there are wave modes, which are sensitive to the damage defined in the composite and metallic part of the structure, represented in Modes 1–3 with respect to the case described here. To be therefore able to separate the damage in the aluminium liner from the damage in the CFRP overwrap, the characteristic interaction of a mode sensitive to the damage only in one part is needed. This has been observed for Mode 4 in the case of the crack in the aluminium liner and for Mode 5 in the case of the delamination at both positions defined in CFRP. One has to be careful here, in the sense that the right modes are used and that a damage in the CFRP part will not be mistaken for a damage in the aluminium liner. This is why it is important to be able to excite modes in different constituent parts of the component, and that chosen modes show characteristic interaction only with the damage in one of the parts. As a remark, in the present work, a case of the CFRP being delaminated at numerous locations across the thickness has not been considered so far, which would represent an impact damage. Supposedly, more modes may interact with such an impact damage, and this may result in higher amplitudes of the reflected modes than in the case of a single delamination considered here. As for micro-cracking of the CFRP overwrap, a different approach is necessary where guided wave modes are analysed based on the change of their phase [10] or group velocity [13], and not the reflection from the damage.

A solution in that regard for the reception as well as the excitation of the desired modes could be an interdigital transducer (IDT) being specifically "tuned" to the frequency and wavelength sensitive to the respective damage to be monitored. A 15 mm pitch between its electrodes may be suitable for the cases described here, which corresponds to Modes 4 and 5 when driven at 400 kHz and 475 kHz, respectively. Excitation of Mode 4 will allow the damage in the aluminium part and excitation of Mode 5 the damage in the CFRP part to be identified. A design of such an IDT based monitoring system could be to get it integrated at the aluminium-CFRP interface. The monitoring system configuration (pattern) should be adapted to as many modes as possible being sensitive to the damage to be monitored. Here, thin and flexible polymer materials such as PVDF can be used. Their thickness may be less than 100 µm, they can be poled to achieve piezo-electric properties, and their electrodes can be structured in the desired manner to excite the wave mode needed [42–47]. A sketch of a proposed arrangement with regard to structural integration is shown in Figure 9.

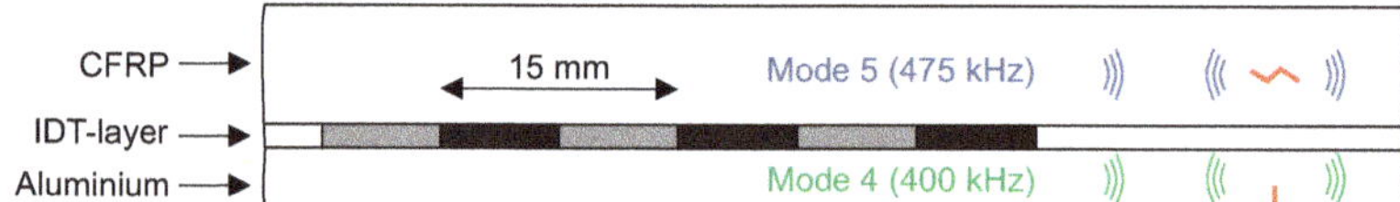

Figure 9. Sketch of an arrangement based on the example considered here for an interdigital SHM system for a composite pressure vessel monitoring.

5. Conclusions

In this contribution, the scaled boundary finite element method was used to calculate dispersion curves and mode shapes, as well as to analyse the propagation of guided waves in a plate consisting of an isotropic metal bonded to anisotropic carbon fibre reinforced layered material. The method allows appropriate modes to be identified and their interaction with different damage types to be analysed. The results show that most of the wave energy can be concentrated in a certain layer depending on the mode used, and by that damage present in this respective layer can be detected. The results obtained help to understand the wave propagation in such a multi-layered structure, which may further help guided wave based techniques to be enhanced in the sense of non-destructive testing and structural health monitoring systems to be developed for hybrid composite structures. Moreover, a concept for an SHM system for composite pressure vessels is proposed.

Author Contributions: Conceptualisation, Y.L., J.B., C.B. and J.P.; investigation, Y.L.; software, J.B.; data curation, Y.L.; writing—original draft preparation, Y.L.; writing—review and editing, Y.L., J.B., C.B. and J.P.; and supervision, C.B. and J.P.

Funding: This research received no external funding.

Conflicts of Interest: The authors declare no conflict of interest.

References

1. *Standard Practice for Guided Wave Testing of Above Ground Steel Pipework Using Piezoelectric Effect Transduction*; ASTM International: West Conshohocken, PA, USA, 2016.
2. Fromme, P.; Wilcox, P.D.; Lowe, M.J.S.; Cawley, P. On the development and testing of a guided ultrasonic wave array for structural integrity monitoring. *IEEE Trans. Ultrason. Ferroelectr. Freq. Control* **2006**, *53*, 777–785. [CrossRef] [PubMed]
3. Wilcox, P.; Evans, M.; Pavlakovic, B.; Alleyne, D.; Vine, K.; Cawley, P.; Lowe, M. Guided wave testing of rail. *Insight Non-Destr. Test. Cond. Monit.* **2003**, *45*, 413–420. [CrossRef]
4. Gao, H.; Rose, J.L. Ice detection and classification on an aircraft wing with ultrasonic shear horizontal guided waves. *IEEE Trans. Ultrason. Ferroelectr. Freq. Control* **2009**, *56*, 334–344. [CrossRef] [PubMed]
5. Rose, J.; Zhu, W.; Zaidi, M. Ultrasonic NDT of titanium diffusion bonding with guided waves. *Mater. Eval.* **1998**, *56*, 535–539.
6. Diamanti, K.; Soutis, C. Structural health monitoring techniques for aircraft composite structures. *Prog. Aerosp. Sci.* **2010**, *46*, 342–352. [CrossRef]
7. Mitra, M.; Gopalakrishnan, S. Guided wave based structural health monitoring: A review. *Smart Mater. Struct.* **2016**, *25*. [CrossRef]
8. Gao, H. Ultrasonic Guided Wave Mechanics for Composite Material Structural Health Monitoring. Ph.D. Thesis, Pennsylvania State University, State College, PA, USA, 2007.
9. Yu, X.; Fan, Z.; Castaings, M.; Biateau, C. Feature guided wave inspection of bond line defects between a stiffener and a composite plate. *NDT E Int.* **2017**, *89*, 44–55. [CrossRef]
10. Castaings, M.; Hosten, B. Ultrasonic guided waves for health monitoring of high-pressure composite tanks. *NDT E Int.* **2008**, *41*, 648–655. [CrossRef]
11. Sause, M.G.; Hamstad, M.A.; Horn, S. Finite element modeling of lamb wave propagation in anisotropic hybrid materials. *Compos. Part Eng.* **2013**, *53*, 249–257. [CrossRef]
12. Lugovtsova, Y.; Prager, J. Structural health monitoring of composite pressure vessels using guided ultrasonic waves. *Insight-Non-Destr. Test. Cond. Monit.* **2018**, *60*, 139–144. [CrossRef]
13. Lugovtsova, Y.; Bulling, J.; Prager, J.; Boller, C. Efficient modelling of guided ultrasonic waves using the Scaled Boundary FEM towards SHM of composite pressure vessels. In Proceedings of the 9th European Workshop on Structural Health Monitoring, Manchester, UK, 10–13 July 2018.
14. Mair, G.W.; Scherer, F.; Duffner, E. Concept of interactive determination of safe service life for composite cylinders by destructive tests parallel to operation. *Int. J. Press. Vessel. Pip.* **2014**, *120–121*, 36–46. [CrossRef]
15. Chou, H.; Mouritz, A.; Bannister, M.; Bunsell, A. Acoustic emission analysis of composite pressure vessels under constant and cyclic pressure. *Compos. Part Appl. Sci. Manuf.* **2015**, *70*, 111–120. [CrossRef]
16. Degrieck, J.; Waele, W.D.; Verleysen, P. Monitoring of fibre reinforced composites with embedded optical fibre Bragg sensors, with application to filament wound pressure vessels. *NDT E Int.* **2001**, *34*, 289–296. [CrossRef]
17. Kang, D.; Kim, C.; Kim, C. The embedment of fiber Bragg grating sensors into filament wound pressure tanks considering multiplexing. *NDT E Int.* **2006**, *39*, 109–116. [CrossRef]
18. Gąsior, P.; Malesa, M.; Kaleta, J.; Kujawińska, M.; Malowany, K.; Rybczyński, R. Application of complementary optical methods for strain investigation in composite high pressure vessel. *Compos. Struct.* **2018**, *203*, 718–724. [CrossRef]
19. Xiao, B.; Yang, B.; Xuan, F.Z.; Wan, Y.; Hu, C.; Jin, P.; Lei, H.; Xiang, Y.; Yang, K. In-Situ Monitoring of a Filament Wound Pressure Vessel by the MWCNT Sensor under Hydraulic Fatigue Cycling and Pressurization. *Sensors* **2019**, *19*, 1396. [CrossRef]
20. Saeter, E.; Lasn, K.; Nony, F.; Echtermeyer, A.T. Embedded optical fibres for monitoring pressurization and impact of filament wound cylinders. *Compos. Struct.* **2019**, *210*, 608–617. [CrossRef]
21. Bulletti, A.; Giannelli, P.; Calzolai, M.; Capineri, L. An Integrated Acousto/Ultrasonic Structural Health Monitoring System for Composite Pressure Vessels. *IEEE Trans. Ultrason. Ferroelectr. Freq. Control* **2016**, *63*, 864–873. [CrossRef]

22. Yaacoubi, S.; McKeon, P.; Ke, W.; Declercq, N.; Dahmene, F. Towards an Ultrasonic Guided Wave Procedure for Health Monitoring of Composite Vessels: Application to Hydrogen-Powered Aircraft. *Materials* **2017**, *10*, 1097. [CrossRef]

23. Yang, B.; Xiang, Y.; Xuan, F.Z.; Hu, C.; Xiao, B.; Zhou, S.; Luo, C. Damage localization in hydrogen storage vessel by guided waves based on a real-time monitoring system. *Int. J. Hydrogen Energy* **2019**. [CrossRef]

24. Lowe, M.J.S. Matrix techniques for modeling ultrasonic waves in multilayered media. *IEEE Trans. Ultrason. Ferroelectr. Freq. Control* **1995**, *42*, 525–542. [CrossRef]

25. Kausel, E. Wave propagation in anisotropic layered media. *Int. J. Numer. Methods Eng.* **1986**, *23*, 1567–1578. [CrossRef]

26. Dong, S.B.; Nelson, R.B. On Natural Vibrations and Waves in Laminated Orthotropic Plates. *J. Appl. Mech.* **1972**, *39*, 739–745. [CrossRef]

27. Dong, S.B.; Huang, K.H. Edge Vibrations in Laminated Composite Plates. *J. Appl. Mech.* **1985**, *52*, 433–438. [CrossRef]

28. Gravenkamp, H.; Song, C.; Prager, J. A numerical approach for the computation of dispersion relations for plate structures using the Scaled Boundary Finite Element Method. *J. Sound Vib.* **2012**, *331*, 2543–2557. [CrossRef]

29. Xu, K.; Ta, D.; Su, Z.; Wang, W. Transmission analysis of ultrasonic Lamb mode conversion in a plate with partial-thickness notch. *Ultrasonics* **2014**, *54*, 395–401. [CrossRef]

30. Fellinger, P.; Marklein, R.; Langenberg, K.; Klaholz, S. Numerical modeling of elastic wave propagation and scattering with EFIT—Elastodynamic finite integration technique. *Wave Motion* **1995**, *21*, 47–66. [CrossRef]

31. Schubert, F.; Peiffer, A.; Köhler, B.; Sanderson, T. The elastodynamic finite integration technique for waves in cylindrical geometries. *J. Acoust. Soc. Am.* **1998**, *104*, 2604–2614. [CrossRef]

32. Leckey, C.A.; Rogge, M.D.; Miller, C.A.; Hinders, M.K. Multiple-mode Lamb wave scattering simulations using 3D elastodynamic finite integration technique. *Ultrasonics* **2012**, *52*, 193–207. [CrossRef]

33. Luchinsky, D.G.; Hafiychuk, V.; Smelyanskiy, V.N.; Kessler, S.; Walker, J.; Miller, J.; Watson, M. Modeling wave propagation and scattering from impact damage for structural health monitoring of composite sandwich plates. *Struct. Health Monit.* **2013**, *12*, 296–308. [CrossRef]

34. Willberg, C.; Duczek, S.; Vivar-Perez, J.M.; Ahmad, Z.A.B. Simulation Methods for Guided Wave-Based Structural Health Monitoring: A Review. *Appl. Mech. Rev.* **2015**, *67*, 010803. [CrossRef]

35. Gravenkamp, H.; Prager, J.; Saputra, A.A.; Song, C. The simulation of Lamb waves in a cracked plate using the scaled boundary finite element method. *J. Acoust. Soc. Am.* **2012**, *132*, 1358–1367. [CrossRef] [PubMed]

36. Gravenkamp, H. Efficient simulation of elastic guided waves interacting with notches, adhesive joints, delaminations and inclined edges in plate structures. *Ultrasonics* **2018**, *82*, 101–113. [CrossRef] [PubMed]

37. Bunsell, A.R. Composite pressure vessels supply an answer to transport problems. *Reinf. Plast.* **2006**, *50*, 38–41. [CrossRef]

38. Scott, A.; Clinch, M.; Hepples, W.; Kalantzis, N.; Sinclair, I.; Spearing, S.M. Advanced micromechanical analysis of highly loaded hybrid composite structures. In Proceedings of the 17th International Conference on Composite Materials, Edinburgh, UK, 27–31 July 2009.

39. Changliang, Z.; Mingfa, R.; Wei, Z.; Haoran, C. Delamination prediction of composite filament wound vessel with metal liner under low velocity impact. *Compos. Struct.* **2006**, *75*, 387–392. [CrossRef]

40. Matemilola, S.A.; Stronge, W.J. Low-Speed Impact Damage in Filament-Wound CFRP Composite Pressure Vessels. *J. Press. Vessel. Technol.* **1997**, *119*, 435–443. [CrossRef]

41. Wilcox, P.D. Lamb Wave Inspection of Large Structures Using Permanently Attached Transducers. Ph.D. Thesis , Imperial College London, London, UK, 1998.

42. Monkhouse, R.; Wilcox, P.; Cawley, P. Flexible interdigital PVDF transducers for the generation of Lamb waves in structures. *Ultrasonics* **1997**, *35*, 489–498. [CrossRef]

43. Ueberschlag, P. PVDF piezoelectric polymer. *Sens. Rev.* **2001**, *21*, 118–126. [CrossRef]

44. Seminara, L.; Pinna, L.; Valle, M.; Basiricò, L.; Loi, A.; Cosseddu, P.; Bonfiglio, A.; Ascia, A.; Biso, M.; Ansaldo, A.; et al. Piezoelectric Polymer Transducer Arrays for Flexible Tactile Sensors. *IEEE Sens. J.* **2013**, *13*, 4022–4029. [CrossRef]

45. Bulletti, A.; Capineri, L. Interdigital Piezopolymer Transducers for Time of Flight Measurements with Ultrasonic Lamb Waves on Carbon-Epoxy Composites under Pure Bending Stress. *J. Sens.* **2015**. [CrossRef]

46. Ren, B.; Lissenden, C.J. PVDF Multielement Lamb Wave Sensor for Structural Health Monitoring. *IEEE Trans. Ultrason. Ferroelectr. Freq. Control* **2016**, *63*, 178–185. [CrossRef] [PubMed]
47. Giannelli, P.; Bulletti, A.; Capineri, L. Charge-mode interfacing of piezoelectric interdigital Lamb wave transducers. *Electron. Lett.* **2016**, *52*, 894–896. [CrossRef]

MDPI

St. Alban-Anlage 66

4052 Basel

Switzerland

Tel. +41 61 683 77 34

Fax +41 61 302 89 18

www.mdpi.com

Applied Sciences Editorial Office

E-mail: applsci@mdpi.com

www.mdpi.com/journal/applsci

9 783039 282982